Technology Application Tools

Karl Barksdale
Technology Consultant
Provo, Utah

Connie M. Forde, Ph.D.
Mississippi State University

Jack P. Hoggatt, Ed.D.
University of Wisconsin, Eau Claire

Mary Ellen Oliverio
Pace University
New York, New York

William R. Pasewark
Office Management Consultant
Lubbock, Texas

Jerry W. Robinson, Ed.D.

Michael Rutter
Brigham Young University

Jon A. Shank, Ed.D.
Robert Morris University
Moon Township, Pennsylvania

Susie H. VanHuss, Ph.D.
University of South Carolina

Bonnie R. White
Auburn University
Auburn, Alabama

Donna L. Woo
Cypress College, California

Australia · Canada · Mexico · Singapore · Spain · United Kingdom · United States

DigiTools: Technology Application Tools

Karl Barksdale, Connie M. Forde, Jack P. Hoggatt, Mary Ellen Oliverio, William R. Pasewark, Jerry W. Robinson, Michael Rutter, Jon A. Shank, Susie H. VanHuss, Bonnie R. White, and Donna L. Woo

VP/Editorial Director:	**VP/Director of Marketing:**	**Copyeditor:**
Jack W. Calhoun	Carol Volz	Tom Lewis
VP/Editor-in-Chief:	**Production Manager:**	**Art Director:**
Karen Schmohe	Patricia Matthews Boies	Stacy Jenkins Shirley
Acquisitions Editor:	**Production Editor:**	**Cover Illustration:**
Jane Congdon	Diane Bowdler	Jon Conrad
Project Manager:	**Technology Project Editor:**	**Internal Design:**
Dave Lafferty	Mike Jackson	Ann Small, a small design studio and Craig LaGesse Ramsdell, www.ramsdelldesign.com
Marketing Manager:	**Manufacturing Coordinator:**	
Mike Cloran	Charlene Taylor	**Cover Designer:**
Marketing Coordinator:	**Production House:**	Ann Small, a small design studio
Georgianna Wright	Cover to Cover Publishing, Inc.	**Printer:**
Consulting Editor:	**Photo Researcher:**	Quebecor World
Dianne Rankin	Darren Wright	Versailles, KY

COPYRIGHT © 2006
Thomson South-Western, a part of The Thomson Corporation. Thomson, the Star logo, and South-Western are trademarks used herein under license.

Printed in the United States of America
1 2 3 4 5 07 06 05 04

ISBN: 0-538-44196-8

ALL RIGHTS RESERVED. No part of this work covered by the copyright hereon may be reproduced or used in any form or by any means—graphic, electronic, or mechanical, including photocopying, recording, taping, Web distribution or information storage and retrieval systems, or in any other manner—without the written permission of the publisher.

For permission to use material from this text or product, submit a request online at http://www.thomsonrights.com.

For more information about our products, contact us at:

Thomson Higher Education
5191 Natorp Boulevard
Mason, Ohio 45040
USA

ASIA (including India)	AUSTRALIA/NEW ZEALAND	CANADA	UK/EUROPE/MIDDLE EAST/AFRICA
Thomson Learning	Thomson Learning Australia	Thomson Nelson	Thomson Learning
5 Shenton Way	102 Dodds Street	1120 Birchmount Road	High Holborn House
#01-01 UIC Building	Southbank, Victoria 3006	Toronto, Ontario	50/51 Bedford Row
Singapore 068808	Australia	M1K 5G4	London WC1R 4LR
		Canada	United Kingdom

The names of all companies or products mentioned herein are used for identification purposes only and may be trademarks or registered trademarks of their respective owners. South-Western disclaims any affiliation, association, connection with, sponsorship, or endorsement by such owners.

Photo Credits: © Getty Images/PhotoDisc pp. i, iv, viii, ix, 1, 2, 24, 34, 60, 69, 70, 108, 109, 127, 190, 217, 226, 227, 262, 282, 283, 292, 294, 311, 319, 349, 351, 371, 372 (top), 375, 378, 388, 389, 406 (both), 434, 456, App-1, App-11, App-15, App-25, App-1; © Jon Conrad pp. iii, 32, 60, 107, 126, 171, 189, 216, 261, 282, 318, 348, 405, 427, 456; © Getty Images/EyeWire pp. 3, 24, 191, 352, 359, 412, 416, 443; © Gyration, Inc. pp. 7, 109; © Acer, Inc. pp. 12, 109, 172 (both), 428; © Jon Feingersh/CORBIS page 50; Pitney Bowes page 143; © 2003 FedEx® Corporation page 152; © 2003 Handspring, Inc. All Rights Reserved page 173; Intuos 2 is a registered trademark of Wacom Company LTD. Photo courtesy of Wacom Technology Co. page 174; © CORBIS page 284; Courtesy of Hewlett Packard pp. 302, 303; Digital Vision pp. 361, 440; EyeCom 9000 Reader/Printer page 372 (bottom right); Courtesy of palmOne Inc. pp. App-26 (top left, center), App-28 (top), App-28 (bottom), App-37; Hewlett-Packard Company. All rights reserved pp. App-26 (top right), App-28 (top right); Comstock Images, page App-26 (bottom); Courtesy of Suunto page App-27; Courtesy of Dell Inc. page App-36; App-42 (left), page App-42 (right); Anthony Bolante/Reuters/Landov page App-47; Courtesy of Canesta Inc. page App-47

Reinforce Learning with Activities for Input Technologies and Computer Applications

Activities for Input Technologies and Computer Applications is a three-ring binder of supplementary activities that can be photocopied for student use. Ideal as a companion for any computer applications course, it can be used with *DigiTools* or *Century 21 Computer Applications and Keyboarding*. This binder contains activities for input technologies, such as keyboarding, word processing, PDAs, and Tablet PCs. In addition, there are activities for word processing, spreadsheets, presentations, database, and the creation of Web sites.

- The binder includes both student and teacher pages.
- Instructor CD is included in the binder and contains student data files, solution files, and Lesson Plans for each project.
- Activities enhance reading and problem-solving skills.
- Over 100 activities are provided for 90+ hours of instruction.

Contents:

Unit 1 Basic Input Technologies
   a Keyboarding Skill Building Projects
   b Word Processing Projects

Unit 2 Advanced Input Technologies
   a Handwriting Recognition and Tablet PC Projects
   b Speech Recognition Projects

Unit 3 Advanced Computer Applications
   a Spreadsheet Projects
   b Database Projects
   c Presentations Projects
   d Web Projects

Unit 4 Capstone Project: Europe Study Tour

Teacher Notes, Lesson Plans, and Solutions on CD-ROM

THOMSON SOUTH-WESTERN

Join us on the Internet at www.swlearning.com

Preface

DigiTools are computerized devices and software programs that help people communicate with each other. DigiTools are important for professional, academic, and economic success. Desktop, laptop, tablet, and handheld computers are obvious examples of DigiTools. Cell phones, calculators, keyboards, headsets, digital pens, scanners, cameras, monitors, and printers are also DigiTools. DigiTools include the software that runs on these high-tech devices. Word processing, spreadsheet, e-mail, Web browser, database, and multimedia applications fill our lives with colorful and exciting new ways to communicate, share, and inspire.

Technology is a driving force that has the power to reshape our learning and business activities. Each new technological advance promotes new ways of thinking, working, communicating, and learning. With each step forward we have the opportunity to be more efficient, better educated, and more prosperous. *DigiTools* is designed to teach you how to be successful in school and on the job using today's powerful digital communication tools.

Three Digital Communication Know-How Skills

You must master three types of skills to be effective in the online world of work and in your academic life.

- Input technology skills
- Software and hardware skills
- Application skills

Input Technology Skills

The way information is input into computers has improved over the last few years. In addition to typing on a keyboard, users can now input information using handwriting. Voice can be used to input text at over 100 words per minute. Scanners and digital cameras are also powerful input devices. Mastering these input technologies will prepare you to compete in the modern digital world.

Software and Hardware Skills

To master using DigiTools, you must go well beyond basic input skills and learn how your computer's hardware and software can create spectacular output to be shared with others. Output can take many forms—reports, press releases, Web pages, database reports,

charts or graphs, multimedia presentations, e-mail or instant messages, photos, movies, graphics, letters, tables, memos, newsletters, brochures, press releases, and advertisements.

Career and Academic Skills

Not only must you learn *about* DigiTools, you must learn *to use* digital communication tools effectively. There is a big difference between knowing the many features of a software program or the parts of a computer and the ability to create dynamic, creative, and meaningful communications.

How do professionals use DigiTools? They apply their digital skills to realistic problems and situations. In this text, you too will apply the skills and knowledge you learn to realistic work situations. You will also learn about 16 career clusters defined by the U.S. Department of Education. Learning about these career clusters will help you as you plan for a future career.

Features

DigiTools is divided into 4 units and 15 chapters. Preview each unit and then examine the other important features described in the following sections.

Unit 1 DigiTools for Academic and Business Success will begin your DigiTools training. You will learn about acceptable computer use policies, login procedures, file management, the *Windows* operating system, basic application program commands, and Web research skills. This unit is designed to get you up and running quickly.

Unit 2 Input Technologies will help you develop your input technology skills. You will learn or review touch keyboarding and 10-key pad skills. Then you will progress to other input technologies—handwriting and speech recognition. By blending these input technologies, you will discover the most efficient ways to work with your computer. You will also develop your word processing skills using *Microsoft Word*.

Unit 3 Increasing Productivity with DigiTools provides training in calculating, analyzing, and managing data using the spreadsheet and database applications *Microsoft Excel* and *Microsoft Access*. You will learn to develop and deliver presentations and to create multimedia slide shows using *Microsoft PowerPoint*. Learning basic HTML and Web site design will allow you to create Web pages to display your work or share messages with others. In this unit, you will also learn to be more productive by managing time and records effectively.

Unit 4 DigiTools and Your Career Pathway will allow you to apply your input, software, and hardware skills in the preparation of a personal employment journal, resume, and employment portfolio. These documents will help you as you plan ways to prepare for a career and choose a career path.

The text has four extensive end-of-unit sections called *Tooling Up!* In each Tooling Up! you will learn about 4 of the 16 career clusters outlined by the U.S. Department of Education. You will complete an exercise related to each career cluster to help you understand the careers in that cluster. You will also learn about business trends and issues, develop your critical thinking skills, and apply communication skills as you complete the exercises in each Tooling Up!

What You Should Know About, Ethics, and *DigiTip* features are scattered throughout the text. You will extend your learning by reading the What You Should Know About features. Ethics features allow you to consider ethical issues related to work and interacting with others on the job. DigiTips provide related information that will help you complete activities. *Peer Checks* and *Self Checks* will help you focus on key points, evaluate your work, and learn from your peers.

Appendices include a reference guide, tips on equipment use, keyboarding skillbuilding, enriching personal productivity with Personal Digital Assistants (PDAs), and a mini-tutorial on Microsoft OneNote.

You will develop teamwork skills as you work with classmates to complete many of the activities in this text. A teamwork icon indicates these activities. Many activities in this text require research on the Internet. An Internet icon indicates these activities.

T E A M WORK

INTER N E T

Data Files

Data files for student use in completing the activities found in the textbook are provided on the *Instructor's Resource CD-ROM*. Your instructor will need to make these files available to you on disk or on your local area network. Some data files are in word processing or spreadsheet format and are to be revised or completed by students. Other files are to be used as source documents or to enhance presentations or documents. Files in HTML format (Web pages) are used as reference sources.

Corporate View Intranet

The *Instructor's Resource CD-ROM* also contains Web pages for a corporate intranet. The intranet is for a fictional company called Corporate View; however, the intranet imitates real corporate intranets. Your instructor may make these files available to you on your computer's hard drive or on your local area network. You can also access the Corporate View intranet on the World Wide Web.

While visiting the Corporate View intranet, you will learn about various departments within the company. These departments relate to the career clusters you will learn about in Tooling Up! sections. For example, you will explore career pathways in Marketing, Sales, and Customer Support. You will also learn about these jobs on the Corporate View intranet. Three Corporate View simulation textbooks are available from South-Western/Thomson Learning. Ask your teacher for more information if you are interested in studying the career areas listed below.

Corporate View: Corporate Communications (ISBN: 0-538-69153-0)

Corporate View: Management & Human Resources (ISBN: 0-538-69978-7)

Corporate View: Marketing, Sales, & Support (ISBN: 0-538-69154-9)

DigiTools Digital Workbook

You have probably used many workbooks that were printed on paper. Because you are learning to use digital tools in this textbook, however, you will use a digital workbook for this course. The *DigiTools Digital Workbook* is provided on the *Instructor's Resource CD-ROM*. Your instructor will need to make these files available to you on disk or on your local area network.

The *DigiTools Digital Workbook* provides activities to review the concepts taught in the chapter, reinforce the vocabulary you have learned, and practice your math skills. You will complete online exercises to review and practice punctuation, grammar, and number and word usage. You will also complete drills to build keyboarding skills. Self Checks and Peer Checks will help you evaluate your work.

Instructor's Materials

Teaching support materials are available for instructors who adopt *DigiTools* for classroom use. These materials are described in the sections that follow.

Instructor's Manual

The *Instructor's Manual* is available to instructors who adopt the textbook for class use. The manual is a comprehensive source for practical ideas in course planning and enrichment. The manual includes teaching and grading suggestions and sample solutions for textbook activities and workbook review questions.

Instructor's Resource CD

An *Instructor's Resource CD* is available to instructors who adopt the textbook for class use. The CD includes:

- Data files for use by students in completing activities for the textbook and for the *DigiTools Digital Workbook*
- Lesson plans for each chapter
- Teaching strategy articles and assessment tools
- Sample solution files for selected student activities
- Electronic slides (in *Microsoft PowerPoint* format) for presenting/reviewing document formats
- The Corporate View intranet
- Supplemental speech recognition lessons for *Dragon NaturallySpeaking*, *IBM ViaVoice*, and *Microsoft Office XP Speech Recognition* in PDF format

ExamView ®

Instructors can purchase a flexible, easy-to-use electronic test bank and test generation software program that contains objective questions. Test bank questions are included for 15 chapter tests and a final exam. The **Exam***View* ® software enables instructors to modify questions from the test bank or add instructor-written questions to create customized tests.

Web Site

South-Western maintains a Web site to support *DigiTools*. Instructors may access the Web site at www.digitools.swlearning.com. The site provides instructor resources selected from the *DigiTools* Instructor's Manual and the Instructor's Resource CD, student resources including data files and *DigiTools Digital Workbook* files, updates (information made available after publication of the text), links to related sites, and information about related products.

Acknowledgements

Comments from reviewers of *DigiTools: Technology Application Tools* have been valuable in the development of this book. Special thanks to the following individuals:

Cindy Agnew
CTE Director
Fife School District
Tacoma, WA

Wanda J. Cadwallader
Business Education Teacher
Wilson High School, Tacoma Public School District
Tacoma, WA

Linda J. Hoff
Business Technology Lab Instructor
Century Career Center
Logansport Community School Corporation
Logansport, IN

Marilyn Horton
Teacher
Deltona High School
Deltona, FL

Linda Mallinson
Digital Design Instructor
Mid Florida Tech
Orlando, FL

Sondra S. Mangan
Curriculum Specialist, Business and Marketing
Puyallup School District
Puyallup, WA

Tina McCloud
Business Teacher
Mt. Vernon High School
Mt. Vernon, IN

Lajuana McKay
Business Education Teacher
Flagler County Schools
Palm Coast, FL

Jane M. McKillip
Business Education Teacher
Olympia High School
Olympia, WA

Donna Raske
Business Education Teacher
Mt. Tahoma High School
Tacoma, WA

Judith P. Sams
Chairman, Business Department
Fuqua School
Farmville, VA

Claudia H. Skinner
Business Education Consultant
NC Department of Public Instruction
Raleigh, NC

Shannon L. Thissen
Computer Technology Teacher
Sumner High School
Sumner, WA

Contents

Unit 1: DigiTools for Academic and Business Success

Chapter 1 Digital Communications Tools 2
- 1-1 Hardware and Network Know-how 5
  - What You Should Know About: Wireless Networking 8
  - Ethics: Computer Crime 9
- 1-2 Computer Operating Systems and Interfaces 10
- 1-3 Exploring *Microsoft Office* 15
- 1-4 File Management in *Windows* 24
- 1-5 Digital Pathways to Career Success 31
  - DigiTools Digital Workbook 32

Chapter 2 Internet and Intranet Basics 34
- 2-1 Browser Software 35
- 2-2 Internet Access and Security 37
  - What You Should Know About: Packets and Firewalls 37
  - What You Should Know About: URLs, Domain Names, and IP Numbers 39
- 2-3 Hyperlinks and Navigation for Web Sites 40
  - What You Should Know About: How Web Addresses and Paths Work 41
  - What You Should Know About: Teamwork 43
- 2-4 Learning from FAQs 44
  - Ethics: Intellectual Property Rights 46
- 2-5 Bookmarks and Favorites Lists 47
- 2-6 Digital Messages and Netiquette 48
- 2-7 E-mail Software and Addresses 52
- What You Should Know About: Computer Viruses 54
- 2-8 Managing E-mail Messages 55
- 2-9 Searching Online Resources 57
  - DigiTools Digital Workbook 60

UNIT 1 TOOLING UP! 60
- Career Clusters 60
  - Education and Training 60
  - Agriculture and Natural Resources 62
  - Architecture and Construction 63
  - Hospitality and Tourism 64
- Business Trends and Issues: Code of Ethics 65
- Critical Thinking: Handling Conflicts at Work 66
- Writing 67
  - Citing Online Sources 67
  - Capitalization Usage 68

Unit 2: Input Technologies

Chapter 3 Keyboarding: Alphabetic Keys 70
- 3-1 Home Keys (fdsa jkl;) 71
- 3-2 New Keys: h and e 75
- 3-3 New Keys: i and r 77
- 3-4 New Keys: o and t 79
- 3-5 New Keys: n and g 81
- 3-6 New Keys: left shift and period (.) 83
- 3-7 New Keys: u and c 85
- 3-8 New Keys: w and right shift 87
- 3-9 New Keys: b and y 89

CONTENTS

Chapter 4 Keyboarding: Number and Symbol Keys   108

What You Should Know About: Repetitive Stress Injury 108

4-1	Figure Keys (8, 1, 9, 4, and 0)	110
4-2	Figure Keys (5, 7, 3, 6, and 2)	112
4-3	Symbol Keys (/, $, !, %, <, and >)	115
4-4	Symbol Keys (#, &, +, @, and ( ))	117
4-5	Symbol Keys (=, _, *, \, and [ ])	118
4-6	Numeric keypad Keys 4/5/6/0	120
4-7	Numeric keypad Keys 7/8/9	122
4-8	Numeric keypad Keys 1/2/3	123
4-9	Subtraction and Multiplication	124
4-10	Division and Math Calculations	125

DigiTools Digital Workbook 126

Chapter 5 Word Processing   127

5-1	Memos	127
5-2	Letters	135

What You Should Know About: Postage Meters 142

5-3	Tables	143

What You Should Know About: Private Mail Delivery Service 152

5-4	Reports	154

Ethics: Protecting Confidential Information 159

5-5	Graphics	165

DigiTools Digital Workbook 171

Chapter 6 Handwriting Recognition   172

6-1	Control the Language Bar or Input Panel	174
6-2	Input with Writing Pad	178
6-3	Improve Writing Skills	182

Ethics: Loyalty 186

6-4	Input with the On-Screen Keyboard	186

DigiTools Digital Workbook 189

Chapter 7 Speech Recognition   190

What You Should Know About: Protecting Your Voice 191

7-1	Initial Training	192

What You Should Know About: Readjusting Your Microphone Settings 194

7-2	Dictating Voice Commands and Text	194

What You Should Know About: The Origins of Speech Recognition 198

7-3	Navigation and Punctuation	198
7-4	Correcting Errors	201

Ethics: Avoiding Gossip 206

7-5	Adding Words to the Dictionary	206
7-6	Changing Case	208
7-7	Formatting Text with Voice Commands	211

DigiTools Digital Workbook 216

UNIT 2 TOOLING UP!   217

Career Clusters	217
Scientific Research and Engineering	217
Health Science	218
Manufacturing	220
Transportation, Distribution, and Logistics	222
Business Trends and Issues: Global Marketplace	223
Critical Thinking: Telecommuting	224
Writing: Possessives, Numbers in Text, and Sentence Structure	225

Unit 3: Increasing Productivity with DigiTools

Chapter 8	**Spreadsheets**	**227**
8-1	Worksheet Basics	227
8-2	Creating and Editing Formulas	236
	What You Should Know About: Company Goals	**241**
8-3	Editing and Formatting Worksheets	243
	Ethics: Stewardship of Company Funds	**245**
8-4	Printing	251
8-5	Charts	257
	DigiTools Digital Workbook	**261**
Chapter 9	**Databases**	**262**
9-1	Creating a Database	263
9-2	Changing Table Structure and Fields	267
9-3	Arranging and Finding Data	272
	Ethics: Protecting Personal Data	**277**
9-4	Forms and Reports	277
	DigiTools Digital Workbook	**282**
Chapter 10	**Presentations**	**283**
10-1	Planning and Preparing a Presentation	283
	Ethics: Honesty in Presentations	**291**
10-2	Creating Visuals	295
	What You Should Know About: Digital Cameras and Scanners	**302**
10-3	Delivering the Presentation	308
	What You Should Know About: Body Language	**311**
	DigiTools Digital Workbook	**318**
Chapter 11	**HTML and Web Site Design**	**319**
11-1	Web Site Interaction Design	320
11-2	HTML Tags	323
11-3	Creating Simple Web Pages	326
11-4	Adding Color and Graphics to Web Pages	332
	Ethics: Graphics from the Web	**338**
11-5	Information Design for Web Pages	339
	What You Should Know About: Work-Related Injuries	**342**
11-6	Web Design Application Programs	346
	DigiTools Digital Workbook	**348**

Chapter 12	**Enhancing Workplace Performance**	**349**
12-1	Time Management	350
12-2	Effective Telephone Communications	359
	What You Should Know About: Voice Mail	**364**
12-3	Records Management	368
	What You Should Know About: Disaster Recovery Plans	**370**
	Ethics: Making Unauthorized Copies	**377**
	DigiTools Digital Workbook	**378**
UNIT 3	**TOOLING UP!**	**378**
	Career Clusters	378
	Arts, Audio/Video Technology, and Communications	378
	Retail/Wholesale Sales and Services (includes Marketing)	380
	Human Services	382
	Information Technology	383
	Business Trends and Issues: Quality Management and Customer Satisfaction	384
	Critical Thinking: Tangible and Intangible Rewards of Work	386
	Writing: Punctuation and Confusing Usage	386

Unit 4: DigiTools and Your Career Pathway

Chapter 13	**Finding and Analyzing Career Choices**	**389**
13-1	Examining Professional Teams in a Business	390
13-2	All About Job Descriptions	392
13-3	Learning About Jobs Online	397
	What You Should Know About: Attitude, Persistence, and Patience	**400**
	Ethics: Job Descriptions	**402**
13-4	Traditional Sources of Job Leads	402
13-5	Manage Job Search Data	404
	DigiTools Digital Workbook	**405**
Chapter 14	**Qualifying and Applying for a Career**	**406**
14-1	Job Search Documents	407

CONTENTS

Ethics: Honesty on Resumes and Applications	410
14-2 Comparing Applicant Qualifications Against Employer Needs	410
14-3 Resume Building: A Lifelong Process	412
What You Should Know About: Resume Myths	414
14-4 Finding E-Mail and Web Hosting Services	416
14-5 Chronological Resume	418
14-6 Letter of Application	420
14-7 Personal Career Web Page	423
14-8 Applying Online and Assessing Your Qualifications	424
What You Should Know About: Resumes and Scanning	425
DigiTools Digital Workbook	427
Chapter 15 Starting Your New Career Successfully	**428**
15-1 Personal Employment Portfolio	429
What You Should Know About: Electronic vs. Hard Copy Portfolios	432
15-2 Job Interviews	440
What You Should Know About: Being Polite	442
What You Should Know About: Smoking, Drugs, Alcohol, and Criminal Activity	448
15-3 Employee Rights and Obligations	452
What You Should Know About: Nepotism	453
Ethics: Respecting Diversity in the Workplace	455
DigiTools Digital Workbook	456
UNIT 4 TOOLING UP!	**456**
Career Clusters	456
Business and Administration	456
Government and Public Administration	458

Finance	460
Law and Public Safety	461
Business Trends and Issues: Work/Life Balance	463
Critical Thinking: Plan and Complete a Project	463
Writing: Ambiguous Pronouns and More Confusing Terms	465

Resources

Appendix A Reference Guide	**App-1**
Proofreaders' Marks	App-1
E-mail Format and Software Features	App-1
Envelope Guides	App-2
State and Territory Abbreviations	App-3
Report Documentation	App-4
Sample Documents	App-5
Appendix B Using Your Equipment and Media Properly and Safely	**App-11**
Operating Electrical Equipment	App-11
Using Your Computer Safely	App-11
Protecting Software and Data	App-11
Working with Printers	App-12
Caring for Peripherals	App-13
Understanding What You Have Learned	App-13
Computer and Media Use Checklist	App-13
Appendix C Skillbuilding	**App-15**
Appendix D Enhancing Personal Productivity with PDAs	**App-27**
Appendix E Microsoft® OneNote®	**App-119**
Index	I-1

UNIT 1

DigiTools for Academic and Business Success

Communicating with people is at the heart of academic and career success. Communication often involves using digital tools—called DigiTools for short. Unit 1 introduces these tools. In this unit, you will learn to use the *Windows* operating system and basic program commands. You will navigate the Internet and use online search tools to find information. You will send e-mail messages and manage computer files and folders. You will also learn tips for protecting computers from unauthorized users and other threats, such as computer viruses.

EXERCISE 4-7

UPDATE A BIOGRAPHY

Jim Plante, who works at Corporate View, submitted a draft of his online biography to Corporate Communications for approval and posting in the intranet. The Corporate Communications editor noticed some ambiguous uses of pronouns in the bio. There were also some terms in the document that were used incorrectly.

1. Start *Word*. Open the data file *Plante Bio*. Read this biography. Identify and correct all mistakes.

2. Save the corrected file as *Exercise 4-7* in the *DigiTools your name\Unit 4* folder.

Peer Check There are several ways to rewrite this document. Your final draft may read differently from the ones prepared by your classmates. Compare your revision with the rewrites completed by a classmate.

CHAPTER 1

Digital Communications Tools

OBJECTIVES — *In this chapter you will:*

1. Explore hardware, software, and input technologies for personal, professional, and academic success.
2. Learn about networks and network security.
3. Explore and use an operating system.
4. Learn about office suites.
5. Use basic commands and enter text in *Microsoft Word*.
6. Practice file management procedures.
7. Examine career pathways.

Communicating with people is at the heart of business, professional, and academic success. Successful communications in our high-tech 21st century require the proficient use of modern digital communications tools—called **DigiTools** *for short.*

TWENTY-FIRST CENTURY DIGITAL COMMUNICATIONS TOOLS

At the center of every DigiTool is a **microprocessor** or "microchip." A microprocessor is a small circuit board that controls all the processing within a computer. Microchips have made communicating more fun and convenient than ever. Almost anything big enough to hold a microprocessor can become a communications device. These devices are referred to as hardware. The physical parts of a computer or related equipment are called **hardware**.

Digital hardware can take many forms, for example, desktop computers, instrument panels in a modern car, or handheld devices. Consider the digital video phone, for instance. In years past when people went on vacation, they mailed postcards of their travels back home to their friends and family. These people could not imagine talking to friends wirelessly while sending live video home with a miniature video phone. This was the stuff of cartoons, like *Dick Tracy, The Jetsons,* or *Dexter's Laboratory.* Today, such magical DigiTools are all around us.

6. Compose a letter to send to each of the businesses in your database. Describe your project and ask for appropriate help (money, volunteers, or a loan of equipment, for instance). Submit the letters to your teacher for review. If you are working on a real project, mail the letters.

7. Prepare a presentation to stimulate support for the project. Include slides for:
- The overall purpose of the project
- Details of what the project involves
- Who the project will benefit and how
- What is needed (money, land, equipment, volunteers) to complete the project
- What you want to listeners to do (give money, donate equipment, volunteer time)
- Questions and answers

Practice the presentation and deliver it to groups such as the Chamber of Commerce, your school board, or the school's Parents Association. If you are working on a simulated project, deliver the presentation to a group of classmates.

8. Prepare a worksheet to record contributions from business or individuals and payments made for items needed to complete the project. If you are working on a simulated project, assume that the first ten businesses in your database from Step 5 made the following donations: $20, $50, $15, $100, $75, $40, $65, $150, $30, and $120. Record the business names and amounts in your worksheet. Calculate the total of the contributions. Write thank-you letters to all the businesses that contributed.

9. Write an advertisement about your project to place in the local newspaper. The purpose of the ad can be to announce the event or to ask for donations or other support.

10. Create a Web page to promote the project. Give complete details about the project similar to what you included in the presentation you prepared in Step 7. Ask readers to attend the event and/or to help provide support for the event as appropriate.

11. Do any other tasks needed to complete your project.

12. Meet with your team members to evaluate the project. Discuss questions such as:
- Did you accomplish the stated purpose of the project?
- What unforeseen problems did the team face in completing the project?
- What changes to the plan did you have to make as the project progressed?
- How could scheduling or tasks be completed more effectively for a long-term project you might do in the future?

WRITING

Ambiguous Pronouns and More Confusing Terms

In Unit 4, you accessed the Corporate View intranet and completed exercises to learn about ambiguous and repetitious pronoun usage and confusing terms. In Exercise 4-7, you will apply what you have learned.

Figure 1-1
A microprocessor controls all processing within a digital device.

These amazing hardware devices work so well because they're digital. **Digital devices** are those that communicate by sending data in electronic form (streams of zeros and ones), which microchips can understand. **Software** is a program that contains instructions for a computer. Software transforms these zeros and ones into pictures, text, or spoken words and transmits them to anywhere in the world. Software programs are also known as **applications**.

In the digital age we have a host of exciting software and hardware digital tools. You might use DigiTools to:

- Send e-mail
- Transmit instant messages
- Write reports
- Create charts and graphs
- Calculate your taxes and income
- Burn (create) CDs and DVDs
- Design Web sites
- Develop tables
- Make phone calls
- Send videos and photographs

A digital communication involves a message, an input method, hardware, software, and output as seen in Figure 1-2. A **message** consists of information—ideas, thoughts, or data that you want to share with others. **Input** refers to the method used to enter or transmit your message to the computer. The message or data itself is also sometimes called input. You might use a keyboard, mouse, speech recognition, handwriting recognition, digital camera, scanner, or some other input device. Your input message is read by software and processed by hardware. After processing, the message is shared as output.

to do the project. You would prepare correspondence but not mail it. Your presentation about the project would be given to classmates instead of members of your community. Follow your teacher's instruction regarding whether you should do a real or a simulated project.

EXERCISE 4-6

TEAM WORK

LONG-TERM PROJECT

Work with three or four classmates to complete this exercise.

1. Read the suggestions listed below to give you ideas for a project.
   - Your school needs a field where students can play soccer (or another sport).
   - Your class or club wants to sponsor an event such as a concert, party, or open house.
   - Your school wants to begin an after-school program to help students with homework and study skills.
   - Your class wants to hold a fund-raising event to raise money for a particular charity or to buy computer, music, or sports equipment.

2. Meet with your teammates and choose a project. Write a short paragraph that describes the purpose of the project clearly. Ask your teacher for approval to work on this project.

3. Work with your teammates to plan the project. Consider the *what, who, when,* and *where* for the project. Write a detailed plan that includes the information noted in the following questions.
   - What is the purpose of the project? State all the goals clearly.
   - What resources will be needed—money, people, equipment, land, etc.?
   - What information do you need to research—costs of renting or buying a field, costs of renting or buying equipment, costs of paying an adult to supervise, etc.?
   - What, if any, approvals will be needed from school officials? What, if any, approvals will be needed from parents so that students can participate?
   - What tasks must be performed to accomplish the project?
   - What software tools can best be used for handling different tasks related to the project?
   - When must each task be done? Develop a timeline/schedule for the project.
   - Who (on the team) will handle each task?
   - Who must be contacted for information or approvals? Who will be contacted to ask for support/money/volunteer time for the project?
   - Where will the event take place?

4. Once your team has planned the project, do the tasks assigned to you. If you are working on a simulated project, prepare any correspondence or write scripts of phone calls you would make. Do research to find information needed to complete the project.

5. Your team has decided to ask for support from local businesses. You can ask for money or for volunteers, speakers, or chaperones who would help in completing the project. Create a database containing information for 12 local businesses. Include the company name, mailing address, phone number, and a contact person at the business. Print mailing labels from the database.

> **DigiTip**
>
> To create mailing labels in *Access*:
> - Open the database. Select **Reports** on the Objects bar.
> - Choose **New** to open the New Report dialog box.
> - Click **Label Wizard.** Select the table that holds your data and click **OK**.
> - Choose options or provide information for the Wizard screens.
>
> When you click **Finish**, the Label report opens in Print Preview. Be sure that labels are loaded in the printer. Print the labels.

Figure 1-2
Digital Flow of Information

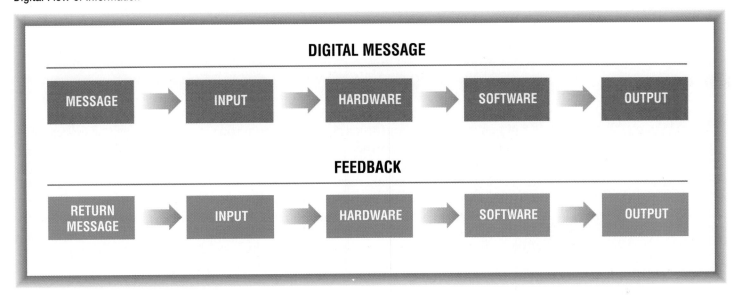

Output is information or data retrieved or received from a computer. The output is what someone sees, hears, reads, or experiences. Examples of output include a Web page, an e-mail message, a phone conversation, or a printed report.

WHY LEARN TO USE DIGITOOLS?

What reasons do you have for learning digital input, hardware, and software skills? The first reason is that using DigiTools is fun! Who doesn't like surfing the Web, using instant messaging, joining a multiuser online game, or downloading a movie trailer? These are all examples of communicating just for fun. A second important reason to study DigiTools is that they are essential for business, professional, and academic success.

WHAT YOU WILL BE DOING

While studying this text, you will learn, step-by-step, the ways you can use DigiTools for academic and professional growth and for personal enjoyment. After you understand the power of your DigiTools, you will be able to apply your skills in a variety of realistic situations.

What You Will Learn

You will start by learning about basic operating systems, Web browsing, intranets, and software tools. Then you will learn or improve keyboarding and word processing skills. From there, you will advance quickly to the new input technologies: handwriting recognition and speech recognition.

In later chapters, you will learn to use spreadsheet and database software. You will create multimedia presentations and use Web page creation tools. You will build skills in time and records management, teamwork, and communications to help you increase

BUSINESS TRENDS AND ISSUES

Work/Life Balance

The term *work/life balance* is commonly used to describe the need workers have to balance work with other aspects of life. In the last 20 years, the number of women in the workforce has increased significantly. This change has created more families with two working parents. Single-parent families are also on the rise. Many of the activities formerly handled by a nonworking parent must now be handled by a working parent.

When you think about a career, consider how your choice will affect all aspects of your life. Different careers make different demands on workers and their families. Some jobs may require much travel, overtime, or a long commute that will reduce time for family or participating in other activities. In many jobs, taking time off to care for a sick child or pursue a personal interest is very difficult.

When employees do not have enough time to take care of their personal matters, they bring stress to the workplace and are less productive. Many companies address this problem by creating a workplace that is supportive of workers' needs. For example, some companies have childcare facilities on site or help pay for childcare. Other companies create positions with flexible work hours. Employees in these positions can choose to arrive and leave work earlier or later than the normal working hours to accommodate their schedules. More companies are also providing benefits for part-time workers. Telecommuting on a part-time or full-time basis is a helpful option for some employees.

Companies find that employee loyalty increases when the company makes accommodations for workers' personal needs. Employees find that these accommodations contribute to their job satisfaction. When choosing an employer, consider whether the company fits your needs as well as whether you fit the needs of the company.

EXERCISE 4-5

LEARN MORE ABOUT WORK/LIFE BALANCE

INTERNET

1. Search the Web or visit the local library to find articles about work/life balance. Identify one or more companies that offer special benefits such as flextime or allowing pets at work. Note the source information for articles or Web pages you use.

2. Write a short unbound report to summarize the information you find. Title the report *Work/Life Balance*. Include a References page to list the source information for articles or Web pages.

3. Save the report as *Exercise 4-5* in the *DigiTools your name\Unit 4* folder. Print the report.

CRITICAL THINKING

Plan and Complete a Project

For this Critical Thinking, you will use all the skills you have learned in your study of *DigiTools* to plan, carry out, and evaluate a long-term project. You may choose a real project that you actually do or a "simulated" project. In a simulated project, you would plan the schedule, resources, and assignments for the project as if you were really going your productivity. You will explore career pathways and learn job search strategies. Throughout the course, you will examine ethical issues and explore business topics to help prepare you for the world of work.

Types of Exercises

As you study this textbook, you will complete activities, applications, and end-of-unit exercises and projects. Activities are learning exercises where you will explore new information or practice new skills. Applications will allow you to apply the new learning to completing a task, such as researching information or writing a report. End-of-unit exercises and projects will extend your learning and allow you to use the information and skills you have learned in the unit.

Self Checks and Peer Checks

On the job, employees are expected to produce a reasonable amount of work in a given time. Employers expect the work to be correct and well done. Do these expectations sound familiar? Your teachers probably expect you and the other students to complete your work correctly and on time. The good work habits you develop during your school years will be helpful to you later when you have a job.

Special instructions, called Self Check and Peer Check, appear in some activities and applications in the textbook. These instructions are designed to help you produce work that is correct and well done. A Self Check usually contains tips or questions that you are to ask yourself about a task or project you have completed. Answering these questions or reading the tip will help you decide whether your work has been done correctly.

On the job, employees sometimes seek input and help from other workers. Coworkers may supply important information, offer advice, or double-check the accuracy of another person's work. You can gain the same type of valuable help from your classmates when you follow the Peer Check instructions. During a Peer Check, you will discuss your completed work with a classmate. The information you gain from these discussions can help you correct or improve your work. Peer Checks also allow you to practice giving feedback on work to others in a tactful and constructive way. This is an important teamwork skill.

LESSON 1-1: HARDWARE AND NETWORK KNOW-HOW

OBJECTIVES *In this lesson you will:*

1. Identify types of digital hardware.
2. Learn about wireless networks.
3. Explore an Acceptable Use Policy.

Digital hardware includes any physical device that contains a microprocessor and is designed to communicate. A hardware device may be the latest Tablet PC or a traditional desktop computer. Whether your DigiTools are new or old, they are all communications

Careers in Law and Public Safety

Position and Yearly Salary		Position and Yearly Salary	
Police Dispatcher	$27,670*	Legal Secretary	$35,970*
Fire Dispatcher	$27,670*	Supervisor of Police and	
Correctional Office Manager	$46,030*	Detectives	$59,940*
Firefighter	$35,880*	Fire Inspector	$44,050*
Bailiff	$32,590*	Detective or Criminal Investigator	$52,150*
Parking Enforcement Worker	$27,480*	Police Officer	$41,950*
Animal Control Worker	$25,670*	Security Guard	$20,460*
Crossing Guard	$20,020	Lawyer	$69,030*
Judge	$79,540	Paralegal or Legal Assistant	$39,220*
Court Reporter	$42,530*	Law Clerk	$32,280*
Correctional Officer	$33,770*		

*Salaries based on 2001 industry averages, Bureau of Labor Statistics, www.bls.gov, downloaded January 2003.

EXERCISE 4-4

RESEARCH JOB SKILL AND TRAINING REQUIREMENTS

Language barriers can present difficult obstacles for professionals who are working to help people. Some professionals in Law and Public Safety must communicate in multiple languages. In any given community in the United States, several languages may be spoken with frequency. The ability to communicate in several languages is a helpful skill for public safety workers. What others skills, education, or training are needed? In this exercise, you will research this information for three careers in this cluster.

1. Choose three of the careers listed in the table for Law and Public Safety.

2. For each career, do research to learn what education, training, or skills are needed for success in the career. Your research may include finding job descriptions posted online for these careers, reading information from the *Occupational Outlook Handbook*, or interviewing persons who work in these careers. Be sure to note any mention of language skills as a job requirement. Note the source information for any articles or Web pages you use.

3. Write a short unbound report to summarize the information you find. Title the report *Career Qualifications*. Use the name of each career as a side heading. Include a References page to list the source information for articles or Web pages.

4. Save the report as *Exercise 4-4* in the *DigiTools your name\Unit 4* folder. Print the report.

Self Check Compare the format of your report with the unbound report on page App-7 in Appendix A. Correct the format, if needed.

tools. They will allow you to create messages and turn your thoughts and ideas into professional-looking outputs. Digital hardware tools include:

- Personal computers (PCs)
- Portable or laptop computers
- Tablet PCs
- Handhelds and personal digital assistants (PDAs): Palm Pilots, Handspring Visors, and Pocket PCs
- Wireless phones
- Digital cameras
- Pagers
- Scanners
- Networking hardware such as hubs, routers, access points, and switches
- Other digital devices

The parts of a traditional desktop computer are shown in Figure 1-3. Other digital devices may look very different from a desktop computer. Although they all contain a central processing unit, they may not have the same type of input and output devices.

Figure 1-3
Parts of a Computer

The numbered parts are found on most desktop computers. The location of some parts will vary.

1. **CPU (Central Processing Unit):** Internal operating unit or "brain" of computer.
2. **Disk drive:** Input/output device that reads data from and writes data to a disk.
3. **Monitor:** Output device that displays text and graphics on a screen.
4. **Mouse:** Input device used for commands and navigation.
5. **Keyboard:** Input device with an arrangement of letter, figure, symbol, control, function, and editing keys and a numeric keypad.

EXERCISE 4-3

TEAMWORK

INTERNET

EXPLORE A CAREER IN FINANCE

In this exercise, you will explore in depth a career in the Finance cluster. You will create a *PowerPoint* presentation to share the information you learn with the class.

1. Work with two classmates to complete this exercise. Your team will create a *PowerPoint* presentation about a career in the Finance cluster. Data about one career area, accountants and auditors, is provided from the *Occupational Outlook Handbook* (OOH) in the data file *Accountants*. You can use this career as the subject of your presentation. With your teacher's approval, you can research a different career related to finance on the OOH Web site. (Go to www.dol.gov and choose the **Occupational Outlook Handbook** link.)

2. Prepare a title slide that includes the career name, the names of your team members, and the current date. On the Notes for the title slide, indicate that the audience will be your classmates. Also note the purpose of the presentation—to inform and educate your listeners about one career in the Finance career cluster.

3. Include one or more slides about each of these topics:
   Nature of Work
   Working Conditions
   Employment
   Training, Other Qualifications, and Advancement
   Job Outlook
   Earnings
   Related Occupations

4. Read the OOH page to find important subpoints for each of the main topics listed in Step 3. Record these details on your slides. List only three or four subpoints on a slide. If you need to present more information, make another slide with the same main title. List the other details on the second slide. The OOH page contains much more information than you can include on your slides. Choose subpoints that you think are the most important. Your goal is to help the listeners understand this career area.

5. Add graphics, motion, or sound to enhance your slides. Save the file as *Exercise 4-3* in the *DigiTools your name\Unit 4* folder. Decide how to divide presenting the content among the team members and practice the presentation.

6. Deliver your presentation to the class or to a small group of classmates.

Law and Public Safety

As its name suggests, careers in the Law and Public Safety cluster involve creating or enforcing laws or helping to ensure the safety of the public. You are probably very familiar with some jobs in this area such as lawyer, police officer, and firefighter. After the attacks on the World Trade Center and the Pentagon on 9/11/01, many people gained a new appreciation of the vital role our firefighters and policemen play in our society. Other individuals with jobs in this cluster protect the rights of people and businesses in court. The court system employs a wide variety of people who help keep our streets safe, our business dealings honest, and our individual liberties protected.

Some jobs in the Law and Public Safety career cluster are shown in the following table. These are only a few of the many jobs in this extremely important area.

ACTIVITY 1-1

DIGITAL HARDWARE

DigiTip
Follow your teacher's instructions regarding when to have Peer Check discussions. Your teacher may set aside a certain time during a class period for discussions.

1. Think about your school, your home, and your workplace. Identify as many digital hardware devices as you can that are found in these places. Make a list of the devices. Examples include desktop computers, digital phones, and handheld computers or scanners. Remember, if the hardware is used to communicate and is a digital device, it qualifies for your list.

2. For each device you identified, give the following information:
   - A brief description of the device and its purpose
   - The input and output methods or equipment used with the device
   - The number of these devices found

Peer Check Discuss your answers with a classmate. What new information about digital hardware did you learn from the discussion?

NETWORKS

Hardware devices are often networked to be more effective. Networks allow data to flow from one computerized device to another through a series of hardware tools. For example, many networks use wires and special connectors to link computers together.

Increasingly, however, wireless networks are being used. Computers connect wirelessly to networks through access points. An **access point** is a piece of hardware that connects to a larger network. The access point routes the signal from wireless devices to that network. For example, you might use an access point to accept signals and connect to the Internet. The **Internet** is a public, worldwide computer network made up of smaller, interconnected networks that span the Earth. Increasingly, Internet connections are linked through fiber optic cables. These cables can transfer billions of communications at the same time.

Figure 1-4
This mouse is a wireless computing device.

2. Create a Web page listing the names of the sites that you have visited. Include a link to each government Web site you list on the page. Include descriptions of the public services each government agency offers.

Example:

U.S. Patent and Trademark Office
www.uspto.gov

"The basic role of the Patent and Trademark Office (PTO) is to promote the progress of science and the useful arts by securing for limited times to inventors the exclusive right to their respective discoveries (Article 1, Section 8 of the United States Constitution)."

Source: http://www.uspto.gov/web/menu/intro.html. April 3, 2003.

The PTO issues patents and registers trademarks. By providing information about patents and trademarks, it promotes an understanding of intellectual property rights.

3. Save your Web page as *Governments* in the *DigiTools your name\Unit 4* folder. Print the Web page.

Finance

Careers in the Finance cluster are varied, however, they all relate to managing money or to reporting or studying how money is used. For example, economists study how society uses money to purchase scarce resources such as land, labor, raw materials, and machinery to produce goods and services. Budget analysts examine, analyze, and seek new ways to improve efficiency and help businesses make more money. Accountants create financial reports such as income statements and balance sheets that report how a company has used its money and other resources. Insurance underwriters establish how much money customers will pay for insurance and write insurance policies.

The year 2002 will be remembered for several serious business failures. These failures resulted, in part, from poor financial reporting. One widely-publicized corporate failure was a company called Enron. Enron was forced to declare bankruptcy. The failure of Enron reminded everyone of the importance of honest, credible, and accurate accounting and earnings reports.

Look at some of the careers in Finance shown in the table below.

Careers in Finance

Position and Yearly Salary		Position and Yearly Salary	
Economist	$72,350*	Accountant	$50,700*
Cost Estimator	$50,450*	Auditor	$50,700*
Financial Manager	$75,430*	Credit Analyst	$62,440*
Benefit Specialist	$45,950*	Personal Financial Advisor	$69,300*
Budget Analyst	$53,040*	Insurance Underwriter	$48,770*
Insurance Appraiser	$42,360*	Financial Examiner	$59,860*
Bank Teller	$20,150*	Loan Officer	$50,060*
Bookkeeping Clerk	$27,820*	Tax Examiner	$45,180*
Payroll Clerk	$29,300*	Tax Preparer	$32,710*

*Salaries based on 2001 industry averages, Bureau of Labor Statistics, www.bls.gov, downloaded January 2003.

WHAT YOU SHOULD KNOW ABOUT
Wireless Networking

Wireless communications require that digital devices speak the same language. These languages are built on standards with funny names, like *802.11*, *WI-FI*, and *Bluetooth*. These standards allow wireless devices to communicate with other devices and their access points. When you move from place to place, your computer can switch automatically from one wireless access point to a closer access point.

Wireless computer networking was first made popular by the *Apple Airport* system. This system uses *802.11* or *WI-FI*. As long as you are within about 150 feet of an *Airport* access point, you can connect to the Internet or any local network.

Windows users can use *802.11*, too. Access points have been installed in many airports, hotels, schools, and other places. If you have the right wireless hardware, you can access the Internet from thousands of new locations. Business offices, government offices, and public parks in some cities are examples. *Bluetooth,* developed by IBM, also connects devices wirelessly that are about 30 feet from each other.

NETWORK SAFETY AND SECURITY

With millions of people using networks, there is an increased chance that someone will accidentally access or destroy the data files of others. Criminals also seek to gain access to information and use it for illegal purposes. Information technology (IT) specialists must manage and protect network accounts. A network account can be compared to a bank account. Only the account owner may access his or her account.

Passwords

To protect network accounts, special networking software keeps track of users and passwords. A **password** is a series of letters and/or numbers that you enter to gain access to a network or program. These passwords protect your data, files, pictures, e-mail, and other information from unauthorized use by others.

Your password is your primary security device. Choose it carefully! Read and follow these rules.

Rule 1 Do not share your password with anyone. Do not leave your password on a paper attached to your monitor, under your keyboard, or some other place where it can be found easily.

Rule 2 Do not create an obvious password that nearly anyone who knows you can easily guess, such as your boyfriend's or girlfriend's name, your birth date, or a pet's name.

Rule 3 Do not create an unnecessarily long password. A well-chosen short password can be just as secure as a long password.

Rule 4 Do not forget your password. Think of a word association to help you remember your password. This is called a mnemonic device. For example, if your password is "ringing," think "Phone call from a friend!"

The administration of government, just like business administration, has been radically changed by DigiTools. Governments are administered at national, regional or state, and county and city levels. Many government departments now have Web sites explaining what they do and how they do it. This is in response to efforts by governments to become more accessible to the public while reducing costs.

For instance, the Internal Revenue Service (IRS) used to be a paper-based tax collection agency. All forms, tax returns, and IRS statements had to be printed on paper. Now you can submit taxes forms electronically in a digital format that saves time and money. Your state, local, or provincial governments may have Web sites informing you of important upcoming elections, government services, and assistance to those in need. Review this list of possible careers in government service.

Careers in Government and Public Administration

Position and Yearly Salary		Position and Yearly Salary	
President of the United States	$400,000**	Community Service Manager	$44,540**
Vice President of the United States	$186,300**	Supervisor and Office Manager	$40,920*
		Administrative Services Manager	$55,470*
Senator	$141,300**	Public Relations Manager	$64,280*
Speaker of the House of Representatives	$181,400**	Compliance Officer	$46,250*
		Emergency Management Specialist	$45,260*
Judge or Magistrate	$79,540*		
Court Clerk	$28,930*		

*Salaries based on 2001 industry averages, Bureau of Labor Statistics, www.bls.gov, downloaded January 2003.
**As of January 2000.

EXERCISE 4-2 — GOVERNMENT ONLINE

1. Use your Web browser to visit government agencies at the national, state or provincial, and local levels. Use a search tool such as *Google* or *Yahoo!* to find Web addresses to complete this task.

- Visit a Web site about elected officials, such as the Congress of United States, the Parliament of Canada, or the Parliament in London.
- Visit a Web site for your state or provincial government.
- Does your city or county have a Web page? Search for it and see what information it contains.
- Visit several government agencies such as the U.S. Internal Revenue Service (www.irs.gov) or the U.S. Patent and Trademark Office (www.uspto.gov).
- Search for a Web page for a local Fire Department. This page might be part of a city or county Web site.
- Search for a Web page for a local Police Department. This page might be part of a city or county Web site.

ETHICS

Computer Crime

Network users are expected to act responsibly. They must be concerned with security and ethics related to information and networks.

Companies and other organizations must be careful to protect their confidential data. They must also protect the data gathered from others. Unauthorized users, called **hackers**, may be able to access and misuse information if proper safeguards are not in place. For example, a hacker might steal a customer's credit card number. Hacking is both unethical and illegal. Penalties for hacking can be to up to 20 years in jail!

A **computer virus** is a destructive program loaded onto a computer. A virus can run without the user's knowledge. Viruses are dangerous because they can quickly destroy data. They can also cause a computer or network to stop working properly. Some viruses can travel across networks and sneak past security systems. You will learn about protecting against computer viruses in Chapter 2.

Never hack or write computer viruses. The consequences could be extremely harsh regardless of your intent. You can learn more about computer crime at www.cybercrime.gov.

For Discussion

1. Have you encountered a computer virus at home, school, or work? What damage did the virus cause?

2. Why is hacking wrong, even if the hacker does not intend to steal or misuse information?

Figure 1-5
The U.S. Department of Justice hosts a Web site related to computer crime.

Source: www.cybercrime.gov.

Acceptable Use Policy

To meet the security and legal demands of employees, students, and customers, organizations must create rules to guide users. These rules are called **Acceptable Use Policies (AUPs)**. These rules are not usually open for debate. If you want to use the network, you must obey the rules. You will review a typical AUP in the following activity.

EXERCISE 4-1

TEAM WORK

SCANS GOALS

In this exercise, conduct a self-evaluation of how your skills compare to the SCANS workplace competency skills. Work with two classmates to complete this exercise.

1. Start *Word*. Open the data file *SCANS*. Create the following table in *Microsoft Word*.

2. Think and talk about the five SCANS skill areas with your team. As a group, think of an example of a career person doing what is described in each of the five SCANS skill areas. Record the examples in your table.

3. Evaluate your current abilities and qualifications for the SCANS categories listed in your table. Use this rating scale:

Poor	1 point
Weak	2 points
Good	3 points
Very Good	4 points
Excellent	5 points

Although 25 points are possible, do not worry if you don't receive a perfect score. Simply be honest in your assessment.

4. Give yourself the challenge to improve your SCANS score. Think about what you can do to improve your SCANS qualifications. In the SCANS table, write suggestions for how you can improve your score for each of the categories in the SCANS table. Save your document as *Exercise 4-1* in the *DigiTools your name\Unit 4* folder.

Government and Public Administration

Governments are made up of three different groups of people:

- Elected officials
- Appointed officials
- Professional career employees

Elected officials. Presidents, prime ministers, members of Congress or Parliament, governors, and state legislators are elected by the people. They serve for a limited time called terms. Their job is to represent the people while leading their governments.

Appointed officials. Judges and heads of government departments, ministries, and agencies are led by appointed leaders. They are appointed by elected officials and can be removed from their appointed positions by those same officials.

Professional career employees. Career government employees may work a lifetime in the same government department. However, governments also consist of employees who work for the government in a professional capacity. Employment for these government workers is unaffected by elections.

ACTIVITY 1-2

ACCEPTABLE USE POLICY

All new employees must take a survey at Corporate View, a company located in Colorado, before they can start using the corporate network. The AUP is in a typical survey format. Read the AUP shown in Figure 1-6 on page 11. Answer the following questions and express yourself on these security issues.

1. What do you think are some of the downsides of letting friends or acquaintances use your user account? What potential problems could arise?

2. Why do you think it is so essential to protect your password? What can happen if someone learns your password?

3. Give three examples of the types of activities that you think may be considered non-official use. What are some of the possible prohibited uses on a network?

4. Use a dictionary to provide an accurate definition of the following words: libel, slander, obscene.

5. What experiences have you had with **spam** (unwanted e-mail, advertising, or Web pages that appear without you requesting them)?

6. List other rules or restrictions specifically for the network you are using that did not appear in the Corporate View network AUP.

7. What are the login procedures for your system? What steps must you follow to enter your network legally and correctly? What are the steps your network administrator/instructor has given you regarding passwords on your local network?

Peer Check Discuss your answers with a classmate. What new information or insights did you gain from the discussion?

LESSON 1-2: COMPUTER OPERATING SYSTEMS AND INTERFACES

OBJECTIVES

In this lesson you will:

1. Learn the importance of a computer operating system.
2. Learn about three types of computer interfaces.
3. Explore a graphical user interface.

Every digital tool using a computer chip requires an operating system. An **operating system (OS)** is a program that controls the operation of a computer. An OS manages all of the digital tools on personal computers and other computing devices. The OS receives input and interprets the instructions of the user. It interacts with your input devices. The OS then sends your commands to the microprocessor.

intranets and networks. Business professionals can now use handheld, tablet, portable, and desktop computers to keep track of contacts, appointments, employees, products, marketing campaigns, financial records, and their public Web sites. Today, executives are likely to create their own business correspondence rather than turn their document preparation over to secretaries or typists.

To improve their effectiveness, business professionals look for faster and more effective DigiTools. For example, some businesses now use wireless networks, speech and handwriting recognition, and mobile computing to improve employee productivity. Examine the following career opportunities in business and administration. As you read through the list, think about how each of these business professionals can use new digital technologies to do their jobs more effectively.

Careers in Business and Administration

Position and Yearly Salary		Position and Yearly Salary	
Chief Executive Officer	$107,670*	HR Employment or	
Receptionist	$21,450*	Recruitment Specialist	$44,320*
Executive Secretary or		General Manager	$73,570*
Administrative Assistant	$33,970*	Human Resource Manager	$61,303*
Business Manager for Artist		HR Administrative Assistant	$30,570*
or Performer	$62,480*	HR Compensation Specialist	$45,950*
Purchasing Agent	$46,090*	Training Specialist	$44,790*
Compliance Officer	$46,250*	Meeting or Convention Planner	$39,680*
		Management Analyst	$64,470*

*Salaries based on 2001 industry averages, Bureau of Labor Statistics, www.bls.gov, downloaded January 2003.

What skills do you need to be successful in business? Certainly an understanding of DigiTools is essential. However, there is a broader set of business skills to consider. SCANS, the Secretary's Commission on Achieving Necessary Skills, identified the following workplace competencies that employers are looking for in their employees. As a potential businessperson, you too will need the following workplace skills:

Effective workers can productively use:

Resources. *They know how to allocate time, money, materials, space, and staff.*

Interpersonal skills. *They can work on teams, teach others, serve customers, lead, and negotiate, and work well with people from culturally diverse backgrounds.*

Information. *They can acquire and evaluate data, organize and maintain files, interpret and communicate, and use computers to process information.*

Systems. *They understand social, organizational, and technological systems; they can monitor and correct performance; they can design or improve systems.*

Technology. *They can select equipment and tools, apply technology to specific tasks, and maintain and troubleshoot equipment.*[1]

---

[1] U.S. Department of Labor Web site. "Secretary's Commission on Achieving Necessary Skills." 3 April 2003. http://wdr.doleta.gov/SCANS/lal/LAL.HTM.

Figure 1-6
Acceptable Use Policy

Acceptable Use Policy

1. Your username and network account are for your professional use only. They are not transferable to others. You may not allow others to use your account. Do you agree not to let others use your network account?

   ☐ I Agree   ☐ I Disagree

2. Your password is the primary way to protect your work, files, folders, and personal information. Do you agree to never share your password with others?

   ☐ I Agree   ☐ I Disagree

3. Access to this network is for official purposes only. You are specifically prohibited from using your access to this network for personal, non-official use. Do you agree to this restriction?

   ☐ I Agree   ☐ I Disagree

4. Employees are prohibited from using the network to publish or distribute libelous, slanderous, obscene, or inappropriate literature, graphics, or other offensive materials. Do you agree to this restriction?

   ☐ I Agree   ☐ I Disagree

5. Disk space on company networks is used to store company data. Space is limited. It is important to delete unwanted e-mail, graphics, and any unused or unnecessary multimedia, music, or video files. Are you willing to delete unnecessary and unwanted files weekly?

   ☐ I Agree   ☐ I Disagree

6. When using e-mail or other Internet software, do not create mass mailings or spam. Do not create any message with instructions to forward the message to multiple addresses. Do not create harassing messages. Comply with any requests to stop sending unwanted messages. Are you willing to follow these instructions?

   ☐ I Agree   ☐ I Disagree

7. Graphics, applications, databases, audio files, video files, and other works that are protected by copyright should not be acquired or distributed over this network without complying with the owners' licensing terms. Only lawful activities are allowed on the network. Do you agree not to acquire or distribute copyrighted information?

   ☐ I Agree   ☐ I Disagree

DIGITOOLS
DIGITAL WORKBOOK — CHAPTER 15

Open the data file *CH15 Workbook*. Complete these exercises in your *DigiTools Digital Workbook* to reinforce and extend your learning for Chapter 15:

- Review Questions
- Vocabulary Reinforcement
- Math Practice: Calculating Person Budget Amounts
- From the Editor's Desk: Confusing Word Usage
- Keyboarding Practice: High-Frequency Words and Speed Building

UNIT 4

Tooling Up!

In this final Tooling Up! Section, you will go online and use a variety of DigiTools to learn, think, and write about career alternatives, business issues, and career trends. The skills and knowledge you have acquired in the previous chapters will help you complete these exercises.

CAREER CLUSTERS

Business and Administration

Workers in the Business and Administration area manage or provide support for others in the company who create the company's products or services. For example, a chief executive officer develops a company's goals and policies and supervises other managers. A general manager manages daily operations of a company and plans the use of materials and labor. A human resource manager is responsible for handling employee benefits questions and recruiting, interviewing, and hiring new employees. Purchasing agents try to obtain the highest-quality material at the lowest possible cost for the company. Business managers of artists, performers, and athletes handle contract negotiations and other business matters for clients.

The administration of business has been deeply changed by modern DigiTools. Computers have revolutionized business communications through the use of corporate

Some operating systems, such as *Windows*, *Linux*, or *Macintosh OS*, have become household names. Other OSs, such as *Windows Pocket PC* or *Palm OS*, are also very popular. These OSs are used to power handheld personal digital assistants. A special OS that runs Tablet PC computers is called the *Windows Tablet PC Edition*.

A **computer interface** is the means by which users communicate and interact with the computer. Early OSs were awkward and input was limited to a keyboard. This type of OS has a text user interface. With a text user interface, users must type text to input commands to a computer. Using such an interface may require memorizing dozens of typed commands.

Other types of OSs are much more user-friendly. In the 1980s, a graphical user interface (GUI) was introduced, first on the *Apple Macintosh* and later in the *Microsoft Windows* operating system. A GUI uses pictures called **icons** to represent commands or programs. With a GUI, users can input computer commands with a mouse, a track pad, or a digital pen. In the late 1990s, a speech user interface (SUI) was also introduced. An SUI allows users to input commands by talking rather than by typing or clicking with a mouse. For example, using an SUI you can say "File" to open the File menu.

Figure 1-7
Microsoft Windows is a GUI OS allowing speech, handwriting, keyboard, or mouse inputs.

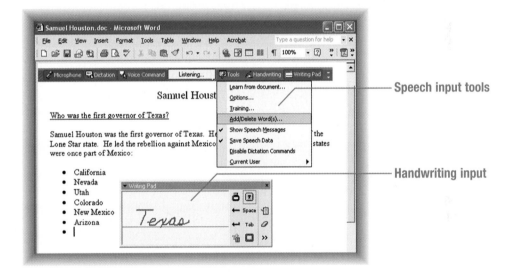

In 2002, Microsoft released a special version of its popular *Windows* OS for Tablet PC computers. A Tablet PC accepts spoken input, handwritten input (by writing directly on the screen), and traditional text input using a keyboard.

Figure 1-8
With a Tablet PC, users can enter text using a keyboard or by writing on the screen with a digital pen.

ETHICS
Respecting Diversity in the Workplace

You have learned that employers are required to treat fairly job applicants and employees of many cultural backgrounds, races, and ages. You, as an employee, must also respect diversity in the workplace. Treating others unfairly because of their background or because they belong to a particular group is unethical.

Diversity, as it relates to organizations, means having a workforce with people from a wide range of ethnic and cultural backgrounds. Many companies seek to have diversity at all levels of the organization. Employees with different backgrounds and new perspectives help businesses meet the needs of diverse customers.

In some companies, a diversity coordinator collects data about the company's hiring and promotion policies. This person also tracks the progress of the company in achieving its diversity goals. Employees are expected to respect coworkers and customers from all backgrounds. Diversity training programs are conducted to help employees become aware of issues related to diversity.

Follow-Up

1. Have you seen a situation where you thought someone was treated unfairly because of his or her age, gender, or ethnic background? Describe the situation.

2. What consequences may an employee who treats others unfairly face?

ACTIVITY 15-19 — REVIEW IMPORTANT EMPLOYEE RIGHTS AND OBLIGATIONS

1. Work in a team with three classmates to learn more about the rights and obligations of employees at a typical corporation.

2. Assign one topic or group of topics below to each team member.
   - At-Will Employment, Equal Opportunity Employment, Other Employment
   - Attendance, Confidentiality Policy, Travel
   - Drug and Alcohol Policy
   - Progressive Discipline

3. Access the Corporate View intranet. Choose the **Regular Features** and **Employee Handbook** links. Choose other links that lead to information about your assigned topic(s). Read the information and make notes about the policies.

4. Write a short summary of each topic you have been assigned. Share these summaries with your team. Discuss them and make sure each team member understands each policy.

5. Save each summary by the name of its link in the *DigiTools your name\Chapter 15* folder. For example, the Travel policy report would be named *Travel*.

In the following activity, you will use your OS and the mouse, touch pad, or digital pen to discover some of the exciting DigiTool applications available to you.

> **DigiTip** Many physically challenged people now have devices that allow them to operate a computer completely with the use of their eyes. By looking at a portion of the screen and blinking, they can open and close programs, send messages, and interact with a computer.

ACTIVITY 1-3

EXPLORE AN INTERFACE

DigiTip
For information on using your computer and related devices properly and safely, see Appendix B on page App-11.

1. Start your computer. When the OS has loaded, usually at the bottom or side of your screen you will have a taskbar. Figure 1-9 shows an example. The taskbar allows you to open and close programs or switch quickly between programs by simply clicking or tapping its buttons.

Figure 1-9
The taskbar in *Windows XP*

DigiTip
The instructions in this textbook direct you to give commands by clicking or double-clicking the mouse. Your teacher may instruct you to use a digital pen or speech recognition tools instead.

2. Start *Microsoft Word*. Click **Start**, (**All**) **Programs**, **Microsoft Word**. You will use *Word* to explore the *Windows* interface. The interface parts are similar in all *Windows* programs.

Figure 1-10
The Parts of a Typical Application Program Window

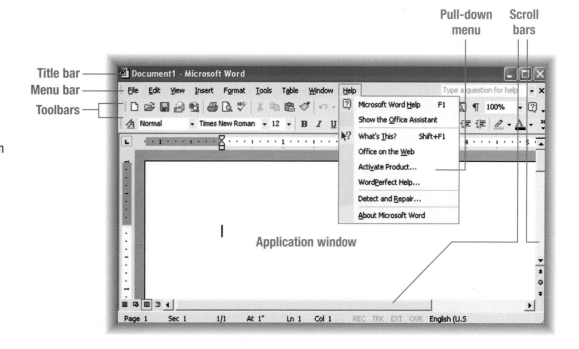

Federal law also seeks to protect workers' safety. The Occupational Safety and Health Act of 1970 was enacted to assure safe and healthful working conditions for workers in the United States. The Occupational Safety and Health Administration (OSHA) is a federal government agency responsible for enforcing the law. In addition, OSHA provides research, information, education, and training on occupational safety and health.

You will be expected to know and follow the rules and regulations regarding any hazardous materials you use on the job. Doing so is important for your safety and that of your coworkers.

Figure 15-12
Employees should know about dangerous substances they may encounter on the job.

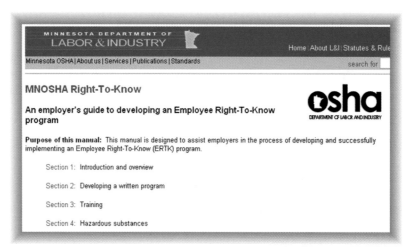

Source: http://www.doli.state.mn.us/rtkgen.html

EMPLOYEE HANDBOOK

After you report to your new employer, you will probably be asked to attend a new employee training orientation. You may also be given an Employee Handbook. An Employee Handbook is a reference guide to your privileges and obligations as an employee.

For instance, you are obligated to help cultivate an atmosphere where harassment is not tolerated. In this next activity, you will review the rules Corporate View has established regarding harassment. In the activities that follow, you will bring your team together and investigate other important issues regarding employee rights, privileges, and obligations.

ACTIVITY 15-18 REVIEW A HARASSMENT POLICY

1. Access the Corporate View intranet. Choose the **Regular Features, Employee Handbook**, and **Harassment Policy** links.

2. Read the information provided on harassment in the Employee Handbook. Write a definition of sexual harassment as explained in this document.

3. What may happen to a Corporate View employee who harasses another employee?

4. What other types of harassment are specifically mentioned in the Harassment Policy?

3. Applications, like a word processing program, appear in windows. When a file holds more information than can be displayed in the window, you can use the scroll bars to move up and down or side to side to see more information. Practice moving the scroll bars. Click or tap on the slider bars or the arrows at the ends of the bars to move them up or down, left or right.

4. The title bar displays the name of the application and document that are currently open. Toolbars allow you to give commands by simply clicking or tapping on toolbar buttons. The menu bar displays drop-down menu choices from which commands can be selected. Click **Help** on the menu bar to view the choices available. Click in the application window on a blank area to close the menu without choosing a menu option.

5. Choosing an option on a menu may execute a command or open a dialog box. A dialog box contains options or settings you can choose or enter. Click **View** on the menu bar. Click **Task Pane**. This command will display the task pane on your screen (or hide the task pane if it is already displayed.) See Figure 1-11. Click **Edit** on the menu bar. Click **Find**. The Find and Replace dialog box displays. See Figure 1-12. In this box, you could enter text in the Find what box that you wish to locate in the document. Click **Cancel** to close the box.

Figure 1-11
Task Pane in *Microsoft Word*

Figure 1-12
The Find and Replace Dialog Box

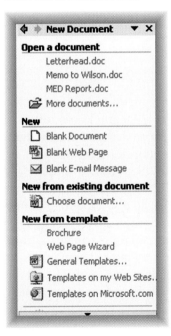

6. Application windows can be minimized, maximized, and restored to their previous conditions. At the right side of the title or menu bar are the Control buttons: Minimize, Close, and Maximize or Restore. Clicking the Minimize button reduces the window to a button on the taskbar. This is useful when you want to use more than one program at once. Click the **Minimize** button. To restore the program, click its button on the taskbar.

Figure 1-13
Windows Control Buttons

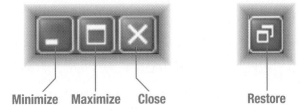

2. Prepare a *PowerPoint* presentation to present key information from this document. Include an appropriate title slide. For each of the laws listed below, include a slide that tells briefly the type of discrimination the law prohibits.
- Title VII of the Civil Rights Act of 1964 (Title VII)
- Equal Pay Act of 1963 (EPA)
- Age Discrimination in Employment Act of 1967 (ADEA)
- Title I and Title V of the Americans with Disabilities Act of 1990 (ADA)
- Sections 501 and 505 of the Rehabilitation Act of 1973
- Civil Rights Act of 1991

3. Include one or two slides that list the discriminatory practices prohibited by these laws.

4. Include a slide or two that lists remedies available when employment discrimination is proved.

5. Insert graphics on some slides to make the show more interesting. For example, you could use a dollar sign graphic on the Equal Pay Act slide. You could use an American flag on the Americans with Disabilities Act slide. Save the slides as *EEOC Show* in the *DigiTools your name\Chapter 15* folder.

6. Deliver your presentation to the class or a small group of classmates.

Peer Check Ask your classmates for feedback on your slides and on how you delivered the presentation.

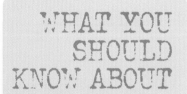

Nepotism is favoritism shown to relatives or close friends by someone in a position of power. For example, nepotism could be the hiring of friends or relatives even though other applicants are more qualified. Nepotism can also be giving raises, promotions, or other special treatment to friends or relatives in the company. Nepotism is simply playing favorites based upon family relationships or friendships.

Many companies have rules against nepotism. These rules do not allow one family member to hire, supervise, or evaluate the performance of other family members. In other words, family members are often required to work in different departments within a company. In some companies, an applicant will not be hired if an immediate family member already works for the company. If you are thinking of applying for a job at a company where an immediate family member already works, ask for information about the company's policy on nepotism.

EMPLOYEE RIGHT TO KNOW

An Employee Right-To-Know Act was passed by the Minnesota Legislature in 1983. Its purpose is to ensure that employees are aware of the dangers associated with hazardous substances, harmful physical agents, or infectious agents they may be exposed to in their workplaces. Many states have these so called "Employee Right-to-Know" laws.

7. Clicking the Maximize button enlarges a window to take up almost the entire screen. After you have maximized a window, the Restore button will replace the Maximize button. Clicking the Restore button restores the window to its previous size and location. Experiment with the Minimize, Maximize, and Restore buttons to become familiar with how they work.

8. An application window can be changed in size or shape. If the program is maximized, click the **Restore** button. Click and drag the bottom corner of the window in and out to make it bigger or smaller.

9. Clicking the Close button closes a window. Click the **Close** button to close *Word*.

LESSON 1-3 EXPLORING *MICROSOFT OFFICE*

OBJECTIVES *In this lesson you will:*

1. Learn about common applications included in software office suites.
2. Open and close programs and documents.
3. Enter text and navigate in *Word*.
4. Use menus and toolbars.
5. Access Help.
6. Print a document.

OFFICE SUITES

An office suite is a bundle of software programs sold together. These programs are designed to meet the needs of people in many walks of life. Some major office suites are:

- *Microsoft Office*
- *Lotus SmartSuite*
- *Corel WordPerfect Office*
- *StarOffice*
- *AppleWorks*

Generally, office suites include word processing, spreadsheet, presentations, database, and personal management programs. *Microsoft Office* is a widely used office suite. The programs in a version of *Microsoft Office* designed for students are:

Microsoft Word

- A word processing program
- Allows the creation, formatting, and printing of documents such as reports, letters, and brochures

LESSON 15-3

EMPLOYEE RIGHTS AND OBLIGATIONS

OBJECTIVES *In this lesson you will:*

1. Review equal opportunity employment hiring practices.
2. Learn about employee rights and obligations.
3. Review company policies in an online Employee Handbook.

HIRING PRACTICES

As an employee, you have certain rights, obligations, and privileges. Some of these rights and obligations relate to hiring and employment practices. Often, job seekers feel at the mercy of interviewers and employers. However, the rules are on your side. All employers must follow certain rules when hiring. For example, employers must provide equal opportunity to employment regardless of race, age, or gender. They must provide a fair and equitable hiring atmosphere.

The most important source of equal opportunity employment information is the **Equal Employment Opportunity Commission** (EEOC). The EEOC is an independent federal agency originally created by Congress in 1964 to enforce Title VII of the Civil Rights Act of 1964. It has authority to establish equal employment policy and provides education and technical assistance activities through field offices serving every part of the United States. You can visit the EEOC online at www.eeoc.gov.

Figure 15-11
The EEOC Online

Source: *http://www.eeoc.gov/*.

ACTIVITY 15-17 — JOB DISCRIMINATION PRESENTATION

1. Start *Word*. Open the data file *Job Discrimination FAQs*. This document has been downloaded from the EEOC Web site. It contains information about job discrimination. Read the entire *Job Discrimination FAQs* file to learn more about job discrimination laws.

Microsoft Excel
- A spreadsheet program
- Used to organize and manipulate numerical and text values in worksheets

Microsoft PowerPoint
- A presentation program
- Used to create multimedia slide shows and visual presentations

Microsoft Outlook
- A personal management and team collaboration program
- Helps individuals and teams plan, schedule meetings, manage appointments, organize contacts, and send e-mail

To appeal to a broad variety of workers, software companies often place additional programs into their office suites. For instance, Microsoft offers a version of *Microsoft Office* that includes *Microsoft FrontPage*, *Microsoft Publisher*, and *Microsoft Access*. *FrontPage* is a Web site creation program. *Publisher* is a desktop publishing program. *Access* is a database application that organizes information so it can be searched, queried, and sorted. Other software developers may add graphics programs or teamwork tools to their office suites.

Using a software suite has advantages over using a group of individual programs. One advantage of using an office suite is that many features and commands that you learn to use in one program can also be used in other programs in the suite. For example, the command used to print a document in *Word* is the same as the *Excel* print command. This makes learning and using the programs easier.

Another advantage of using an office suite is that the programs in a suite are designed to work together. This means that you can share data among programs easily. For example, a graph created in *Excel* can be placed in a report created with *Word*. An *Access* database containing names and addresses can be merged with a *Word* document to create personalized letters. An outline for a speech created in *Word* can be used in *PowerPoint* to create electronic slides. You will learn how to share data among programs as you use the *Microsoft Office Suite* and complete the lessons in this textbook.

OPENING AND CLOSING PROGRAMS

You will begin by learning to open and close programs. The procedure is the same for all programs in the *Microsoft Office* suite. Later, you will open and close documents while leaving *Word* open.

ACTIVITY 1-4　　OPEN AND CLOSE PROGRAMS

1. Turn on the computer and the monitor. Enter a password to access your computer or network, if required, to display the *Windows* desktop.

2. Click the **Start** button at the bottom of the screen. Point to (**All**) **Programs** to display the Programs menu. Click a program name (in this case, **Microsoft Word**). The *Word* application window displays (as seen earlier in Figure 1-10).

WHEN YOU DON'T GET THE JOB

Always be prepared mentally for the possibility that you may not be selected for a specific job. Don't let a rejection upset you! Instead, use the interview experience to help you prepare for your next interview. Think about what you did right and what you can do better to improve your chances the next time. You can improve your chances by keeping the following thoughts in mind:

- Keep applying! Remember, it may take a dozen interviews or more to get a job you really want.

- Keep practicing your interviewing skills.

- Keep reviewing, rewriting, and targeting your resume toward the skills needed in the job for which you are applying.

- Keep improving your employability skills. Take classes, study, and take advantage of government retraining programs for new job seekers.

Rarely does anyone fit an employer's job description perfectly. If you do have gaps in your preparation, what can you do to eliminate them or minimize their effect? Do you need to take some classes or get into a training program? Simply participating in a training program may give you the edge you need.

Federal, state, and local governments often provide assistance for those looking for work. They sometimes provide financial support and tax breaks for expenses incurred while improving job skills. To learn about training, scholarships, and assistance packages, talk with personnel at local Job Service and Unemployment Offices about programs that may apply to you.

ACTIVITY 15-16 ENHANCING YOUR QUALIFICATIONS

In this activity, take some time to think about how you can improve your qualifications for employment.

1. Start *Word* and open a new blank document. Think of a job you would like to have sometime in the future. Key the name of the job and the career cluster in which it falls.

2. List possible courses, training, schools, and self-improvement programs that may help you obtain this job.

3. List skills you need to learn that are related specifically to this job.

4. List general work skills, like organization or problem-solving ability, that you need to improve as you prepare for a job.

5. Save the file as *Enhancing My Qualifications* in your *DigiTools your name\Chapter 15* folder.

3. The *Windows* OS allows multitasking. This means you can have two or more applications open at the same time. Open *Excel*. (Click the **Start** button. Point to (**All**) **Programs** to display the Programs menu. Click **Microsoft Excel**.) The *Excel* screen displays.

Figure 1-14
Excel Application Window

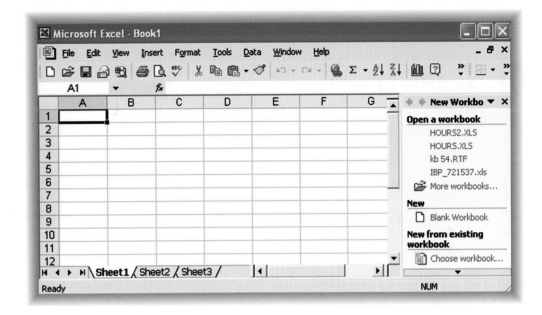

4. The *Excel* application window may appear on top of the *Word* application window. To bring a window to the front or to restore it from the taskbar, click the document name on the taskbar. Practice switching between the *Word* and *Excel* windows by clicking the document names on the taskbar.

Figure 1-15
Click a document name on the taskbar to restore it.

5. You can close a program by choosing a command from the menu bar or by clicking the Close Control button in the upper right corner of the screen. Switch to the *Excel* program. Click **File** from the menu bar. Click **Exit**. (If you do not see the Exit command on the menu, click the double arrow at the bottom of the menu to display more choices.)

6. Now close *Word*. (Restore the program to the screen if it is not displayed.) Click the **Close** button at the right of the title bar.

CHAPTER 1   Lesson 1-3   17

3. Conclude your letter by expressing your continued interest in working for Corporate View. Keep your letter to one page in length. See page 470 in Appendix A to review block style letter format. Include your return address above the date or create a personal letterhead to appear at the top of the page.

4. Save your letter as *Follow-up Letter* in your *DigiTools your name\Chapter 15\Portfolio* folder. Print the letter.

ACCEPTING EMPLOYMENT

Waiting is always difficult. It is especially difficult waiting for a job offer. You will probably be contacted by phone, e-mail, or registered letter to be offered a job or to be told when a job has been filled. If a reasonable time has passed, it is acceptable to call and check on the status of the hiring.

After you have received a job offer, you will need to accept or reject the offer. If you decide to reject the job offer, do so politely. Give reasons why you have decided not to accept the job and thank the company for its consideration. For example, perhaps you think the salary you have been offered is not acceptable.

If you decide to accept the job, respond to the employer promptly by phone and work out the details of when you will begin work. Also, accept the job formally in writing using an acceptance letter. Depending upon the nature of the job, you may be asked to sign an employment contract.

ACTIVITY 15-15 — WRITE A LETTER OF ACCEPTANCE

Imagine you have received a letter offering you a job. You must prepare a letter of acceptance.

1. Write a sample letter of acceptance to place into your personal employment journal for future reference. Pretend you have been hired by Melissa Kim at Corporate View for the Web and Intranet News Writer and Editor position. You are to start work on the first day of next month. Your salary is $30,000 per year.

2. The letter should:
- State your acceptance with appreciation.
- Briefly reaffirm agreed-on terms of employment, such as the day you will begin work and your salary.
- Express enthusiasm about joining the company.

3. Keep your letter to one page in length. See page App-6 in Appendix A to review block style letter format. Include your return address above the date or create a personal letterhead to appear at the top of the page.

4. Save your letter as *Acceptance Letter* in your *DigiTools your name\Chapter 15\Portfolio* folder. Print the letter.

ENTERING TEXT

When you enter text using your keyboard or handwriting or speech recognition, it is entered at the insertion point (the blinking vertical bar). In *Word*, when a line is full, the text automatically moves to the next line. This feature is called wordwrap. To begin a new paragraph, press the **Enter** key.

ACTIVITY 1-5

ENTER TEXT IN *WORD*

1. Open the *Word* program. Locate the blinking vertical line in the application window. This is the insertion point.

Figure 1-16
The insertion point is where text will appear on the screen.

2. Using your keyboard, key the text that follows. Press the **Space Bar** at the bottom of the keyboard to enter a space between words. Hold down a **Shift** key as you strike a letter to key a capital letter. Press **Enter** only at the end of a paragraph. (Don't worry if you don't know how to type. You will learn to key properly by touch in Chapter 3. For now, looking at the keys is okay.)

```
Text is keyed in the application window.  To change or
edit text, you must move the insertion point around
within the document.  You can move to different parts
of the document by using the mouse or the keyboard.

To use the mouse, move the pointer to the desired
position and click the left mouse button.  You can also
use the arrow keys on the keyboard to move the
insertion point to a different position.
```

3. Practice using the arrow keys and the mouse or digital pen to move to different locations in the text you have keyed. Move the insertion point to just before *application* in the first line. Key `Word`.

4. Move the insertion point to just after the word *different* in the second paragraph. Press the **Backspace** key several times to delete *different*. Key the word `all`.

5. Press the **Home** key to move quickly to the beginning of the line. Press the **End** key to move quickly to the end of the line.

DigiTip
The instructions in this textbook often direct you to enter text by keying (typing) it. Your teacher may instruct you to use handwriting or speech recognition tools instead.

Figure 15-10
The CDC and other government agencies provide information about smoking and the workplace.

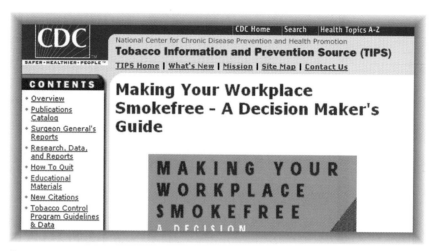

Source: http://www.cdc.gov/tobacco/research_data/environmental/etsguide.htm

After the Interview

After the interview, find a quiet spot and take notes on anything you think you should remember. Analyze your performance. Think about how you could have done better, thereby preparing yourself for future interviews.

Every contact with a potential employer is an opportunity to make a good impression. Always write a follow-up letter after the interview. Thank the interviewer for seeing you and mention again your most relevant qualifications for the job.

Keep a record of every interview in which you participate. Make a note of the names of the people you met or talked with. Note your impressions of how you did on the interview. Also note any questions you want to ask if you have an opportunity to talk with the interviewer again.

ACTIVITY 15-14   WRITE A FOLLOW-UP LETTER

Imagine that you have just interviewed for a job you really want! In this activity, you will write a follow-up letter to increase your chances of being considered.

1. Pretend that you have been interviewed by Melissa Kim at Corporate View for a Web and Intranet News Writer and Editor position. Access and read the description for this job on the Corporate View intranet. Choose **Human Resources & Management, Current Job Openings at TeleView, Corporate Communications**. Then choose **Web and Intranet News Writer & Editor**.

2. Write a thank you follow-up letter addressed to Melissa. Start by thanking the interviewer for a pleasant interview experience. In a separate paragraph, list the skills and experiences that you think match what the interviewer seems to be looking for in a new hire. Consider the general work skills Corporate View values (listed on page 444) as well as the information in the job description.

6. Move the insertion point to just after *also* in the second paragraph. Press the **Delete** key five times to delete the word.

7. Press the **Page Up** key to move quickly to the top of the screen. Press the **Page Down** key to move quickly to the end of the page.

8. Leave your document on the screen to use in the next activity.

SAVING A FILE

Saving a document preserves it so that it can be used again. If a document is not saved, it will be lost once the program is closed or the computer is shut down. It is a good idea to save a document before printing. The first time you save a document, you must give it a filename. Filenames should accurately describe the document. For example, in this course you might use the activity number as the filename (*Activity 1-6*).

The Save As command on the File menu is used to save a new document or to rename an existing document. The Save As dialog box may either be blank or display a list of files that have already been saved.

Word makes it easy to create a new folder when a file is saved. A folder would be created for storing related files. The Create New Folder button is located near the top of the Save As dialog box.

ACTIVITY 1-6 — SAVE A FILE

1. *Word* should be open with the document you keyed in the previous activity displayed on the screen. Click **File** on the menu bar; then click **Save As**. The Save As dialog box displays.

Figure 1-17
Save As Dialog Box

3. In the second and third mock interviews, different team members will play the roles of job applicant and interviewer. When you are the applicant, behave as if this were a real interview. Answer the questions seriously and completely. Remember the points listed on page 445 about how to behave during an interview.

4. When you are an interviewer, use a rating scale of 5 to 1 (similar to the one in Figure 15-9) to rate the performance of the applicant. Assume that the applicant's education and skills are adequate for the job. Your rating will be based on how well the applicant answers your questions. Also consider the points listed on page 445 about how to behave during an interview when rating the applicant. Write a few brief notes to explain your rating. What did the applicant do well; what needs improvement?

5. After all the interviews have been completed, discuss them with your team members. Explain the rating you gave each applicant and why. Discuss how each person can improve his or her interviewing skills.

WHAT YOU SHOULD KNOW ABOUT
Smoking, Drugs, Alcohol, and Criminal Activity

Once you have landed a good job, you will want to avoid problems that may put your job in jeopardy. Using alcohol or drugs or stealing from the company can lead to your dismissal from a job.

Alcohol and illegal drug use are prohibited on the job. Employers are allowed to order drug testing to test employees for illegal drug use. Employees who come to work under the influence are usually given a warning and sent home. Repeated occurrences can lead to the loss of the job.

Employers have an interest in promoting the good health of their employees. Employers are concerned about the health risks of smoking, both to smokers and to nonsmokers from second-hand smoke. For this reason, smoking or other use of tobacco is not allowed on many jobs. In some jobs, smoking is allowed during breaks in specified smoking areas, which are often outside the building. Laws or regulations in some counties or cities do not allow smoking in any public building or enclosed work area. You should be prepared to follow your employer's rules regarding smoking.

Stealing from the company is also grounds for dismissal. Employers lose millions of dollars each year in the form of shrinkage. The term **shrinkage** means that the supplies and products at a company disappear through pilfering and theft. Never leave the office with company equipment, computers, and supplies unless you specifically have permission to do so. If you accidentally leave your place of work with company supplies, be sure to return them immediately.

Employers also have the right to check your background for a criminal record. If you have a criminal record, it's best to talk openly about the situation if the topic comes up in the interview. If your employer finds that you have given false information on your job application or during your interview, you may be dismissed from the job.

2. If necessary, change the folder or drive in the Save in box. Click the down arrow and browse to locate the desired drive and folder. (You will learn more about creating and managing files and folders in the next lesson. For now, choose a drive and folder as directed by your teacher.)

3. Key the filename `Activity 1-6` in the File name text box. **Word Document (*.doc)** should be displayed in the Save as type box. Click the **Save** button. *Word* automatically adds the file extension *doc* to the filename when the document is saved. The file extension may or may not display as part of the filename, depending on your computer's settings.

> **HELP KEYWORDS**
> **Save As**
> Save a document

4. Note that in Figure 1-17, both the filename (*Memo to Wilson*) and the extension (*doc*) are shown. Your program may show only the filename. This depends on your system settings. You will learn how to change the View settings in a later lesson.

5. Keep this document on screen for use in the next activity.

CLOSING AND OPENING DOCUMENTS

The Close command clears the screen of the document. If you have not saved the document or have made a change to it since you saved it, you will be prompted to save the document before closing.

When all documents have been closed, *Word* displays a blank screen. You can create a new blank document (using the New command) or open a document (using the Open command). Any documents that have been saved can be opened and used again. When you open a file, a dialog box displays the names of folders or files within a folder.

ACTIVITY 1-7

CLOSE AND OPEN DOCUMENTS

1. *Word* should be open with the document you saved in the previous activity displayed on the screen. Click **File** on the menu bar; then click **Close**.

2. To create a new document, click the **New Blank Document** button on the Standard toolbar.

3. Key the text below in the document.

> **DigiTip**
> If the Standard toolbar is not displayed, click **View**, **Toolbars**. Then click **Standard**, the name of the toolbar.

```
The Exit command closes all documents that are on the
screen and then closes the software.  When you exit Word,
you close both the document and the program window.  You
will be prompted to save before exiting if you have not
already saved the document or if you have made changes to
it since last saving.
```

> **HELP KEYWORDS**
> **Blank document**
> Create a document

4. Save the document using the filename *Activity 1-7*. Close the document.

5. Now practice opening the two documents you have saved. Click **File** on the menu bar, then click **Open**. The Open dialog box displays.

> **DigiTip** You can also open a file by clicking the Open button on the Standard toolbar.

Figure 15-9
Interview Rating Scale

Interview Rating Scale

Points

5 Excellent
All major elements of the job were addressed in the candidate's answers to the interview questions. Their qualifications greatly exceed job requirements. The candidate answered each question very clearly.

4 Very Good
Most major criteria for this job were addressed by the candidate in their answers to the interview questions. Few if any major deficiencies exist in the candidate's qualifications. The candidate answered questions clearly.

3 Good
Some of the major and minor criteria for the job were addressed in the answers given by the candidate. Some deficiencies exist in the candidate's answers and qualifications, but none is a major concern. The candidate answered questions adequately.

2 Weak
Few of the major criteria for the job were addressed by the candidate in the answers given. Major deficiencies exist and these deficiencies are of some concern. The candidate struggled with some questions.

1 Poor
Few of the criteria for the job were addressed by the candidate in their answers to interview questions. Many deficiencies exist in the candidate's qualifications that could lead to major problems if the candidate were hired. The candidate had trouble answering most of the questions effectively.

ACTIVITY 15-13 — MOCK INTERVIEW WITH TEAM MEMBERS

TEAM WORK

In this activity, you will use the sample interview questions you wrote earlier and practice answering questions in a mock interview.

1. Work with two classmates to conduct mock interviews. Allow 12 to 15 minutes for each interview.

2. In the first mock interview, one team member will be a job applicant. The other two team members will be interviewers. The interviewers will ask questions from those your team created in Activity 15-12. (**Hint:** You may also use the questions concerning drive on page 444.)

Figure 1-18
The Open Dialog Box

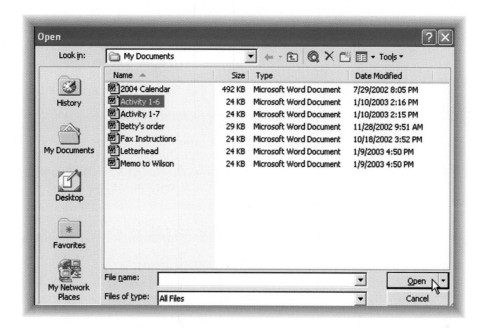

6. In the Look In box, click the down arrow; then click the drive where your files are stored. If necessary, double-click the folder name to display the filenames. Click the desired filename, *Activity 1-6*. Click **Open**.

7. Open your other file, *Activity 1-7*. Key the text below at the end of the document.

```
Make a habit of saving documents often as you work on
them.  To save a document again using the same name,
click the Save button.
```

8. Click the **Save** button on the Formatting toolbar to save the document using the same name. You will practice saving a document using a different name in a later lesson.

9. Close both documents.

Self Check Open *Activity 1-7*. Does the new paragraph you keyed appear at the bottom of the document? If yes, you saved the document correctly. If not, try Steps 7 and 8 again. Close the document.

PREVIEWING AND PRINTING DOCUMENTS

The Print Preview command enables you to see how a document will look when it is printed. Use Print Preview to check the layout of your document, such as margins, line spacing, and tabs, before printing.

You can print a document by clicking Print on the File menu or by clicking the Print button on the Standard toolbar. Clicking the Print button immediately prints the document using all of the default settings. To view or change the default settings, you can click Print on the File menu or use the keyboard shortcut **Ctrl+P** to display the Print dialog box.

During the Interview

Interviewers may begin evaluating you even before you have been introduced. Follow these guidelines as you enter the interview room:

- Shake hands with each member of the interview team.
- Make eye contact with each interviewer.
- Introduce yourself, and listen for the names of the interviewers. Try to address each interviewer by his or her name.

Follow these guidelines during the interview and when answering questions:

- Sit up straight.
- When an interviewer asks a question, look at him or her intently. Show interest in the question.
- Answer questions completely and clearly; however, do not draw out the answers. Do not spend too much time answering any one question.
- Be (or at least act) confident.
- Do not hem and haw. Avoid phrases like, "and a," "like," and "ugh."
- Avoid slang terms.
- Use technical and job-related terms when appropriate.
- Never use profanity—ever.
- Do not take lengthy notes. If you need to write something down, write it quickly and then make eye contact again.
- Avoid making negative comments about present or former coworkers or supervisors.

Usually, at the end of the interview, you will be asked if you have any questions for your potential employer. If you are confused about anything, ask questions. Only ask one or two questions if possible. (**Note:** If you are offered the job, you will have other chances to ask questions later.) Thank each interviewer for his or her time and shake hands with each one, if convenient.

Interview Rating Scale

Many corporations use a **rating scale** to evaluate candidates interviewing for a new job. A sample scale based on several real corporate interview rating scales is shown in Figure 15-9 on page 447. Understanding how interviewers rate candidates can help you do a better job during an interview. Notice in the sample rating scale that how well you answer interview questions is considered important as well as your qualifications for the job.

ACTIVITY 1-8 — PRINT A DOCUMENT

1. Open the *Word* program. Open the document you created earlier, *Activity 1-6*.

2. Click **File** on the menu bar. Click **Print Preview**. In Print Preview, a special toolbar displays with additional options for viewing the document. For example, when you click on the Magnifier button, the mouse pointer changes to a magnifying glass. When you click the magnifying glass on the page, you can see a portion of the document at 100%.

Figure 1-19
The Print Preview Toolbar

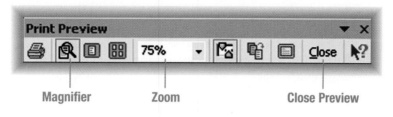

DigiTip
Follow any special instructions regarding printing that your teacher may give you. For example, you may be instructed to enter your name on the first line of a document before printing.

3. Click **Close** to close Print Preview. At this point, you can make any changes that might be needed to the document before printing. Always use Print Preview to check your documents before printing. Doing so will help prevent having to print a document more than once, saving time and supplies.

4. Check to be sure that your printer is turned on and has paper. Click **File** on the menu bar. Click **Print**. The Print dialog box displays. Compare your dialog box with the one below. The printer name may be different, but other choices should be the same. Verify that the Number of copies box is set to 1. Then click **OK**.

HELP KEYWORDS
Print
   Print a document

Figure 1-20
The Print Dialog Box

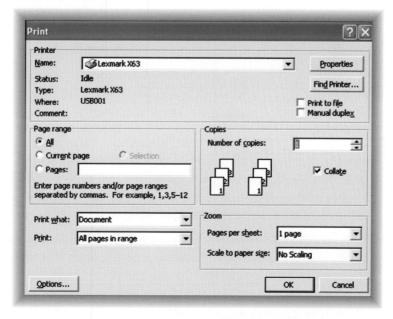

5. Close the document. Retrieve your document from the printer.

Possible interview questions regarding drive might include:

1. Why do you want to work?
2. What personal job-related goals do you have for your future?
3. Give an example of how you have worked to achieve your goals.
4. Have you ever set a goal that became impossible to achieve? Explain.
5. What process do you go through to update and revise your goals?
6. Describe a situation where you increased the productivity of others.
7. What motivates you to do a good job?
8. Explain a situation where you motivated others to work hard and get a job done effectively.
9. Why is work important to you?

When you are preparing for an interview, review the job description carefully. Scan the company's Web site if one is available. Try to anticipate what skills, talents, or attitudes the company values. Then try to anticipate questions you may be asked.

ACTIVITY 15-12

ANTICIPATE INTERVIEW QUESTIONS

TEAMWORK

Work with a team of two classmates to complete this activity.

1. Each member of your team has been asked to interview for a job at Corporate View. You need to anticipate interview questions that you may be asked about the general skills Corporate View values. You have already reviewed some questions you might be asked about *drive*.

2. Each member of your team should choose two of the six other work skills listed below. Each person should choose different skills.
   - Problem-solving skills
   - Technical communication skills
   - Flexibility
   - Interpersonal skills
   - Project management skills
   - Teamwork skills

3. In a word processing file, define the two skills you have chosen. Use a dictionary or search the Web to find the meaning of the terms. List several phrases that describe a person who demonstrates these skills.

4. Write three or four possible interview questions for each skill. Print and share copies of these definitions and questions with the other team members. You will need these questions for the next activity.

5. Save your definitions and questions as *Interview Questions* in the *DigiTools your name\Chapter 15* folder.

GETTING HELP

The **Help** feature provides you with quick access to information about commands, features, and screen elements. To access it, click **Help** on the menu bar. Then click **Microsoft Word Help**. This option includes Contents, Answer Wizard, and Index.

The *Contents* tab displays a list of topics you can click on to display helpful information.

The *Answer Wizard* tab lets you ask a question. When you click Search, the Wizard displays a list of topics relating to your answer. Click on a topic to display additional information.

The *Index* tab enables you to key a word or select a keyword to display a list of topics pertaining to the keyword. You can click Search and then click on the topic to display the information in the window to the right.

In many places in this textbook, Help keywords appear to the left of software instructions. Using these words will help you quickly find a Help entry related to the topic in the instructions. Key the first keyword in the Type keywords box. Choose the second keyword phrase from the Choose a topic list. You will practice this procedure for the Print keywords in Activity 1-9.

ACTIVITY 1-9   ACCESS HELP

1. Open the *Word* program. Open the document you created earlier, *Activity 1-6*.

2. Click **Help** on the menu bar. Click **Microsoft Word Help**. The Help window displays.

3. Click the **Index** tab. In the Type keywords box, key `print`. Click the **Search** button. A list of Help topics related to printing appears in the Choose a topic list.

4. Click **Print a document**. The Help entry appears in the right window. Use the scroll bar, if needed, to view all the Help entry. Click the **Close** button in the Help window to close Help when finished reading the entry.

Figure 1-21
The Help Window

3. Take notes of the answers you are given. Summarize the answers in an unbound report titled *Professional Interview*. Give the name and professional title of the person you interviewed. Include the name of the organization where this person works. Describe the person's job briefly and tell which career cluster it fits into. List the questions you asked and summarize the answers.

4. Save the report as *Professional Interview* in the *DigiTools your name\Chapter 15* folder. Print the report.

Self Check Does the format of your report match the unbound report shown in Appendix A on page App-7?

Interview Questions

There are two general categories of questions that you may be asked in any job interview. The first set of questions will relate specifically to the skills listed in the job description. The second category of questions deals with general work habits and skills. For example, at Corporate View, the seven skills listed below are considered vital for every employee. They include:

- Problem-solving skills
- Technical communication skills
- Flexibility
- Drive
- Interpersonal skills
- Project management skills
- Teamwork skills

Knowing that the company thinks these skills are important, you should consider questions related to these skills that you might be asked in an interview. To anticipate interview questions, you must first know the definition of each of these skills. For example, *drive* can be defined in a variety of ways. In business, drive means *to work hard, as to reach a goal*. A person with drive:

- Is excited about the job
- Works to accomplish goals
- Constantly revises his or her goals
- Finds ways to be more productive
- Is motivated to do a good job
- Can motivate others
- Desires to succeed

LESSON 1-4

FILE MANAGEMENT IN *WINDOWS*

OBJECTIVES *In this lesson you will:*

1. Learn about internal and external storage devices for electronic data.
2. Use *Windows Explorer* to see how files and folders are organized on a computer.
3. Create folders for storing computer files.
4. Rename computer files and folders.
5. Copy and move computer files.
6. Arrange files and view details in *Windows Explorer*.
7. Delete computer files.

When you save information you have input on a computer, it is stored on a variety of devices. Some devices are internal (housed inside the computer) and some are external. A **hard drive** is a common internal storage device. Compact discs (CDs), digital video discs (DVDs), and floppy disks are popular external devices on which to save data. A hard drive can also be an external device.

Figure 1-22
Floppy disks and CDs are common external storage devices.

To understand the various sizes of storage devices, compare each of them to different places where you might store printed documents such as reports or letters. For example, you can put data in:

Device	Compare To:
Floppy disk	Several folders in a file cabinet
CD	One drawer in a file cabinet
Hard drive on a computer	Four or five entire file cabinets
Hard drive on a network	A room full of file cabinets

Figure 15-8
Professional dress is essential to making a good impression at an interview.

ACTIVITY 15-11

INTERVIEW A PROFESSIONAL

There is no better way to prepare for a job interview than to interview a professional in your chosen career field. If you do not know what a career entails or are unsure of what an interview will be like, ask someone who knows.

1. Interview someone who works at a job that interests you. You may interview a parent, a friend, teacher, relative, or neighbor. You may politely contact the Human Resources department of a company and ask to speak to someone in a job that interests you.

2. Prepare ten questions to ask the person you interview. Here are some samples to get you started:
- What tasks does an employee in this job perform?
- What is a typical career path for entry-level employees in this field?
- What educational background will you need to work in this career?
- What work experience does someone need to secure a job in this career path?
- What specific things does an employer look for on a resume?
- What questions might someone be asked if he or she were to interview for a job in this career area?
- What advice do you have for someone who is about to interview for a job in this career area?

Information is stored in files. You can save text, pictures, video, or even sound files. Files are stored and organized in folders. Creating a logical and easy-to-use file management system will help you organize files and find them quickly and easily. You can manage files on the desktop or in your file management program, *Windows Explorer*. This feature may look somewhat different on your computer, depending on your *Windows* version and setup.

NAMING FILES AND FOLDERS

Good file organization begins with giving your files and folders names that are logical and easy to understand. For example, you might create a folder for your English assignments called *English*. In this folder, you might have a journal that you add to each day (called *Journal*); monthly compositions (e.g., *Composition 10-12, Composition 3-01*); and occasional essays (such as *Essay Sports* or *Essay Ethics*). A system like this would make finding files simple.

UNDERSTANDING THE FILE SYSTEM

You can use *Windows Explorer* to see how files and folders are organized on your computer. *Explorer* shows files and folders in a hierarchical or tree view. At the top is *Desktop*. The *Desktop* folder contains all the items that appear on the desktop of your computer. An item in *Desktop*, *My Computer*, contains the files and folders on your computer, organized by drives.

Figure 1-23
The *My Computer* folder contains the files and folders on the computer.

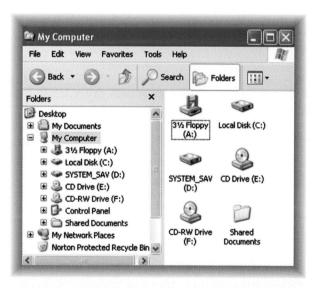

ACTIVITY 1-10

NAVIGATE THE FILE SYSTEM

1. Click the **Start** button. Point to (**All**) **Programs** (then to **Accessories** if you have *Windows 2000* or *XP*), then click **Windows Explorer**.

> **DigiTip** In *Windows XP*, click **My Computer** on the desktop or the Start menu to open *Windows Explorer* quickly. Folders may not be displayed in the left pane. To display the folders, click **View** on the menu bar. Click **Explorer Bar**. Click **Folders**.

WHAT YOU SHOULD KNOW ABOUT
Being Polite

There was once an extremely qualified candidate (whom we will call Sally) who had preliminary interviews for a position at a major software firm. In the second round of interviews, Sally was asked to give a multimedia presentation highlighting what she would do if she obtained the position. Her presentation was flawless, and she seemed sure to be offered her dream job.

After the interview, the interviewers began discussing the strengths and weaknesses of the final three candidates. The administrative assistant who had taken calls from Sally remarked that the candidate had been rude, insistent, pushy, and had put her on the defensive in more than one phone call. When Sally visited the Human Resources office to check on the status of her application, she behaved rudely. She said Sally made her feel like "a servant."

After the interviewers heard this report, the position was given to another person.

Always be aware of what you say and how you say it. Be polite to everyone with whom you talk. If you do this, you should not, like Sally, miss an opportunity for a great job.

DEVELOPING YOUR INTERVIEW SKILLS

Often, many qualified people fail to get the job they want because their interview skills are lacking. Knowing how to interview properly is important.

Before an Interview

Before any interview, you should practice how you will answer interview questions. This often means inventing possible questions that you will be expected to answer in a real interview. As the day of the interview approaches, you should:

- Carefully consider the clothes you will wear. Professional dress is essential to making a good impression. Read more about dress in the next section.

- Prepare extra copies of your resume.

- Prepare printed copies of your portfolio as needed. (Not every job requires a portfolio, but you should be ready just in case.)

- Arise early. It's better to arrive 15 minutes early than it is to arrive one second late. Plan plenty of time to arrive at the interview site early so you won't feel rushed or pressured because you arrived late.

- As the time of your interview approaches, turn off your cell phone or pager so these devices will not interrupt your interview.

Dressing for an Interview

Dressing appropriately is important for any interview. It is better to be overdressed than to appear less than professional. For women, a business suit or pants or a skirt, blouse and sweater or jacket, polished shoes, and matching accessories are appropriate. For men, a business suit or pressed pants, neutral colored shirt, tie, and polished shoes would be considered proper attire. Hair should be trimmed and neatly combed. You should also know that many employers are uncomfortable with body piercing and tattoos.

2. Click the plus sign beside a drive or folder to display below it a list of any folders that it contains. Click the minus sign to hide the list.

3. Click a folder (one without a plus sign) in the left pane. All its contents (files) will be displayed in the right pane.

4. Double-click a folder with a plus sign (double-click the icon or name, not the plus sign). Any folders inside the folder will be listed below, and all the contents of the folder (files and/or folders) will be displayed in the right pane.

5. Practice Steps 2–4 with other folders.

CREATING FOLDERS

You will want to create folders to store files. You can do so in *Windows Explorer* or on the desktop. In addition to putting files in your folders, you can create folders within folders if you need to.

- In *Windows Explorer*, click the drive or folder that will contain the new folder. Click the **File** menu, point to **New**, and click **Folder**.

- On the desktop, double-click the drive or folder that will contain the new folder. (If the drive or folder is not on the desktop, you can access it by double-clicking **My Computer**.) In the window that opens, click the **File** menu, point to **New**, and click **Folder**.

- To create a folder on the desktop itself, right-click in a blank area of the desktop, point to **New**, and click **Folder**.

ACTIVITY 1-11   CREATE FOLDERS

1. Start *Windows Explorer* if it is not already open. In the left pane of *Windows Explorer*, click **Desktop** (you may need to scroll up a little to find it).

2. Click the **File** menu, point to **New**, and click **Folder**. A new folder called **New Folder** will appear in both panes of the window. In the right pane, the name will be highlighted.

Figure 1-24
Creating a New Folder in *Windows Explorer*

ACTIVITY 15-10 — INTERVIEW ACCEPTANCE LETTER

1. Imagine you have just received an e-mail message telling you that you have been selected from hundreds of possible candidates for an interview. Read this e-mail message before you prepare a reply.

E-mail: Interview

From: Melissa.Kim@corpview.com
Date: Current
To: Student Name
Subject: Interview

Congratulations! You have been selected from among over 140 other interested people to interview for a job at Corporate View. Interviews will be held between 8 a.m. and 5 p.m. next Thursday at Corporate View headquarters. We tentatively have your interview scheduled for 9:45 a.m. Please reply to let me know whether this time is acceptable for your interview. Also, let me know if you have any other questions.

Ms. Melissa Kim
Human Resources Department
Corporate View
One Corporate View
Boulder, CO 80303

---

2. Using *Word*, draft a reply to the e-mail message you will send to accept this interview opportunity. Ask politely for all of the information you will need to attend the interview. Thank Ms. Kim for this opportunity.

3. Save your message as *Interview Acceptance* in the *DigiTools your name\Chapter 15\Portfolio* folder. Print the message.

> **DigiTip** After you have received all of the information you need regarding an interview, record this information carefully. Use your PDA or PIM software and set alarms or reminders so you are sure to leave for your interview in plenty of time.

3. Key a name for the folder, *DigiTools*. Press **Enter**.

4. Minimize *Windows Explorer* to the taskbar. Right-click in a blank area of the desktop. Point to **New** and click **Folder**. Key a name for the folder (*Workbook*) and press **Enter**.

Figure 1-25
Creating a New Folder on the Desktop

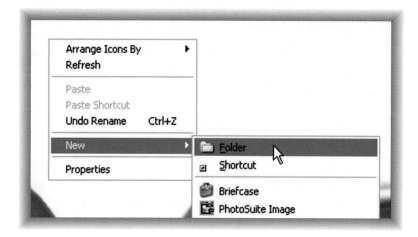

RENAMING FILES AND FOLDERS

You can rename a file or folder in one of these ways:

- In *Windows Explorer*, click the file or folder. Choose **Rename** from the File menu. Key the new name and press **Enter**.

- Right-click the file or folder and choose **Rename**. Key the new name and press **Enter**.

In the filename *Lesson 1.doc*, the *doc* extension indicates that the file is a *Microsoft Word* document. When you rename a file, be sure to include the file extension used in the original filename. Otherwise, you may not be able to open the file.

ACTIVITY 1-12 — RENAME FOLDERS

1. Display your desktop. Right-click the **Workbook** folder on the desktop. Choose **Rename**, key `Digital Workbook`, and press **Enter**.

2. Start or maximize *Windows Explorer*, if it is not already displayed. If necessary, click the **DigiTools** folder once to highlight it. Choose **Rename** from the File menu. Key `DigiTools Chapter 1`, and press **Enter**.

Figure 1-26
You can rename a folder on the desktop or in *Windows Explorer*.

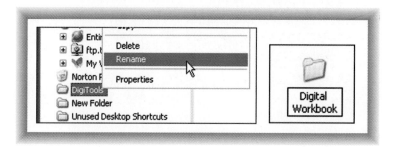

LESSON 15-2

JOB INTERVIEWS

OBJECTIVES *In this lesson you will:*

1. Write acceptance messages and other career-related correspondence.
2. Develop your interview skills.
3. Interview a professional.
4. Think about ways you can improve your skills and qualifications.

Your resume, portfolio, and letter of application are designed to open the door to an interview. If your resume indicates that you have the qualifications needed, you may be selected by an employer for an interview.

ACCEPTING AN INTERVIEW

DigiTip

If you have not heard about a possible interview after a few days, call to inquire about the status of your application. Whenever possible, try to schedule interviews in the morning, when people tend to be fresh and at their best.

The first step is actually setting up the appointment time and place when you have been offered an interview. Your invitation to be interviewed may come through a phone call, from an e-mail message, by fax, or even by registered mail. You must reply quickly, accept the interview, and make sure you know the following information:

- Location of interview
- Time of interview
- Date of interview
- Anything you need to bring (i.e., extra copies of your resume, a copy of your portfolio, or multimedia presentation)

Figure 15-7
You may be notified that you have been selected for an interview by phone or by mail.

MOVING AND COPYING FILES AND FOLDERS

You can move or copy files or folders in *Windows Explorer* or on the desktop.

- To move a file or folder, click and drag it to its new location.
- To copy a file or folder, hold down the **Ctrl** key while dragging. The pointer icon will change to include a plus sign to indicate that you are copying.

Drag your file or folder on top of the destination drive or folder. You will know you are doing it correctly if the destination drive or folder is darkened, just as when you click it. If you are moving or copying to the open window for a drive or folder, drag the item anywhere inside the window.

When you are moving or copying files or folders, selecting (clicking) several items at once can save time.

- To select consecutive items, click the first item, hold down the **Shift** key, and click the last item.
- To select items in different places, hold down the **Ctrl** key while you click each item.

ACTIVITY 1-13   COPY AND MOVE FILES

1. Start *Windows Explorer* if it is not already open. In the left pane of *Windows Explorer*, locate and click the drive or folder where you saved the files you created earlier (*Activity 1-6* and *Activity 1-7*). The files will be displayed in the right pane.

2. If necessary, scroll in the left pane until you can see the *Digital Workbook* folder.

3. Hold down **Ctrl** and drag *Activity 1-6* to the *Digital Workbook* folder. You now have two copies of this file stored in two different folders.

4. After thinking about the file arrangement, you realize that the *Activity 1-6* file should be stored in the *DigiTools Chapter 1* folder. In *Windows Explorer*, double-click the *Digital Workbook* folder to display its contents.

5. If necessary, scroll in the left pane until you can see the *DigiTools Chapter 1* folder. Click and drag *Activity 1-6* from the *Digital Workbook* folder in the right pane to the *DigiTools Chapter 1* folder in the left pane.

Figure 1-27
Drag a file from the right pane to a new folder in the left pane.

ACTIVITY 15-8 PERSONAL WEB PAGE

1. An online Web page gives you the opportunity to showcase yourself as a prospective employee. A personal Web page is like a personal introduction. Web biographies can be less formal than a resume. Review your existing personal Web page and think about ways to improve it.

2. Suggestions for additional material that you can link to in your personal page are listed below. You can link to these documents in whatever electronic format you now have them. However, users may need the software you used to view the files. As an alternative, you could create Web pages to contain this material and link to these Web pages.
- Your resume
- Samples of your work
- Letters or certificates related to honors or awards
- Electronic copies of diplomas or certificates to highlight educational experience
- Sound or video clips or an online project you have completed
- An electronic copy of a letter of recommendation

You could also create links to the Web sites of any schools you have attended.

4. Update your Web page and create links to the files or other Web pages you created. Remember not to use your personal contact information on any online site. If an employer wants to contact you, e-mail should be sufficient. Also, never include the contact information of others. Indicate that references will be provided upon request in the electronic copy of your resume. Review all your electronic documents and remove contact information for yourself and your references wherever it appears.

5. Save this new and enhanced version of your Web page as *yourname2.htm* in the *DigiTools your name\Chapter 15\Portfolio* folder. Use appropriate names for other Web pages if you created them.

ACTIVITY 15-9 FINALIZE YOUR PORTFOLIO

In this activity, you will put together a final version of your portfolio.

1. Organize all electronic files you have collected. Organize hard copies of all your documents. Place the hard copies in a three-ring binder or report cover for a professional-looking presentation.

2. Use your word processing skills to make a cover for your hard copy portfolio. Include your name, the title *Employment Portfolio*, and the current date. You may wish to include an appropriate graphic that relates to your chosen career area.

Peer Check Present your portfolio to two or three of your classmates for their review and inspection. Discuss ways to improve the portfolio and to make it more interesting. Make changes based on this discussion and publish a final copy of your portfolio.

6. Note that in Figure 1-27 the filename extension (*doc*) is displayed. You can choose to view filenames only (the default) or filenames and extensions. Follow the steps below a couple of times to practice displaying and hiding file extensions. Note the difference in how the file list appears each time.

- Click **Tools** on the menu bar.
- Click **Folder Options**.
- Click the **View** tab.
- Scroll down the list of options and locate **Hide Extensions for Known File Types**. Select (click) the box beside this option if you want to hide filename extensions. Deselect (click to remove the check mark) this option if you want to display the filename extensions. Click **OK**.

ARRANGING FILES AND VIEWING DETAILS

You can arrange the icons in a window by name, type, size, or date. You can also view details about files such as the file size and the date the file was last modified.

ACTIVITY 1-14 — ARRANGE FILES AND VIEW DETAILS

1. Start *Windows Explorer* if it is not already open. In the left pane, click a folder to display its contents. (Choose a folder that has several files in it.) Click the **View** menu, point to **Arrange Icons**, and select **by Name**.

2. With no files selected, note that the status bar (at the bottom of the *Explorer* window and the taskbar) shows how many files the folder contains and the total file size. (If you do not see the status bar, click **Status Bar** on the View menu.)

Figure 1-28
Windows Explorer Status Bar

3. Select one file. What does the status bar tell you about it? (You should see the type of file, the author, the date the file was last modified, and the file size.) What does it tell you about a group of selected files? (You should see the number of files selected and the total size.)

4. Select **Details** from the View menu. If necessary, scroll to see the information this view provides.

5. Click the **View** menu, point to **Arrange Icons**, and choose **by Date** (or **Modified**). When might this view be useful?

ACTIVITY 15-5 USE A TEMPLATE TO CREATE A RESUME

1. Start *Word*. Open and print *Resume 1* (or the most up-to-date copy of your resume). You will need to reference this information in the steps that follow.

2. Choose **File**, **New**, and choose **General Templates**.

3. Select the **Other Documents** tab. Choose one of the resume templates that is available: **Contemporary**, **Elegant**, or **Professional**. (**Note:** If you like, you may use the Resume Wizard.)

4. Reference the information from *Resume 1* and transfer all of the key information to your new resume. (Delete the sample data in the template.) Try to keep this resume to one-page in length if possible.

5. Save your new resume as *Template Resume* in the *DigiTools your name\Chapter 15\ Portfolio* folder.

HELP KEYWORDS
Template
  Create a resume

ACTIVITY 15-6 CREATE AN ORIGINAL RESUME DESIGN

1. Start *Word*. Open a new blank document.

2. Design a multipage resume that is more comprehensive than your current resume. Create a tri-fold brochure or print on larger sheets of paper folded to make a four-page presentation from a single sheet. Plan how you want to fold the paper and identify the information you want to place on each page. Use *Word's* column, table, or text box features to help you position the text.

DigiTip A tri-fold brochure has six pages if you use all panels and sides of the paper. The brochure can be folded in different ways depending on the order in which you want the reader to view the pages. You might want to fold a piece of paper and number the pages to help you plan the brochure.

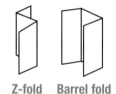

Z-fold   Barrel fold

DigiTip You can use *Microsoft Publisher* or *Adobe PageMaker* if you know these programs to design a spectacular version of your resume.

3. Include your picture and other exciting information not found in a traditional resume. Use all of the information from *Resume 1* and then refer to your growing portfolio or your personal employment journal for additional items you may include.

4. Save your original design *Original Resume* in the *DigiTools your name\Chapter 15\ Portfolio* folder. Print the resume.

ACTIVITY 15-7 LETTER OF APPLICATION

1. Review the letter of application that you created in Chapter 14. What can you do to improve the format or content of the letter?

2. Update the letter of application. For example, you might create a personal letterhead for the letter. You might rewrite a sentence that sounds awkward or correct an error you have found on the letter.

3. Save the document as *Application Letter* in the *DigiTools your name\Chapter 15\Portfolio* folder. Print the updated letter.

DELETING FILES AND FOLDERS

You can delete files and folders. If you delete a folder, you automatically delete any files and folders inside it. You can select and delete several files and folders at once, just as you can select several items to move or copy. Two ways to delete a file or folder are described below.

- In *Windows Explorer* or in a drive or folder window, select the file or folder and choose **Delete** from the File menu. Answer *Yes* to the question about sending the item to the *Recycle Bin* folder.
- Right-click the file or folder and choose **Delete**. Answer *Yes* to the question about sending the item to the *Recycle Bin*.

When you delete a file or folder, the item goes to the *Recycle Bin* folder. You can restore files and folders from the *Recycle Bin*. To restore a file or folder:

- In *Windows Explorer* or on the desktop, open the *Recycle Bin* folder.
- Right-click the name of the file or folder and click **Restore**.

Recycle Bin

ACTIVITY 1-15 — DELETE FILES

1. Start *Windows Explorer* if it is not already open. Click the *DigiTools Chapter 1* folder to open it.

2. Right-click the *Activity 1-6* filename. Choose **Delete**, and answer *Yes* to send the file to the *Recycle Bin*.

3. Delete the *DigiTools Workbook* folder you created earlier. Locate the folder in the left pane. Right-click the folder name and choose **Delete**. Answer *Yes* to send the file to the *Recycle Bin*.

Self Check The *DigiTools Chapter 1* folder should now be empty. One copy of the file *Activity 1-6* and the folder *DigiTools Workbook* should appear in the *Recycle Bin* folder.

APPLICATION 1-1 — CREATE FOLDERS FOR YOUR FILES

1. Start *Windows Explorer* if it is not already open.

2. Rename the *DigiTools Chapter 1* folder to *DigiTools* and *your name*, for example, *DigiTools Karl Barksdale*.

3. Move the *DigiTools your name* folder to the drive and/or folder where your teacher instructs you to save work for this class.

4. Click the *DigiTools your name* folder to open it. Within this folder, create a new folder for each chapter of the *DigiTools* textbook: *Chapter 1, Chapter 2*, etc. The book has 15 chapters, so you should create 15 folders. You will use these folders to store the files you create for each chapter.

EMPLOYMENT PORTFOLIO DOCUMENTS

How long do you suppose an employer looks at the average resume and application? It may surprise you to learn that most employers will spend less than a minute looking at a resume. In some situations, a computer may actually scan your resume. A human being may never read it. This is why you may need printed copies of your resume at an interview. During the interview may be the only time when your interviewers will see your resume.

Your first resume, written in Chapter 14, was plain, simple, and easy to scan. The resume you take to an interview, however, should be formatted for an attractive, professional appearance. As with your earlier resume efforts, it must be clear, state your skills, and be professionally written. When you go to an interview, you will most likely have a chance to give your resume to the interviewers. They may only glance at your resume during the interview, but after the interview, they may take more time and look over your resume carefully.

Your resume should be attractive, flawless, and concise. If your resume is rambling, contains errors, or is sloppy, it will convey a poor lasting impression. You may not be considered for the job—even if you are qualified.

You may want to use the templates in *Microsoft Word* to help you create a resume with an attractive format. A **template** determines the basic structure for a *Word* document and contains settings such as fonts and special formatting. However, these templates are widely available to millions of people. Most interviewers have probably seen them over and over again. You may want to design your own creative resume format. To make sure you have an attractively formatted resume, you will try it both ways. First, you will create a resume using a template. Then you will format your resume using a design of your own creation.

Figure 15-6
Resume created with *Word's* Professional Resume template

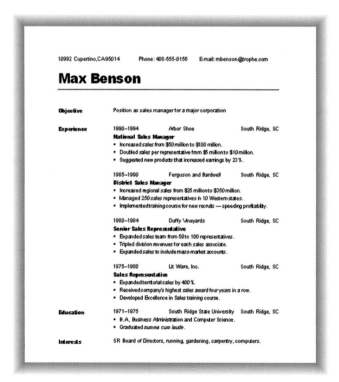

5. Within the *DigiTools your name* folder, create a new folder for each unit in the *DigiTools* textbook: *Unit 1, Unit 2,* etc. The book has four units, so you should create four unit folders. You will use these folders to store the files you create for end-of-unit exercises.

6. Move your original *Activity 1-6* and *Activity 1-7* files to the *DigiTools your name\Chapter 1* folder. Close *Windows Explorer*.

Self Check Does your *DigiTools your name* folder contain 19 subfolders? You should have 15 chapter folders and 4 unit folders. Does your *Chapter 1* folder contain two files?

LESSON 1-5: DIGITAL PATHWAYS TO CAREER SUCCESS

OBJECTIVES

In this lesson you will:

1. Learn about career cluster pathways.
2. Identify jobs in various career clusters.
3. List technology or software that might be used in various jobs.

As an adult, you will have many terrific careers to choose from. The number of job possibilities doubles every few years. Exhilarating opportunities are available in careers as different as medicine, tourism, sports marketing, fast food, the law, or international banking. Each individual can choose a unique path and everyone can succeed at what he or she has chosen to do.

The U.S. Department of Education identified the career clusters or pathways in the list that follows. Down each pathway are many great jobs and careers. You will investigate these career cluster pathways in later activities in this textbook. Success in all of these paths will be enhanced by DigiTools—including success in many careers that have not even been thought of yet!

Career Clusters and Pathways to Professional Success

Agriculture and Natural Resources
Architecture and Construction
Arts, Audio/Video Technology, and
    Communications
Business and Administration
Education and Training
Hospitality and Tourism
Manufacturing
Finance

Government and Public Administration
Health Science
Human Services
Information Technology
Law and Public Safety
Retail/Wholesale Sales and Services
    (includes Marketing)
Transportation, Distribution, and Logistics
Scientific Research and Engineering

Figure 15-5
You may be asked to deliver a multimedia presentation about your qualifications.

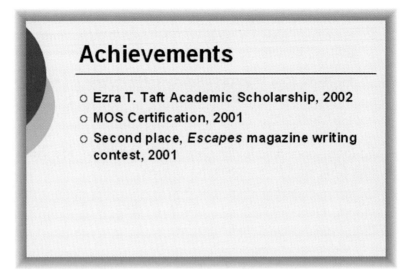

Your multimedia presentation should include information about you, your qualifications, and your vision of what you can do professionally. Include such things as your definition of teamwork, your strengths as they relate to the job, and how you can make any organization that you join more effective.

ACTIVITY 15-4   PREPARE A *POWERPOINT* PRESENTATION

1. Start *PowerPoint*.

2. Prepare a multimedia slide show about yourself. Your slide shows should take about five to seven minutes to deliver. Follow these guidelines when creating the show:
- Use the slides to introduce yourself. Include photos and perhaps an audio or video clip about you if possible.
- Allow your personality to be reflected in the show, but keep the tone professional.
- Highlight your skills, experience, and achievements that are most relevant to the career you want to pursue.
- Include information about what you can contribute to the efforts of any business or organization.

3. Practice delivering your slide show. As you do, proofread carefully and make any necessary corrections.

4. Save your slide show as *Personal Slide Show* in your *DigiTools your name\Chapter 15\ Portfolio* folder.

Peer Check Ask two or three classmates to watch you deliver your presentation and give you feedback on how it can be improved.

To learn more about career clusters, visit the U.S. Department of Education Web site at www.ed.gov and search for *career clusters*.

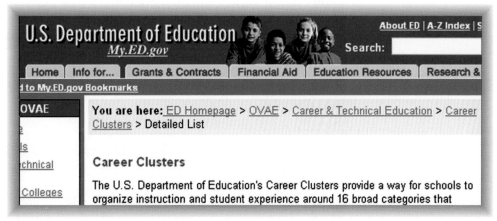

Figure 1-29
Information on career clusters can be found on the U.S. Department of Education Web site.

Source: http://www.ed.gov/offices/OVAE/clusters/occlist.html.

ACTIVITY 1-16

EXPLORE CAREER PATHWAYS

TEAM WORK

Work in a team with two other students to complete this activity.

1. Consider how digital communications tools may be used in a typical career. Look at this example:

   Pathway: Agriculture and Natural Resources
   Career: Farmer
   Application: Spreadsheet or financial application
   Input: Farm expenses and crop yields
   Output: Profit and loss statement

2. List six career pathways of your choice. For each one, list a career in that pathway. Then list a technology or software application that might be used in that career. Give the input and output methods used by that technology or software application.

DIGITOOLS
DIGITAL WORKBOOK — CHAPTER 1

You have probably used many workbooks in various classes during your school years. Usually, a workbook is printed on paper. Because you are learning to use digital tools in this textbook, however, you will use a digital workbook for this course. The *DigiTools Digital Workbook* is stored online for your use. Your teacher will tell you where to access the files for the workbook.

For each chapter, you will open a workbook file in *Word*. For Chapters 1 and 2, you may want to print the workbook pages and complete the activities by handwriting the

A good letter of recommendation need not contain all the information listed above; however, it should present you in a positive light and discuss your qualifications. Only positive letters of recommendation should be included in your portfolio. If you receive a letter that you think does not present you in a positive light, thank the writer for his or her time. File the letter with your other employment journal documents, but do not use it. Do not ask the writer to rewrite the letter. You want to use only people who can give you honest, positive recommendations as references. When you begin actively seeking a job, your employment portfolio should contain at least three strong, positive letters of recommendation. For now, obtain one good letter of recommendation to help build your portfolio.

ACTIVITY 15-3 — ACQUIRE A LETTER OF RECOMMENDATION

1. Review the list of contacts you created for your Personal Employment Journal in Chapter 14. Find one person you think will be willing to write a positive letter of recommendation for you.

2. Ask this person to write a one-page letter of recommendation that you can use during the interview process in a future job search. Give the writer your resume and other helpful information such as a transcript of courses you have taken. Tell the writer the career area in which you are interested.

3. Remember that writing letters of recommendation is time-consuming and never easy. You should write a thank-you letter to a person who prepares a letter of recommendation for you. Write a sample thank-you letter addressed to your contact thanking him or her for this effort on your behalf. See page App-6 in Appendix A to review letter format. When you receive the letter of recommendation, send your thank-you letter to the writer.

4. Save your thank-you letter as *Thank You* in your *DigiTools your name\Chapter 15\Portfolio* folder. Print the letter.

PERSONAL MULTIMEDIA SHOW

Some jobs require several interviews. Often preliminary interviews will be scheduled to narrow down the potential candidates to a select few. These preliminary interviews are often conducted in person, but sometimes they are held over the phone.

A second interview is often organized to choose from among the top two to five potential candidates. When you are selected for a second round of interviews, you may be asked to deliver a multimedia presentation about your qualifications to a group of interviewers. You should prepare a slide show in advance to have ready in case you need it. Then, just before an interview, you will have time to practice delivering your show and modify it to fit the specific job you're applying for. Your electronic *PowerPoint* show will become a part of your personal employment portfolio.

answers. After you learn to key properly in Chapter 3 (or if you already know how to key), you can complete the activities online by keying in the workbook file. After you learn to use handwriting or speech recognition, you may want to use these tools also.

For every chapter, you will complete workbook activities to review the concepts taught in the chapter, reinforce the vocabulary you have learned, and practice your math skills. Beginning in Chapter 2, you will complete online exercises called *From the Editor's Desk*. These exercises provide review and practice in punctuation, grammar, and number and word usage. Beginning in Chapter 3, you will also complete drills to build keyboarding skills. Self Checks and Peer Checks will help you evaluate your work.

Open the data file *CH01 Workbook*. Complete these exercises in your *DigiTools Digital Workbook* to reinforce and extend your learning for Chapter 1:

- Review Questions
- Vocabulary Reinforcement
- Math Practice: Calculating Averages

LETTERS OF RECOMMENDATION

Most employers require references. References are listed either on an application form or at the end of a resume. Human resource specialists and interviewers will contact these references, usually by phone or e-mail. Sometimes, however, employers require written references, usually in the form of a **letter of recommendation**.

You will want to have letters of recommendation in your portfolio. Individuals from your contacts list will need to write the letters for you. This is not something you can do for yourself. When thinking about whom to choose, consider someone who knows you well and can evaluate your work ethic, abilities, and commitment to do a good job. Who do you know that will be willing to write a positive letter of recommendation for you?

> **DigiTip** Make copies of letters of recommendation and other valuable documents and place the originals in a safe place. Obtain electronic copies of letters if possible. If not, scan the letters. You do not want to lose a letter and have to ask the author to write another.

Figure 15-4
Ask someone who knows you well to write a letter of recommendation for your portfolio.

You can help the person you ask to write a letter of recommendation by providing him or her with a copy of your resume and information about a specific job or the career area in which you are interested. A transcript or list of relevant courses you have taken can also be helpful. Tell the writer the full name and address of the person to whom the letter should be sent if the letter is for a particular potential employer. Request the letter at least two weeks before it is needed. Give the writer more time, if possible, so he or she will not feel rushed or pressured when writing the letter.

Once you receive a letter of recommendation, review the letter to make sure it is one that will be helpful to you in getting the job you want. The letter should contain information such as:

- The writer's relationship to you and the length of time he or she has known you
- Details about your skills, talents, responsibilities, and work projects that the writer is familiar with
- Your personal characteristics such as honesty, initiative, or a positive attitude
- Your interpersonal skills such as the ability to get along with others and work well in a team
- How your strengths, skills, training, or characteristics will contribute to your job performance
- The writer's contact information where he or she can be reached for questions

CHAPTER 2

Internet and Intranet Basics

OBJECTIVES *In this chapter you will:*

1. Use browser software to access, navigate, and bookmark Web pages.
2. Learn how networks are protected with firewalls and virus protection software.
3. Learn about the Internet's Domain Name System, Web addresses, and paths.
4. Locate, explore, and apply online learning resources.
5. Learn about intellectual property rights and plagiarism.
6. Search the Web and research information online.
7. Send, receive, forward, and delete e-mail.

Nearly everyone is familiar with the Internet. The Internet is known by several names: the World Wide Web, the Web, or just the Net. This amazing digital communications system connects computers, phones, TVs, and other devices to Web sites, databases, e-mail, online games, and other valuable services.

A **Web site** is a collection of related Web pages. For example, if you visit the Web site of a company such as The Thomson Corporation, you can find the latest product news, information about the company, and a host of customer services. This Web site is for public display. Thomson Learning wants its customers to see, enjoy, learn from, and participate with this site. The Web site is open to everyone who wants to visit.

Figure 2-1
Many companies and organizations provide a public Web site.

Source: *http://www.thomson.com/bizinfo/bizinfo.jsp*

- Word processing document
- Spreadsheet document or assignment
- Database document or assignment
- *PowerPoint* show
- Keyboarding assignment, test score, or timing
- Speech recognition assignment, test score, or timing
- Handwriting recognition assignment
- Web page

2. Create a subfolder called *Portfolio* in the *DigiTools your name\Chapter 15* folder.

3. Save each sample from Step 1 as a file in your *DigiTools your name\Chapter 15\Portfolio* folder. Use an appropriate name for each file such as *Word Processing Sample* for the word processing document. Make a note of the names you use for each file for later reference.

4. Print each file so you will have a copy of the work sample to include in your hard copy portfolio.

ACTIVITY 15-2

ACHIEVEMENTS SCAVENGER HUNT

In this activity, you will document all of your achievements that should be represented in your portfolio.

1. Read your resume. Examine each part, and think of any sort of physical evidence that you may have that will provide proof that you actually accomplished this task, acquired this skill, or won this award. Examine each part and make a list of items that can be included in a portfolio. List any awards, degrees, certifications, or achievements that you have received for items listed in any of the following sections of your resume.
- Personal Web Address
- Related Experience and Employable Skills
- Educational Experience
- Work Experience
- Relevant Personal Achievements

2. Collect the items you listed and decide how you can represent them as a "snapshot" in your portfolio. For example, you could include a printed copy of your personal employment Web page in your hard copy portfolio. Are there any documents or achievements you can collect, scan, and add to your electronic portfolio?

3. Prepare the items you have chosen for both your hard copy and your electronic portfolio. For the electronic portfolio, scan "snapshot" examples of all the items you have collected and create files for your electronic portfolio. Use appropriate filenames and note the names for later reference. Save the files in your *DigiTools your name\Chapter 15\Portfolio* folder. Make photocopies of documents such as awards, certificates, or diplomas for your hard copy portfolio. Place the originals in a safe place.

> **DigiTip**
> Always use copies of original documents and place the originals in your personal employment journal filing cabinet or storage box. Take pictures of trophies, medals, or bulky awards and save the pictures digitally.

Another level of network activity takes place on intranets, outside the view of most people. An **intranet** is a computer network created by a company or organization. It is meant for the use of its employees or members. Using an intranet, employees or members share information. They work as teams, communicate with each other, and stay informed. Because companies want to protect their company secrets, strategies, and new product research, intranets are not open to the public.

LESSON 2-1 BROWSER SOFTWARE

OBJECTIVES *In this lesson you will:*

1. Use browser software.
2. Access an intranet locally.

Both public Internet Web sites and private intranets make use of many of the same technologies and tools. If you can use the Internet, you will be able to use an intranet.

A **browser** is a program that allows users to view Web pages and online information. Browsers are Web surfing tools with names like *Internet Explorer, Opera, Mozilla, Safari,* or *Netscape Navigator*. Browsers are popular because these DigiTools are easy to use and provide easy access to the Web.

Figure 2-2
The parts of a Web browser

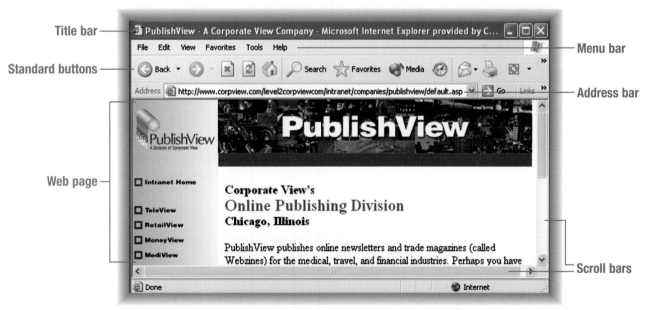

In Activity 2-1, you will access the intranet of a company called Corporate View. (This is not a real company. The company intranet and other information have been created as a learning tool for students.) This intranet is modeled after the intranets of major companies. A copy of the intranet may have been installed to your computer or your local area network. Begin by seeing if you have access to a local copy of the intranet.

Figure 15-3
An employment portfolio can show "snapshots" of your work.

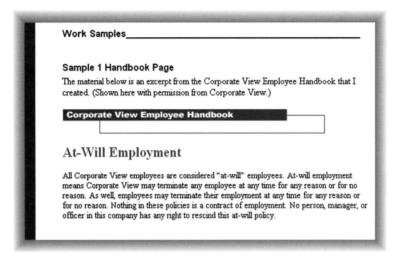

WHAT YOU SHOULD KNOW ABOUT
Electronic vs. Hard Copy Portfolios

In the days before computers, scanners, and Web pages, the only way to prepare portfolios was to print hard copy examples and place them in some sort of book or binder. A portfolio was the employment equivalent of a scrapbook. Today, in the age of DigiTools, photos, certificates, awards, cover pages, and all sorts of items can be scanned and converted into digital images. These items can be placed into an electronic portfolio. This portfolio can be viewed online and copies can be printed as needed.

You will want printed copies of your portfolio to take to interviews or to show interested family members. You may even want to place the printed portfolio in a binder. Digital portfolios have the added advantage of allowing you to display them online. Digital files can also be sent to interested people via e-mail or using file transfer protocol (FTP). A digital format gives you flexibility in how you share your portfolio information.

Digital files are vulnerable, however, as are all the data on your computer's hard drive. You should back up your digital portfolio on a CD or some other safe place so your electronic portfolio will not be lost due to a crashing hard drive, a virus, or some other unforeseen circumstance.

ACTIVITY 15-1 — DIGITOOLS ASSIGNMENTS SCAVENGER HUNT

What have you accomplished during this course that you might want to include in your portfolio? You will collect samples of your work in this activity.

1. Search for the best examples of your work for each of the following categories. Find at least one sample assignment that you are particularly proud of in each category.

ACTIVITY 2-1

ACCESS AN INTRANET LOCALLY

Note: If you do not have access to a local copy of the Corporate View intranet, you can access it live via the Web. Skip to Lesson 2-2.

If the intranet program has been installed on your computer:
Choose **Start, (All) Programs, Corporate View, Corporate View Management and HR**.

If the intranet program has been installed on a local network:
Start your Web browser. Choose **Start, (All) Programs, Internet Explorer** (or your browser name). Open the *index.html* file from your local copy of the Corporate View intranet.

- Choose **Open** from the File menu.
- Browse and open the intranet folder. (Your instructor or network administrator will provide information on where Corporate View intranet files are stored on your local network or stand-alone computer.)
- Select the *index.html* file.
- Choose **OK** to open the file.

HELP KEYWORDS
Open
Learning how to browse the Web...
To open a Web page or folder

DigiTip Your instructor will provide directions if you are using a different Web browser.

1. When a local copy of the Corporate View intranet opens, you will be greeted by a splash or welcome screen as seen in Figure 2-3.

DigiTip A splash screen is the first page or "welcome page" that visitors view when entering a Web site.

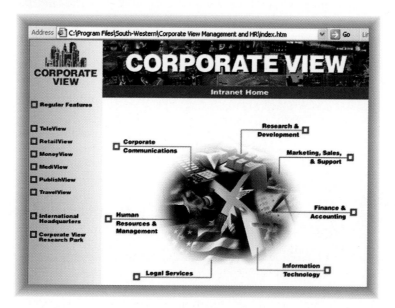

Figure 2-3
The Corporate View Intranet Home page welcomes visitors.

2. Later you will explore parts of the Corporate View Web site. For now, click the **Close** button in the upper right corner of the browser window to close the browser and the Corporate View intranet.

Figure 15-2
Portion of a Resume

Related Experience
- Currently a Communications Coordinator in the TeleView Division
- Wrote press releases, coordinated fund-raising events, and worked closely with the public relations team for the U.S. Olympic Ski Team
- Worked part-time at a public radio station writing copy for fund-raising spots and public service announcements
- Rewrote and edited Corporate View's online Employee Handbook
- Edited eight manuals from other writers within the company
- Created 40 percent of the Internet pages on TeleView's Web site
- Created training on how to write effective press releases
- Know Microsoft Office extremely well. Obtained MOS certification
- Proficient with Dragon NaturallySpeaking and Microsoft speech and handwriting software
- Proficient with Web design software including Dreamweaver, Flash, Illustrator, and Photoshop

Achievements
- Ezra T. Taft Academic Scholarship, 2002
- MOS Certification, 2001
- Second place, *Escapes* magazine writing contest, 2001

PORTFOLIO SNAPSHOTS

DigiTip
The personal employment portfolio, whether saved digitally on your computer, in a filing cabinet, or in a big box, is a place to deposit everything. All of the successes, achievements, and awards related to employment should be preserved. However, only the best of your work should be found in your portfolio.

Carrying a huge box of career-related documents and items to an interview would, of course, be impractical. So think back to our snapshots analogy. How can Maria take snapshots of work samples or each important achievement for her portfolio?

- Perhaps she could scan the cover and title pages of the manuals she has edited and include them in her portfolio.
- She might include the best one or two samples of Web pages she has created for TeleView.
- Perhaps she can include one or two press releases—those she is most proud of—for placement in her portfolio.
- Possibly a sample page from the training materials she created on how to write press releases could be included.

In this way, Maria can create "snapshots" that represent her work. These items could be placed in her personal employment portfolio. A portion of Maria's electronic portfolio that includes work samples is shown in Figure 15-3 on page 432.

LESSON 2-2: INTERNET ACCESS AND SECURITY

OBJECTIVES *In this lesson you will:*

1. Use browser software.
2. Access an intranet Web site on the World Wide Web.

An intranet can be used at the company location or from other locations. An employee may need information from the intranet while working at home or traveling. **Remote access** allows users to connect to an intranet from a distant location. The connection is often made by using a modem and a computer. A **modem** is a hardware device that allows electronic data to be sent by telephone and cable service lines.

Security features are used to protect intranets. Many intranet sites are protected by password and login procedures. These procedures help ensure that only authorized users can get data from the network.

As you log in during the next activity, your data will pass through a firewall. A **firewall** is a security system designed to prevent unauthorized entry to a network. A firewall can be made up of hardware and/or software. A firewall can scan incoming and outgoing messages for viruses. A firewall can also help prevent unauthorized users from getting on an intranet. Once you are connected, your movements on the intranet can be tracked by IT (Information Technology) employees. All these efforts are designed to protect a company's private information.

WHAT YOU SHOULD KNOW ABOUT Packets and Firewalls

Data sent over the Internet is broken into little digital bundles called **packets**. All packets contain information to tell the network where the data is going. These packets are sent separately across the Internet. Packets fly across the Net in all different directions heading for their destinations. The TCP/IP (Transmission Control Protocol/Internet Protocol) makes sure packets are addressed and arrive properly.

When all the packets have successfully arrived at their destination, they are reassembled like a jigsaw puzzle and displayed by the Web browser. If a packet is missing, TCP/IP lets the browser know so that new packets can be sent. Without TCP/IP and a Web browser, packets would sit in a digital pile with nothing to do, waiting to be erased.

Firewalls can be hardware, software, or both working together. Every message entering or leaving an intranet will pass through the firewall. The firewall looks at each message. It blocks those packets that seem suspicious or that do not meet the network's security screening criteria.

Firewall tools can hide network e-mail and Web addresses so hackers and viruses cannot attack them directly. Firewalls can also block obscene, libelous, or slanderous material.

A firewall can also protect a single computer connected to the Internet. This type of firewall monitors all the messages sent to the computer. It helps prevent outsiders from accessing your computer.

What else should go into a portfolio? The answer will depend on the career cluster you are interested in. For example, someone pursuing a career in the Architecture and Construction cluster would include drawings or photographs of the buildings he or she has designed or helped to build. Someone in Information Technology would need to show industry-recognized certification documents. Those pursuing a career in the Retail/Wholesale Sales or Marketing cluster should include pictures of any product displays or samples of marketing campaigns they have worked on. For each career cluster, a different set of portfolio items should be included.

Identify the career cluster from the table that follows that best matches your career goals at this time. Think for a few minutes about what achievements should be found in a typical portfolio for professionals in this cluster. Make a few mental notes and continue with the lesson.

Career Clusters

Agriculture and Natural Resources	Government and Public Administration
Architecture and Construction	Health Science
Arts, Audio/Video Technology, and Communications	Human Services
	Information Technology
Business and Administration	Law and Public Safety
Education and Training	Retail/Wholesale Sales and Services (includes Marketing)
Hospitality and Tourism	
Manufacturing	Transportation, Distribution, and Logistics
Finance	Scientific Research and Engineering

EVIDENCE TO SUPPORT YOUR RESUME

You should collect examples or evidence to support your resume. For instance, Maria Bravo works in a Corporate Communications department, which fits into the Arts, Audio/Video Technology, and Communications cluster. What type of items do you think Maria should provide as evidence for the items listed in the Related Experience and Achievements sections of her resume shown in Figure 15-2 on page 431.

Maria should have samples of the press releases she wrote for the U.S. Olympic Ski Team. She should also have copies of the fund-raising spots and public service announcements she wrote while working at the public radio station. She should retain a copy of the Employee Handbook that she helped edit. Somewhere she should store copies of the eight manuals that she edited. Screen captures, electronic copies, or printed copies of the Internet Web pages she created should also be stored away with her personal employment journal. She should have a copy of her MOS certification and a copy of the letter notifying her of her Ezra T. Taft Academic scholarship. She should also have a copy of the award she received from *Escapes* magazine.

Locating all of the items needed for a portfolio may require some effort. After all of the honors, work samples, certificates, awards, and achievements have been collected, they should be stored away in a filing cabinet or a clearly marked storage box with a name like "Personal Employment Journal Items." Do you have all of your items like these stored in a place where you can locate them quickly?

Figure 2-4
A firewall helps prevent unauthorized access to a computer or network.

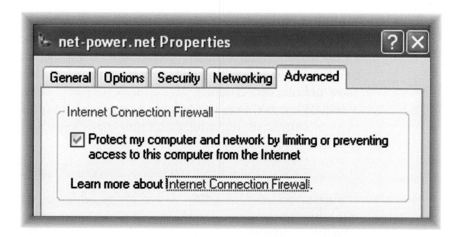

ACTIVITY 2-2: ACCESS AN INTRANET ON THE WEB

INTER N E T

DigiTip
You must be a registered user to log in on a secure intranet. If you do not have access rights, the firewall will block your entry into the system.

Note: If you do not have access to the Internet, you can use a local copy of the Corporate View intranet to complete many of the exercises in this text. If you are not connected to the Net, review Lesson 2-1 and then skip to Lesson 2-3.

1. Start your Web browser.

2. Enter the Web address `www.corpview.com` on the address bar. Press **Enter**.

3. Click the **Intranet Employee Login** link to access the intranet Employee Login page. (See Figure 2-5.)

Figure 2-5
The Corporate View Intranet Login Link

DigiTip
A different username must be created for each person with a user account on a network. If another person has already chosen a name you want, you will need to pick another name.

4. To register, you must select a username and a password. To access the registration screen, enter `register` in the Username text box. Enter `level2` in the Password text box and click **OK**. Follow the instructions on the screen to complete the registration process.

and abilities. A portfolio is not a full feature film or personal documentary detailing every aspect of your life. Those types of details are better placed in your personal employment journal for reference purposes. Items that cannot be recorded digitally should be stored in a filing cabinet or a well-marked storage box.

The Internet is a good source of information about developing employment portfolios. America's Career InfoNet is a government-sponsored Web site that provides links to articles with guidelines for developing a portfolio.

Figure 15-1
Many Web sites provide information about employment portfolios.

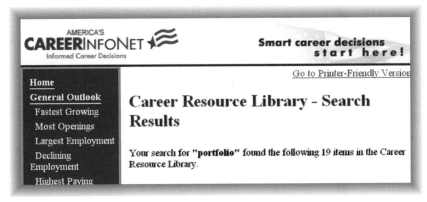

Source: http://www.acinet.org/acinet/library_search.asp?Keyword=portfolio&x=13&y=10

LESSON 15-1: PERSONAL EMPLOYMENT PORTFOLIO

OBJECTIVES

In this lesson you will:

1. Create an employment portfolio.
2. Demonstrate key accomplishments in a portfolio.
3. Acquire a letter of recommendation.
4. Develop a *PowerPoint* presentation to introduce yourself and highlight your job qualifications.
5. Improve your resume and other employment documents.
6. Organize, design, and print your final portfolio.

You have already created some important pieces of a professional portfolio. As you complete this chapter, you will improve these documents. You will also add to this important collection of personal portfolio information. You already have:

- A resume
- A letter of application
- A personal employment Web page

5. To access the Corporate View intranet, enter the username and password you selected during the registration process. (See Figure 2-6.) Click **OK**.

Figure 2-6
Log in with a username and password.

6. After you have successfully completed the login, click the **Corporate View Level 2 Intranet** link on the Employee Portfolio page. The Corporate View Intranet Home page appears.

7. Later you will explore parts of the Corporate View Web site. For now, click the **Close** button in the upper right corner of the browser window to close the browser and the Corporate View Intranet Web site.

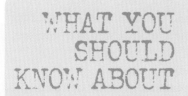

WHAT YOU SHOULD KNOW ABOUT

URLs, Domain Names, and IP Numbers

A Web address, such as www.corpview.com/intranet/local.html, is called a **Uniform Resource Locator (URL)**. URLs are part of the uniform system of naming that allows Web visitors to locate the exact "resource" they are looking for. Resources can include Web pages, graphics, text files, video clips, or other files.

URLs contain domain names. For example, *corpview.com* is a domain name. The domain name *nasa.gov* will take you to the National Aeronautics and Space Administration Web site.

The Internet works digitally, that is, by using numbers, not names. Internet Protocol (IP) numbers identify computers on the Web. These numbers are made up of four sets of digits, such as 198.80.140.11. However, people can remember domain names much easier than IP numbers. The Internet's Domain Name System (DNS) converts the names people can remember into the IP numbers computers work with. Several special computers around the world convert domain names, like corpview.com, into their corresponding IP numbers. For example the IP number for corpview.com is 198.80.140.11.

CHAPTER 2   Lesson 2-2

CHAPTER 15

Starting Your New Career Successfully

OBJECTIVES *In this chapter you will:*

1. Prepare a personal employment portfolio.
2. Develop a multimedia presentation to introduce yourself and highlight your job qualifications.
3. Update your resume and other portfolio documents.
4. Digitize key awards, diplomas, certifications, and other personal achievements.
5. Prepare for an interview and practice answering interview questions.
6. Interview a professional.
7. Learn about hiring practices and the rights and obligations of employment.
8. Improve your career qualifications.

In Chapter 13, you surveyed several of the career opportunities available. In Chapter 14, you prepared the documents necessary to apply for jobs. In this chapter, you will learn how to start your new career by completing the interview and hiring process successfully.

After you have applied for a job and submitted a resume, the next step is to prepare for an interview. Some interviews are informal, but most are very formal. They may involve a team of interviewers and many applicants applying for the same job. Interviews are important events for which you must be totally prepared. Highly competitive job interviews may require more than a simple resume. They may require candidates to present a portfolio. To prepare, you can create your portfolio in advance and have it ready to take with you to a job interview.

A **portfolio** is a collection of items that demonstrates your skills and abilities. If you are confused about what a portfolio is, think of it this way: "What would an artist take to an interview?" Obviously, samples of his or her artwork. What can *you* take to an interview? Evidence of your skills, including a flawless resume, awards or certifications, samples of your work, letters of recommendation, and anything else that you can think of that may give you an edge in a competitive job market.

You must be selective, however, when choosing items for a portfolio. A portfolio is a place to display the best of your work samples. Not everything should go into an employment portfolio. Think of a portfolio as a series of snapshots displaying your skills

What You Should Know About

URLs, Domain Names, and IP Numbers

continued

No two domain names are alike, and there are no duplicate numbers. This is important because IP numbers must correspond directly with the domain name for the Web site you are locating.

The last few letters following the period in a domain name is called a top-level domain (TLD). Most people are familiar with .com. This top-level domain name is assigned to *commercial* business domains and Web sites.

Many other TLDs have been created by InterNIC (Internet Network Information Center). InterNIC is the organization responsible for Internet naming and numbering. You can look up top-level domain names at the organization's Web site at www.icann.org.

Top-level domain names can give you some insight into the purpose of the site you are visiting. Here are a few examples:

.edu	United States educational institutions	.gov	United States government
.mil	United States military	.net	Network providers
.org	Organizations	.us	United States country code
.biz	Businesses	.mx	Mexico country code.

LESSON 2-3: HYPERLINKS AND NAVIGATION FOR WEB SITES

OBJECTIVES *In this lesson you will:*

1. Use hyperlinks to navigate Web pages.
2. Learn to understand Web addresses and paths.
3. Explore the primary departments found in many businesses.
4. Review a company mission statement.

USING HYPERLINKS

Hyperlinks are the tools that first made the Web popular. This system of interlinking pages is so easy to use that people can't resist clicking or tapping to see what is behind interesting hyperlinks. You can even speak any hyperlink aloud using speech recognition tools to go to a new Web resource.

A **hyperlink** can be any word, phrase, or picture that, when clicked, take the user to another location. Hyperlinks can link to information on the same page, to information on another Web page, or to another Web site thousands of miles away.

APPLICATION 14-1 COMPARE YOUR QUALIFICATIONS AGAINST EMPLOYER NEEDS

1. Collect these documents that you printed earlier:
   - The job description that you selected in Activity 14-8
   - Your resume
   - Your letter of application
   - Your online resume form
   - Your personal Web page

2. Start *Word* and open a new blank document. Create a table in which to compare your skills and qualifications against the job description you chose in Activity 14-8. Use a format similar to the table you created in Activity 14-2. Record at least seven requirements listed in the job description. Record all qualifications you have to meet these requirements.

3. Review your selected job description and your resume one more time. Are there any requirements in the job description that you do not meet? List them below your table. Are there any weaknesses you should be aware of as you apply for this position? List them in your document.

4. What can you do to better qualify yourself for the particular job you have chosen? What educational goals do you have that could help you achieve success in the career you have chosen? Record these answers in your document. Then list five things you can do to better prepare yourself for a career in which you are interested.

5. Save your work as *My Career Analysis* in your *DigiTools your name\Chapter 14* folder. Print and close the document.

DIGITOOLS
DIGITAL WORKBOOK CHAPTER 14

Open the data file *CH14 Workbook*. Complete these exercises in your *DigiTools Digital Workbook* to reinforce and extend your learning for Chapter 14:

- Review Questions
- Vocabulary Reinforcement
- Math Practice: Calculating Comparative Salaries
- From the Editor's Desk: Confusing Word Usage
- Keyboarding Practice: Number Review and Speed Building

Figure 2-7
Hyperlink Options on a Web Page

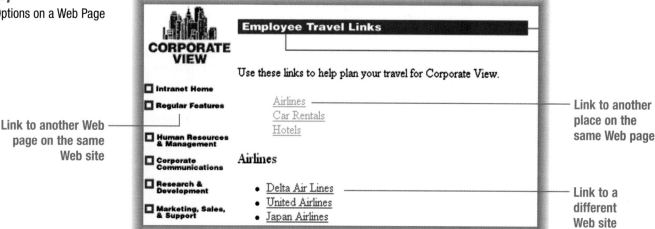

Link to another Web page on the same Web site

Link to another place on the same Web page

Link to a different Web site

WHAT YOU SHOULD KNOW ABOUT

How Web Addresses and Paths Work

How does your Web browser know exactly where to go to find the specific Web resource you are looking for? Your Web browser uses an URL as a map to follow a well-defined path directly to whatever resource you are requesting. A **path** is the URL or the drive, folder names, and filename that tell the exact location of stored data. Consider the following URL for instance:

http://www.corpview.com/Mission-CriticalFunctions/HumanResources/default.htm

When you request a page by choosing a hyperlink, your Web browser immediately locates the computer where the resource is located. In this case, the host computer can be found by its domain name and IP number:

corpview.com = 198.80.140.11

As you learned earlier, computers store files in folders, also called directories. In Chapter 1 you created *Chapter 1, Chapter 2, Chapter 3*, and other folders in which to store your work. Inside folders you may find other folders called subfolders or subdirectories. In our example, the primary folder or directory is called */Mission-CriticalFunctions/*.

Folders are separated by slashes (/). Inside this first folder or directory is a subfolder or subdirectory called */HumanResources/*.

Finally, inside this subfolder or subdirectory is the exact file you are looking for:

default.htm

A similar path can be followed on your local computer. If you access the same Web page from the local copy of your corporate intranet, the path will look something like the sample below. Although the slashes are reversed, the principle of following the path is the same. To get to the file, you must pass from one folder to the next until you reach the *default.htm* file. When the *default.htm* file opens, you will see the specific page that you requested.

C:\Program Files\South-Western\Corporate View Management and HR\Mission-CriticalFunctions\HumanResources\default.htm

3. Prepare your resume for use in an online employment application form. Open the file you saved earlier as *Letter of Application 1*. Edit your letter to a format that can be scanned successfully. Eliminate any bold or italicized words and left-align all of the text. Save the document as *Letter of Application 2* in the *DigiTools your name\Chapter 14* folder. Close the document.

4. Access the Corporate View intranet. Choose **Human Resources & Management, Corporate View Employment Application Form**.

5. Fill out the online application form as completely as you can. To move to a field, click in the field or press **Tab** to move to the next field. The form is divided into sections. Enter your personal data and your educational data. Assume that you will complete high school this year if you have not already graduated.

6. Open the file *Resume 2*. Copy and paste your resume into the resume section of the online form. Follow the instructions on the form advising you to use the **Enter** or **Return** key at the end of each line when submitting your resume. Insert a hard return at the end of each line in your resume on the form. Close *Resume 2*.

7. Open the file *Letter of Application 2*. Select all of the text from the salutation to the complimentary close of the letter. Copy and paste the text into the online form. Close *Letter of Application 2*.

8. Click the **Print Form** button at the bottom of the form. Proofread your application carefully. Correct all mistakes or Print the form again and place a copy with your hard copy resume, letter of application, and personal Web page.

9. Submit the form by clicking the **Submit** button at the end of the form. Close the Corporate View intranet and your browser.

ASSESSING YOUR QUALIFICATIONS

What do you have to offer an employer? Self-evaluation is an important early step in your job search and resume preparation. Be honest in assessing your strengths and weaknesses. If you know your strengths and weaknesses, you can focus on your assets and make plans to acquire the qualifications you may need.

In Activity 14-2, you compared Maria Bravo's qualifications against the job description for which she was applying. Now it is time for you to make a similar comparison of your skills to a job description. Can you compete with other possible candidates? Complete Application 14-1 to find out.

EXPLORING TEAMWORK IN ORGANIZATIONS

To be effective, people in businesses and other organizations must work well in teams. There are many kinds of teams. In general, a **team** is a group of people who share common tasks and goals within a company or organization. The biggest teams are often organized into departments. Departments are responsible for critical functions within the company such as the research, marketing, accounting, human resources, and legal services. Working together, the teams try to fulfill the company mission.

In the next activity, you will use hyperlinks to explore the primary departments found in many businesses. The Corporate View Web site displays these important departments on its Home page. You will also read the Corporate View mission statement. A **mission statement** gives the goals of a company or other organization in broad terms.

Figure 2-8
Essential Departments and Functions at Corporate View

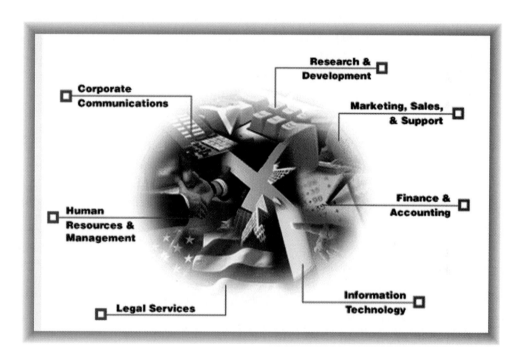

ACTIVITY 2-3   EXPLORE CORPORATE DEPARTMENTS AND MISSION STATEMENT

1. Open the local or online Corporate View intranet as explained in either Lesson 2-1 or 2-2.

2. Choose the **Research & Development** link. Then choose **About Research & Development**. Notice how the path has changed in the address/location window. Write the URL for the new address.

3. Read the short report about Research & Development. Write a short summary of the information explaining what this department does and the duties for which it is responsible.

DigiTip
To choose a hyperlink, click it with your mouse, tap it with your digital pen, or speak the link name when using speech recognition.

In an online application form, you should fill out all sections as completely as you can. Identify your strongest skills and experiences and make sure they appear somewhere on the form. Proofread carefully before clicking the Submit or Send button, because you will not be able to retrieve the form once it has been sent.

For a company, there are advantages to online applications. Data from an online form can be placed in a database, allowing employers to sort the applications looking for the skills and qualities they want in future employees. Human resource specialists can search the database looking for the best matches to the company's job descriptions. Also, when applicants complete resumes online, it helps prove to the employer that the applicant is computer literate and has some experience using the Web.

WHAT YOU SHOULD KNOW ABOUT
Resumes and Scanning

After you submit your resume, employers are likely to scan it. Employers use software to scan your resume for keywords and phrases that match their needs as listed in job descriptions. This process helps determine which candidates to drop and which to select for further consideration.

When you analyze a job description, identify the keywords that employers are likely to scan for. To give yourself a good chance of making the scanners work for you, include as many keywords as you can in your resume. For example, "Microsoft Office," "Dreamweaver," and "Bachelor of Science" are keywords that a company may look for in resumes for a Web designer position.

Describe your skills and experience using concrete rather than big words. For example, say "proficient with *Microsoft Word*" as opposed to "I enjoy working with software." Do not abbreviate terms unless the abbreviation is very common. For instance, IBM is the well-known acronym for International Business Machines, and B.S. and BS are understood to mean Bachelor of Science degree. However, FBLA is not a term that would be understood by some. You should spell out Future Business Leaders of America rather than using the acronym.

ACTIVITY 14-12   COMPLETE AN ONLINE APPLICATION

1. Prepare your resume for use in an online employment application form. Open the file you saved earlier as *Resume 1*. Edit your resume to a format that can be scanned successfully.
- Change all of the bulleted items to regular text.
- Make all bold text uppercase and then remove all bold from the text.
- Remove all indents and centering to left-justify all of the text, including your contact information.

2. Save this document as *Resume 2* in the *DigiTools your name\Chapter 14* folder. Close the document.

4. Return to Corporate View Intranet Home page. Click the **Back** button several times or choose the **Intranet Home** link at the left of the screen in the navigation bar.

> **DigiTip** A **navigation bar** is a listing of hyperlinks. It is usually at the side or the top of a Web page. This bar allows users quick access to important locations.

Figure 2-9
Options on a navigation bar may change depending on your location.

5. Use links to find information about two of the other departments (of your choice). Write a short summary of the information explaining what each department does and the duties for which it is responsible.

6. Return to the Intranet Home page. Choose the **Regular Features** link. Choose the **All About Corporate View** link. Choose the **Mission Statement** link. Write in your own words a list of the goals that make up Corporate View's mission.

7. Choose one of the departments you read about earlier. Describe how you think this team could help achieve one or more of the goals in the mission statement.

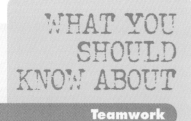

WHAT YOU SHOULD KNOW ABOUT Teamwork

Frequently, people must work in teams to complete tasks. Teamwork involves combining the efforts of two or more people to accomplish a task or achieve a goal. For a team to function well, each team member must understand the purpose or goals of the team. Each member of the team must accept responsibility for completing his or her duties and communicate clearly with other team members. For remote teams, in which members may be located around the world rather than down the hall, communication is especially important.

2. Look at Maria's resume on her Web page. Notice that personal information, such as her home address and phone number, has been removed.

3. Look at Maria's work samples. She has provided two samples of her work that are mentioned on her resume. Close the Web page and your browser.

4. Write information for your own personal career-oriented Web page. Include only information similar to Maria's About Me section. You will update your resume for electronic format and gather work samples in Chapter 15.

5. Use *Notepad*, *Word*, or a Web page editor such as *FrontPage* or *Dreamweaver* to create your Web page. Scan your picture and save it as a .gif or .jpeg file. Place your picture on your Web page.

6. Save the document as *yourname.htm* in the *DigiTools your name\Chapter 14* folder.

7. Open your personal employment Web page in your Web browser. Print your Web page.

LESSON 14-8: APPLYING ONLINE AND ASSESSING YOUR QUALIFICATIONS

OBJECTIVES *In this lesson you will:*

1. Convert your resume and letter of application to a format appropriate for scanning.
2. Complete an online application.
3. Compare your qualifications against employer needs.
4. Think and write about how you can better qualify yourself for a chosen career.

ONLINE APPLICATIONS

Many employers, especially high-tech companies, only accept resumes entered into forms on their Web site. An online application form often requires information that would be found in an application, resume, and letter of application.

Figure 14-8
Online Resume Form

What You Should Know About Teamwork *continued*

When you have a job, you will probably be expected to work in teams. You will want to be successful in this part of your job. Consider these guidelines for working effectively in teams:

- Set clear goals for the team and create an action plan for achieving the goals.
- Define the responsibilities of each team member in achieving team goals.
- Identify how success will be measured. How will the team know its goals have been accomplished?
- Identify obstacles to achieving the team's goals and discuss ways to overcome them.
- Communicate clearly and often with all team members. Be open to all feedback and ideas. Schedule regular meetings or reports to track the progress toward achieving team goals.
- Discuss how differences will be resolved. Understand that all members of a team may not have the same level of authority.
- Build on the strengths of individual members. Encourage all members of the team to participate. Each team member has different skills and ideas that can be valuable to the team.
- Recognize accomplishments of team members and the team as a whole.
- As an individual team member, develop trust by doing your part of the work well. Show a positive attitude when discussing team activities.

Follow Up

As part of their ongoing training program, Corporate View often publishes articles related to teamwork. Find and read the current article. Access the Corporate View intranet. On the Corporate View Intranet Home page, choose **Regular Features**. Next choose **Corporate View Archive** and then **Productivity Through Collaboration (Article)**.

LESSON 2-4 LEARNING FROM FAQS

OBJECTIVES *In this lesson you will:*

1. Locate, explore, and apply online learning resources.
2. Copy and paste information from Web pages.
3. Learn about intellectual property rights and plagiarism.

Using a browser is easy. However, you may need help sometimes when working with a browser or other software. A reference book or an instructor may not always be available to help you. Companies often find that employees need the same kind of help. To provide the help needed, companies often place employee training materials online. These materials can be accessed remotely by any employee.

ACTIVITY 14-10 — WRITE A LETTER OF APPLICATION

1. Start *Word* and open a new blank document.

2. Create a letter of application for the job description you chose earlier. Follow the guidelines discussed earlier in this lesson. Use the format and include the elements shown in Figure 14-7.

3. Spell-check and proofread the document carefully; correct all errors. Print the letter of application.

4. Save the letter as *Letter of Application 1* in the *DigiTools your name\Chapter 14* folder. Close the document.

Self Check Does your letter include all the letter parts shown in Figure 14-7?

LESSON 14-7 — PERSONAL CAREER WEB PAGE

OBJECTIVES — *In this lesson you will:*

1. Preview a personal career Web page.
2. Create a personal career Web page.

A personal career Web page can be a great icebreaker. Reading your Web page gives future interviewers a chance to learn about you. This may make you and the interviewer feel more at ease during an interview. Your Web page can also demonstrate some of your technical skills or display samples of your work. For example, Maria Bravo is applying for a job related to Web page development and design. Her Web page allows her to demonstrate her Web design skills.

Some information should not be shared online. For example, did you notice that there is no personal contact information listed on Maria Bravo's personal Web page? At most, you may wish to post an e-mail address on such a page, but never give your personal phone number or home address.

ACTIVITY 14-11 — CREATE A PERSONAL CAREER WEB PAGE

1. Start your browser. Open the data file *Bravo*. Review Maria Bravo's personal career Web page again. Notice that it starts with a quote from Maria related to her love of words. This sets the tone for the paragraphs that follow where she talks about her work and educational experiences. She also includes information about sports that she enjoys.

Training materials often take the form of FAQs (frequently asked questions). **FAQs** are questions and their answers provided online to help users. In this activity, you will use FAQs to learn to copy and paste text from a Web page to *Microsoft Word*.

ACTIVITY 2-4

COPY AND PASTE WEB INFORMATION

1. Open your Web browser and the Corporate View intranet. On the Corporate View Intranet Home page, choose the **Regular Features** link.

2. Choose the **Intranet FAQs** link. Then choose **Copying and Pasting Information from the Web and the Intranet**.

3. Choose the **Copy Text** link and read the instructions.

4. Select and copy the information about copying text from a Web page.

HELP KEYWORDS

Copy
Saving pictures or text from a Web page

Figure 2-10
Copy Text from a Web Page

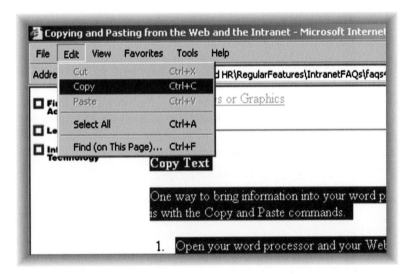

5. Start *Microsoft Word*. Open a new blank document. Paste the information from the Web page into the document by clicking the Paste button. You can also paste by choosing **Paste** from the Edit menu.

6. Save your *Microsoft Word* document in your *DigiTools your name\Chapter 2* folder. Use the filename *Activity 2-4*. Print the document.

CHAPTER 2   Lesson 2-4

Figure 14-7
Sample Letter of Application

1 inch or default side margins

1 inch top margin or center page vertically

Return address
Date

2818 River Bottom Road
Boulder, CO 80303
April 12, 20—

Letter address

Ms. Robin Mills
Human Resource Department
Corporate View
One Corporate View Drive
Boulder, CO 80303

Salutation

Dear Ms. Mills

State the job and ask to be considered an applicant

Please consider me as an applicant for the position of Web and Intranet News Writer and Editor. As a member of the Corporate Communications team in Corporate View's TeleView Division, I have seen this position develop in importance within our company. I think I would find this position both challenging and exciting.

Body

Introduce your qualifications and mention your enclosed resume

For the past year, I have worked as a Communications Coordinator at TeleView. In this position I have created 40 percent of the intranet pages on TeleView's Web site. I have also edited technical manuals from other writers within the company. My editing abilities have been greatly enhanced through the completion of a Bachelor of Arts degree in English at Colorado State University. Please see my enclosed resume for more details about my skills, education, and experience.

Ask for an interview

Please contact me for an interview with you at your convenience. I would like to discuss the many ways I can benefit the company in this job. You may reach me by phone or e-mail as shown on my resume. Thank you for your consideration.

Complimentary close

Sincerely

Signature

Maria Bravo

Writer's name and title

Maria Bravo
Communications Coordinator

Enclosure notation

Enclosure

CHAPTER 14   Lesson 14-6

ETHICS
Intellectual Property Rights

The United States Congress has created the U.S. Patent and Trademark Office (PTO) and passed laws to preserve intellectual property rights. **Intellectual property rights** grant certain privileges to persons, companies, and organizations that create something new, different, useful, and potentially profitable.

To protect intellectual property in the form of an invention, the PTO can issue a patent. A patent gives exclusive rights for an invention to the patent holder, in many cases for up to 17 years or more.

A product name or logo is often considered intellectual property. A registered trademark can be obtained to secure exclusive rights to a product name or logo. These names or logos often carry the ® symbol. For example, Pepsi is the registered trademark of a soft drink. So are Coca-Cola, Dr. Pepper, and 7Up. Because these product names are protected, others cannot use them. To do so would be illegal as well as unethical.

To protect intellectual property in the form of works such as books, articles, music, plays, movie scripts, and artwork, a copyright can be granted. Software also falls into this category and is often copyrighted. Legally copyrighted materials usually carry the © symbol. This symbol informs the public that the material is copyrighted.

The term **plagiarism** refers to using material another person has created and claiming it as your own. This practice is highly unethical and may be illegal depending on the circumstances.

Copyright laws regulate how copyrighted documents and other works can be legally copied or used. You should be very careful not to copy or download copyrighted information from the Web or other sources unless you have the owner's permission. You may be able to use a small portion of a copyrighted work for educational purposes under the "fair use" doctrine. To learn more about the fair use doctrine, access the U.S. Copyright Office Web site at www.loc.gov/copyright/. Search the site using the term *fair use*.

Follow Up

To learn more about copyright laws, do some research on the Corporate View intranet.

- Open your Web browser and the Corporate View intranet. On the Intranet Home page, choose the **Legal Services** link.

- Choose the **Patent, Trademark, and Copyright Resources** link. Choose the **Copyright Basics** link. Use this document to find answers to the following questions:
  1. Who can claim a copyright?
  2. The owner of a copyright has the exclusive right to do and to authorize others to do what five things?
  3. What eight categories of works are protected under copyright law?
  4. What are some examples of material not protected under copyright law?

DigiTip
In Chapter 5 you will learn to cite the source of quotes taken from copyrighted material.

- State that you want to be considered an applicant for a particular job
- Introduce your strongest and most relevant qualifications for the job
- Invite the prospective employer to review your qualifications in more detail in your accompanying resume
- Ask for an interview
- Thank the reader for his or her time

When writing your letter, consider your audience—the prospective employer. The employer wants to fill a job that requires specific skills. To be effective, your letter of application should introduce the two or three skills you have that best fit the job requirements. Your goal is to create interest so your prospective employer will read your resume carefully. Then your resume can do its job of selling your qualifications in more detail. This may lead you to the next step in the selection process—an interview. Read the sample letter of application shown in Figure 14-7 on page 421.

STYLE AND FORMAT

Your letter of application (or e-mail) is likely to be the prospective employer's first contact with you. To make a good impression, your format, style, and mechanics must be appropriate for a business letter or e-mail message.

If you are submitting a hard copy document, remember this is a formal letter. Use a traditional business format, such as the block style shown in Figure 14-7. Do not use a less formal style, such as a memo format.

You may send your correspondence in electronic format, such as in an e-mail or by posting your letter in an online application form. When this is the case, prepare the letter using block style, default fonts, minimal formatting, and no special characters or graphics. Keeping the format simple will increase the chances that your e-mail or electronic letter will arrive in a readable format. If you are creating a paper letter, then you can apply more intricate styles.

Letters of application should be one page only. Remember, this correspondence is just a preview of your resume. You are trying to quickly capture the interest of your reader, not explain everything in detail. Leave that to your resume. The same length rule applies to letters or e-mail in electronic form. Employers want to see at a glance whether reading your resume is worth their time.

The **tone** of your correspondence is its personality, style, or mood. You want the letter or e-mail to have a professional tone. You want it to inspire confidence and convey self-assurance, not arrogance. This is not the time to be overly assertive or cute.

Mechanics are important. Proofread! Check the document carefully and correct it until there are no typos, voice mistakes, misspellings, or grammatical errors. Your communication skills are on display. Mechanical errors look unprofessional and may give the impression that you are careless.

Use a good-quality bond paper if you are sending a hard copy document. Make sure it is at least 20-pound weight. Many companies scan letters and resumes into computers for analysis. Use a light neutral color, such as cream or white, to make the letter easy to scan.

Figure 2-11
The U.S. Copyright Office Web site provides information on fair use of copyrighted material.

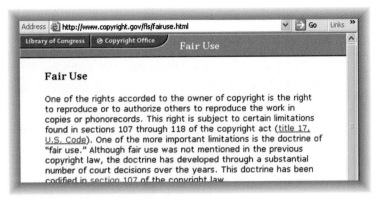

Source: *http://www.copyright.gov/fls/fairuse.html.*

LESSON 2-5: BOOKMARKS AND FAVORITES LISTS

OBJECTIVES

In this lesson you will:

1. Locate and apply online learning resources.
2. Bookmark Web pages for a Favorites list.

Web sites and intranets are full of useful information. You may find that you refer to the same sections of a Web site often. Browsers allow you to create bookmarks on Favorites lists. A **bookmark** is a link that takes you directly to a location in a document or on a Web site. Links can be organized into folders to make finding a particular link easier.

Figure 2-12
A Favorites List on Internet Explorer

ACTIVITY 2-5

CREATE BOOKMARKS

HELP KEYWORDS

Favorites
Add a page to your list of favorites

1. Open your Web browser and the Corporate View intranet.

2. On the Corporate View Intranet Home page, choose the **Regular Features** link. Next choose the **Intranet FAQs** link. Then choose the **Bookmarking and Making Favorites Lists** link.

CHAPTER 2   Lesson 2-5   47

6. Review the work experience section of your *Qualifications* file. In your resume, enter the side heading `Work Experience`. Under the heading, list the three to five most relevant jobs you have had. To add variety to your resume, use a bulleted list for this information. Include volunteer jobs as well as paid jobs. List the jobs in chronological order with the most recent first. List the job position first, the employer second, and the date of employment third. Do not use periods at the end of each line.

7. Review the personal achievements listed in your *Qualifications* file. In your resume, enter the side heading `Achievements`. Under the heading, list three to five relevant awards or achievements you have received. Use a bulleted list for this information. Do not use periods at the end of each line.

DigiTip
You should ask permission from each person listed in your reference section before you place them on your resume.

8. Open the file *References*. Scan through your contacts list. Choose three people who can give employers information about your skills, abilities, education, or work experience. You should choose three people with different backgrounds or occupations. Generally, do not list members of your immediate family as references. In your resume, enter the side heading `References`. Under the heading, enter the information for three references. For each reference, list the person's name, title, complete mailing address, daytime phone number, and e-mail address. Set tabs or create a table to align the references in an attractive format similar to Figure 14-2.

9. Spell-check and proofread your resume carefully; correct all errors. Save the document as *Resume1* in the *DigiTools your name\Chapter 14* folder. Print and close the document.

Peer Check Ask a classmate to proofread your resume as you proofread his or her resume. Discuss any possible errors and make corrections, if needed.

LESSON 14-6 LETTER OF APPLICATION

OBJECTIVES *In this lesson you will:*

1. Review a letter of application.
2. Prepare a letter of application.

PURPOSE

The purpose of your resume is to highlight your skills and create interest so you will be granted an interview. The purpose of a letter of application (or cover letter) is to interest the prospective employer in looking at your resume. A letter of application (or an e-mail message) almost always accompanies a resume. Your cover letter or e-mail message will demonstrate your communication skills. Follow these guidelines when writing a letter of application:

3. Instructions for *Microsoft's Internet Explorer (IE)* and *Netscape's Navigator and Communicator (NN)* are provided. Locate the information for your browser. Read the instructions. You may want to print the Web page if you think you will have trouble remembering the instructions.

4. Choose the **Corporate Communications** link in the navigation bar. Then choose **Style Guide**. The Style Guide contains information on preparing various documents. You will refer to it often in later lessons.

5. Add this page to your Favorites as you learned to do in Step 3.

Figure 2-13
Add Favorite Dialog Box

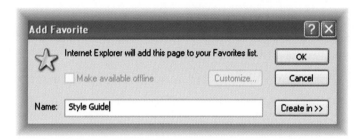

6. Click the **Back** button to return to the Corporate Communications page. Then choose the **From the Editor's Desk** link. You will use this section of the intranet often to improve your writing skills.

7. Bookmark or add this page to your Favorites as you learned to do in Step 3.

Self Check Choose **Favorites** on the menu bar. Do links appear for Style Guide and From the Editor's Desk? Click each link in turn to test the links.

LESSON 2-6 DIGITAL MESSAGES AND NETIQUETTE

OBJECTIVES   *In this lesson you will:*

1. Learn about types of digital messages.
2. Research e-mail netiquette.

E-MAIL

DigiTip
To review the parts of a digital message, see page 3 in Chapter 1.

E-mail is a shortened term for electronic mail. E-mail is a popular type of digital message. E-mail was invented in 1971. Since then, e-mail has become an important way of communicating for people, businesses, and organizations. E-mail allows electronic messages to be sent to inboxes. Once in an inbox, messages are stored until they can be opened, answered, or deleted. You will learn to use e-mail in a later lesson.

Figure 14-6
Job Postings on the Corporate View Intranet

2. Print a copy of this job description so you can refer to it in the next few activities.

ACTIVITY 14-9   WRITE YOUR PERSONAL RESUME

1. Start *Word*. Open a new blank document. Refer to Figure 14-2 on page 408 frequently as you complete the steps of this activity. Format your resume similar to the format used in Figure 14-2.

2. Enter your contact information. Include your name, address, phone number, e-mail address, and a URL to your personal Web page. If you do not have an e-mail or Web page address, leave that blank for now. Your contact information should look like part A of Figure 14-2.

3. Open the file *Career Objectives*. Review the information you entered and choose (or rewrite) a career objective statement for your resume. In your resume, enter the side heading `Career Objective` followed by your objective. Do not use a period or punctuation mark at the end of the career objective statement.

4. Open the file *Related Experience*. Review the information you entered. Choose six to ten items that you think would most interest an employer. In your resume, enter the side heading `Related Experience` followed by a bulleted list of experiences. List the most recent and/or important experiences toward the top of the list. Use a single sentence for each item and keep each statement as brief as possible. Do not use periods or punctuation marks at the end of each item.

5. Open the file *Qualifications*. Review the information you entered about your education. In your resume, enter the side heading `Educational Experience`. Under the heading list the last two or three schools you attended. List the name of the school first, location second, and any diploma, degree, or certificate received and the date last. List the schools in chronological order with the most recent school at the top of the list. You may also list any special training programs that you have attended. Do not use periods at the end of each line.

> **DigiTip**
> You can copy information from your personal employment journal files and paste it into your resume. You may need to adjust the formatting in your resume after pasting text from other files.

INSTANT MESSAGING

The success of e-mail led to other types of digital messages. In the 1990s, instant messaging became popular. **Instant messaging (IM)** allows people to have a conversation through the Internet by typing messages. Both people must be online at the same time. Typed messages are exchanged almost instantly. An on-screen message lets the user know when the other person is online and can respond to a message. Instant text messages can also be sent to some pagers and mobile phones. Software such as *Windows Messenger* (shown in Figure 2-14) is used for instant messaging.

Figure 2-14
Windows Messenger Instant Message Window

INTERNET TELEPHONY

Another type of digital message is Internet telephony. **Internet telephony** is using the Internet to connect for a voice conversation. Users call from a PC to another PC or a telephone. With PC-to-PC service, calls are sent over the Internet. The caller and the person called need the proper software, a sound card, a microphone and speakers or a headset, and a connection to the Internet. Because callers are connected via the Internet, there is no charge, even for long distance calls.

With the PC-to-phone service, calls are sent over the Internet and through phone lines to the person receiving the call. The caller often must sign up with a voice provider. The user pays a charge for this service. The service is not expensive, however.

VIDEO CONFERENCES

Digital messages may contain video as well as voice. Communicating using voice and video is called **video-conferencing**. A digital video camera (also called Web camera or NetCam) and a computer can be used to send streaming video over the Internet. **Streaming** means playing video and sound in real time as it is received (rather than waiting for it to be stored on your computer.) Streaming video allows a caller to see the person he or she is talking with during the conversation.

LESSON 14-5

CHRONOLOGICAL RESUME

OBJECTIVES *In this lesson you will:*

1. Choose a job description for reference.
2. Write a chronological resume.

A **chronological** resume lists your work experience, education, skills, and achievements in order by date. The most recent information should be listed first. For example, in the work experience section, your current or most recent job will be listed first. The first job you had will be listed last. This format is sometimes called the traditional resume, because it has been around for quite some time. This format is an old standard, and is the type of resume you should start with. The chronological resume usually has seven parts, as seen in Figure 14-2: contact information, career objective, related experience, education, work experience, achievements, and references.

The exact order of the parts can vary. If you wish to move your work experience above your educational experience, or move your related experience section down, you can do so. The order should be determined by which section you think is the most powerful and will interest your reader the most.

ACTIVITY 14-8

CHOOSE A JOB DESCRIPTION

When writing resumes, letters of application, or personal career Web pages, it always helps to have a job description in front of you for reference purposes. A job description will help guide your writing. After all, every job you apply for will have a specific description of needs. There are several job descriptions you can choose from to help you with the activities that follow:

1. Choose one job description from the following list:
   - The Corporate View job description you selected in Activity 13-1
   - The job description for which you gave the highest score in Chapter 13
   - Any of the four other job descriptions you researched in Chapter 13
   - Any job description that you like from the Corporate View intranet (Choose **Human Resources & Management**, **Current Job Openings @ TeleView** on the Corporate View intranet to view job descriptions.)
   - A new job description that you find from the Web, a newspaper, or other source

> **DigiTip**
>
> Imagine you are applying for the job in the job description you selected in Activity 14-8. Keep it in front of you at all times, and see how well you can write your resume to meet the requirements of this job description.

Individual users can hold video conferences. Both the caller and the person called must have the proper software, a sound card, a microphone and speakers or a headset, and a Web camera connected to the PC. If the person you call does not have a camera, you will not see that person. He or she will be able to see you, however. A fast connection to the Internet, such as a cable modem or digital subscriber line (DSL) is best. Using video in a call with a dial-up Internet connection is possible. A fast connection, however, makes video-conferencing work much better.

In the business world, a special conference room may be used for video conferences. Such a room is equipped with microphones, cameras, screens, and computers. People attending the conference can see and hear others at distant locations. Data and documents can be viewed or exchanged electronically during the meeting.

Figure 2-15
Video conferences allow people at different locations to see and hear each other and to exchange information electronically.

NETIQUETTE

Communicating successfully requires following certain rules of behavior. For example, you would not answer a phone and start shouting at someone. This would be considered rude. You should follow rules of polite behavior when communicating electronically. These rules are called **netiquette**. The word *netiquette* is a hybrid of the words *etiquette* (the requirements for proper social behavior) and *Net* (from the word Internet).

In the next activity, you will review the rules for e-mail netiquette that Corporate View has established for its employees. These rules are generally accepted and are good rules for you to follow.

When creating a new career-oriented e-mail account, choose a professional-sounding name. E-mail addresses like *spoiledbrat@corpview.com* or *iamlazy@corpview.com* are not good choices. Try to use your name, or invent a clever career-oriented e-mail address.

Every e-mail name must be different from all others. Your first choice for an e-mail name may already be taken. For example, Daniel.Jones@corpview.com might not be available. However, djones@corpview.com may be available. Sometimes you may have to try several names before you find one you like that is not already taken.

WEB HOSTING

For many of today's high-tech jobs, it is important that you post your own personal Web site. Your site can display your qualifications and skills in an online resume. These Web pages can be very creative documents. You can use graphics, text, and even multimedia to demonstrate your skills and abilities for particular careers.

If you have an ISP, space to post a Web site may be covered in your monthly fee. If not, you can rent space from an ISP. As is the case with e-mail, some sites offer free Web site hosting for job seekers.

ACTIVITY 14-7 — RESEARCH E-MAIL AND WEB HOSTING SERVICES

In this activity, you will research Web site and e-mail hosting services that will fit your budget and your needs as a job seeker.

1. Start *Word*. Open a new blank document.

2. Write five appropriate e-mail names that you can use for a career-oriented account and list on your resume. Use the @trophe.com extension for each of these addresses. List each one in your *Word* document.

INTER N E T

3. Access the Internet and start your browser. Search the Web looking for three e-mail providers. Find two e-mail providers where you would pay a monthly fee. Find another e-mail provider that offers free or very low-cost accounts. List the three providers in your word processing document.

DigiTip
Once you create a new career-oriented e-mail address, you may want to use it only for business purposes. Create a separate e-mail account for friends and personal business.

4. Search the Web to find three providers that will allow you to post your Web pages. Find two sites where you would pay a fee for the service and one site that offers free service. List each one in your *Word* document.

5. Save your document as *E-mail and Web Site* in your *DigiTools your name\Chapter 14* folder.

CHAPTER 14   Lesson 14-4

ACTIVITY 2-6

RESEARCH INFORMATION ON E-MAIL NETIQUETTE

1. Open your Web browser and the Corporate View intranet. Access the **Regular Features** page.

2. Choose the **Intranet FAQs** link followed by **Netiquette on the Corporate View Intranet**. Then choose the **Email Netiquette** link. Read this section of the FAQs.

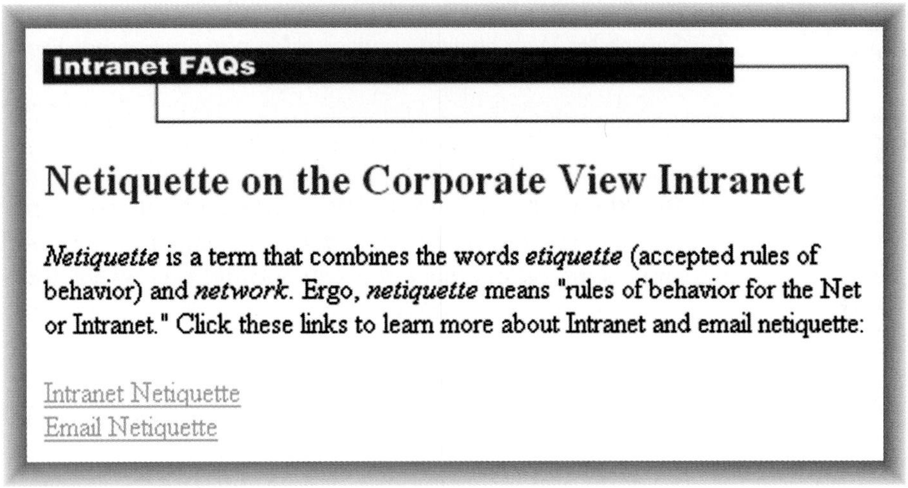

Figure 2-16
Netiquette FAQs Page

> **Intranet FAQs**
>
> ## Netiquette on the Corporate View Intranet
>
> *Netiquette* is a term that combines the words *etiquette* (accepted rules of behavior) and *network*. Ergo, *netiquette* means "rules of behavior for the Net or Intranet." Click these links to learn more about Intranet and email netiquette:
>
> Intranet Netiquette
> Email Netiquette

3. According to the Intranet FAQs on e-mail netiquette, what is the proper length for an e-mail message?

4. What is a possible problem with the "Reply to All" feature in e-mail?

5. Is personal e-mail at Corporate View allowed? If so, how much?

6. Do you think it is right or fair that IT employees at Corporate View have the authority to review personal employee e-mail sent over their e-mail system? Why or why not?

7. What are "flames"?

8. At Corporate View, employees should answer their e-mail within how many hours?

9. Why should using all capital letters in an e-mail be avoided?

10. Why is proper spelling and grammar in e-mail important?

11. What are emoticons? Give an example.

LESSON 14-4

FINDING E-MAIL AND WEB HOSTING SERVICES

OBJECTIVES *In this lesson you will:*

1. Evaluate the need for an e-mail address for your job search.
2. Create acceptable job search e-mail names.
3. Search for Web hosting services for a personal career Web page.

E-MAIL SERVICE

You should list an e-mail address in your resume. E-mail connects the working world, and you need to be connected, too. Having an e-mail address that you can check at any time is important. For instance, you might be on vacation when a message arrives notifying you of an important job interview. You do not want to miss this message. Several e-mail options can give you the flexibility to check your e-mail almost anywhere.

You might choose to open a paid e-mail account. If you have an Internet service provider (ISP), such as America Online, Earthlink, or Microsoft Network (MSN), you probably already have an e-mail account. These services charge a monthly fee. Most provide local access numbers to the Internet so you can check your e-mail from any Internet connection—at home, a local library, Internet café, or hotel room.

You might choose to sign up for a free e-mail account. You might not have the money for a paid account or you might not have your own computer. You can obtain a free e-mail account from one of dozens of free e-mail service providers on the Web. You can then use any Internet connection to check your e-mail. You can borrow a friend's computer or use one at your local library or at a job placement center.

Figure 14-5
Check your e-mail frequently for messages related to your job search.

LESSON 2-7: E-MAIL SOFTWARE AND ADDRESSES

OBJECTIVES *In this lesson you will:*

1. Learn about client software and e-mail addresses.
2. Compose and send an e-mail message.

CLIENT SOFTWARE

To use e-mail or instant messaging, you must have the appropriate software. End user software, like e-mail, instant messaging, and Web browser applications, are often called clients. Clients communicate with servers. **Servers** are fast, powerful computers. Servers store e-mail messages being sent to and from users. Servers also store information for Web sites. Servers then serve (send) e-mail and Web pages to requesting clients just like a waiter or server will serve food to customers at a restaurant.

To use e-mail, you must first establish an e-mail account. You may also need to set up an account with an Internet service provider. An **Internet service provider (ISP)** is a company or organization that provides customer connections to the Internet and related services. America Online and Earthlink are popular ISPs.

Several e-mail programs are available. Fortunately, most e-mail programs work in very similar ways. In this lesson, Microsoft's *Outlook Express* will be used to demonstrate how e-mail works. Don't worry if you have a different e-mail program. You should be able to apply what you learn easily to your software.

DigiTip
You can access some online e-mail services by following this hyperlink path on the corporate intranet: **Regular Features**, **Intranet FAQs**, **Email Services**.

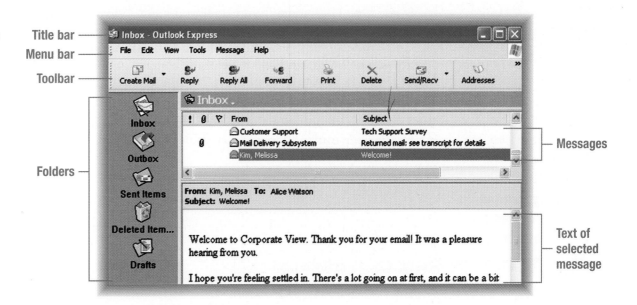

Figure 2-17
Outlook Express is a typical e-mail program.

CHAPTER 2 Lesson 2-7 52

3. List each school or training program you have attended in chronological order. Organize your list of schools and training programs by date attended. Place the most recent school or training program at the top of the list and your earlier educational experiences at the bottom.

4. You should list your work experience on a resume. (See Figure 14-2, part E.) To help you think about this section of your resume, list any jobs that you have had, including volunteer work. List everything—full- or part-time jobs, paid or unpaid work—no matter how small or large. For each job, list the following details as completely as possible:
- Name of your employer
- Dates you worked for each employer
- Specific jobs or duties that you had while at this job
- Address or location of each job that you had

5. List each job you have had in chronological order. Organize your list by date with the most recent job at the top of the list, and your earlier employment experiences at the bottom of the list.

6. Listing your achievements on a resume may give you an edge over other job candidates. (See Figure 14-2, part F.) To help you think about this section of your resume, list any awards or special achievements you have received.
- Name each achievement, scholarship, recognition, or award you have received.
- Include the dates you received, achieved, or completed the achievement.

7. List each award in chronological order. Organize your list by date with the most recent achievement at the top of the list.

8. Save your answers as *Qualifications* in your *DigiTools your name\Chapter 14* folder. Close the document.

ACTIVITY 14-6 — RECORD REFERENCES AND CONTACTS

Many good job leads come from people you know. Friends, family, acquaintances, teachers, and counselors can become part of your personal employment network. These people can provide a rich source of job-related information and referrals. In the next activity, you will begin to build a list of job contacts. Consider it a running list. Add names to it later as you think of them. It's time to create a personal job network.

1. Start *Word*. Open a new blank document. Create a list of individuals who can help you find or prepare for a job. Include an address and phone number for each person you list. Also include e-mail addresses if available. Your list should include:
- Your very best friends
- Other friends, coworkers, neighbors, and businesspeople you know
- Teachers and guidance counselors
- Adults in your immediate and extended family
- People you know who are in the career field that interests you

2. Save the document as *References* in your *DigiTools your name\Chapter 14* folder. Print the list. Close the document.

3. Save a backup copy of your employment contacts list. For example, save your list on a floppy disk or a CD. If you have a PDA, place a backup copy of your list there for easy reference.

> **DigiTip**
> You may want to create a special folder called *Employment Journal* to store these files that you created: *Career Objectives, Related Experience, Qualifications,* and *References.*

E-MAIL ADDRESSES

People with whom you communicate by e-mail are called *contacts* (sometimes *buddies*). All contacts are accessed by their e-mail names. Before you can use any e-mail or instant messaging software, you must first have an e-mail address. Each e-mail address must be different from all others on the Web. Otherwise, e-mail messages may be delivered to the wrong person. An e-mail address contains a username and a domain name separated by the at sign (@). For example:

Kim@swep.com

Kim.Miller@thomsonlearning.com

MKim@speakingabout.com

Maria_Bravo@speakingsolutions.com

maria_bravo@corpview.com

Spaces are not used in e-mail addresses. Sometimes an underscore or a period is used to separate first and last names. You need not worry about capitalization when entering e-mail addresses. Maria_Bravo@corpview.com is the same address as maria_bravo@corpview.com.

ACTIVITY 2-7   SEND E-MAIL

INTERNET

HELP KEYWORDS
Send
Sending e-mail messages

1. Open your e-mail software. (If you don't have e-mail, visit **Email Services** under **Intranet FAQs** on the Corporate View intranet. There you will learn about various free e-mail options available on the Internet.)

2. Open a new e-mail message screen. Click **Create Mail**. (Similar options are Compose Mail or New Message.) A new e-mail message form will appear. (See Figure 2-18.)

3. Melissa Kim is a member of the Human Resources team at Corporate View. Imagine that you have decided to accept an internship opportunity at the company. Write to Melissa and let her know of your intention to join the company for a semester of study and work. In the *To:* text box, enter Melissa's e-mail address: `mkim@corpview.com`.

4. In the *Subject:* text box, enter a descriptive title, such as `Internship`.

5. In the *Message:* window, enter a message. Tell Melissa that you plan to take part in the Corporate View internship program. Explain why you are excited about this opportunity.

ACTIVITY 14-4 — RECORD RELATED EXPERIENCE AND ABILITIES

1. Start *Word*. Open the data file *Related Experience*.

2. Resumes usually include job-related experience statements. (See Figure 14-2, part C.) To help you think about writing this section, answer questions about the valuable skills that you currently have. Record any experiences you have had that can relate to a future career.

3. Save your document using the same name in your *DigiTools your name\Chapter 14* folder. Close the document.

WHAT YOU SHOULD KNOW ABOUT Resume Myths

Four of the common myths about resumes are described below. Do not let these myths cause you to write an ineffective resume.

True/False?   An effective resume lands you the job you want.
              False. An effective resume gets your foot in the door.

True/False?   An effective resume lists all your qualifications and skills.
              False. An effective resume highlights your key skills and abilities.

True/False?   A resume will be thoroughly read.
              False. If you're lucky, a resume will be glanced at.

True/False?   An effective resume has lots of information.
              False. An effective resume has enough data to pique your reader's interest.

ACTIVITY 14-5 — RECORD EDUCATION, WORK EXPERIENCE, AND ACHIEVEMENTS

1. Start *Word* and open a new blank document.

2. All resumes should include educational experience statements. (See Figure 14-2, part D.) To help you think about this section of your resume, list any schools and special training (classes) you have attended. For each school or training opportunity, list the following details as completely as possible:
   - The name of the school or training program
   - The address or location of each school or training center you have attended
   - The specific dates you attended the school or training program
   - Any specific courses you have taken that would help prepare you for a job

Figure 2-18
New Message Window in *Outlook Express*

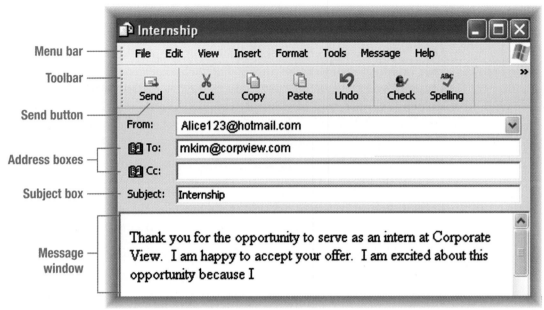

6. Click the **Send** button to send your message.

7. Locate your sent message in the *Sent* folder. Open the message by double-clicking the message in the Sent list. Click the **Print** button to print the message.

Peer Check Compare your message with a classmate's message. Discuss how both messages could be improved.

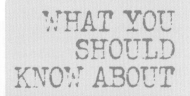

WHAT YOU SHOULD KNOW ABOUT
Computer Viruses

One of the major problems with e-mail is that messages can carry computer viruses. A computer virus is a program that can be loaded onto your computer or network without your knowledge. Viruses run against your wishes. They can cause serious damage to the data on your computer. Computer viruses are created and spread by criminals who want to disrupt and damage computers, data, networks, intranets, and the Internet at large.

Thousands of computer viruses are known to exist. Antivirus programs have been written to detect, isolate, and destroy viruses before they can do any damage. Antivirus programs can scan e-mail as it comes into your inbox and scan your computer's drives looking for viruses.

To protect yourself from viruses, do the following:

- Run virus protection scanning software.
- Set your virus software to scan all incoming e-mail messages.
- Scan your computer drives for viruses at least once a week.

PERSONAL EMPLOYMENT JOURNAL

Many experiences from your life (education, training, experience) should be listed on your resume. Learn to make a record of such experiences so that when you are writing a resume, you can look up the details quickly. A **personal employment journal** is a written self-assessment of your interests, skills, work-related experience, and other job qualifications. Preparing a personal employment journal can help you focus your resume-writing efforts on a job that is right for you. In a personal employment journal you will analyze your:

- **Career Objectives and Interests.** Examine what you want from a job.
- **Career Related Experience and Abilities.** Determine the personal qualities, experiences, and skills you have that would make you a good employee.
- **Educational Experience.** List any schools and special training classes you have taken.
- **Work Experience.** List any jobs you have had including volunteer work.
- **Achievements.** List any special awards or recognitions you have received.
- **References and Contacts.** Create a list of contacts that can help you obtain a job, advance your career, and expand your career opportunities.

Once you have created a personal employment journal, update it through all of the working years of your life. As you change jobs, receive awards, accomplish new things, or make new contacts, constantly record your new career-related experiences and qualifications in your journal. You will find the personal employment journal an important tool in your job searches.

ACTIVITY 14-3 — RECORD CAREER OBJECTIVES AND INTERESTS

Think about your future work-related goals and objectives. Why do you want to work in a certain career? What are your career goals? Even if you are not exactly sure what you want to do, spend some time writing your thoughts. Perhaps answering questions about these topics can help you clarify your goals.

1. Start *Word*. Open the data file *Career Objectives*. Answer the questions related to your employment interests and career objectives.

2. Resumes require a career objective statement. (See Figure 14-2, part B.) The statements do not have to be complicated. However, they should reflect your career goals. Three examples are shown below. Write three career objectives of your own at the end of your document.

Examples:
- A summer internship tracking Internet sales trends
- A position in one of the major accounting firms in Tallassee, Florida
- A teaching position for third or fourth grades

3. Save your document using the same name in your *DigiTools your name\ Chapter 14* folder. Close the document.

What You Should Know About Computer Viruses *continued*

- Make backup copies of important files. Store the copies on separate drives, floppy disks, or CDs.
- Do not share unlicensed software, data disks, or CDs with other users. Do not allow others to open their homemade CDs on your computer. If you must load data from an unfamiliar CD, always scan it first.
- Beware of e-mail attachments. **Attachments** are files that are loaded onto e-mail messages and travel with them to their recipients. If you receive an attachment from an unknown source, do not open it!
- Commercial software is often prone to attack by viruses. Download any updates or service releases to your software to ensure that any potential security breaches have been fixed.
- Update your virus scanning software frequently.
- Inform your teacher or network administrator if you receive any viruses.

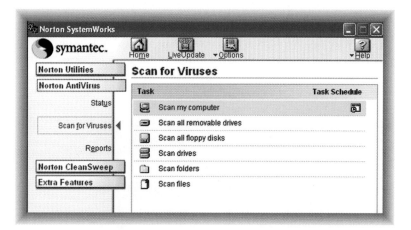

Figure 2-19
Scan your computer for viruses regularly with a program such as *Norton AntiVirus*.

LESSON 2-8 MANAGING E-MAIL MESSAGES

OBJECTIVES *In this lesson you will:*

1. Receive and open e-mail.
2. Forward e-mail.
3. Delete e-mail.

Sending and receiving e-mail is a popular activity for many Internet users. In business, e-mail is an important aid in getting work done. However, one of the problems with e-mail is that it can take a lot of time to read, reply to, and organize. Some people spend too many hours each day sending and answering their e-mail.

LESSON 14-3

RESUME BUILDING: A LIFELONG PROCESS

OBJECTIVES *In this lesson you will:*

1. Create a personal employment journal.
2. Think and write about career-related objectives, interests, and experiences.
3. Think and write about educational and work experiences.
4. Think and write about personal achievements.
5. Develop a list of contacts and references.

JOB CHANGES

For many people, career preparation and job searching are a lifelong process. Before you focus on a specific career, you will probably work at a number of other jobs. These jobs will pay the bills while you are in school and will provide valuable experience that you can list on your resume. Think of all the classes you take and courses you attend as potential items that can be listed on your resume.

Figure 14-4
You may work in a number of jobs as you gain work experience or complete your education.

A generation ago it was common for employees to work for a single company throughout their entire careers. This type of stability is rare today. If you are like the average worker, you will make a half-dozen major job changes or more in a lifetime of work! In addition, you may compete for promotions at your company, which often requires an updated resume. Throughout your career, you will probably update your resume and write letters and applications many times. Although you may write or update a resume in a few short hours, the resume reflects skills and experiences gained throughout your life.

E-mail messages arrive in an inbox. You should manage your e-mail messages to prevent your inbox from becoming overcrowded. After you have finished reading an e-mail message, you can:

- **Reply** to an individual
- **Reply to All** of the people listed in the *To:* and *Cc:* fields
- **Forward** the message to another person
- **Delete** the message
- **File** or **save** the message

Manage your e-mail effectively by handling each e-mail message the first time you read it. When you open a message, reply to it if that is appropriate. Then delete the message or save it in an appropriate folder if it contains information you may need later. In this lesson, you will manage e-mail by reading, forwarding, and deleting messages.

ACTIVITY 2-8

RECEIVE AND FORWARD E-MAIL

INTER N E T

DigiTip
You can also open an e-mail message by selecting it and choosing **Open** from the File menu.

Figure 2-20
Messages from others appear in the *Inbox* folder.

1. Open your e-mail program.

2. After a reasonable wait, you should have received a reply from Melissa Kim about your internship. This message will arrive in your *Inbox* folder. Double-click the reply message to open it. Read the reply.

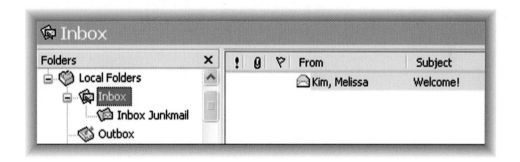

3. Forward this good news to a classmate, close friend, or family member who also has e-mail. Click the **Forward** button.

4. In the *To:* address box, enter the e-mail address of the person to whom you are forwarding the message. Key a brief note in the *Message:* window. Click the **Send** button.

5. Select the message from Melissa Kim. Delete it by clicking the **Delete** button.

6. Open your **Deleted Items** folder and see if the message appears there. If so, you may need to select the message again and delete it to clear the *Deleted Items* folder.

HELP KEYWORDS
Forward
Forwarding e-mail messages

ACTIVITY 14-2

MATCH QUALIFICATIONS TO NEEDS

Maria Bravo believes she has been doing a great job since she started working in Corporate Communications a little over a year ago. During this time, she has been working on her Bachelor of Arts (B.A.) degree in English part time in the evenings. She is hoping this degree will help her achieve a **promotion**—a new job with higher pay and increased responsibilities.

This week, a great new job opening was listed that interests Maria very much. Can she compete with other possible candidates? Let's find out.

1. Open your browser and access the Corporate View intranet. Choose **Human Resources & Management**, **Current Job Openings @ TeleView**, **Corporate Communications.** Choose the fifth job in the list called **Web and Intranet News Writer & Editor**. This job description describes the new position for which Maria is interested in applying.

2. Open *Word* and a new blank document. Create a table with three columns and at least eleven rows. Enter numbers and column heads as shown in the sample table below. In the second column of the table, enter all the skills, attributes, abilities, and requirements listed in the job description.

	REQUIREMENTS IN JOB DESCRIPTION	MARIA S SKILLS AND EXPERIENCE
1.	Required to live near Boulder, Colorado	Maria already lives near Boulder, Colorado
2.	HTML background is a big plus	Proficient with Dreamweaver
3.		

3. Read Maria's resume shown in Figure 14-2 on page 408. In the third column of your table, list the training, education, abilities, and skills Maria has that most closely match the requirements found in the job description. Two examples are shown in the sample table above.

4. Review the job description and Maria's resume one more time. Are there any requirements in the job description that Maria does not meet? Are there any weaknesses Maria should be aware of as she applies for this position?

5. What do you think Maria's chances are of getting this job? Is she qualified?

6. Save your table and responses as *Activity 14-2* in the *DigiTools your name\Chapter 14* folder.

LESSON 2-9: SEARCHING ONLINE RESOURCES

OBJECTIVES *In this lesson you will:*

1. Learn about search engines.
2. Research information on the Internet.
3. Learn about technology certifications.
4. Find information on ethics related to computer use.

SEARCH ENGINES

The Internet contains vast amounts of information. Articles and reports can be found on almost any topic. Finding the exact information you need, however, can be a challenge. Search engines can help you meet this challenge. A **search engine** is a program that performs keyword searches for information on the Internet. The search may cover titles of documents, URLs, indexed words in documents, or the full text of documents. *Google, Yahoo!, AltaVista, Lycos, Northern Light,* and *WebCrawler* are some popular search engines.

To use a search engine, log on to the Internet and open your browser program. Enter the URL of a search engine in the address bar. For example, you could enter *www.google.com* to access *Google*. The search engine will have a text box where you can enter the keywords (a word, phrase, or title) that you want to find. For example, you might enter *inkjet printer* to find information about printers. After the keywords are entered, click a button to start the search. Common names for this type of button are *Search, Go,* or *Find*.

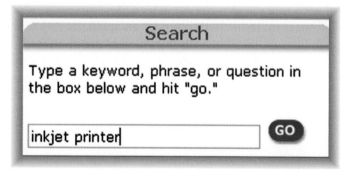

Figure 2-21
Search engines help users find information on the Internet.

When a search is completed, the search engine will display a list of results. For each result (also called a hit), information such as the title of the document or Web site, the URL, the file size, the language, and a short extract is displayed. Search engines typically display 10 to 15 results at a time. You can move to another page of results if more are available. Click on a result to move to that Web site or document.

ETHICS

Honesty on Resumes and Applications

Some people have a natural tendency to exaggerate. If you are telling a story to your friends and the story isn't particularly interesting, exaggeration is a fun way to enliven a conversation. Using exaggerations is fine as long as everyone knows the story is mostly fiction. For example, a wild fishing story about the one that got away is fine among good fishing buddies who know not to take the tall tale seriously.

When completing an application, resume, or letter of application, however, honesty must be the watchword. Employers appreciate honesty above everything else. Giving a potential employer inaccurate or exaggerated information is unethical. If an employer finds that an employee has given false information on an application or resume, the employee may be dismissed from the job. When writing a resume, tell the truth, the whole truth, and nothing but the truth. Honesty is always the best policy.

Follow-Up

What problems might you have on the job as a result of having given false information on your job application?

LESSON 14-2: COMPARING APPLICANT QUALIFICATIONS AGAINST EMPLOYER NEEDS

OBJECTIVES *In this lesson you will:*

1. Learn why an effective resume is important for the job search process.
2. Compare requirements on a job description to an applicant's resume.

Job applicants must learn what employers want and need. Generally, these needs are explained in a job description, which you learned to analyze in Chapter 13. An employer is more likely to consider an applicant if that applicant's qualifications, skills, and abilities (as listed in a resume) closely match those listed in a job description. For this reason, an effective resume is very important for a successful job search.

If an applicant falls short in his or her qualifications, steps must be taken to improve the qualifications. For example, Maria Bravo thought that she needed a Bachelor of Arts degree in English to qualify for the type of job in which she is truly interested.

Figure 14-3
An applicant's qualifications and skills must match employer needs.

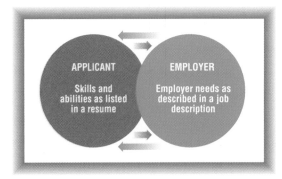

ANALYZING SEARCH RESULTS

The results list may contain dozens or thousands of entries. How do you decide which results are valid or useful? With many search engines, the results are sorted in order by relevance. This means that the Web sites or documents most likely to contain the information you want are listed first. Sometimes a results list will have a number, such as 50%, 80%, or 100%, by each entry. The entries with the highest numbers are the most likely to have the information you want.

You may see the heading *Sponsored Sites* (or a similar name) at the top of the list of search results on some search engines. These sites are listed at the top because their owners have paid the search engine to list them first. These sites may not be the most relevant to your search terms. Ethical search engine providers will let you know when a site is sponsored.

To learn more about search engines, search the Web using the term *search engine*. Web sites such as Search Engine Showdown (http://searchengineshowdown.com) provide descriptions and comparisons of various search engines.

Many Web sites have a search feature that allows you to search for information on that Web site. Look for a Search link if you don't see a search text box on the main page. Enter words in the search text box just as you would in a search engine like *Google*. The search results may or may not be ranked in order of relevance.

ACTIVITY 2-9   SEARCH THE WEB

Some workers who deal with computers, networks, and software take tests to prove their skills and knowledge. This process is called certification. People who pass certification tests may have a better chance at getting a job or may receive a raise or promotion at their current job. In this activity, you will search the Internet to find information about various certification programs.

Work in a team with four other classmates to complete this activity.

TEAMWORK

INTERNET

1. Open your Web browser and complete any logon procedures necessary to access the Internet.

2. Each team member, choose one of the following search tools. Enter any one of these Web addresses in the address bar. Press **Enter** to open the search engine. If the engine you choose does not respond, try another one.

HELP KEYWORDS

Find
Finding the information you want on the Internet

www.google.com   www.yahoo.com   www.askjeeves.com
www.altavista.com   www.northernlight.com   www.webcrawler.com

3. When the search engine opens, enter one of the following search terms in the search engine text box. Each team member, use a different search term. Click **Go, Find,** or **Search** depending on the prompt given you by the search tool.

Microsoft certification
Cisco certification
Oracle certification
Macromedia certification
Adobe certification

Figure 14-2
Sample Resume

1 inch top margin or center page vertically

A — Bold for name and side headings

Maria Bravo
2818 River Bottom Road
Boulder, CO 80303

(506) 555-0123
mbravo@trophe.com
www.trophe.com/bravo.html

B — **Career Objective**

A promotion to the position of *Web and Intranet News Writer & Editor*

1 inch or default side margins

C — **Related Experience**

- Currently a Communications Coordinator in the TeleView Division
- Wrote press releases, coordinated fund-raising events, and worked closely with the public relations team for the U.S. Olympic Ski Team
- Worked part-time at a public radio station writing copy for fund-raising spots and public service announcements
- Rewrote and edited Corporate View's online Employee Handbook
- Edited eight manuals from other writers within the company
- Created 40 percent of the Internet pages on TeleView's Web site
- Created training on how to write effective press releases
- Know Microsoft Office extremely well. Obtained MOS certification
- Proficient with Dragon NaturallySpeaking and Microsoft speech and handwriting software
- Proficient with Web design software including Dreamweaver, Flash, Illustrator, and Photoshop

Indent text under side headings .5 inch

D — **Educational Experience**

Colorado State University, Fort Collins, CO. B.A. Major: English; Minor: Business Management, May 2003, GPA 3.72

Walla Walla Community College, Walla Walla, WA. Associate's Degree: Mass Media Communication, May 1998, GPA 3.5

Granger High School, West Valley City, UT. 1996

E — **Work Experience**

- Communications Coordinator, Corporate View's TeleView Division, Boulder, CO 80303. 2003-Present
- Public Relations Writer, U.S. Olympic Ski Team, Park City, UT 84663. 1998-1999
- Copy Writer, Campus Public Radio Station, Walla Walla, WA 99362. 1996-1997

F — **Achievements**

- Ezra T. Taft Academic Scholarship, 2002
- MOS Certification, 2001
- Second place, *Escapes* magazine writing contest, 2001

G — **References**

Dr. Sally Rogers	Mr. Robert Petrie	Ms. Laura Moore
Professor of English	Director	HR Manager
Department of English	Corporate View	Corporate View
Colorado State University	One Corporate View Drive	One Corporate View Drive
Fort Collins, CO 80523	Boulder, CO 80303	Boulder, CO 80303
(802) 555-0177	(303) 555-0177	(303) 555-0177
dr.s.rogers@co.csu.edu	rpetrie@corpview.com	lmoore@corpview.com

4. Scan the first page of search results. Pick three results that seem the most promising. Access and scan the Web pages for these three sites.

5. Return to the Web page that best describes the certification program. You need to print this page to discuss with your teammates. The Web site may have many pages. If you click the Print button, you may print several pages that you do not need. Instead, choose **Print** from the File menu. In the Page Range section, choose the **Pages** option. In the Pages box, enter **1** to print one page or **1-2** to print pages one and two. Then click **Print**.

6. Using your printed Web page, discuss the information you found with your teammates. Prepare a list that tells what type of hardware, software, or networking tools or skills each certification program relates to. For each type of certification, give the name and the URL of the page where you found the information.

APPLICATION 2-1   RESEARCH COMPUTER ETHICS

You have learned netiquette rules to help you behave properly and politely when sending e-mail messages. In this application, you will search for rules of conduct for the proper use of information technology as a whole.

INTER N E T

1. Open your Web browser and complete any logon procedures necessary to access the Internet.

2. Access a search engine such as *Google, Yahoo!,* or *Northern Light*.

3. Information on computer ethics can be found on various Web sites such as the one for the Computer Ethics Institute. Use the search engine to find addresses for Web sites that provide information on computer ethics. Review your list of hits. Access those hits that seem most valid.

4. Once you reach a site, look at the links on the home page. Do you see a link that seems to provide rules for computer ethics? If yes, follow the link. If not, use the Web site's search feature if one is available. (You may need to scroll to the bottom of the page to see the Search link.) Use a search term such as *computer ethics*.

5. When you find information or a list of rules for computer ethics, print the Web page. Read the page and keep it for reference.

Peer Check Compare the information you found about computer ethics with the information found by a classmate. Discuss how the Web pages are similar and how they are different.

2. Review a typical online application. Open your browser and access the Corporate View intranet. On the Corporate View Intranet Home page, choose **Human Resources & Management**. Choose **Corporate View Employment Application Form**.

3. Do not enter data in the online application. Just scan the form from top to bottom. This is an input form for a database. The rectangles indicate fields where an applicant can enter data. Take a look at the various fields. Take notes about what you need to know to fill in each field completely. For example, do you have an e-mail address? Do you know a place where you can post a personal Web page? Can you remember all of your educational information?

4. Return to the Corporate View Intranet Home page.

5. A letter of application is a formal document. It introduces you and asks that you be considered for a job. Examine a typical letter format that you can use for a letter of application. To view a sample letter format, choose **Corporate Communications, Style Guide, Letters, Sample Letter**.

6. Review the sample letter. Take notes about the pieces of information you will need. For example, you will need the recipient's name and title, company name, street address, city, state, and ZIP code. Notice the Enclosures notation at the end of the letter. Read the last paragraph of the letter to learn the purpose of this notation. Close the Corporate View intranet.

7. Start your browser. Open the data file *Bravo*. Review Maria Bravo's personal Web page. Maria uses this page to explain about herself, her interests, and her career. Can you tell from reading this page what her interests are, what she likes to do, and if she would most likely be an outstanding team member? Do you know how to create a Web page like this? Do you know where to post a personal Web page so it can be read online? Take notes about anything you find confusing. Close the Web page and your browser.

8. Save your notes as *Activity 14-1* in the *DigiTools your name\Chapter 14* folder. Print and close the document.

TEAM WORK

9. In later activities, you will create all four of these documents that are important for a job search. Meet with a team of classmates or your instructor and ask questions about anything you found confusing in the documents you reviewed. Use your notes to guide the discussion. Talk about any portions of the documents that you do not understand. Discuss the key parts of these documents and how to create them. Talk about how you can research and find the required information.

DIGITOOLS
DIGITAL WORKBOOK CHAPTER 2

Open the data file *CH02 Workbook*. Complete these exercises in your *DigiTools Digital Workbook* to reinforce and extend your learning for Chapter 2:

- Review Questions
- Vocabulary Reinforcement
- Math Practice: Converting Fractions, Decimals, and Percents
- From the Editor's Desk: Capitalization Rules

UNIT 1

Tooling Up!

In these unit exercises, you will learn about four career path clusters. You will go online to learn, think, and write about a variety of DigiTools, business issues and trends, and career paths. The skills and knowledge you have learned in the previous chapters will help you complete these exercises.

CAREER CLUSTERS

Education and Training

When you first think of the education and training career cluster, you probably think of your school teachers. Indeed, over 7.5 million teachers are employed in the United States alone. Teachers are a sizable part of the American workforce.

For those with a love of teaching and who possess great interpersonal skills, education can be a very rewarding career path. Teachers specialize by age and subject area. There are preschool, kindergarten, elementary, secondary, college, and adult education teachers. Content areas, especially at the high school and college levels, cover a wide range of areas such as:

- Science
- Mathematics
- Language arts
- History

LESSON 14-1

JOB SEARCH DOCUMENTS

OBJECTIVES *In this lesson you will:*

1. Preview a sample resume, application, and personal career Web page.
2. Preview a sample letter format.
3. Discuss the documents needed to compete for employment.

A **resume** is a summary of your work qualifications and skills. It is an organized collection of data that will help "sell" your skills to an employer. A resume may also contain a list of references. **References** are individuals who can give a potential employer information about your job qualifications or your character. A resume is usually accompanied by a letter of application. A **letter of application** is a document that introduces you to a potential employer and asks that you be considered an applicant for a job. The letter should typically be limited to one page. You may also be asked to complete an application form. An **application form** is a document that has space for an applicant to enter personal contact information and information related to education, skills, and work experience.

You may wish to create a personal Web page that potential employers can visit to learn more about you. Such a Web page will help you project a professional image. Think of these documents as advertisements that get you noticed in the crowded, competitive job market. They represent you and pave the way for a job interview. These documents can literally open doors of career opportunity for you.

ACTIVITY 14-1

PREVIEW JOB SEARCH DOCUMENTS

In this activity, you will preview important job search documents. This will help you create flawless job search documents in later activities. Take notes listing anything you find confusing about resumes, applications, Web pages, or letters of application. You can use several software options to take notes for this activity. If you have a PDA, use the *Notes* feature. If you have a Tablet PC, use either *Journal* or *Sticky Notes*. If you use a traditional PC, use *Word* or *Notepad*.

DigiTip

Generally, a resume should be one page in length. If the References section makes the resume exceed one page, the references may be placed on a separate page or printed on the back of a one-page resume.

1. Review the sample resume in Figure 14-2 on page 408. If there is any piece of information required in the resume that you currently do not know or that you find confusing, make a note. Examine each part carefully. Identify the following parts, which are labeled with the letters shown here.

   A. Contact Information
   - Name
   - Address
   - Phone Number
   - E-Mail Address
   - Personal Web Address

   B. Career Objective or Job Statement

   C. Related Experience and Skills
   D. Educational Experience
   E. Work Experience
   F. Relevant Personal Achievements
   G. References

CHAPTER 14 Lesson 14-1

- Computers and technology
- Music
- Art
- Drama
- Physical education, health, and sports

Another group of educators work directly with adults in businesses. An estimated $60 billion is spent each year training employees. Corporate trainers teach many different skills and subjects. Some topics are general, such as workplace safety. Other training is for specific job skills. Consider training for workers at a company that makes microchips. Workers are taught the details of computer chip design and production. They learn corporate secrets and skills that cannot be learned anywhere else.

A bachelor's degree is usually required to work in education and training. However, about 1.18 million teacher assistants serve in schools. Many of the assistants have an associate's degree or some college training but no degree. Review a few sample jobs and potential salaries in education and training in the table below.

Careers in Education and Training

Position and Yearly Salary		Position and Yearly Salary	
Teacher Assistant	$19,430*	Kindergarten Teacher	$41,100*
Librarian	$43,750*	Elementary Teacher	$43,320*
Museum Curator	$38,160*	Secondary School Teachers	$43,580*
Special Education Teacher	$44,890*	Postsecondary Teachers	$49,770*
Vocational Education Teacher	$45,050*	(Varies based on subject taught)	$69,620*
Corporate Technology Trainer	$47,747**	Adult Literacy Instructor	$40,230*
School Principal	$60,078*	Coach	$34,680*
Preschool Teacher	$20,940*		

*Salaries based on 2001 industry averages, Bureau of Labor Statistics, www.bls.gov, downloaded January, 2003.

**Salaries estimated from a Midwestern city median averages using data from www.salary.com calculated in January, 2003. Salary estimates will vary greatly from city to city, state to state, and from year to year.

EXERCISE 1-1

RESEARCH SEMINARS

You have learned that Corporate View offers general training on its intranet, such as the Style Guide and From the Editor's Desk. Corporate View also offers classes, called seminars, to its employees. Investigate these courses provided for employees.

1. Access the Corporate View intranet. On the Corporate View Intranet Home page, choose the **Regular Features** link. Choose the **Employee Training & Evaluations** link. Scroll down the page until you see the seminars list.

2. Choose five of the seminar topics that would interest you the most. Write a brief description of each seminar.

3. Which of the five seminar topics would you find most interesting to take as a seminar class?

CHAPTER

CHAPTER 14

Qualifying and Applying for a Career

OBJECTIVES *In this chapter you will:*

1. Preview a sample resume, application, personal career Web page, and letter of application.
2. Compare applicant qualifications against the skills required by an employer.
3. Create a personal employment journal.
4. Research e-mail and Web hosting services.
5. Write a chronological resume, letter of application, and personal career Web page.
6. Fill out an application online.
7. Compare your qualifications against the skills required by an employer.

In Chapter 13, you explored career choices. You saw the wide diversity of careers available. While reading a variety of job descriptions, you were able to think about what you might want in a career. In this chapter, you will learn how to present your skills and other qualifications to employers.

Job openings are doorways to opportunity. They open for a short time and then close as jobs are filled. To get a great job, you must compete for it. Often, there are many more applicants than jobs. You will need to impress a prospective employer with your job-related skills. If you can communicate your skills in a powerful resume, an application, a personal career Web page, and a letter of application, you will be noticed by employers and increase your chances of being hired.

Figure 14-1
You want prospective employers to be interested in reading your resume.

Agriculture and Natural Resources

Workers in agriculture and natural resources careers help provide food, clothing, heat, light, and other support services for us all. Workers in this area might grow corn in Nebraska, cotton in Mississippi, or tomatoes in California. They may harvest paper trees in Maine, mine coal in West Virginia, or discover new oil reserves in Wyoming. We are all dependent upon the millions of Americans employed in farming, fishing, forestry, and mining.

New areas in the natural resources cluster have emerged recently. These areas involve biotechnology, environmental conservation, and other areas of agricultural and mining science. Review some sample jobs and potential salaries in this highly diverse and varied career cluster in the table below.

Careers in Agriculture and Natural Resources

Position and Yearly Salary		Position and Yearly Salary	
Agricultural Inspector	$29,630*	Hazardous Material Remover	$34,510*
Greenhouse Worker	$15,730*	Rotary Drill Operator	
Forest and Conservation Worker	$21,440*	(Oil and Gas)	$35,640*
Logging Equipment Operator	$27,240*	Continuous Mining Machine	
Farm Labor	$15,730*	Operator	$34,320*
Mining and Geological Engineer/		Bioagricultural Engineer	$64,330*
Mining Safety Engineer	$64,370*	Agricultural and Food Scientist	$52,290*
Environmental Scientist	$50,700*	Biological Technician	$34,030*
Landscaper	$20,880*	Landscape Architect	$51,640*
Environmental Engineer	$64,330*	Veterinarian	$69,150*
Supervisor/Manager of			
Mining Operations	$49,430*		

*Salaries based on 2001 industry averages, Bureau of Labor Statistics, www.bls.gov, downloaded January, 2003.

EXERCISE 1-2 — RESEARCH EDUCATIONAL RESOURCES

Workers are employed in many different jobs in the agriculture and natural resources career cluster. Most related pathways require considerable education and scientific research. A few examples are:

- Agricultural science
- Conservation and environmental science
- Landscape architecture and horticulture
- Mining and geology
- Veterinary science

Your task is to use your search tools to find a school, university, college, or technical training center that provides education or does research in each of these areas.

7. Sort the records in ascending order by salary. Which job description has the highest salary? Which job description has the lowest salary?

8. Close the database file and exit *Access*.

Peer Check Compare your plan for fields for the database with a classmate. Discuss data types for each field. Make changes, if needed, after your discussion.

ORGANIZING LINKS FOR CAREER RESOURCES

While studying this chapter you have visited several Web sites and online sources for career information. Keeping track of all of these resources can be difficult unless you organize this information somehow. A Web page is a great tool for cataloging sites. You can create a Web page with links to online career sources, job boards, and employment Web sites.

APPLICATION 13-2 — EMPLOYMENT RESOURCES WEB PAGE

1. Using your HTML skills, create a Web page to list and link to career sites such as the ones you have visited. Include links to at least ten career or employment information Web sites.

2. Save your Web page as *Employment.htm* in the *DigiTools your name\Chapter 13* folder. (If you want to enhance your Web pages with graphics, create a subfolder named *joblinks* in the *DigiTools your name\Chapter 13* folder. Save your Web page in this subfolder.)

3. Open your Web page in your browser and test the links to make sure each of them work.

DIGITOOLS
DIGITAL WORKBOOK — CHAPTER 13

Open the data file *CH13 Workbook*. Complete these exercises in your *DigiTools Digital Workbook* to reinforce and extend your learning for Chapter 13:

- Review Questions
- Vocabulary Reinforcement
- Math Practice: Scoring Employee Evaluations
- From the Editor's Desk: Ambiguous Pronoun References
- Keyboarding Practice: Finger Reaches and Speed Building

TEAMWORK

INTERNET

1. Work as a team with four other students on this project. Each team member, choose a different career area from the list on the previous page.

2. For each career area, identify a school, university, college, or technical training center that provides education or does research in this area. To find information, search the Internet. Use a search engine such as *Google* (www.google.com) or *Lycos* (www.lycos.com). Use the name of the career area and *education*, *research*, or *training* as the search terms. For example, *veterinary science training*.

Hint: Start close to home. Visit the Web sites of universities, technical schools, and colleges in your community or state. Discover if they have programs related to any of these areas. Then expand out to other states as necessary.

3. Write a short description of each school and what it offers to students in each of the areas being researched. Note the source information (author's name, if given; Web site name; URL; and date you accessed the site) for each Web site you use.

4. Each team member, discuss your findings with the team. After you listen to each member of your team, offer comments or questions to help improve the school description.

5. Save your work for use later. In another activity your team will prepare a report to present the information you found.

Architecture and Construction

Construction and architectural occupations are highly complex and highly rewarding. Workers in this career area design, construct, inspect, and maintain homes, offices, recreational facilities, and buildings large and small.

Some workers in this career cluster must know the strengths and weaknesses of thousands of new construction products. They must know how to engineer buildings for safety from natural disasters such as earthquake and from any other hazards. These workers must understand how to use electricity, natural gas, and ventilation systems, and how to build state-of-the-art fire, anti-asphyxiation, and security systems.

Millions of North American workers build homes, offices, and public buildings. They also design, construct, and maintain recreational facilities, theme parks, malls, and restaurants. Many involved in these career paths say that there is nothing more rewarding than looking back upon a building or facility they have built. Review a few sample jobs and potential salaries from this interesting career cluster in the table below.

Careers in Architecture and Construction

Position and Yearly Salary		Position and Yearly Salary	
Architect	$59,590*	Landscape Architect	$51,640*
Surveyors	$41,510*	Electrical Drafter	$43,200*
Civil Engineer	$61,000*	Carpenter	$36,110*
Carpet Installer	$34,290*	Construction Laborer	$27,790*
Electrician	$43,160*	Painter	$30,850*
Roofer	$32,350*	Highway Maintenance Worker	$28,490

*Salaries based on 2001 industry averages, Bureau of Labor Statistics, www.bls.gov, downloaded January 2003.

LESSON 13-5: MANAGE JOB SEARCH DATA

OBJECTIVES — *In this lesson you will:*

1. Create a database to store job description data.
2. Sort the database by career cluster, salary, and total score.
3. Create a Web page to organize online career resources.

ORGANIZING JOB SEARCH DATA

As you go online looking for job opportunities, you will collect information on dozens, even hundreds of possible jobs. You have already collected data for five jobs and prepared job description analysis forms. As your searching continues, you will need to organize the information you find in a convenient way. One of the best tools to use for this purpose is a database.

As you plan the database, create fields that are appropriate for your data. After you have created the database, you will be able to sort your database by fields such as score, career cluster, or employer.

APPLICATION 13-1: JOB ANALYSIS DATABASE

1. Start *Access*. Create a database that will allow you to record information from all of the job descriptions that you have collected. Name your new database *Job Analysis*. Save the file in the *DigiTools your name\Chapter 13* folder.

2. Design and create a table to hold your job data. Name the table *Jobs*. Use the Job Description Analysis Form as a guide for planning the fields for the database. For the salary information, create two fields, one for the lowest salary and one for the highest salary. Use yearly salary figures. You should have 11 fields in total to record data from each category on the form and the total score.

3. Use appropriate data types for the fields. For example, use Currency as the data type for the salary fields. Use Memo as the data type for the job environment information.

4. Use the forms in your *Job 1, Job 2, Job 3, Job 4,* and *Job 5* files as source documents. Enter the data for each job into your database. Summarize, including only the main points, in memo fields. You may wish to create a form based on the Jobs table to make entering the data easier. Save your updates to the database.

5. Use your database to analyze the information you have entered. Sort your records in ascending order by score. Which job description did you give the highest score? Which job description did you give the lowest score?

6. Sort the records by career cluster. Which career cluster has the most job descriptions?

EXERCISE 1-3

INTER N E T

RESEARCH SOFTWARE TOOLS

1. Choose three positions in the architecture and construction career cluster.

2. Use Internet search tools to find software that can help professionals in these three positions do their jobs more effectively. Use a search engine such as *Google* (www.google.com) or *Lycos* (www.lycos.com). Use the position and the word *software* as the search terms. For example: *landscape architect software*.

3. Try to find at least one software for each of the three job positions you chose. List the name of each program. Give a short description of what the software helps the user do.

Hint: Some companies allow their architectural, landscape, or construction software to be downloaded free of charge for evaluation purposes. If you find such an application, download it and see what you can make it do! Be sure to delete the software after your trial period is over. Who knows? You may get hooked on their software and wish to purchase it at some future time for your own professional use.

Hospitality and Tourism

Some people work in careers helping others enjoy their leisure time. Careers in hospitality and tourism can take you anywhere in the world. These careers can be rewarding and exciting. An employee in this industry might work for an amusement park in Orlando, a cruise ship in the Caribbean, or a ski lodge in Park City, Utah.

A big part of the hospitality industry relates to eating out. Over 10 million people are involved in the food preparation industry alone. Millions more work for hotels and recreational services of all kinds including spas, golf courses, and amusement parks. Review a variety of sample jobs and potential salaries in this career cluster in the table below.

Careers in Hospitality and Tourism

Position and Yearly Salary		Position and Yearly Salary	
Housekeeping Cleaner	$16,900*	Janitor	$19,800*
Groundskeeper	$20,880*	Hotel Resort Clerk	$17,640*
Chef	$30,330*	Waiter or Waitress	$45,310*
Fast Food Cook	$14,530*	Dishwasher	$15,080*
Food Service Manager	$38,290*	Amusement and Recreation Attendant	$15,970*
Coatroom Attendant	$17,230*		
Tour Guide	$20,340*	Bellhop	$22,030*
Travel Coordinator	$34,622	Flight Attendant	$46,880*
Lodging Manager	$36,820*	Housekeeping Manager	$29,500*

*Salaries based on 2001 industry averages, Bureau of Labor Statistics, www.bls.gov, downloaded January 2003.

**Salaries estimated from a Midwestern city median averages using data from www.salary.com calculated in January 2003. Salary estimates will vary greatly from city to city, state to state, and from year to year.

Tooling Up 1

School Career Centers and Placement Services. Most colleges and universities maintain lists of internships, part-time, and full-time jobs. An **internship** is a program of temporary, supervised work designed to give workers experience in a particular job or career. Visit your school's placement office, or that of a local college, and look for new postings. The placement office might post jobs on the school's Web site as well.

Professional organizations. Unions and other professional organizations frequently list jobs in their publications, on their Web sites, and at their offices. Ask your school's placement office or search the Web or local phonebook for professional organizations that advertise jobs in your career area.

Employment Agencies. Employment agencies are government or private agencies that help people find jobs. Employers often register job openings with employment agencies that search for suitable candidates. Government agencies provide free services, but private agencies charge a fee. If you are considering a private employment agency, choose one that charges the employer, not you.

Community Agencies. Some local nonprofit organizations provide job placement services for specific groups, such as minorities, women, young people, or other unemployed persons.

Local Employers. Employers in your local community usually have offices in the community. Visit a local employer and ask about employment opportunities. Request a job description or ad for a position in which you are interested.

In the next activity, you will look for job openings using these traditional sources of employment information.

ACTIVITY 13-9 SEARCH TRADITIONAL JOB INFORMATION SOURCES

1. Use one or more of the sources of employment information listed below to locate a job description for a job in which you are interested.
   - Help-Wanted Ads
   - School Career Centers and Placement Services
   - Professional Organizations
   - Employment Agencies
   - Community Agencies
   - Local Employers

2. Obtain a copy of the job ad or description if possible. Read and analyze the job information.

3. Start *Word*. Open the data file *Analysis Form*. Complete this form using information for the job description. Fill in the information in Column 3 as completely as you can, including the job title, an appropriate career cluster, the working conditions, and the primary responsibilities. Don't forget the educational level, skills, and experience sections. If the salary is listed, place that information in your form calculated to an annual rate.

4. Score this job in Column 4 using the scoring guide from Activity 13-5. Save the completed form as *Job 5* in the *DigiTools your name\Chapter 13* folder. Print and save the document.

EXERCISE 1-4

RESEARCH JOBS IN HOSPITALITY AND TOURISM

Finding a career you're interested in is one thing. Finding a place to work at your chosen profession is quite another. This exercise will give you a chance to explore companies and organizations in the hospitality and tourism industry. You will investigate the types of job opportunities available.

INTERNET

1. Choose a large company or organization that is in the hospitality and tourism business. You can search the Web for companies or use one of the ones listed below:
   - Universal Studios Theme Parks (themeparks.universalstudios.com)
   - Disney Corporation (www.disney.com)
   - Six Flags Theme Parks (www.sixflags.com)
   - Marriott Hotels (www.marriott.com)
   - Wendy's Restaurant (www.wendys.com)
   - AAA (American Automobile Association) (www.aaa.com)

2. Visit the Web site for the company or organization you chose. Look for career opportunities by following links such as *Careers, Employment,* or *Jobs*. These links may appear in small print at top or bottom of the Web page.

3. Locate at least one job opening at one company. List the following information about the job:
   - Company or organization name
   - Job or position title
   - Location of the job
   - Brief description of the job
   - Salary (if given)
   - Benefits (if given)
   - Educational or experience requirements for the job
   - How you may apply for the job (online, by telephone or fax, in person)

BUSINESS TRENDS AND ISSUES

Code of Ethics

Ethical standards require honesty and fairness in all business dealings. These qualities lead to trust. Companies want to build trust with their employees, their customers, other companies, and the public.

Many companies and organizations have developed standards of conduct for their employees or members. These standards are called codes of ethics or codes of conduct. Such codes are given to all workers or members, usually when they first join the organization.

A code of ethics helps employees know what to do in certain situations. For example, a company code of ethics may state that employees are not to accept gifts from other businesses with which the company deals.

Companies usually have steps to follow when someone does not follow the code of ethics. A person who violates the code ethics may be given a warning, fined, or suspended from work without pay for a period of time. For very serious violations, employees may be dismissed from their jobs.

ETHICS
Job Descriptions

Misrepresenting the duties involved in a job on a job ad or description is unethical. Companies should be honest with potential employees about what a job involves. Job applicants should realize, however, that a job ad or description usually does not describe every detail or duty of a job. Consider this situation:

Beth is the new office assistant to Mr. Hope, supervisor of the Customer Service department. This morning Mr. Hope called Beth into his office and asked her to arrange for dinner reservations for him and two clients at a local restaurant. When Beth returned from Mr. Hope's office, she looked upset. "I don't think I was hired to be a social secretary," she said. "I guess rank has all the privileges. Don't you agree?"

Beth obviously thinks that she is being treated unfairly. Perhaps this type of task was not discussed when she interviewed for her job. Employees must be flexible and willing to accept reasonable tasks, even if they are not listed as part of their job descriptions.

Business meetings are often held after regular business hours and away from the company offices. As part of a work team, you may be expected to arrange for these meetings or to attend the meetings. You should not, however, be expected to handle personal social arrangements for your employer unless that is part of your stated job duties.

Follow-Up

1. Do you think Mr. Hope is making an unreasonable request of Beth? Why or why not?

2. What are some tasks that Mr. Hope might ask Beth to handle that would reflect unethical behavior on his part?

LESSON 13-4: TRADITIONAL SOURCES OF JOB LEADS

OBJECTIVES   *In this lesson you will:*

1. Examine occupations using traditional, offline sources.

2. Search for and select an interesting job description from an offline source.

3. Complete a career analysis form for a job description.

You can learn about job openings from a wide variety of sources other than the Internet. Job ads, people (teachers, counselors, friends, and family), and employment agencies are all good sources of information. When you are looking for the right job, explore all avenues open to you.

Help-Wanted Ads. Read the help-wanted ads, or classifieds, in your local newspapers. Many ads will be for entry-level jobs. You may find a job that is appropriate for you. At the least, ads can provide useful information about job qualifications, duties, and pay.

EXERCISE 1-5

COMPARE CODES OF ETHICS

Many organizations and companies post their code of ethics or conduct online. In this exercise you will research and compare codes from three organizations.

INTER N E T

1. Access the Internet and open a search engine such as *Google* (www.google.com) or *Lycos* (www.lycos.com).

2. Perform a search using terms such as *code of ethics* or *code of conduct*.

3. Follow links for the search results to find a code of ethics or conduct for three companies or professional organizations. Read each code of ethics that you find.

4. List the companies or organizations for which you found a code of ethics. What rules or standards do the three codes have in common? What rules are unique to each company or organization?

Peer Check With two classmates, discuss the codes of ethics each of you found. What new information did you learn from the discussion?

CRITICAL THINKING

Handling Conflicts at Work

Many employees spend 40 hours or more per week on the job. Making these hours as stress-free as possible is to everyone's benefit. You may need to be patient and use your critical thinking skills to avoid or resolve conflicts with coworkers. Remember that you do not have to be friends with your coworkers. You do, however, need to be able to work with them productively.

A **conflict** is a disagreement, quarrel, or controversy. Because human beings are not perfect, those with whom you work will have a variety of weaknesses and problems—just as you do. At times, problems may arise that hurt relations among coworkers. Responsible employees take steps to deal with conflicts in a mature and constructive way. These strategies can be helpful in resolving conflicts at work:

1. Communicate. Listen and talk with your coworkers to be sure you all have the same understanding of the situation.

2. Analyze the situation. Determine the real or underlying problems that may be leading to the conflict. Try to resolve a conflict at the earliest stage possible so that a small problem does not become a big problem.

3. Be objective. Focus on the issue—not the person. Do not let your personal feelings for the people involved stand in the way of resolving the problem.

4. Look inward. Objectively examine your role in the situation. Are you contributing to the problem or to a solution? Be willing to admit your mistakes and apologize when your behavior or comments hurt others.

Figure 13-7
The Career link describes job opportunities at Thomson.

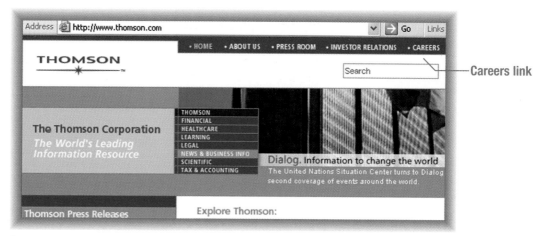

Source: http://www.thomson.com/

Major corporations have jobs for every career cluster. In the next activity, you will locate major corporations that offer job information directly on their Web sites.

ACTIVITY 13-8 — SEARCH FOR JOBS ON CORPORATE WEB SITES

INTERNET

1. Choose five of the career clusters listed in step 1 in Activity 13-1 on page 391. For each of your choices, locate a major business in each career cluster area that advertises jobs and career information on its Web site. For example:

   Career Cluster: Information Technology
   Company: Microsoft Corporation
   URL: www.microsoft.com

2. Visit each of the five Web sites. Examine job opportunities at each one. Choose one job description for a job you really like. Print the job description.

3. Read and analyze your corporate job description.

4. Start *Word*. Open the data file *Analysis Form*. Complete this form using information for the job description. Fill in the information in Column 3 as completely as you can, including the job title, an appropriate career cluster, the working conditions, and the primary responsibilities. Don't forget the educational level, skills, and experience sections. If the salary is listed, place that information in your form calculated to an annual rate.

5. Score this job in Column 4 using the scoring guide from Activity 13-5. Save the completed form as *Job 4* in the *DigiTools your name\Chapter 13* folder. Print and close the document.

DigiTip

If salary information is not provided, use online tools as you did in Activity 13-3 to find this information. If the salary is reported hourly, convert it to a yearly salary as you did in Activity 13-4.

5. Look for solutions. Brainstorm with coworkers, if appropriate, to find ways to resolve the conflict.

6. Be tactful when suggesting possible solutions. Focus on finding a solution rather than on placing blame. Explain how resolving the conflict is of benefit to others.

7. Compromise. When appropriate, be willing to give up demands or make adjustments to reach a settlement.

EXERCISE 1-6

ENCOUNTER WITH A MANAGER

A staff assistant, Ana, had quietly taken on extra responsibilities when another assistant, Kay, was on an extended leave. By working extra hours and through lunch, she has just managed to do her regular work and the absent staff member's work, too. Near the end of Kay's leave, the manager stopped by Ana's desk and said: "Because you are handling both jobs so well, I'm going to recommend that Kay be assigned to another department. We really don't need her." At that point, the manager walked away.

Ana was stunned. Up to this time, she had not commented about what she was doing. She assumed that her manager was aware of the great efforts she had made to do both jobs. Ana has asked to meet with the manager to discuss the situation.

1. Assume you are Ana. Describe clearly in two or three sentences what you want the outcome of the meeting with your manger to be.

2. List two or three of the main points you will make in the meeting with your manager in an effort to achieve the outcome you want.

3. What kinds of supporting materials might you take to the meeting to support your points?

4. Describe the attitude with which you will approach this meeting.

5. What might you (Ana) have done during the past few weeks to help avoid the conflict you now face with your manager?

WRITING

Citing Online Sources

Using the Internet and other resources, you can learn about almost everything online. You will often use information you find online or elsewhere in reports or other documents. You should give credit to those that provide this information. Fortunately, Corporate View has provided training on how to give credit to online sources you quote. In this exercise, you will review reference citations for various types of online material.

UNIT 1 Tooling Up 1

DigiTip

If salary information is not provided, use online tools as you did in Activity 13-3 to find this information. If the salary is reported hourly, convert it to a yearly salary as you did in Activity 13-4.

2. Print the job description. Remember to look for "Printer Friendly" links before printing. Close your browser. Read and analyze the job description.

3. Start *Word*. Open the data file *Analysis Form*. Complete this form using information for the job description. Fill in the information in Column 3 as completely as you can, including the job title, an appropriate career cluster, the working conditions, and the primary responsibilities. Don't forget the educational level, skills, and experience sections. If the salary is listed, place that information in your form calculated to an annual rate.

4. Score this job in Column 4 using the scoring guide from Activity 13-5. Save the completed form as *Job 3* in the *DigiTools your name\Chapter 13* folder. Print and close the document.

WHAT YOU SHOULD KNOW ABOUT
Attitude, Persistence, and Patience

Finding a good job takes some work. Determine now that, no matter what happens, you're going to stay positive—even in the tough times. Do not allow yourself to give in to negative thinking if you do not find the job you want right away.

For most people, job searching takes time. Employment counselors suggest it takes a month (sometimes two or three months) of searching for every $10,000 you want to earn on the job.

If you are willing to move to a different part of the country, you may increase your employment opportunities and earnings potential. Be patient and flexible. Consider most jobs carefully before ruling out an opportunity. Great jobs can sometimes pop up in the most unlikely places.

While looking for a full-time job, you might consider temporary employment. A **temporary employment agency** is a company that hires individuals to fill temporary or short-term jobs. These jobs may last from a single day to many months. Many businesses contract with employment agencies to use temporary workers on a periodic basis. By taking temporary jobs, young workers can gain a variety of experiences. These experiences can help them understand better which types of jobs will be most appealing. These jobs can also provide experience to help you qualify for a full-time job. In some instances, temporary workers are asked to accept permanent, full-time positions.

JOB DESCRIPTIONS ON CORPORATE WEB SITES

Many companies and professional organizations list employment opportunities directly on their Web sites rather than (or in addition to) posting them on employment Web sites. Many major corporations have employment links that can be found on their Web site welcome pages. For example, The Thomson Corporation provides a link to job opportunities as shown in Figure 13-7 on page 401.

EXERCISE 1-7 — REVIEW CITATIONS FOR ONLINE MATERIAL

1. Access the Corporate View intranet.

2. Use the Favorites bookmark you created earlier to go to the Style Guide page. On the Style Guide page, choose the **Citing Electronic Sources** link. (If you do not have a bookmark, access the Corporate View intranet. Choose **Corporate Communications, Style Guide.**)

3. Read this page carefully and review the sample citations near the end of the Web page.

4. Select and copy the list of sample citations. Open *Word*. Paste the list into a new blank document in *Word*.

5. At the bottom of your *Word* document, enter a citation for the page you copied. Review the example given for a corporate Web site. (In the citation, use today's date, or the day you actually downloaded the information.)

6. Save the document in your *DigiTools your name\Unit 1* folder. Use the name *Exercise 1-7*. Print the document.

Capitalization Usage

You have completed some online training exercises designed to help you prepare well-written documents. In this exercise you will apply what you have learned in the online training.

EXERCISE 1-8 — PROOFREADING A MEMO

1. Open the data file *Reading Group*. Print the document.

2. Read the memo carefully. Circle all errors in capitalization that you find in the document.

Government agencies have a commitment to help their citizens find employment. For this reason, government agencies also post jobs online. For example, the U.S. Department of Labor maintains America's Job Bank, which lists thousands of jobs. Individual states also will often post job banks online. Most search engines, such as *Yahoo!* and *Google*, have career information sections that can help you find state, province, and city job boards. If you're looking for a job locally, these online job boards will be the right place to start.

Branches of the U.S. military post Web sites to help you learn about careers in the military. For example, the U.S. Army Web site has tools to help you explore the Army's job bank and discover which jobs might be best for you.

Figure 13-6
The U.S. Army Web Site

Source: http://www.goarmy.com/index07.htm

ACTIVITY 13-7 — SEARCH EMPLOYMENT WEB SITES

INTER N E T

1. Use a search engine to find the URL for two or more of the following sites. Visit several of the sites. While visiting, search for and choose one job description in which you may be interested.
    - CareerBuilder.com
    - Monster
    - FlipDog.com
    - America's Job Bank
    - National Association of Colleges and Employers Job Web
    - U.S. Department of Labor's Employment and Training Administration
    - U.S. Department of Labor's Women's Bureau
    - U.S. Army and U.S. Army Recruitment
    - U.S. Navy
    - U.S. Air Force
    - U.S. Marine Corps

UNIT 2

Input Technologies

To create a document or message using many DigiTools, you must first enter information into a software program. Unit 2 introduces a variety of ways to input data. Entering data using a keyboard is still the most commonly used input method. Proficiency in keyboarding and word processing software is essential. In Chapters 3 and 4, you will learn or review how to touch-key letter, number, and symbol keys. You will also learn how to perform simple math calculations with the numeric keypad. In Chapter 5, you will learn to use *Microsoft Word* to create documents such as memos, letters, reports, and tables. In Chapters 6 and 7, you will learn to use handwriting and voice to input information and create documents.

Figure 13-5
The *Occupational Outlook Handbook* Online

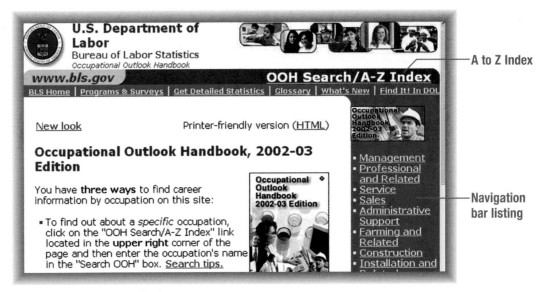

3. Select one job from the *OOH* that interests you. Print the job information so you can refer to it easily. Choose the **Printer-friendly version HTML** link before printing.

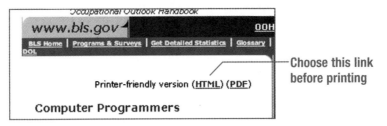

4. Read and analyze your *OOH* job description.

5. Start *Word*. Open the data file *Analysis Form*. Complete this form using information from the OOH job description. Fill in the information in Column 3 as completely as you can, including the job title, an appropriate career cluster, the working conditions, and the primary responsibilities. Don't forget the educational level, skills, and experience sections. If the salary is listed, place that information in your form as an annual rate.

6. Score this job in Column 4 using the scoring guide from Activity 13-5. Save the completed form as *Job 2* in the *DigiTools your name\Chapter 13* folder. Print and close the document.

DigiTip

If salary information is not provided, use online tools as you did in Activity 13-3 to find this information. If the salary is reported hourly, convert it to a yearly salary as you did in Activity 13-4.

SEARCHING FOR JOBS ONLINE

In some ways, searching for job opportunities has never been easier. The World Wide Web allow you to find job descriptions, locate current job openings, distribute your personal resume, and send applications to potential employers. Information from the Web can also help you build a winning resume and prepare for a job interview.

CHAPTER 3

Keyboarding: Alphabetic Keys

OBJECTIVES *In this chapter you will:*

1. Learn to arrange your work area.
2. Learn to use correct keying position and posture.
3. Master correct touch-keying for letter keys and common punctuation keys.
4. Learn to use command and function keys.
5. Build keyboarding skill.

Professional workers in many settings use a computer keyboard. Keying is a skill that you are likely to use in your career and in your academic and personal activities. You will use a keyboard to enter data, retrieve information, and communicate with others. Learn to use this DigiTool effectively to improve your productivity. Refer to Figure 3-1 and locate the parts listed on your keyboard.

Figure 3-1
Computer Keyboard.

1. **Alphanumeric keys:** Letters, numbers, and symbols.
2. **Numeric Keypad:** Keys at the right side of the keyboard used to enter numeric copy and perform calculations.
3. **Function (F) keys:** Used to execute commands, sometimes with other keys. Commands vary with software.
4. **Arrow keys:** Move insertion point up, down, left, or right.
5. **ESC (Escape):** Closes a software menu or dialog box.
6. **TAB:** Moves the insertion point to a preset position.
7. **CAPS LOCK:** Used to make all capital letters.
8. **SHIFT:** Makes capital letters and symbols shown at tops of number keys.
9. **CTRL (Control):** With other key(s), executes commands. Commands may vary with software.
10. **Windows:** Display the Windows Start menu.
11. **ALT (Alternate):** With other key(s), executes commands. Commands may vary with software.
12. **Space Bar:** Inserts a space in text.
13. **ENTER (RETURN):** Moves insertion point to margin and down to next line. Also used to execute commands.
14. **DELETE:** Removes text to the right of insertion point.
15. **NUM LOCK:** Activates/ deactivates numeric keypad.
16. **INSERT:** Activates insert or typeover.
17. **BACKSPACE:** Deletes text to the left of insertion point.

COMMAND (Mac® keyboards only): With another key, executes commands that vary with software.

2. Add the score numbers to find a final score. Enter the total in the Total Score cell. Save the form using the same name. Print the completed form. Close the document.

LESSON 13-3 LEARNING ABOUT JOBS ONLINE

OBJECTIVES *In this lesson you will:*

1. Examine a variety of occupations using the online *Occupational Outlook Handbook*.
2. Search for and select job descriptions from the *Occupational Outlook Handbook*, employment Web sites, government Web sites, and corporate Web sites.
3. Complete career analysis forms for job descriptions.

The job openings listed at Corporate View represent one single company among the thousands of possible employers around the world. You probably did not find the exact job you are looking for on the Corporate View list. In this lesson, you will examine job descriptions from other sources.

OCCUPATIONAL OUTLOOK HANDBOOK

Many career explorers are unaware of the choices available in the world of work. Instead of looking at all of the possibilities, they settle for what they see around them in their own community. The **Occupational Outlook Handbook (OOH)** is a publication of the U.S. Bureau of Labor and contains information about jobs. The *OOH* is one of the best sources of career information. It can help you broaden your career horizons. The *OOH* describes thousands of jobs that people do. Somewhere in this listing, you will find several jobs that appeal to you.

ACTIVITY 13-6 SEARCH THE *OCCUPATIONAL OUTLOOK HANDBOOK* ONLINE

INTERNET

1. Access the Internet and open your browser. Access the *OOH* online by entering the following address in your Web browser: www.bls.gov/oco.

2. Browse through the A to Z index or the navigation bar options (see Figure 13-5 on page 398) to get an idea of the wide variety of career choices. Spend at least 20 minutes searching for interesting jobs.

LESSON 3-1

HOME KEYS (fdsa jkl;)

OBJECTIVES *In this lesson you will:*

1. Learn control of home keys (**fdsa jkl;**).
2. Learn control of **Space Bar** and **Enter** key.

3-1A • Work Area Arrangement

Arrange work area as shown at the right.

- alphanumeric (main) keyboard directly in front of chair; front edge of keyboard even with edge of table or desk
- monitor placed for easy viewing
- disk drives placed for easy access and disks within easy reach (unless using a network)
- book behind or at side of keyboard; top raised for easy reading

Properly arranged work area

3-1B • Keying Position

The features of proper position are shown at right and listed below:

- fingers curved and upright over home keys
- wrists low, but not touching keyboard
- forearms parallel to slant of keyboard
- body erect, sitting back in chair
- feet on floor for balance

Proper position at computer

CHAPTER 3 Lesson 3-1 71

Peer Check Compare your answers with a classmate's answers. If you have different answers, compare the data and formulas in your worksheets to find why the two are different.

ACTIVITY 13-5 — SCORE JOB DESCRIPTIONS

In the process of looking for a job, you may have to read and analyze dozens of job descriptions. Use a scoring system to help you identify those jobs that interest you and for which you are prepared. A scoring system will help compare one job description to another and screen out jobs that are impractical.

1. Start *Word*. Open your *Job 1* file. Score your job description in the last column using this scoring guide.

	CATEGORY	SCORING GUIDE
4.	Job Environment	Would you be comfortable working in this job environment, under these working conditions? Rank from 0-10. 10 points = very comfortable 0 points = not at all comfortable
5.	Responsibilities or Duties	Would you be comfortable meeting all of the duties, assignments, and responsibilities of this job? Would you like the nature of this work as described in this job description? 10 points = like the work very much 0 points = like the work not at all
6.	Education	How close are you to meeting the educational requirements of this position? 10 points = fully meet or exceed requirements 5 points = nearly meet the requirements 0 points = do not meet requirements
7.	Skills or Training	How close are you to having the skills or training required for this job? 10 points = fully meet or exceed requirements 5 points = nearly meet the requirements 0 points = do not meet requirements
8.	Experience	How close are you to meeting the years of experience requirement? 10 points = meet or exceed requirement 5 points = nearly meet the requirement 0 points = do not meet requirement
9.	Salary Range	Would you be happy with the minimum salary or hourly wage that is being offered for this job? 10 points = very happy with salary 5 points = moderately happy with salary 0 points = salary unacceptable

3-1C • Home-Key Position

1. Find the home keys on the chart: **f d s a** for left hand and **j k l ;** for right hand.
2. Locate and place your fingers on the home keys on your keyboard with your fingers well curved and upright (not slanting).
3. Remove your fingers from the keyboard; then place them in home-key position again, curving and holding them *lightly* on the keys.

3-1D • Techniques: Home Keys and Spacebar

1. Read the hints and study the illustrations at the right.
2. Place your fingers in home-key position as directed in 1C above.
3. Strike the key for each letter in the first group below the illustration.
4. After striking ; (semicolon), strike the *Space Bar* once.
5. Continue to key the line; strike the *Space Bar* once at the point of each arrow.
6. Review proper position (1B); then repeat Steps 3–5 above.

TECHNIQUE HINTS

Keystroking: Strike each key with a light tap with the tip of the finger, snapping the fingertip toward the palm of the hand.

Spacing: Strike the Space Bar with the right thumb; use a quick down-and-in motion (toward the palm). Avoid pauses before or after spacing.

Space once.

fdsajkl; f d s a j k l ; ff jj dd kk ss ll aa ;;

3-1E • Technique: Hard Return at Line Endings

1. Read the information and study the illustration at the right.
2. Practice the ENTER key reach several times.

Hard Return

To return the insertion point to the left margin and move it down to the next line, strike ENTER.

This is called a **hard return**. Use a hard return at the end of all drill lines. Use two hard returns when directed to double-space.

Hard Return Technique

Reach the little finger of the right hand to the ENTER key, tap the key, and return the finger quickly to home-key position.

ACTIVITY 13-3

COMPARE SALARIES ONLINE

INTERNET

DigiTip
If you do not find the exact name or job title you are looking for, look for similar jobs. For example, the job *Senior Corporate Securities Counsel* is comparable to a *Senior (Sr) Legal Compliance Officer.*

1. Access the Internet and open your Web browser. Search for salary information about the job you choose in Activity 13-1. Search for sites such as Salary.com or PayScale! (www.payscale.com) that compare salaries for various jobs. You can also find information by entering the job name and *salary range* as a search term in a search engine. For example, use the search term *computer programmer salary range* to find salary information for that job.

2. Compare salaries for the job from different locations, one near your home and at least one other in a major city in a different part of the country. From all the information you find, choose what you think are typical low and high salaries for your area for the job.

3. Start *Word*. Open *Job 1* that you created earlier. Record the salary information on your job description analysis form.

4. Save your updated form using the same name. Print and close the document.

ACTIVITY 13-4

CONVERT SALARIES

Salaries are listed either by hourly rates or annual rates of pay. You need to convert hourly salaries into yearly salaries so you can compare jobs accurately. For example, which is more, $10 per hour or $20,000 per year?

For each job described below, convert the hourly rate into a yearly salary. Create an *Excel* worksheet to record data about each job and calculate the annual salary. (See Figure 13-4.)

Example: A cashier job pays an hourly rate of $10. The workday is 8 hours. Employees work 5 days per week, 48 weeks per year.

Figure 13-4
Sample Salary Worksheet

	A	B	C	D	E
			Hourly	Hours per	Weeks per	Annual
1			Rate	Week	Year	Salary
2	Job					
3	Cashier	$10.00	40	48	$19,200.00

(E3 formula: =B3*C3*D3)

1. Job: Commercial Artist
   Salary: $24.95 per hour for 48 weeks at 8 hours per day, 5 days a week

2. Job: Certified Teacher Assistant (Assisting English as a second language students)
   Salary: $17 an hour, 20 hours per week, 36 weeks

3. Job: Retail Stocker (Placing products on store shelves)
   Salary: $6.25 per hour, 38 hours per week for 50 weeks. The stocker works 8 hours per day. The job also offers 2 paid holidays (calculated at 8 hours each).

4. The company advertising for the retail stocker position from Problem 3 above is offering a $.50 per hour raise for 38 hours per week for 50 weeks. Remember to calculate the two paid holidays (calculated at 8 hours each). How much more per year will be paid on an annual basis?

5. Save the file as *Activity 13-4* in the *DigiTools your name\Chapter 13* folder.

3-1F • Home-Key and [Spacebar] Practice

1. Place your hands in home-key position (left-hand fingers on **f d s a** and right-hand fingers on **j k l ;**).
2. Key the lines once: single-spaced (SS) with a double space (DS) between 2-line groups. Do not key line numbers.

Fingers curved and upright

Down-and-in spacing motion

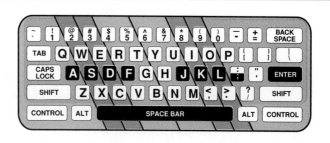

Strike Space Bar once to space.

```
1  j jj f ff k kk d dd l ll s ss ; ;; a aa jkl; fdsa
2  j jj f ff k kk d dd l ll s ss ; ;; a aa jkl; fdsa
```
Strike the ENTER key twice to double-space (DS).
```
3  a aa ; ;; s ss l ll d dd k kk f ff j jj fdsa jkl;
4  a aa ; ;; s ss l ll d dd k kk f ff j jj fdsa jkl;
                                                     DS
5  jf jf kd kd ls ls ;a ;a fj fj dk dk sl sl a; a; f
6  jf jf kd kd ls ls ;a ;a fj fj dk dk sl sl a; a; f
                                                     DS
7  a;fj a;sldkfj a;sldkfj a;sldkfj a;sldkfj a;sldkfj
8  a;fj a;sldkfj a;sldkfj a;sldkfj a;sldkfj a;sldkfj
```
Strike the ENTER key 4 times to quadruple-space (QS).

3-1G • Technique: [Enter] Key Practice

Key the lines once: single-spaced (SS) with a double space (DS) between 2-line groups. Do not key line numbers.

SPACING CUE

When lines are SS, strike ENTER twice to insert a DS between 2-line groups.

```
1  a;sldkfj a;sldkfj
2  a;sldkfj a;sldkfj
                     DS
3  ff jj dd kk ss ll aa ;;
4  ff jj dd kk ss ll aa ;;
                     DS
5  fj dk sl a; a; as df ;l kj;
6  fj dk sl a; a; as df ;l kj;
                     DS
7  fj dk sl a; jf kd ls ;a fdsa jkl;
8  fj dk sl a; jf kd ls ;a fdsa jkl;
                     DS
9  k; fa kl ds ak dl fj s; lafj ksd; dlj
10 k; fa kl ds ak dl fj s; lafj ksd; dlj
                     DS
11 fa sd j; kl ak sj fl d; akdj s;lf sfk; djl
12 fa sd j; kl ak sj fl d; akdj s;lf sfk; djl
                     QS
```

Reach with little finger; tap ENTER key quickly; return finger to home key.

CHAPTER 3   Lesson 3-1   73

JOB SALARIES

A **salary** is the money a worker receives in exchange for doing a job. Salaries are not always listed in job descriptions. When this is the case, it's important to learn the salaries for similar jobs. This will help you know whether you want to pursue a particular job. If no salary is listed, you can go online and see what other companies in your area are paying for a job similar to the one that interests you.

You need to ask yourself seriously if doing a job is worth the salary being offered. When you think a salary is not appropriate, you can make a counteroffer. A **counteroffer** is an alternate proposal, in this case for a salary amount you would be satisfied with. You can find out what employers are paying locally for similar jobs by checking the classified ads in a local newspaper. You can learn about salaries for jobs locally or nationwide by searching the Internet.

Companies use job descriptions to help them decide how much they should pay for a particular job. Employers establish either a yearly or an hourly salary range for each job. The range is the difference between the initial low salary and the highest possible salary that a company will pay for a job. For example, for a Technical Writer position at Corporate View the salary information may be something like:

- Salary Range: $38,000–$52,000 per year, depending on qualifications

or

- Hourly Wage: $22–$42 per hour, depending on qualifications

Often, a worker starts at the low end of the salary range and receives higher pay when he or she gains experience or additional training. Offering a competitive salary helps companies attract the most qualified applicants. Salaries for the same job will vary from state to state and city to city.

Figure 13-3
The U.S. Office of Personnel Management provides salary information for some government jobs online.

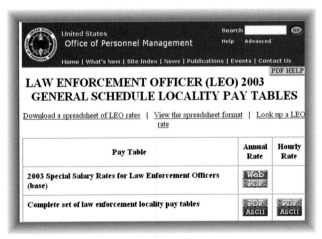

Source: http://www.opm.gov/oca/03tables/indexLEO.asp

By comparing the responsibilities, working conditions, location of employment, and qualifications, Human Resources (HR) employees can compare jobs against each other to decide which jobs should be paid at higher or lower wages. HR departments can use the Internet to help make these comparisons.

3-1H •
Home-Key Mastery

1. Key the lines once (without the numbers); strike the ENTER key twice to double-space (DS).
2. Rekey the drill at a faster pace.

TECHNIQUE CUE
Keep fingers curved and upright over home keys, right thumb just barely touching the Space Bar.

SPACING CUE
Space once after ; used as punctuation.

Correct finger alignment

```
1 aa ;; ss ll dd kk ff jj a; sl dk fj jf kd ls ;a jf
                                                    DS
2 a a as as ad ad ask ask lad lad fad fad jak jak la
                                                    DS
3 all all fad fad jak jak add add ask ask ads ads as
                                                    DS
4 a lad; a jak; a lass; all ads; add all; ask a lass
                                                    DS
5 as a lad; a fall fad; ask all dads; as a fall fad;
```

3-1I •
End-of-Lesson Routine

1. Exit the software.
2. Remove disk from disk drive.
3. Turn off equipment if directed to do so.
4. Store materials as the instructor directs.

Disk removal

3-1J •
Enrichment

1. Key the drill once as shown to improve control of home keys.
2. Key the drill again to quicken your keystrokes.

SPACING CUE
To DS between single-spaced lines, strike ENTER twice.

```
1 ja js jd jf f; fl fk fj ka ks kd kf d; dl dk dj a;
                                                    DS
2 la ls ld lf s; sl sk sj ;a ;s ;d ;f a; al ak aj fj
                                                    DS
3 jj aa kk ss ll dd ;; ff fj dk sl a; jf kd ls ;a a;
                                                    DS
4 as as ask ask ad ad lad lad all all fall fall lass
                                                    DS
5 as a fad; as a dad; ask a lad; as a lass; all lads
                                                    DS
6 a sad dad; all lads fall; ask a lass; a jak salad;
                                                    DS
7 add a jak; a fall ad; all fall ads; ask a sad lass
```

CHAPTER 3   Lesson 3-1

description will help applicants screen themselves so that they only apply for jobs they find interesting and for which they are qualified. Corporations often receive hundreds of applications for the same position. Job descriptions provide criteria that can help eliminate unqualified candidates.

PARTS OF A JOB DESCRIPTION

For each of the professional teams who work in a company, there are many separate jobs. As you learned, these jobs are detailed in documents called job descriptions. A **job description** lists the duties to be performed in a job as well as the education, experience, and skills required. As you read your job description in Lesson 13-1, you probably noticed several important parts.

- Job title
- Job environment or working conditions (including the location of employment)
- Main responsibilities or primary duties
- Education, skills, and experience required

You will need to analyze carefully each part of the job description for jobs you consider. This process will help you determine how well you qualify for various jobs. In the next activity, you will use a job description analysis form to help you organize all the factors you need to consider.

ACTIVITY 13-2 — COMPLETE A JOB DESCRIPTION ANALYSIS FORM

1. Start *Word*. Open the data file *Analysis Form*. This form contains questions you will answer to analyze a job description. Place your name in a header. Save the form as *Job 1* in the *DigiTools your name\Chapter 13* folder.

2. Access the Corporate View intranet. Return to the job description you chose in the last activity, for example, Corporate Communications: Art Director.

3. Reread and summarize the job description you chose in Activity 13-1 in the third column of your Job Description Analysis Form. Fill out Sections 1 to 8 as completely as you can. Leave the salary information blank for now. You will find this information in Activity 13-3. You will also add the score information later.

4. Save your updated form using the same name. Close the document. Close your browser.

Self Check Did you enter all the duties and responsibilities for the job? Did you enter all the education and experience requirements?

LESSON 3-2

NEW KEYS: h AND e

OBJECTIVES *In this lesson you will:*

1. Learn reach technique for **h** and **e**.
2. Combine smoothly **h** and **e** with all other learned keys.

3-2A •
Get Ready to Key

At the beginning of each practice session, follow the *Standard Plan* given at the right to get ready to key the lesson.

Standard Plan for Getting Ready to Key

1. Arrange work area similar to the illustration on p. 71.
2. Check to see that the computer, monitor, and printer (if any) are plugged in.
3. Load the computer software specified by your instructor.
4. Align the front of the keyboard with the front edge of the desk or table.
5. Position the monitor and the textbook for easy reading.
6. Sit back in chair, body erect, feet on floor for balance.

3-2B •
Plan for Learning New Keys

All keys except the home keys (**fdsa jkl;**) require the fingers to reach in order to strike them. Follow the *Standard Plan* given at the right to learn the proper reach for each new key.

Standard Plan for Learning New Keys

1. Find the new key on the keyboard chart given on the page where the new key is introduced.
2. Look at your own keyboard and find the new key on it.
3. Study the reach-technique picture at the left of the practice lines for the new key. (See p. 76 for illustrations.) Read the statement below the illustration.
4. Identify the finger to be used to strike the new key.
5. Curve your fingers; place them in home-key position (over **fdsa jkl;**).
6. Watch your finger as you reach it to the new key and back to home position a few times (keep it curved).
7. Refer to the set of three drill lines at the right of the reach-technique illustration. Key each line twice SS (single-spaced):
   - once slowly, to learn new reach;
   - then faster, for a quick-snap stroke. DS (double-space) between 2-line groups.

3-2C •
Home-Key Review

Key each line twice single-spaced (SS): once slowly; again, at a faster pace; double-space (DS) between 2-line groups.

All keystrokes learned

```
1 a;sldkfj a; sl dk fj ff jj dd kk ss ll aa ;; fj a;
2 a a as as ask ask ad ad lad lad add add fall falls
3 as as ad ad all all jak jak fad fad fall fall lass
4 jj kk; jj kk; as jak; as jak; ask a lad; ask a lad
5 a jak; a fad; as a lad; ask dad; a lass; a fall ad
6 a sad fall; all fall ads; as a lass asks a sad lad
7 a fad; ask dad; ask a lass; add a lad; a sad lass;
```

Strike ENTER 4 times to quadruple-space (QS) between lesson parts.

4. Read your chosen job description. Think about the following questions as your read, but don't write down any information yet. You will do that in Activity 13-2.
- What is the job environment like? What are the working conditions for this job? (This includes the location of employment if known.)
- What are the main responsibilities or primary duties of this job?
- What education, skills, and experience are required for this job?
  - What degree, diploma, or certification is required or preferred?
  - What skills are required?
  - How much experience is needed?

5. Add your job description choice to your Favorites list in your Web browser so you can find it quickly later. (If you cannot save Favorites on your system, note the department and job names.)

6. In Lesson 1-5 you were introduced to 16 career clusters representing thousands of different jobs. These career clusters are listed in the table that follows. Which career cluster do you think best relates to your job description? For example, Corporate Communications: Art Director would fit into the Arts, Audio/Video Technology, and Communications cluster.

CAREER CLUSTERS

Agriculture and Natural Resources	Government and Public Administration
Architecture and Construction	Health Science
Arts, Audio/Video Technology, and Communications	Human Services
	Information Technology
Business and Administration	Law and Public Safety
Education and Training	Retail/Wholesale Sales and Service (Includes Marketing)
Hospitality and Tourism	
Manufacturing	Transportation, Distribution, and Logistics
Finance	Scientific Research and Engineering

7. Close the Corporate View intranet and your browser.

LESSON 13-2 — ALL ABOUT JOB DESCRIPTIONS

OBJECTIVES *In this lesson you will:*

1. Learn the purpose of job descriptions.
2. Examine the parts of a typical job description.
3. Use a job description analysis form to analyze and score a job description.
4. Compare salaries using online sources.
5. Convert hourly wages into yearly salaries.

Corporations use job descriptions as screening devices. You can use them to screen employment opportunities, too. Screening is a process of elimination. A well-written job

3-2D •
New Keys: H and E

1. Use the *Standard Plan for Learning New Keys* (p. 75) for each key to be learned. Study the plan now.

2. Relate each step of the plan to the illustrations below and text at the right. Then key each line twice SS; leave a DS between 2-line groups.

h *Right index* finger

e *Left middle* finger

Do not attempt to key line numbers, the vertical lines separating word groups, or the labels (home row, h/e).

Learn h

1 j j hj hj ah ah ha ha had had has has ash ash hash
2 hj hj ha ha ah ah hah hah had had ash ash has hash
3 ah ha; had ash; has had; a hall; has a hall; ah ha

Strike ENTER twice to double-space (DS) below the set of lines.

Learn e

4 d d ed ed el el led led eel eel eke eke ed fed fed
5 ed ed el el lee lee fed fed eke eke led led ale ed
6 a lake; a leek; a jade; a desk; a jade eel; a deed

Combine h and e

7 he he he|she she she|shed shed|heed heed|held held
8 a lash; a shed; he held; she has jade; held a sash
9 has fled; he has ash; she had jade; she had a sale

Strike ENTER 4 times to quadruple-space (QS) between lesson parts.

3-2E •
New-Key Mastery

1. Key the lines once SS with a DS between 2-line groups.

2. Key the lines again with quick, sharp strokes at a faster pace.

Space once after ; used as punctuation.

 Note:

Once the screen is filled with keyed lines, the top line disappears when a new line is added at the bottom. This is called **scrolling**.

Fingers curved

Fingers upright

home row
1 ask ask|has has|lad lad|all all|jak jak|fall falls
2 a jak; a lad; a sash; had all; has a jak; all fall
DS

h/e
3 he he|she she|led led|held held|jell jell|she shed
4 he led; she had; she fell; a jade ad; a desk shelf
DS

all keys learned
5 elf elf|all all|ask ask|led led|jak jak|hall halls
6 ask dad; he has jell; she has jade; he sells leeks
DS

all keys learned
7 he led; she has; a jak ad; a jade eel; a sled fell
8 she asked a lad; he led all fall; she has a jak ad

ACTIVITY 13-1
INVESTIGATE A BUSINESS CAREER PATH

1. Access the Corporate View intranet. On the Corporate View Intranet Home page, click the link for a department, such as **Finance & Accounting**. Choose the link to the section that describes the department, such as **About Finance and Accounting**. Read the department description to learn what this team does. Repeat the process for all the departments:
   - Research & Development
   - Marketing, Sales, & Support
   - Finance & Accounting
   - Information Technology
   - Legal Services
   - Human Resources & Management
   - Corporate Communications

2. Choose the **Human Resources & Management** link followed by **Current Job Openings @ TeleView** link. The Web page displays career teams or departments at TeleView that currently have job openings. Choose one department from the list. (See Figure 13-2.)

Figure 13-2
TeleView Departments with Current Job Openings

3. Under each link you will find one or more jobs in a list. Choose the job listing that interests you the most. (**Note:** If you wish to change your choice, click the **Back** button to return to the list of departments and choose another job description.) Keep looking until you find a job description that interests you.

LESSON 3-3

NEW KEYS: i AND r

OBJECTIVES *In this lesson you will:*

1. Learn reach technique for **i** and **r**.
2. Combine smoothly **i** and **r** with all other learned keys.

3-3A • 3'
Get Ready to Key

Follow the steps in the *Standard Plan for Getting Ready to Key* on p. 75.

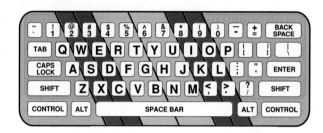

3-3B • 5'
Conditioning Practice

Key each line twice SS; DS between 2-line groups.

PRACTICE CUE

- Key each line at a slow, steady pace, but strike and release each key quickly.
- Key each line again at a faster pace; move from key to key quickly.

home keys   1   a;sldkfj a;sldkfj as jak ask fad all dad lads fall

                                                                                         Strike ENTER twice to DS.

h/e   2   hj hah has had sash hash ed led fed fled sled fell

                                                                                       DS

all keys learned   3   as he fled; ask a lass; she had jade; sell all jak

                                      Strike ENTER 4 times to quadruple-space (QS) between lesson parts.

3-3C • 5'
Speed Building

Key each line once DS.

SPACING CUE

To DS when in SS mode, strike ENTER twice at end of line.

SPEED CUE

In lines 1–3, quicken the keying pace as you key each letter combination or word when it is repeated within the line.

1   hj hj|ah ah|ha ha|had had|ash ash|has has|had hash

2   ed ed|el el|ed ed|led led|eke eke|lee lee|ale kale

3   he he|she she|led led|has has|held held|sled sleds

4   he fled; she led; she had jade; he had a jell sale

5   a jak fell; she held a leek; he has had a sad fall

6   he has ash; she sells jade; as he fell; has a lake

7   she had a fall jade sale; he leads all fall sales;

8   he held a fall kale sale; she sells leeks as a fad

After taking time to research your future, set some goals for yourself. Plan the education options you must pursue to qualify for a special career opportunity that's perfect for you. Be open-minded. In your search, you may discover an opportunity you haven't even thought about and find a new career path that will be surprisingly fulfilling.

LESSON 13-1: EXAMINING PROFESSIONAL TEAMS IN A BUSINESS

OBJECTIVES *In this lesson you will:*

1. Learn about various teams or departments in a business.
2. Choose a job description from among several listed by business departments.
3. Classify your selected job description according to its career cluster.

In every business, many jobs must be done. One person cannot do all of the tasks that are required. Teamwork is essential. Many professional people working together are needed for a business to reach its potential.

If you choose to work in a business, you will likely be assigned to a team with members that perform similar types of jobs. These professionals who work together are often called departments, groups, or teams. Perhaps you can see yourself becoming a member of one of these essential teams.

- Research and Development
- Marketing, Sales, and Customer Support
- Finance and Accounting
- Information Technology
- Legal Services
- Human Resources and Management
- Corporate Communications
- Grounds and Maintenance
- Manufacturing, Warehousing, and Shipping

> **DigiTip**
> Most of these department or team functions must also be accomplished by government organizations, schools, and nonprofit organizations.

The different departments listed above have very different job duties. For example, a Grounds and Maintenance Team maintains a comfortable, aesthetically pleasing, clean, and safe working environment for all employees. A Manufacturing, Warehousing, and Shipping Team makes, stores, or transports the products that a company sells. You will learn about the other departments on the Corporate View intranet in the next activity. You will also review job descriptions for jobs in some of these departments.

3-3D • 18'
New Keys: I and R

Key each line twice SS (slowly, then faster); DS between 2-line groups; if time permits, key lines 7–9 again.

Technique Goals:
- curved, upright fingers
- finger-action keystrokes
- eyes on copy

i *Right middle* finger

r *Left index* finger

Follow the *Standard Plan for Learning New Keys* outlined on p. 75.

Learn i

1 k k ik ik is is if if did did aid aid kid kid hail
2 ik ik if if is is kid kid his his lie lie aid aide
3 a kid; a lie; if he; he did; his aide; if a kid is

Learn r

4 f f rf rf jar jar her her are are ark ark jar jars
5 rf rf re re fr fr jar jar red red her her far fare
6 a jar; a rake; a lark; red jar; hear her; are dark

Combine i and r

7 fir fir|rid rid|sir sir|ire ire|fire fire|air airs
8 a fir; if her; a fire; is fair; his ire; if she is
9 he is; if her; is far; red jar; his heir; her aide

Quadruple-space (QS) between lesson parts.

3-3E • 19'
New-Key Mastery

1. Key the lines once SS with a DS between 2-line groups.
2. Key the lines again at a faster pace.

Technique Goals:
- fingers deeply curved
- wrists low, but not resting
- hands/arms steady
- eyes on copy as you key

reach review
1 hj ed ik rf hj de ik fr hj ed ik rf jh de ki fr hj
2 he he|if if|all all|fir fir|jar jar|rid rid|as ask
DS

h/e
3 she she|elf elf|her her|hah hah|eel eel|shed shelf
4 he has; had jak; her jar; had a shed; she has fled
DS

i/r
5 fir fir|rid rid|sir sir|kid kid|ire ire|fire fired
6 a fir; is rid; is red; his ire; her kid; has a fir
DS

all keys learned
7 if if|is is|he he|did did|fir fir|jak jak|all fall
8 a jak; he did; ask her; red jar; she fell; he fled
DS

all keys learned
9 if she is; he did ask; he led her; he is her aide;
10 she has had a jak sale; she said he had a red fir;

CHAPTER 13

Finding and Analyzing Career Choices

OBJECTIVES *In this chapter you will:*

1. Search a variety of sources for career and employment information.
2. Categorize job descriptions by their career clusters.
3. Explore the parts of a job description.
4. Create a job description analysis form to analyze, score, and rank job descriptions.
5. Analyze and compare salaries by career and location.
6. Convert hourly wages into yearly salaries.
7. Search for job opportunities through Web sites, online government sources, and from traditional offline sources.
8. Create a database to categorize and sort job description information.
9. Create a Web page to organize online employment sites.

CAREER CHOICES

What do you want to do with the rest of your life? It's a challenging question. Choosing a career is never easy and cannot be taken lightly. The decision is so important that you should take the time to study the possibilities, search the latest employment information, think about what you want to do, and evaluate your career choices.

Figure 13-1
Articles and Web sites about career planning can be found on the Internet.

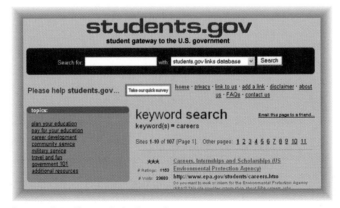

Source: *http://www.students.gov/*

LESSON 3-4

NEW KEYS: o AND t

OBJECTIVES *In this lesson you will:*

1. Learn reach technique for **o** and **t**.
2. Combine smoothly **o** and **t** with all other learned keys.

3-4A • 8'
Conditioning Practice

Key each line twice SS (slowly, then faster); DS between 2-line groups.

In Lessons 4–8, the time for the *Conditioning Practice* is changed to 8'. During this time, you are to arrange your work area, prepare your equipment for keying, and practice the lines of the *Conditioning Practice* as directed.

Fingers curved

Fingers upright

home row 1 `a sad fall; had a hall; a jak falls; as a fall ad;`
3d row   2 `if her aid; all he sees; he irks her; a jade fish;`
all keys learned 3 `as he fell; he sells fir desks; she had half a jar`

3-4B • 20'
New Keys: O and T

Key each line twice SS (slowly, then faster); DS between 2-line groups; if time permits, key lines 7–9 again.

o *Right ring* finger

t *Left index* finger

Follow the *Standard Plan for Learning New Keys* outlined on p. 75.

Learn o

1 `l l ol ol do do of of so so lo lo old old for fore`
2 `ol ol of of or or for for oak oak off off sol sole`
3 `do so; a doe; of old; of oak; old foe; of old oak;`

Learn t

4 `f f tf tf it it at at tie tie the the fit fit lift`
5 `tf tf ft ft it it sit sit fit fit hit hit kit kite`
6 `if it; a fit; it fit; tie it; the fit; at the site`

Combine o and t

7 `to to|too too|toe toe|dot dot|lot lot|hot hot|tort`
8 `a lot; to jot; too hot; odd lot; a fort; for a lot`
9 `of the; to rot; dot it; the lot; for the; for this`

DIGITOOLS AND YOUR CAREER PATHWAY

DigiTools and Your Career Pathway

UNIT 4

In Unit 4, you will think about what you might choose as a future career as you explore jobs in various career clusters. In Chapter 13, you will search for job descriptions from a variety of online and traditional sources. You will learn to analyze and rank job opportunities. In Chapter 14, you will learn to present your skills and other qualifications to employers. You will prepare a resume, application, personal career Web page, and letter of application. You will also compare your qualifications against the skills required by an employer. In Chapter 15, you will learn how to start a new career by completing the interview and hiring process successfully. You will prepare a personal employment portfolio and learn strategies for communicating in an interview. You will also learn about hiring practices and rights and obligations of employees that will help you be successful in the workplace.

3-4C • 22'
New-Key Mastery

1. Key the lines once SS; DS between 2-line groups.
2. Key the lines again at a faster pace.

PRACTICE CUE

In lines 3, 5, and 7, speed up the second keying of each word.

- curved, upright fingers
- wrists low, but not resting
- eyes on copy as you key

reach review
1 hj ed ik rf ol tf jh de ki fr lo ft hj ed ol rf tf
2 is led fro hit old fit let kit rod kid dot jak sit

h/e
3 he he|she she|led led|had had|see see|has has|seek
4 he led|ask her|she held|has fled|had jade|he leads

i/t
5 it it|fit fit|tie tie|sit sit|kit kit|its its|fits
6 a kit|a fit|a tie|lit it|it fits|it sits|it is fit

o/r
7 or or|for for|ore ore|fro fro|oar oar|roe roe|rode
8 a rod|a door|a rose|or for|her or|he rode|or a rod

space bar
9 of he or it is to if do el odd off too for she the
10 it is|if it|do so|if he|to do|or the|she is|of all

all keys learned
11 if she is; ask a lad; to the lake; off the old jet
12 he or she; for a fit; if she left the; a jak salad

3-4D •
Enrichment

1. Key the drill once SS at an easy pace to gain control of all your reach-stroke motions. DS between 2-line groups.
2. Key the drill again to speed up your motions and build continuity.

reach review
1 hj ed ik rf jhj ded kik frf a;sldkfj a;sldkfj fja;
2 if led ski fir she ire sir jak has did jar kid rid

o/t
3 ol ol|old old|for for|oak oak|ode ode|doe doe|does
4 tf tf|it it|to to|kit kit|the the|fit fit|sit sits

i/r
5 ik ik|if if|it it|fir fir|ski ski|did did|kid kids
6 rf rf|or or|for for|her her|fir fir|rod rod|or for

h/e
7 hj hj|he he|ah ah|ha ha|he he|she she|ash ash|hash
8 ed ed|el el|he he|her her|elk elk|jet jet|she|shed

all keys learned
9 of hot kit old sit for jet she oak jar ore lid lot
10 a ski; old oak; too hot; odd jar; for the; old jet

all keys learned
11 she is to ski; is for the lad; ask if she has jade
12 he sold leeks to her; she sells jade at their lake

EXERCISE 3-7

TV COMMERCIAL

1. Start *Word*. Open the data file *TV Commercial*.

2. Proofread the script and identify errors in the use of commas, semicolons, colons, and confusing terms. You should find 20 grammatical errors in this TV script. Do not correct fragmented sentences describing sound effects. Fragments are permitted in a script in order to save space. Also, because this is a technical document, it requires precise comma usage.

3. Edit the document to correct errors. Save the document as *Exercise 3-7* in the *DigiTools your name\ Unit 3* folder. Print the document.

Peer Check Compare your corrected document with a classmate's document. What did you learn from this comparison? Make corrections, if needed, after your discussion.

LESSON 3-5

NEW KEYS: n AND g

OBJECTIVES *In this lesson you will:*

1. Learn reach technique for **n** and **g**.
2. Combine smoothly **n** and **g** with all other learned keys.

3-5A • 8'
Conditioning Practice

Key each line twice SS (slowly, then faster); DS between 2-line groups.

home row 1 has a jak; ask a lad; a fall fad; had a jak salad;
o/t   2 to do it; as a tot; do a lot; it is hot; to dot it
e/i/r   3 is a kid; it is far; a red jar; her skis; her aide

3-5B • 20'
New Keys: N and G

Key each line twice SS (slowly, then faster); DS between 2-line groups; if time permits, key lines 7–9 again.

n *Right index* finger

g *Left index* finger

Follow the *Standard Plan for Learning New Keys* outlined on p. 75.

Learn n

1 j j nj nj an an and and end end ant ant land lands
2 nj nj an an en en in in on on end end and and hand
3 an en; an end; an ant; no end; on land; a fine end

Learn g

4 f f gf gf go go fog fog got got fig figs jogs jogs
5 gf gf go go got got dig dig jog jog logs logs golf
6 to go; he got; to jog; to jig; the fog; is to golf

Combine n and g

7 go go|no no|nag nag|ago ago|gin gin|gone gone|long
8 go on; a nag; sign in; no gain; long ago; into fog
9 a fine gig; log in soon; a good sign; lend a hand;

3-5C • 5'
Technique: Enter **Key**

Key each line twice SS; DS between 2-line groups.

PRACTICE CUE

Keep up your pace to the end of line, strike the ENTER key quickly, and start the new line without pause.

1 she is gone;
2 she got an old dog;
3 she jogs in a dense fog;
4 she and he go to golf at nine;
5 he is a hand on a rig in the north;

Reach out and tap ENTER.

CHAPTER 3 Lesson 3-5 81

CRITICAL THINKING

Tangible and Intangible Rewards of Work

As you have learned, high productivity and effective use of resources are common goals of businesses. Productivity depends in part on the work ethic of employees. **Work ethic** is a general term that combines a deep belief in the value of work in one's life and a willingness to meet the demands of work. Persons with a strong work ethic value both tangible and intangible rewards of work. Tangible rewards, such as pay and benefits, are important to most workers. Persons without a strong work ethic may not place much value on intangible rewards, such as enjoyment of the work performed or pride in a job well done. Persons with a strong work ethic tend to define job satisfaction differently from those without a strong work ethic.

EXERCISE 3-6 RANK REWARDS OF WORK

Both tangible and intangible rewards of work will contribute to your job satisfaction. Which type of reward is most important to you? You will identify and rank tangible and intangible rewards of work in this exercise.

1. Create a list of ten or more tangible rewards of work such as salary, stock options, company-paid life insurance, and so on.

2. Create a list of ten intangible rewards of work such as a feeling of pride in work done well, the enjoyment of socializing with coworkers, or a feeling that your work contributes to the well-being of others.

3. Think about a job or career that interests you. Place the name of this job or career at the top of your two lists. Rank the tangible and intangible rewards you have listed in order of their importance to you.

4. Key a paragraph that explains how the job or career you identified in Step 3 will allow you to experience the tangible and intangible rewards of work. Print the document. Save the document as *Exercise 3-6* in the *DigiTools your name\Unit 3* folder.

WRITING

Punctuation and Confusing Usage

In Unit 3 you learned about the proper usage for commas, semicolons, colons, and confusing terms. In the next exercise, you will apply what you have learned as you edit a script of a television commercial for a mountain hotel called the Corporate View Lodge. This type of script would be written to guide the filming of a commercial. The script also guides the creation of a storyboard. Storyboards are a series of sketches and notes that will help a director and a camera operator visualize what they want to create.

3-5D • 17'
New-Key Mastery

1. Key the lines once SS; DS between 2-line groups.
2. Key the lines again at a faster pace.

Technique Goals:
- curved, upright fingers
- wrists low, but not resting
- quick-snap keystrokes
- down-and-in spacing
- eyes on copy as you key

reach review
1 a;sldkfj ed ol rf hj tf nj gf lo de jh ft nj fr a;
2 he jogs; an old ski; do a log for; she left a jar;

n/g
3 an an|go go|in in|dig dig|and and|got got|end ends
4 go to; is an; log on; sign it; and golf; fine figs

space bar
5 if if|an an|go go|of of|or or|he he|it it|is is|do
6 if it is|is to go|he or she|to do this|of the sign

all keys learned
7 she had an old oak desk; a jell jar is at the side
8 he has left for the lake; she goes there at eight;

all keys learned
9 she said he did it for her; he is to take the oars
10 sign the list on the desk; go right to the old jet

3-5E •
Enrichment

Key each line twice SS; DS between 2-line groups; QS after lines 3, 7, and 12.

lines 1–3:
- curved, upright fingers
- steady, easy pace

lines 4–7:
- space immediately after each word
- down-and-in motion of thumb

lines 8–12:
- maintain pace to end of line
- strike ENTER key quickly
- start new line immediately

lines 13–16:
- speed up second keying of each repeated word or phrase
- think words, not each letter

Reach review
1 nj nj gf gf ol ol tf tf ik ik rf rf hj hj ed ed fj
2 go fog an and got end jog ant dog ken fig fin find
3 go an on and lag jog flag land glad lend sign hand

Space Bar
4 if an it go is of do or to as in so no off too gin
5 ah ha he or if an too for and she jog got hen then
6 he is to go|if it is so|is to do it|if he is to go
7 she is to ski on the lake; he is also at the lake;

Enter key
8 he is to go;
9 she is at an inn;
10 he goes to ski at one;
11 he is also to sign the log;
12 she left the log on the old desk

Short words and phrases
13 do do|it it|an an|is is|of of|to to|if if|or or or
14 he he|go go|in in|so so|at at|no no|as as|ha ha ha
15 to do|to do|it is|it is|of it|of it|is to|is to do
16 she is to do so; he did the sign; ski at the lake;

goods and services. All personnel, from the president to staff in the mailroom, are asked to view their work with an awareness of TQM. Many companies have developed slogans such as "Quality is everybody's business" or "We want to be the best in all we do" to highlight their quality goals.

Over time, the policies and procedures used by a company may become outdated or inefficient. Companies seek to avoid this problem by applying the concept of continuous improvement. **Continuous improvement** means being alert at all times to ways of working more productively. Continuous improvement is a concept that overlaps the principles of TQM. All employees are encouraged to participate in continuous improvement efforts.

Customer satisfaction is a focus for many successful companies. These companies are described as having a customer-service based culture. "We are here to serve customers" is a message that all kinds of organizations send to employees. Having a **customer-service attitude** means thinking through what you do in relation to what it will mean to customers. Companies often conduct surveys to see if they are meeting the standards customers expect. They study the results of such surveys and then make changes to improve customer service.

EXERCISE 3-5   A CUSTOM-SERVICE ATTITUDE

You were standing at a desk of a coworker when her telephone rang. This is what you heard her say:

Who do you want?

A Mr. Ted Wells? Are you sure he works for this company?

Gee, I really don't know who the executives are. I don't work for any of them. I work for the director of catering services.

Oh, you work for Johnson Corporation. Well, you know how hard it is to know your own job, let alone know what is going on in the company.

You say our operator gave you this extension? Possibly, the operator doesn't know much more about the company than I do.

If I knew the extension for the president's office, I'd transfer you because I'd guess the president's secretary knows where everyone is—but, I don't know the number offhand and I could never find my directory on this messy desk... Let me transfer you back to the operator. Is that okay? I so wish I could be helpful.

Just hold on. But first, where are you calling from? Why don't you call when you aren't busy, and we can have a chat. Do you have my number? It's 513-555-0192, extension 344.

Hold on. Good luck in finding Mr. Wells. Goodbye.

1. Describe the impression you think the caller has of your coworker's knowledge of the company and of her way of working. Does she have a customer-service attitude?

2. If your coworker maintained an orderly desk, what would she have done as soon as it was clear that the caller had the wrong extension? What might she have said instead of the comments shown here?

3. What steps could she take as part of a continuous improvement effort to improve the way she handles calls?

UNIT 3   Tooling Up 3

LESSON 3-6

NEW KEYS: left shift AND period (.)

OBJECTIVES — In this lesson you will:

1. Learn reach technique for **left shift** and **.** (period).
2. Combine smoothly **left shift** and **.** (period) with all other learned keys.

Finger-action keystrokes

Down-and-in spacing

Quick out-and-tap ENTER

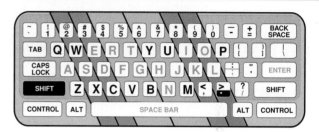

3-6A • 8'
Conditioning Practice

Key each line twice SS (slowly, then faster); DS between 2-line groups.

reach review 1 ed ik rf ol gf hj tf nj de ki fr lo fg jh ft jn a;
space bar   2 or is to if an of el so it go id he do as in at on
all letters learned 3 he is; if an; or do; to go; a jak; an oak; of all;

3-6B • 20'
New Keys: Left [Shift] and [.]

Key each line twice SS (slowly, then faster); DS between 2-line groups; if time permits, repeat lines 7–9.

Left shift *Left little* finger

. (period) *Right ring* finger

SHIFTING CUE

Shift, strike key, and release both in a quick 1-2-3 count.

Learn left shift key

1 a a Ja Ja Ka Ka La La Hal Hal Kal Kal Jae Jae Lana
2 Kal rode; Kae did it; Hans has jade; Jan ate a fig
3 I see that Jake is to aid Kae at the Oak Lake sale

Learn . (period)

4 l l .l .l fl. fl. ed. ed. ft. ft. rd. rd. hr. hrs.
5 .l .l fl. fl. hr. hr. e.g. e.g. i.e. i.e. in. ins.
6 fl. ft. hr. ed. rd. rt. off. fed. ord. alt. asstd.

Combine left shift and . (period)

7 I do. Ian is. Ola did. Jan does. Kent is gone.
8 Hal did it. I shall do it. Kate left on a train.
9 J. L. Han skis on Oak Lake; Lt. Haig also does so.

SPACING CUE

Space once after . following abbreviations and initials. Do not space after . within abbreviations. Space twice after . at end of a sentence except at line endings. There, return without spacing.

Careers in Information Technology

Position and Yearly Salary		Position and Yearly Salary	
Computer Scientist	$76,970*	Network Administrator	$56,440*
Computer Programmer	$62,890*	Web Art Director	$80,379**
Software Engineer	$72,370*	Graphic Design Specialist	$36,646**
Computer Support Specialist	$41,920*	Webmaster	$49,914**
Database Administrator	$58,420*	Technology Trainer	$47,747**
Level 1 Web Designer	$40,864**		

*Salaries based on 2001 industry averages, Bureau of Labor Statistics, www.bls.gov, downloaded January 2003.

**Salaries estimated from a Midwestern city median averages using data from www.salary.com calculated in January 2003. Salary estimates will vary greatly from city to city, state to state, and from year to year.

EXERCISE 3-4

TEAM WORK

INTERNET

IT ACRONYMS

Information technology professionals must learn a number of very technical terms to do their jobs effectively. Some terms are confusing. Others seem very similar to each other. Acronyms are abbreviations for commonly used terms. There are so many IT acronyms that employees in IT often sound like they speak a different language. Sort out some of these confusing terms in this exercise.

Work with a classmate to answer the following questions and learn about important IT acronyms. Use your Internet search tools, such as *Yahoo* or *Google,* to help you answer the questions.

1. What is the difference between RAM and ROM?

2. What are a bit, a byte, a kilobyte (KB), a megabyte (MB), a gigabyte (GB), and a terabyte?

3. What does the abbreviation MHz stand for and what does the term mean?

4. The world of CD standards is very confusing. Find definitions for the terms CD-ROM, CD-R, CD-RW, and DVD-ROM.

5. What do these acronyms for IT certification titles represent: CCIE, MCSE, MOS, OCP, DBA, and CNE? What skills do these certifications involve?

6. Network terms can be confusing. Explain briefly what these network term relate to: Ethernet, 802.11, and Bluetooth.

BUSINESS TRENDS AND ISSUES

Quality Management and Customer Satisfaction

Issues such as quality management, customer satisfaction, and teamwork affect how successful an organization is in achieving its goals. These issues must be the concerns of all employees. The company expects all employees to be reliable and cooperative in efforts to increase productivity and meet company goals in a global marketplace.

The primary goal of all businesses is to make profits. In an effort to increase profits, many companies have adopted **total quality management** (TQM) plans. TQM involves setting high standards in how work is done and in the creation and delivery of

3-6C • 17'
New-Key Mastery

1. Key the lines once SS; DS between 2-line groups.
2. Key the lines again at a faster pace.

Technique Goals:
- curved, upright fingers
- finger-action keystrokes
- out-and-down shifting

TECHNIQUE CUE
Eyes on copy except when you lose your place.

abbrev./initials
1 He said ft. for feet; rd. for road; fl. for floor.
2 Lt. Hahn let L. K. take the old gong to Lake Neil.

3d row emphasis
3 Lars is to ask at the old store for a kite for Jo.
4 Ike said he is to take the old road to Lake Heidi.

key words
5 a an or he to if do it of so is go for got old led
6 go the off aid dot end jar she fit oak and had rod

key phrases
7 if so|it is|to do|if it|do so|to go|he is|to do it
8 to the|and do|is the|got it|if the|for the|ask for

all letters learned
9 Ned asked her to send the log to an old ski lodge.
10 J. L. lost one of the sleds he took off the train.

3-6D • 5'
Technique: Spacebar and Enter

1. Key each line once SS; DS at end of line 7.
2. Key the drill again at a faster pace if time permits.

SPACING CUE
Quickly strike Space Bar *immediately* after last letter in the word.

1 Jan is to sing.
2 Karl is at the lake.
3 Lena is to send the disk.
4 Lars is to jog to the old inn.
5 Hanna took the girls to a ski lake.
6 Hal is to take the old list to his desk.
7 Lana is to take the jar to the store at nine.

Strike ENTER quickly and start each new line immediately.

3-6E •
Enrichment

1. Key each line once SS; DS between 3-line groups.
2. Rekey the drill at a faster pace if time permits.

Spacing/Shifting (Use down-and-in spacing; use out-and-down shifting.)
1 K. L. Jakes is to see Lt. Hahn at Oak Lake at one.
2 Janet Harkins sent the sales sheet to Joel Hansen.
3 Karla Kent is to go to London to see Laska Jolson.

Keying easy sentences (Keep insertion point moving steadily—no stops or pauses within the line.)
4 Kae is to go to the lake to fish off an old skiff.
5 Joel is to ask his good friend to go to the shore.
6 Lara and her dad took eight girls for a long hike.
7 Kent said his dad is to sell the oak and ash logs.

TEAMWORK

INTERNET

1. Work with a classmate to complete this activity. Find four organizations in your county, state, or province that provide human services. If you have trouble identifying organizations, look in the government section of your local phone book or search the Web. The U.S. Department of Health and Human Services Web site (www.hhs.gov) provides links to information about many human services organizations. You may find organizations such as:
   - Local health department
   - Environmental health department
   - Health promotion authority
   - Substance abuse facilities
   - Public health nursing authority
   - Homeless shelter
   - Charities involved with human services such as the American Red Cross
   - Local food banks or soup kitchens
   - Faith-based service groups

2. Call, visit in person, or look at each organization's Web site to find the main purpose of the organization. Learn whether the organization is private or part of a government agency or office. List the main services each organization provides. You should find information about at least four organizations.

3. Create a *PowerPoint* slide show to present the information you found about the human services organizations. Include a title slide, one or two slides that describe the human services area in general, and a slide for each of the organizations you researched. Save the presentation as *Exercise 3-3* in the *DigiTools your name\Unit 3* folder.

4. Deliver your presentation to the class or to another team.

Information Technology

Careers in Information Technology (IT) have grown in recent years. Many businesses, government agencies, educational systems, and other organizations need skilled IT personnel. IT workers perform tasks as varied as:

- Repairing, updating, and networking desktop, portable, handheld, and tablet computers

- Creating and maintaining communications networks

- Creating and maintaining Web pages, Web sites, intranets, and Web services such as e-mail, instant messaging, and videoconferencing

- Programming computers and other computerized objects such as toys, cars, cell phones, video game consoles, microwave ovens, and other equipment that can make use of a microchip

Work in the IT industry generally requires college-level education and industry-recognized certifications. Certification courses are taught at technical centers, community colleges, universities, and private career training institutes.

UNIT 3   Tooling Up 3

LESSON 3-7

NEW KEYS: u AND c

OBJECTIVES *In this lesson you will:*

1. Learn reach technique for **u** and **c**.
2. Combine smoothly **u** and **c** with all other learned keys.

3-7A • 8'
Conditioning Practice

Key each line twice SS (slowly, then faster); DS between 2-line groups.

reach review 1 `nj gf ol rf ik ed .l tf hj fr ki ft jn de lo fg l.`
space bar   2 `an do in so to go fan hen log gin tan son not sign`
left shift   3 `Olga has the first slot; Jena is to skate for her.`

3-7B • 20'
New Keys: U and C

Key each line twice SS (slowly, then faster); DS between 2-line groups: if time permits, repeat lines 7–9.

Follow the *Standard Plan for Learning New Keys* outlined on p. 75.

u *Right index* finger

c *Left middle* finger

Learn u

1 `j j uj uj us us us jug jug jut jut due due fur fur`
2 `uj uj jug jug sue sue lug lug use use lug lug dues`
3 `a jug; due us; the fur; use it; a fur rug; is just`

Learn c

4 `d d cd cd cod cod cog cog tic tic cot cot can cans`
5 `cd cd cod cod ice ice can can code code dock docks`
6 `a cod; a cog; the ice; she can; the dock; the code`

Combine u and c

7 `cud cud cut cuts cur curs cue cues duck ducks clue`
8 `a cud; a cur; to cut; the cue; the cure; for luck;`
9 `use a clue; a fur coat; take the cue; cut the cake`
10 `Jake told us there is ice on the road to the lake.`
11 `Jack asked us for a list of all the codes he used.`
12 `Louise has gone to cut the cake on the green cart.`

Human Services

Some of the most rewarding careers involve helping people and are in the Human Services career cluster. Workers in human services careers assist people of all ages from preschoolers to the elderly. Some careers in this cluster involve helping people with physical or mental handicaps and other special needs. If you care about helping people, this is the career cluster for you. Although the pay is often below the salaries found in many other career clusters, the jobs can be personally rewarding. Well-educated professionals in this career cluster can make a good living while helping others. There are many areas of human service to consider:

- Helping people overcome addiction
- Providing psychological care for the mentally ill
- Assisting the handicapped with physical rehabilitation exercises
- Developing after-school programs for teenagers
- Immunizing schoolchildren
- Assisting the aged

Review this list of possible career alternatives in the human services cluster.

Careers in Human Services

Position and Yearly Salary		Position and Yearly Salary	
Home Healthcare Aide	$18,510*	Health and Safety Specialist	$42,750**
Occupational Therapist	$33,130*	Psychiatric Aid	$23,760*
Foodservice Manager	$38,290*	Physical Therapist	$35,740*
Social and Community Services Manager	$44,540*	Funeral Director	$48,410*
		Preschool Manager	$36,988*
Social Worker: Elementary and Secondary Schools	$40,170*	Social Worker: Local Government	$35,780*
Social Worker: Individual and Family Services	$29,730*	Social Worker: Nursing Care Facilities	$31,580*
		Environmental Engineer	$53,210*

*Salaries based on 2001 industry averages, Bureau of Labor Statistics, www.bls.gov, downloaded January 2003.

EXERCISE 3-3 THE HUMAN SERVICES COMMUNITY

Organizations that provide human services have products and services that people need to know about. For example, consider a health and human services organization in your community. What messages do they have that need to be shared with the public? What services do they provide that people who have special needs should know about? You will learn about these messages and services in this exercise.

3-7C • 17'
New-Key Mastery

1. Key the lines once SS; DS between 2-line groups.
2. Key the lines again at a faster pace.

- Reach up without moving hands away from body.
- Reach down without moving hands toward body.
- Use quick-snap keystrokes.

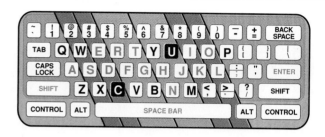

3d/1st rows
1 in cut nut ran cue can cot fun hen car urn den cog
2 Nan is cute; he is curt; turn a cog; he can use it

left shift and .
3 Kae had taken a lead. Jack then cut ahead of her.
4 I said to use Kan. for Kansas and Ore. for Oregon.

key words
5 and cue for jut end kit led old fit just golf coed
6 an due cut such fuss rich lack turn dock turf curl

key phrases
7 an urn|is due|to cut|for us|to use|cut off|such as
8 just in|code it|turn on|cure it|as such|is in luck

all keys learned
9 Nida is to get the ice; Jacki is to call for cola.
10 Ira is sure that he can go there in an hour or so.

3-7D • 5'
Technique: Spacebar and Left Shift

Key the lines once SS; DS between 3-line groups. Keep hand movement to a minimum.

space bar
1 Ken said he is to sign the list and take the disk.
2 It is right for her to take the lei if it is hers.
3 Jae has gone to see an old oaken desk at the sale.

left shift
4 He said to enter Oh. for Ohio and Kan. for Kansas.
5 It is said that Lt. Li has an old jet at Lake Ida.
6 L. N. is at the King Hotel; Harl is at the Leland.

3-7E •
Enrichment

1. Key each line once SS; DS between 2-line groups.
2. If time permits, key the lines again at a faster pace.

PRACTICE CUE
Try to reduce hand movement and the tendency of unused fingers to fly out or follow the reaching finger.

u/c
1 uj cd uc juj dcd cud cut use cog cue urn curl luck
2 Huck can use the urn for the social at the church.

n/g
3 nj gf nj gin can jog nick sign nigh snug rung clog
4 Nan can jog to the large sign at the old lake gin.

all keys learned
5 nj gf uj cd ol tf ik rf hj ed an go or is to he 1.
6 Leona has gone to ski; Jack had left here at nine.

all keys learned
7 an or is to he go cue for and jak she all use curt
8 Nick sells jade rings; Jahn got one for good luck.

CHAPTER 3   Lesson 3-7   86

Careers in Retail/Wholesale Sales and Service (Includes Marketing)

Position and Yearly Salary		Position and Yearly Salary	
Market Research Analyst	$58,230*	Sales Representative (Wholesale)	$48,240*
Wholesale and Retail Buyer	$44,200*	Advertising Sales Agent	$45,700*
Advertising Manager	$64,950*	Securities and Financial Services Sales Agent	$73,430*
Marketing Manager	$78,410*		
Sales Manager	$77,000*	Insurance Sales Agent	$48,560*
Salesperson (retail)	$20,920*	Telemarketer	$21,840*
Parts Salesperson	$25,940*	Real Estate Broker	$65,410*
Demonstrator and Product Promoter	$25,910*		

*Salaries based on 2001 industry averages, Bureau of Labor Statistics, www.bls.gov, downloaded January 2003.

Everyone is familiar with the retail side of business, sales, and marketing. Just visit your neighborhood mall or local store to see *retail* stores in action. Retail sales bring products close to customers. For example, when you enter JCPenney or Wal-Mart, you can see, hold, try on, sample, and buy products on the spot. Wal-Mart and JCPenney are among the most successful retail stores in the world.

Wholesalers supply retail stores and outlets. Products must be invented, designed, built, and manufactured. For example, imagine a Tablet PC built in Taiwan. The manufacturing plant cannot reach customers around the world, so they *wholesale* their products to buyers and distributors. Wholesalers take the finished computers and deliver and sell them to retail outlets.

The role of marketing is to inform customers of the many goods, products, and services that are available for purchase. Marketers use advertising to convince customers that their products are better than the products of their competitors. Some of the most famous brands in the world include Coca-Cola, Wal-Mart, Microsoft, JCPenney, and the Green Bay Packers. Do you know what type of products or services these companies provide?

In this next exercise, you will examine the differences between the roles of marketers, salespersons, and customer support staff.

EXERCISE 3-2 — ALL ABOUT MARKETING, SALES, AND SUPPORT

1. Access the Corporate View intranet. Choose **Marketing, Sales, & Support**, followed by **About Marketing, Sales, and Customer Support**.

2. Create a short unbound report about marketing, sales, and customer support. See page App-7 in Appendix A to review unbound report style.

3. Use an appropriate title for the report. In the first section of the report, describe what marketing, sales, and customer support personnel do as explained on the Corporate View intranet. Focus on how their jobs differ from one another.

4. In the second section of your report, explain how Corporate View believes that the marketing, sales, and support teams can work together to improve the sale of their products.

5. Save your report as *Exercise 3-2* in the *DigiTools your name\Unit 3* folder. Print the report.

LESSON 3-8

NEW KEYS: w AND right shift

OBJECTIVES *In this lesson you will:*

1. Learn reach technique for **w** and **right shift**.
2. Combine smoothly **w** and **right shift** with other learned keys.

3-8A • 8'
Conditioning Practice

Key each line twice SS (slowly, then faster); DS between 2-line groups.

reach review 1 rf gf de ju jn ki lo cd ik rf .l ed hj tf ol gf ft
u/c 2 us cod use cut sue cot jut cog nut cue con lug ice
all letters learned 3 Hugh has just taken a lead in a race for a record.

3-8B • 20'
New Keys: W and Right Shift

Key each line twice SS (slowly, then faster); DS between 2-line groups; if time permits, repeat lines 7–9.

w *Left ring* finger

Right shift *Right little* finger

SHIFTING CUE
Shift, strike key, and release both in a quick 1-2-3 count.

Follow the *Standard Plan for Learning New Keys* outlined on p. 75.

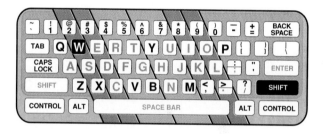

Learn w

1 s s ws ws sow sow wow wow low low how how cow cows
2 sw sw ws ws ow ow now now row row own own tow tows
3 to sow; is how; so low; to own; too low; is to row

Learn right shift key

4 A; A; Al Al; Cal Cal; Ali or Flo; Di and Sol left.
5 Ali lost to Ron; Cal lost to Elsa; Di lost to Del.
6 Tina has left for Tucson; Dori can find her there.

Combine w and right shift

7 Dodi will ask if Willa went to Town Center at two.
8 Wilf left the show for which he won a Gower Award.
9 Walt will go to Rio on a golf tour with Wolf Lowe.
10 Wilton and Donna asked to go to the store with us.
11 Walter left us at Willow Lake with Will and Frank.
12 Ted or Walt will get us tickets for the two shows.

- Use HTML, or a program like *Dreamweaver* or *FrontPage*, to create a Web site that communicates your message.
- Make a film. Digitize your movie, edit it, and copy your final movie to a CD for later viewing.
- Use word processing software to plan and write a transcript of a radio broadcast. Then record your broadcast, complete with music and sound effects, on your computer. Copy your radio broadcast onto a CD for later playback.
- Create traditional artwork using the traditional tools of art. Then scan your masterpieces, converting the images into digital files. Display the art online on a Web page.
- Write a short story and then format and publish it.
- Use software to design a new line of fashion clothing.

2. Think about your message. Write a clear, concise statement that explains your project. Clearly state the main points of the message you would like to communicate to others. You can add the details later.

Example project: A television commercial to be viewed in North America

Example message main points:
- Convince people to abandon their favorite soft drink and to try a new brand of bottled water.
- The bottled water comes from Aucuabueissimo, a small village in Italy at the base of the Alps.
- The spring has the purest water in all of Europe.
- The water is more refreshing and better for your health than diet soda.

3. Get approval for your message and project idea from your instructor.

4. After your message idea and project have been approved, choose and list the DigiTools you will use to create and share your message. This list should include all software you will use as well as devices such as a computer, printer, scanner, or digital camera.

5. Create your project. Share your finished project with your classmates.

Peer Check Ask a classmate to give you feedback on your project. Make adjustments or improvements to your project after the discussion.

Retail/Wholesale Sales and Services (includes Marketing)

Just like artists and moviemakers, businesspeople in marketing, retail sales, wholesale sales, and customer support services also have messages to share. Marketing and sales messages usually concern products. Products are commodities, goods, or services that are bought, sold, traded, and exchanged in wholesale and retail outlets. Examine some of the careers in this fast-paced and exciting career cluster.

3-8C • 17'
New-Key Mastery

1. Key the lines once SS; DS between 2-line groups.
2. Key the lines again at a faster pace.

PRACTICE CUE

Key at a steady pace; space quickly after each word.

Goal: finger-action reaches; quiet hands and arms

w and right shift
1. Dr. Rowe is in Tulsa now; Dr. Cowan will see Rolf.
2. Gwinn took the gown to Golda Swit on Downs Circle.

n/g
3. to go|go on|no go|an urn|dug in|and got|and a sign
4. He is to sign for the urn to go on the high chest.

key words
5. if ow us or go he an it of own did oak the cut jug
6. do all and for cog odd ant fig rug low cue row end

key phrases
7. we did|for a jar|she is due|cut the oak|he owns it
8. all of us|to own the|she is to go|when he has gone

all keys learned
9. Jan and Chris are gone; Di and Nick get here soon.
10. Doug will work for her at the new store in Newton.

3-8D • 5'
Technique: Spacing with Punctuation

Key each line once DS.

SPACING CUE

Do not space after an internal period in an abbreviation; space once after each period following initials.

No space / Space once.

1. Use i.e. for that is; cs. for case; ck. for check.
2. Dr. Wong said to use wt. for weight; in. for inch.
3. R. D. Roth has used ed. for editor; Rt. for Route.
4. Wes said Ed Rowan got an Ed.D. degree last winter.

3-8E •
Enrichment

1. Key each pair of lines once SS.
2. Key each even-numbered line again to increase speed.

Technique Goals:
- steady hands/arms
- finger-action keystrokes
- unused fingers curved, upright over home keys
- eyes on copy as you key

u/c
1. uj cd uc cut cut cue cue use use cod cod dock dock
2. Jud is to cut the corn near the dock for his aunt.

w and right shift
3. Don and Willa|Dot or Wilda|R. W. Gowan|Dr. Wilford
4. Dr. Wold will set the wrist of Sgt. Wills at noon.

left shift and .
5. Jane or Karl|Jae and Nan|L. N. Hagel|Lt. J. O. Hao
6. Lt. Hawser said that he will see us in New London.

n/g
7. nj gf ng gun gun nag nag got got nor nor sign sign
8. Angie hung a huge sign in front of the union hall.

o/t
9. ol tf to too dot dot not not toe toe got gild gild
10. Todd took the tool chest to the dock for a worker.

i/r
11. ik rf or ore fir fir sir sir ire ire ice ice irons
12. Risa fired the fir log to heat rice for the girls.

h/e
13. hj ed he the the hen hen when when then then their
14. He was with her when she chose her new snow shoes.

Commercial artists rely on computer graphics software. Movies are digitally enhanced. Digital television signals are bounced off satellites to receivers that convert the zeros and ones into high-definition images. The Internet has become a major outlet for artistic expression. Output from radio and television signals, once limited to a few hundred miles, now streams digitally to listeners around the world.

Artists, radio and television professionals, Web designers, and corporate communications specialists must learn to use new DigiTools to communicate their messages effectively. These messages may be factual, entertaining, serious, or frivolous. Such messages are crafted in the minds of artists, writers, musicians, producers, dancers, designers, and Web designers who create the messages to be shared with viewers and fans. Take a look at some careers in the art, audio, video, and communications industries in the table below.

Careers in Arts, Audio/Video Technology, and Communications

Position and Yearly Salary		Position and Yearly Salary	
Art Director	$65,570*	Technical Writer	$51,650*
Fine Artist or Painter	$36,330*	Interpreter or Translator	$34,680*
Sculptor	$36,330*	Actor	$36,790*
Illustrator	$36,330*	Dancer	$28,770*
Multimedia Artist	$46,700*	Choreographer	$32,750*
Animator	$46,700*	Musician	$46,690*
Fashion Designer	$56,340*	Singer	$46,690*
News Analyst, Reporter, or Correspondent	$37,800*	Photographer	$27,940*
		Camera Operator	$34,180*
Editor	$44,910*	Film and Video Editor	$42,020*
Author	$48,120*	Level 1 Web designer	$40,864**
Web Art Director	$80,379**		

*Salaries based on 2001 industry averages, Bureau of Labor Statistics, www.bls.gov, downloaded January 2003.

**Salaries estimated from a Midwestern city median averages using data from www.salary.com calculated in January 2003. Salary estimates will vary greatly from city to city, state to state, and from year to year.

EXERCISE 3-1

A PROJECT WITH A MESSAGE

When you have a message to share—a story, performance, or concept to express—DigiTools can help you create and communicate your message. In this exercise, you will use your skills and DigiTools to complete a communications project with a strong message.

1. Read the list of project suggestions below. Think about a message you wish to communicate. What would you like to say, write, draw, design, compose, or perform? Choose a project from the list or think of a similar project you would like to complete.
   - Use your word processor to develop a newsletter, brochure, flyer, handout, or some other document communicating a message; for example, a program for a school play, sports event, concert, or cultural event.
   - Use *PowerPoint* to create a slide show expressing a message; for example, a report for a class in school. Use a spreadsheet to calculate and chart information that helps communicate your message. Integrate the graph into your *PowerPoint* show.

LESSON 3-9

NEW KEYS: b AND y

OBJECTIVES *In this lesson you will:*

1. Learn reach technique for **b** and **y**.
2. Combine smoothly **b** and **y** with all other learned keys.

Fingers curved

Fingers upright

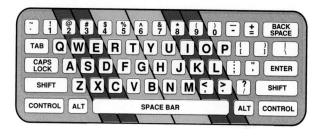

3-9A • 7'
Conditioning Practice

Key each line twice SS (slowly, then faster); DS between 2-line groups.

reach review 1 uj ws ik rf ol cd nj ed hj tf .l gf sw ju de lo fr
c/n 2 an can and cut end cue hen cog torn dock then sick
all letters learned 3 A kid had a jug of fruit on his cart in New Delhi.

3-9B • 5'
Technique: Spacebar

Key each line once.

Technique Goal:
Space with a down-and-in motion immediately after each word.

1 He will take an old urn to an art sale at the inn.
2 Ann has an old car she wants to sell at this sale.
3 Len is to work for us for a week at the lake dock.
4 Gwen is to sign for the auto we set aside for her.
5 Jan is in town for just one week to look for work.
6 Juan said he was in the auto when it hit the tree.

3-9C • 4'
Technique: Enter

1. Key each line once SS; at the end of each line quickly press the ENTER key and immediately start new line.
2. On line 4, see how many words you can key in 30 seconds (30").

A **standard word** in keyboarding is five characters or any combination of five characters and spaces, as indicated by the number scale under line 4 at the right. The number of standard words keyed in 1' is called **gross words a minute** *(gwam)*.

1 Dot is to go at two.
2 He saw that it was a good law.
3 Rilla is to take the auto into the town.
4 Wilt has an old gold jug he can enter in the show.

gwam 1' | 1 | 2 | 3 | 4 | 5 | 6 | 7 | 8 | 9 | 10 |

To find 1-minute (1') *gwam*:

1. Note on the scale the figure beneath the last word you keyed. That is your 1' *gwam* if you key the line partially or only once.

2. If you completed the line once and started over, add 10 to the figure determined in Step 1. The result is your 1' *gwam*.

To find 30-second (30") *gwam*:

1. Find 1' *gwam* (total words keyed).
2. Multiply 1' *gwam* by 2. The resulting figure is your 30" *gwam*.

DIGITOOLS
DIGITAL WORKBOOK — CHAPTER 12

Open the data file *CH12 Workbook*. Complete these exercises in your *DigiTools Digital Workbook* to reinforce and extend your learning for Chapter 12:

- Review Questions
- Vocabulary Reinforcement
- Math Practice: Calculating Folder Amounts
- From the Editor's Desk: Confusing Word Usage
- Keyboarding Practice: Reach Technique and Speed Building

UNIT 3

Tooling Up!

In these unit exercises, you will go online and use a variety of DigiTools to learn, think, and write about career alternatives, business issues, and technology trends. The skills you have acquired in the previous chapters will help you complete these exercises.

CAREER CLUSTERS

Arts, Audio/Video Technology, and Communications

In years gone by, artists created with paintbrush and easel. Movies were captured on film. Television was broadcast from towers sending high-frequency waves to antennas on rooftops. In recent years, however, the tools used in this career cluster have become digital. Professionals in this line of work must pay special attention to the digital tools of the trade and how a digital message is created.

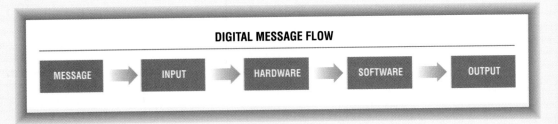

DIGITAL MESSAGE FLOW

MESSAGE → INPUT → HARDWARE → SOFTWARE → OUTPUT

UNIT 3 Tooling Up 3

3-9D • 19'
New Keys: B and Y

Key each line twice SS (slowly, then faster); DS between 2-line groups; if time permits, key lines 7–9 again.

b *Left index* finger

y *Right index* finger

Follow the *Standard Plan for Learning New Keys* outlined on p. 75.

Learn b

1 f f bf bf fib fib rob rob but but big big fib fibs
2 bf bf rob rob lob lob orb orb bid bid bud bud ribs
3 a rib; to fib; rub it; an orb; or rob; but she bid

Learn y

4 j j yj yj jay jay lay lay hay hay day day say says
5 yj yj jay jay eye eye dye dye yes yes yet yet jays
6 a jay; to say; an eye; he says; dye it; has an eye

Combine b and y

7 by by buy buy boy boy bye bye byte byte buoy buoys
8 by it; to buy; by you; a byte; the buoy; by and by
9 Jaye went by bus to the store to buy the big buoy.

3-9E • 15'
New-Key Mastery

1. Key the lines once SS; DS between 2-line groups.
2. Key the lines again at a faster pace.

PRACTICE CUE

- Reach up without moving hands away from your body.
- Reach down without moving hands toward your body.
- Use quick-snap keystrokes.

reach review
1 fg sw ki gf bf ol ed yj ws ik rf hj cd nj tf .l uj
2 a kit low for jut led sow fob ask sun cud jet grow

3d/1st rows
3 no in bow any tub yen cut sub coy ran bin cow deck
4 Cody wants to buy this baby cub for the young boy.

key words
5 by and for the got all did but cut now say jut ask
6 work just such hand this goal boys held furl eight

key phrases
7 to do|can go|to bow|for all|did jet|ask her|to buy
8 if she|to work|and such|the goal|for this|held the

all letters learned
9 Becky has auburn hair and wide eyes of light jade.
10 Juan left Bobby at the dog show near our ice rink.

| gwam | 1' | 1 | 2 | 3 | 4 | 5 | 6 | 7 | 8 | 9 | 10 |

ETHICS

Making Unauthorized Copies

Using company equipment and supplies, such as copiers and paper, for your personal use is unethical. Some companies permit employees to make one or two copies occasionally using company equipment and supplies. Any personal use of the company copier beyond this is not appropriate. If you want to make copies for some community event or charitable cause, always get permission to do so from a company manager.

Companies often track copying costs by the individual or department making the copies. This helps control costs from making unnecessary copies or from making copies for personal use. Devices that monitor the copier use may be placed on each copier or a copier log book may be used to record information about each copy job. This information may include the employee name or department and the number of copies made.

Follow-Up

1. Why is using the company copier to make copies for personal use unethical?

2. What other company supplies or equipment might employees be tempted to use for personal reasons?

ACTIVITY 12-9 — RESEARCH DIGITAL COPIERS

1. Access the Internet and open your browser. Use a search engine to find information about three different models of digital copiers. Search using the term *color digital copier*.

2. Open *Excel*. For each copier, record:
   - The copier name or model number
   - The number of black copies per minute the copier can make
   - The number of color copies per minute the copier can make
   - At least three features such as those shown in the table of Common Copier Features on page 375

3. Create a column chart that compares the number of black and the number of color copies per minute the copiers can make. Format the chart attractively. Title the chart *Copier Comparison*. Save the worksheet as *Activity 12-9* in your *DigiTools your name\Chapter 12* folder. Print the chart.

LESSON 3-10

NEW KEYS: m AND x

OBJECTIVES — *In this lesson you will:*

1. Learn reach technique for **m** and **x**.
2. Combine smoothly **m** and **x** with all other learned keys.

3-10A • 7'
Conditioning Practice

Key each line twice SS (slowly, then faster); DS between 2-line groups.

reach review
1 bf ol rf yj ed nj ws ik tf hj cd uj gf by us if ow

b/y
2 by bye boy buy yes fib dye bit yet but try bet you

all letters learned
3 Robby can win the gold if he just keys a new high.

3-10B • 20'
New Keys: M and X

Key each line twice SS (slowly, then faster); DS between 2-line groups; if time permits, key lines 7–9 again.

Follow the *Standard Plan for Learning New Keys* outlined on p. 75.

m *Right index* finger

x *Left ring* finger

Learn m

1 j j mj mj am am am me me ma ma jam jam ham ham yam
2 mj mj me me me may may yam yam dam dam men men jam
3 am to; if me; a man; a yam; a ham; he may; the hem

Learn x

4 s s xs xs ox ox ax ax six six fix fix fox fox axis
5 xs xs sx sx ox ox six six nix nix fix fix lax flax
6 a fox; an ox; fix it; by six; is lax; to fix an ax

Combine m and x

7 me ox am ax ma jam six ham mix fox men lax hem lox
8 to fix; am lax; mix it; may fix; six men; hex them
9 Mala can mix a ham salad for six; Max can fix tea.
10 Mary will bike the next day on the mountain roads.
11 Martin and Max took the six boys to the next game.
12 Marty will go with me on the next six rides today.

Special features designed to meet specific copying needs and to increase the user's productivity are available on many copiers. Most of the copiers you will use will offer several of the features, which are only a few of those available.

COMMON COPIER FEATURES

Automatic duplexing	Allows you to copy on both sides of the paper. Saves paper and postage costs
Automatic image shift	Creates a margin on one or both sides of the copy paper to allow space for three-hole punching or for binding the copies
Image enlargement and reduction	Allow you to make a photocopy larger or smaller than the original document
Sorter	Automatically collates the copies, arranging the copies in order or sets
Automatic document feed	Automatically feeds the originals into the machine
Self-diagnosis feature	Detects common problems (a paper jam, for example) and displays words or symbols to alert the user
Automatic exposure control	Adjusts the darkness or lightness of copies after sensing the density of the original
Color	Color images on the original are reproduced on the copies

Operating Procedures

Employees need to be knowledgeable about the proper use of copiers and selection of supplies. You will find that the selection of paper, toner, and other materials can affect the cost of making copies and the operation of the machines. All employees are expected to follow closely the recommendations of the vendor or manufacturer and company guidelines when using copier supplies in order to control costs.

Many companies post guidelines near each copier for employees who use copiers. As a responsible employee, you should adhere to these general guidelines:

- Be cost conscious when planning to use the copier. Use the copier's economy features, such as duplexing, and do not make more copies than you need.

- Comply with copyright laws that describe those documents that cannot be legally copied.

- Do not use company resources to make copies for personal use.

- Follow good housekeeping rules. Always clean up the area after you have completed your copying project. Deal with any copier problems, such as a paper jam, or notify the appropriate person of the problem.

- Practice common courtesy when using the copier. If you have a long copy job and another worker needs a priority copy, stop at a convenient point and let the other person have access to the machine. If you need a few copies and someone else is near the end of a long copying job, wait until the other person is finished to make your copies.

3-10C • 17'
New-Key Mastery

1. Key lines once SS; DS between 2-line groups.
2. Key the lines again at a faster pace.

- Reach up without moving hands away from body.
- Reach down without moving hands toward body.
- Use quick-snap keystrokes.

Goal: finger-action keystrokes; quiet hands and arms

3d/1st rows	1	by am end fix men box hem but six now cut gem ribs
	2	me ox buy den cub ran own form went oxen fine club
space bar	3	an of me do am if us or is by go ma so ah ox it ow
	4	by man buy fan jam can any tan may rob ham fun guy
key words	5	if us me do an sow the cut big jam rub oak lax boy
	6	curl work form born name flex just done many right
key phrases	7	or jam|if she|for me|is big|an end|or buy|is to be
	8	to fix|and cut|for work|and such|big firm|the call
all keys learned	9	Jacki is now at the gym; Lex is due there by four.
	10	Joni saw that she could fix my old bike for Gilda.

3-10D • 6'
Technique: Spacing with Punctuation

Key each line once DS.

SPACING CUE

Do not space after an internal period in an abbreviation, such as Ed.D.

1 Mrs. Dixon may take her Ed.D. exam early in March.
2 Lex may send a box c.o.d. to Ms. Fox in St. Croix.
3 J. D. and Max will go by boat to St. Louis in May.
4 Owen keyed ect. for etc. and lost the match to me.

3-10E •
Enrichment

1. Key each line twice SS (slowly, then faster); DS between 2-line groups.
2. Key each line once more at a faster pace.

PRACTICE CUE

Keep the insertion point moving steadily across each line (no pauses).

m/x	1	Max told them that he will next fix the main axle.
b/y	2	Byron said the boy went by bus to a bayou to hunt.
w/right shift	3	Wilf and Rona work in Tucson with Rowena and Drew.
u/c	4	Lucy cut a huge cake for just the four lucky boys.
./left shift	5	Mr. and Mrs. J. L. Nance set sail for Long Island.
n/g	6	Bing may bring a young trio to sing songs at noon.
o/t	7	Lottie will tell the two little boys a good story.
i/r	8	Ria said she will first build a large fire of fir.
h/e	9	Chet was here when the eight hikers hit the trail.

COPYING PAPER RECORDS

Reprographics is the process of making copies of graphic images, such as hard-copy documents, and also includes other image processing such as scanning images into computer files. Reprographics plays an important role in a records management system. Although technology is bringing changes to records management systems, paper is still the most common medium for storing documents and for sharing information with others. Copies of paper records are often needed during both the use and maintenance phases of the record life cycle.

Businesses have different needs for reprographic services. Their needs depend on the size of the company and the types of documents to be reproduced. In some businesses, you may leave your original document and a request for copies at a central copy center. Workers in the copy center make the copies and send them to you. In other companies, employees make their own copies as needed.

Office Photocopiers

Photocopiers, often simply called copiers, produce copies directly from an original document. The original can be handwritten, printed, or drawn. The quality of the copy is excellent if the machine is in good condition and the original is of high quality. Many copiers reproduce onto one or both sides of a sheet of paper and can copy onto letterhead paper, mailing labels, and colored paper. Some machines copy in color as well.

Figure 12-14
Photocopiers play an important role in records management.

Electronic copier/printers, sometimes called intelligent copiers, can receive, transmit, store, print, and copy data. These copiers can produce copies from sources such as a computer file, graphic scanners, or even pictures. For example, you may key material at your computer, proofread the copy, and then transmit it electronically to the copier/printer in a nearby location, where the copies will be printed.

LESSON 3-11

NEW KEYS: p AND v

OBJECTIVES *In this lesson you will:*

1. Learn reach technique for **p** and **v**.
2. Combine smoothly **p** and **v** with all other learned keys.

Fingers upright

Fingers curved

Hard return

3-11A • 7'
Conditioning Practice

Key each line twice SS (slowly, then faster); DS between 2-line groups.

one-hand words
1 in we no ax my be on ad on re hi at ho cad him bet

phrases
2 is just|of work|to sign|of lace|to flex|got a form

all letters learned
3 Jo Buck won a gold medal for her sixth show entry.

3-11B • 20'
New Keys: P and V

Key each line twice SS; DS between 2-line groups; if time permits, key lines 7–9 again.

p *Right little* finger

v *Left index* finger

Follow the *Standard Plan for Learning New Keys* outlined on p. 75.

Learn p

1 ; ; p; p; pa pa up up apt apt pen pen lap lap kept
2 p; p; pa pa pa pan pan nap nap paw paw gap gap rap
3 a pen; a cap; apt to pay; pick it up; plan to keep

Learn v

4 f f vf vf via via vie vie have have five five live
5 vf vf vie vie vie van van view view dive dive jive
6 go via; vie for; has vim; a view; to live; or have

Combine p and v

7 up cup vie pen van cap vim rap have keep live plan
8 to vie; give up; pave it; very apt; vie for a cup;
9 Vic has a plan to have the van pick us up at five.

A variety of filing systems are used in offices today. The positioning of guides and folders within filing systems will vary from office to office. Regardless of the system used, the guides and folders should be arranged so they are easy to see and in a logical order. You can see that the arrangement in Figure 12-13 allows your eye to move easily from left to right to locate guides and folders.

Before placing records in folders, you should index and code each record. **Indexing** is the process of deciding how to identify each record to be filed—either by name, subject, geographic location, number, or date. Many companies use standard filing rules recommended by the Association of Records Managers and Administrators, Inc. (also known as ARMA International).

Coding is the process of marking a record to indicate how it was indexed. You may retrieve and refile a record many times. By coding a record, you help ensure that it will be filed correctly each time it is returned to the files.

Employees need an orderly way to retrieve records. When a record is removed from the file, information such as the following is usually recorded: the name and department of the worker who is taking the record, the date the record was retrieved, and the date it will be returned. This information is kept in case someone else must locate the record. A retrieval procedure may also indicate whether all workers or only certain staff members have free access to the records.

APPLICATION 12-3   FILE RECORDS

Business such as publishers, mail-order houses, radio and television advertisers, and real estate agencies often file records geographically. Practice your geographic filing skills in this activity.

1. You work for Philips Associates, a real estate agency that files records geographically. Open and print the data file *Philips*. This file shows the names and addresses of companies and individuals whose records you will sort for filing.

2. Open your browser and access the data file *Manual*. This Web page contains part of the online office procedures manual that employees at Philips Associates have been asked to follow. Read Rules 1–8 and scan the examples for filing names of individuals and organizations.

3. Determine the filing order for the records. The records should first be arranged alphabetically by city. The records for each city should be arranged alphabetically by the company or person's name. Refer to the alphabetic filing rules as needed to determine the filing order of names for persons, businesses, and other organizations.

4. Number the records on your printout to indicate the correct filing order.

Self Check Did you group records first by city and then by individual or organization name?

3-11C • 17'
New-Key Mastery

1. Key the lines once SS; DS between 2-line groups.
2. Key the lines again at a faster pace.

Technique Goals:
- Reach up without moving hands away from your body.
- Reach down without moving hands toward your body.
- Use quick-snap keystrokes.

Goal: finger-action keystrokes; quiet hands and arms

reach review
1. vf p; xs mj ed yj ws nj rf ik tf ol cd hj gf uj bf
2. if lap jag own may she for but van cub sod six oak

3d/1st rows
3. by vie pen vim cup six but now man nor ton may pan
4. by six but now may cut sent me fine gems five reps

key words
5. with kept turn corn duty curl just have worn plans
6. name burn form when jury glad vote exit came eight

key phrases
7. if they|he kept|with us|of land|burn it|to name it
8. to plan|so sure|is glad|an exit|so much|to view it

all letters learned
9. Kevin does a top job on your flax farm with Craig.
10. Dixon flew blue jets eight times over a city park.

3-11D • 6'
Technique: [Shift] and [Enter] Keys

Key each 2-line sentence once SS as "Enter" is called every 30 seconds (30"). DS between sentences.

Goal:
To reach the end of each line just as the 30" guide ("Enter") is called.

The 30" *gwam scale* shows your gross words a minute if you reach the end of each line as the 30" guide is called.

Eyes on copy as you shift and as you strike ENTER key

	gwam 30"
1. Marv is to choose a high goal	12
2. and to do his best to make it.	12
3. Vi said she had to key from a book	14
4. as one test she took for a top job.	14
5. Lexi knows it is good to keep your goal	16
6. in mind as you key each line of a drill.	16
7. Viv can do well many of the tasks she tries;	18
8. she sets top goals and makes them one by one.	18

3-11E •
Enrichment

1. Key each line once at a steady, easy pace to master reachstrokes.
2. Key each line again at a faster pace.

Technique Goals:
- Keep fingers upright.
- Keep hands/arms steady.

m/p
1. mj p; me up am pi jam apt ham pen map ape mop palm
2. Pam may pack plums and grapes for my trip to camp.

b/x
3. bf xs be ax by xi fix box but lax buy fox bit flax
4. Bix used the box of mix to fix bread for six boys.

y/v
5. yj vf buy vow boy vie soy vim very have your every
6. Vinny may have you buy very heavy silk and velvet.

STORAGE EQUIPMENT AND SUPPLIES

Storage equipment, such as filing cabinets, should be chosen with specific storage media in mind. For example, if your records are on paper, you might use a different filing cabinet than if your records are on microfilm. You should keep certain especially valuable records in fireproof cabinets or vaults.

Each drawer or shelf in a file contains two different kinds of filing supplies: guides and file folders. The guides divide the drawer into sections and serve as signposts for quick reference. They also provide support for the folders and their contents. File folders hold the papers in an upright position in the drawer and serve as containers to keep the papers together. Labels are attached to file folders to identify the contents of each folder. Labels are also attached to file cabinet drawers to identify the contents of each drawer. The words or numbers written on labels are called captions.

FILING PROCEDURES

Filing is the process of storing office records in an orderly manner within an organized system. The procedure you follow to file records will vary according to the storage media used and the manner in which the files are organized. For example, paper records are often stored alphabetically according to names of individuals, organizations, businesses, subjects, or geographic locations. Files may also be organized numerically or by date. As a company employee, you will be expected to understand your organization's filing system so that you can file and retrieve records correctly.

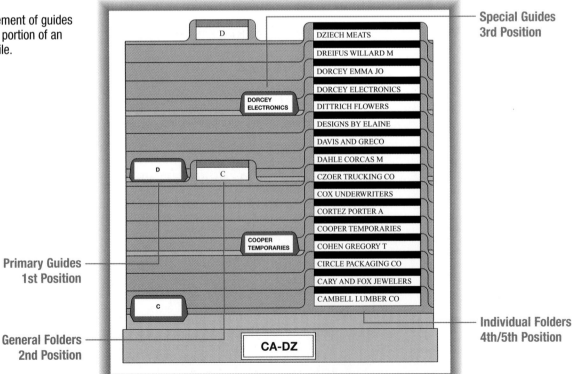

Figure 12-13
Notice the arrangement of guides and folders in this portion of an alphabetic name file.

CHAPTER 12   Lesson 12-3   373

LESSON 3-12

NEW KEYS: q AND comma (,)

OBJECTIVES *In this lesson you will:*

1. Learn reach technique for **q** and **,** (comma).
2. Combine smoothly **q** and **,** (comma) with all other learned keys.

3-12A • 7'
Conditioning Practice

Key each line twice SS (slowly, then faster); DS between 2-line groups; if time permits, key the lines again.

all letters learned 1 do fix all cut via own buy for the jam cop ask dig
p/v   2 a map; a van; apt to; vie for; her plan; have five
all letters learned 3 Beth will pack sixty pints of guava jam for David.

3-12B • 20'
New Keys: Q and ,

Key each line twice SS; DS between 2-line groups; if time permits, key lines 7–9 again.

q *Left little* finger

, (comma) *Right middle* finger

SPACING CUE

Space once after , used as punctuation.

Follow the *Standard Plan for Learning New Keys* outlined on p. 75.

Learn q

1 a qa qa aq aq quo quo qt. qt. quad quad quit quits
2 qa quo quo qt. qt. quay quay aqua aqua quite quite
3 a qt.; pro quo; a quad; to quit; the quay; a squad

Learn , (comma)

4 k k ,k ,k kit, kit; Rick, Ike, or I will go, also.
5 a ski, a ski; a kit, a kit; a kite, a kite; a bike
6 Ike, I see, is here; Pam, I am told, will be late.

Combine q and , (comma)

7 Enter the words quo, quote, quit, quite, and aqua.
8 I have quit the squad, Quen; Raquel has quit, too.
9 Marquis, Quent, and Quig were quite quick to quit.
10 Quin, Jacqueline, and Paque quickly took the exam.
11 Rob quickly won my squad over quip by brainy quip.
12 Quit, quiet, and quaint were on the spelling exam.

Optical Media

A **compact disc** (CD) is an optical storage form. Information is put on the disc by laser and read by a CD drive in the computer. These discs are in many ways better than most magnetic media, such as floppies, because they can hold more information than any but a hard disk. CDs should be handled carefully and kept in a protective jacket or case to prevent the surface from being scratched or getting dirty. Scratches or dirt on the surface and warping, which can be caused by exposure to extreme heat, may make a CD unreadable.

Figure 12-11
Many records are stored on optical media.

Microforms

You may use micrographics in different forms, collectively called microforms. Roll microfilm and microfiche are popular microforms. **Microfiche** is a small rectangular sheet of microfilm that contains a series of records arranged in rows and columns. Micrographics are used when paper or computer files would be less practical. For instance, an automobile dealer usually will keep parts lists for past-year vehicles on microfiche. Because the list is unchanging, keeping the data on magnetic media that can be updated is not necessary. Because the fiche is less bulky, it is easier to store and retrieve than paper records. Libraries often keep back issues of magazines and newspapers on microfilm because storing rolls of films is much easier and less costly than storing huge stacks of periodicals.

Figure 12-12
Microfilm and microfiche are used to store records that would be bulky in paper form.

3-12C · 17'
New-Key Mastery

1. Key the lines once SS; DS between 2-line groups.
2. Key the lines again at a faster pace.

Technique Goals:
- Reach up without moving hands away from your body.
- Reach down without moving hands toward your body.
- Use quick-snap keystrokes.

Goal: finger-action keystrokes; quiet hands and arms

reach review
1 qa .l ws ,k ed nj rf mj tf p; xs ol cd ik vf hj bf
2 yj gf hj quo vie pay cut now buy got mix vow forms

3d/1st rows
3 six may sun coy cue mud jar win via pick turn bike
4 to go|to win|for me|a peck|a quay|by then|the vote

key words
5 pa rub sit man for own fix jam via cod oak the got
6 by quo sub lay apt mix irk pay when rope give just

key phrases
7 an ox|of all|is to go|if he is|it is due|to pay us
8 if we pay|is of age|up to you|so we own|she saw me

all letters learned
9 Jevon will fix my pool deck if the big rain quits.
10 Verna did fly quick jets to map the six big towns.

3-12D · 6'
Technique: Spacing with Punctuation

Key each line once DS.

Space once after , and ; used as punctuation.

Space once.

1 Aqua means water, Quen; also, it is a unique blue.
2 Quince, enter qt. for quart; also, sq. for square.
3 Ship the desk c.o.d. to Dr. Quig at La Quinta Inn.
4 Q. J. took squid and squash; Monique, roast quail.

3-12E ·
Enrichment

1. Key each line once at a steady, easy pace to master reachstrokes.
2. Key each line again at a faster pace.

Technique Goals:

lines 1–3: fingers upright

lines 4–6: hands/arms steady

lines 7–9: two quick taps of each double letter

Adjacent keys
1 re io as lk rt jk df op ds uy ew vc mn gf hj sa ui
2 as ore ask opt buy pew say art owe try oil gas her
3 Sandy said we ought to buy gifts at her new store.

Long direct reaches
4 ce un gr mu br ny rv ym rb my ice any mug orb grow
5 nice curb must brow much fume sync many dumb curve
6 Brian must bring the ice to the curb for my uncle.

Double letters
7 all off odd too see err boo lee add call heed good
8 door meek seen huff less will soon food leek offer
9 Lee will seek help to get all food cooked by noon.

What You Should Know About

Disaster Recovery Plans

continued

Plans will vary depending on the needs of the organization. The disaster recovery plan should be reviewed periodically and updated as needed.

Many organizations and companies promote awareness and education about disaster recovery for businesses. The Disaster Recovery Institute International (www.drii.org) has a professional certification program for business disaster recovery planners.

STORAGE MEDIA

An organization may keep records on a variety of media: paper, magnetic media such as computer disks or tape, optical media such as compact discs, and **micrographics** (documents reduced and placed on film). Employees may be expected to work with all of these media.

Paper

The most common storage medium is paper. The advantage of keeping paper records is that you can immediately read the information. Although paper records will remain a major part of filing systems for years, businesses are recording more information on magnetic media, optical media, and micrographics. These systems require less space to store the records and allow them to be accessed more quickly.

Figure 12-10
The most common records storage medium is paper.

Magnetic Media

Magnetic media are reusable and contain information that is stored electronically. The most frequently used forms of magnetic media are computer hard disks (hard drives), flexible (floppy) disks, and tapes. Magnetic tape is used primarily for backing up (making a copy of the files on) hard drives and for holding large amounts of information that is not used on a regular basis. Because tape may be of great length, it has a large storage capacity.

LESSON 3-13

NEW KEYS: z AND colon (:)

OBJECTIVES *In this lesson you will:*

1. Learn reach technique for **z** and **:** (colon).
2. Combine smoothly **z** and **:** (colon) with all other learned keys.

3-13A • 7'
Conditioning Practice

Key each line twice SS; then key a 1' writing on line 3; determine *gwam*.

```
all letters   1 Jim won the globe for six quick sky dives in Napa.
learned
spacing       2 to own|is busy|if they|to town|by them|to the city
easy          3 She is to go to the city with us to sign the form.
gwam 1' | 1 | 2 | 3 | 4 | 5 | 6 | 7 | 8 | 9 | 10 |
```

3-13B • 18'
New Keys: Z and :

Key each line twice SS (slowly, then faster); DS between 2-line groups; if time permits, key lines 7–10 again.

z Left little finger

: (colon) Left shift and strike *:* key

Follow the *Standard Plan for Learning New Keys* outlined on p. 75.

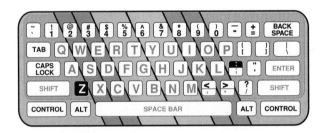

Learn z

```
1 a a za za zap zap zap zoo zoo zip zip zag zag zany
2 za za zap zap zed zed oz. oz. zoo zoo zip zip maze
3 zap it, zip it, an adz, to zap, the zoo, eight oz.
```

Learn : (colon)

```
4 ; ; :: :: Date:  Time:  Name:  Room:  From:  File:
5 :: :: To:  File:  Reply to:  Dear Al:  Shift for :
6 Two spaces follow a colon, thus:  Try these steps:
```

Combine z and :

```
7 Zelda has an old micro with : where ; ought to be.
8 Zoe, use as headings:  To:  Zone:  Date:  Subject:
9 Liza, please key these words:  zap, maze, and zoo.
10 Zane read:  Shift to enter : and then space twice.
```

LANGUAGE SKILLS CUE

- Space twice after : used as punctuation.
- Capitalize the first word of a complete sentence following a colon.

Figure 12-9
These phases make up the life cycle of a record.

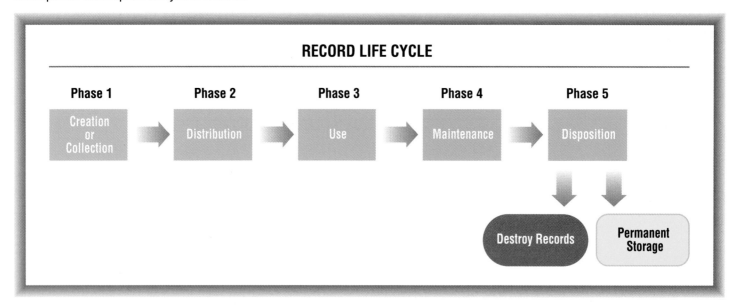

ACTIVITY 12-8

CREATE A RETENTION SCHEDULE

Each company creates its own records retention schedule. Your manager has sent you an e-mail message with some notes that you will use to create a records retention schedule for your company. In determining the retention times for each record, she considered how long the records will be used, how frequently they will be used, the form in which the records will be kept, and laws that pertain to records retention.

1. Start *Word*. Open and print the data file *Retention E-mail*, which contains your manager's e-mail. Close the document and *Word*.

2. Start *Access*. Create a new database as requested by your manager. Name the database file *Records Retention* and save it in your *DigiTools your name\Chapter 12* folder.

3. Sort and query the records as requested by your manager. Print the sorted datasheet and the query results datasheet. Close the database file and *Access*.

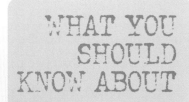

WHAT YOU SHOULD KNOW ABOUT

Disaster Recovery Plans

A disaster is an event that causes serious harm or damage. A disaster recovery plan provides procedures to be followed in case of such an event. The terrorist attacks on the World Trade Center in New York City and the Pentagon are extreme examples of the need organizations have for disaster recovery plans. Hundreds of offices and records were destroyed in these disasters.

While every business may not be involved in a major national crisis, every business does need a disaster recovery plan. A disaster may be caused by natural events such as a hurricane, tornado, or earthquake. A disaster can also be caused by accidental or intentional human acts such as fires, computer viruses, sabotage, bombs, or even human error.

3-13C • 15'
New-Key Mastery

1. Key the lines once SS; DS between 2-line groups.
2. Key the lines again at a faster pace.

- curved, upright fingers
- quiet hands and arms
- steady keystroking pace

q/z
1 zoo qt. zap quo zeal quay zone quit maze quad hazy
2 Zeno amazed us all on the quiz but quit the squad.

p/x
3 apt six rip fix pens flex open flax drop next harp
4 Lex is apt to fix apple pie for the next six days.

v/m
5 vim mam van dim have move vamp more dive time five
6 Riva drove them to the mall in my vivid lemon van.

easy
7 Glen is to aid me with the work at the dog kennel.
8 Dodi is to go with the men to audit the six firms.

alphabet
9 Nigel saw a quick red fox jump over the lazy cubs.
10 Jacky can now give six big tips from the old quiz.

3-13D • 10'
Block Paragraphs

1. Key each paragraph (¶) once SS; DS between them; then key them again faster.
2. If your instructor directs, key a 1' writing on each ¶; determine your *gwam*.

Paragraph 1   gwam 1'

The space bar is a vital tool, for every fifth or   10
sixth stroke is a space when you key. If you use   20
it with good form, it will aid you to build speed.   30

Paragraph 2

Just keep the thumb low over the space bar. Move   10
the thumb down and in quickly toward your palm to   20
get the prized stroke you need to build top skill.   30

gwam 1' | 1 | 2 | 3 | 4 | 5 | 6 | 7 | 8 | 9 | 10 |

3-13E •
Enrichment

1. Key each line once at a steady, easy pace to master reaches.
2. Key each line again at a faster pace.

Technique Goals:
- Keep fingers upright.
- Keep hands/arms steady.

x/:
1 xs :;|fix mix|Max: Use TO: and FROM: as headings.
2 Read and key: oxen, exit, axle, sixty, and sixth.

q/,
3 qa ,k|aqa k,k|quo quo,|qt. qt.,|quite quite,|squat
4 Quen, key these: quit, aqua, equal, quiet, quick.

p/z
5 p; za|;p; zaza|zap zap|zip zip|size size|lazy lazy
6 Zip put hot pepper on his pizza at the zany plaza.

m/v
7 mj vf|jmj fvf|vim vim|vow vow|menu menu move movie
8 Mavis vowed to move with a lot more vim and vigor.

A **records management system** is a set of procedures used to organize, store, retrieve, remove, and dispose of records. The main purpose of a records management system is to make sure records are available when needed. To make an intelligent decision or complete a task well, you need accurate, current information. You must be able to access needed records easily and quickly. An effective records management system will help you to be more productive because you will not waste valuable time searching for information that should be easily available.

RECORD LIFE CYCLE

Records come from many sources. Some records, such as letters from clients, come from outside the organization. Others are created within the organization. Examples of these records include interoffice memos, records of sales and purchases, reports, and copies of outgoing correspondence. Records are categorized according to their usefulness and importance.

- **Vital records** are essential to the company. These records are often not replaceable. Examples include original copies of deeds, copyrights, and mortgages.

- **Important records** are needed for the business to operate smoothly and would be expensive to replace. Examples include tax returns, personnel files, and cancelled checks.

- **Useful records** are convenient to have but are replaceable. Examples include correspondence, purchase orders, and the names and addresses of suppliers.

- **Nonessential records** have one-time or very limited usefulness. Examples include meeting announcements and advertisements.

The usefulness of each record has a beginning and an end. Therefore, each record has a life cycle.

1. **Creation or collection.** The cycle begins when you create or collect the records.

2. **Distribution.** During this phase, records are sent to the people who need them.

3. **Use.** Records are commonly used in making decisions or answering questions. Records may be kept to satisfy legal requirements.

4. **Maintenance.** When records are kept for later use, they must be stored, retrieved as needed, and protected from damage or loss.

5. **Disposition.** Records are disposed of either by destroying the records or by transferring them to permanent storage, as stated on the records retention schedule. A **retention schedule** is a document that identifies how long particular types of records should be kept.

LESSON 3-14

NEW KEYS: caps lock AND question mark (?)

OBJECTIVES *In this lesson you will:*

1. Learn reach technique for **Caps Lock** and **?** (question mark).
2. Combine smoothly **Caps Lock** and **?** (question mark) with other learned keys.

3-14A • 7'
Conditioning Practice

Key each line twice SS; then key a 1' writing on line 3; determine *gwam*.

```
alphabet  1  Lovak won the squad prize cup for sixty big jumps.
     z/:  2  To:  Ms. Mazie Pelzer; From:  Dr. Eliza J. Piazzo.
    easy  3  He is to go with me to the dock to do work for us.
   gwam  1' |  1  |  2  |  3  |  4  |  5  |  6  |  7  |  8  |  9  | 10 |
```

3-14B • 16'
New Keys: Caps Lock and ?

Key each line twice SS (slowly, then faster); DS between 2-line groups; if time permits, key lines 7–9 again.

Caps Lock
Left little finger

? (question mark)
Left shift; then *right little* finger

Depress the Caps Lock key to key a series of capital letters. To release the Caps Lock to key lowercase letters, press the Caps Lock key again.

Learn Caps Lock

1 Hal read PENTAGON and ADVISE AND CONSENT by Drury.
2 Oki joined FBLA when her sister joined PBL at OSU.
3 Zoe now belongs to AMS and DPE as well as to NBEA.

Learn ? (question mark)

Space twice.

4 ; ; ?; ?; Who? What? When? Where? Why? Is it?
5 Who is it? Is it she? Did he go? Was she there?
6 Is it up to me? When is it? Did he key the line?

Combine Caps Lock and ?

7 Did he join a CPA firm? I will stay on with NASA.
8 Is her dad still CEO at BSFA? Or was he made COB?
9 Did you read HOMEWARD? If so, try WHIRLWIND next.
10 Did Julie fly to Kansas City, MISSOURI, or KANSAS?
11 Did Dr. Sylvester pay her DPE, PBL, and NBEA dues?
12 Did you say go TWO blocks EAST or TWO blocks WEST?

- Plan your calls so the time spent during a long-distance or any other call is used efficiently.
- If possible, call when long-distance rates are least expensive.
- Notify the operator immediately after reaching a wrong number so you can receive credit for the call.
- Be an informed consumer of telephone services. Compare rate plans and promotional offerings.
- Learn how to use the equipment and features of your telephone system.

APPLICATION 12-2

INTERNET

NEW LONG-DISTANCE CARRIER

Your company, Dee-Lite's Chocolates, is dissatisfied with its current long-distance service provider. Prices have risen on domestic calls over the last six months. A coworker, Timothy, has researched several plans and recorded the domestic rates for long-distance calls. You will complete the research and update the worksheet. You will also integrate the worksheet table into a document with your recommendation for which plan to choose.

Open the data file *Phone Rates* and read Timothy's e-mail to you. Complete the project as requested by Timothy. Save the completed document as *Application 12-2* in your *DigiTools your name\Chapter 12* folder. Print and close the integrated document.

DigiTip
See Appendix A, page App-5 to review memo format.

LESSON 12-3 RECORDS MANAGEMENT

OBJECTIVES *In this lesson you will:*

1. Identify the purpose of a records management system.
2. Explore the phases of the record life cycle.
3. Prepare a records retention schedule.
4. Describe storage media, equipment, and supplies for records management.
5. Learn basic filing procedures.
6. Learn guidelines for using office copiers effectively.

RECORDS MANAGEMENT SYSTEM

A business cannot operate without records. A **record** is any information—text, image, voice, or other data—kept for future reference. For example, each time an item or service is purchased or sold by an organization, a record of the transaction is made and kept in the files.

3-14C • 18'
New-Key Mastery

1. Key the lines once SS; DS between 2-line groups.
2. Key the lines again at a faster pace.
3. Key a 1' writing on line 11 and then on line 12; determine *gwam* on each writing.

TECHNIQUE CUE

- Reach *up* without moving hands away from your body.
- Reach *down* without moving hands toward your body.
- Use Caps Lock to make ALL CAPS.

To determine 1' *gwam*:

Add 10 for each line you completed to the scale figure beneath the point at which you stopped in a partial line.

Goal: finger-action keystrokes; quiet hands and arms

caps lock/?
1 Did she join OEA? Did she also join PSI and DECA?
2 Do you know the ARMA rules? Are they used by TVA?

z/v
3 Zahn, key these words: vim, zip, via, zoom, vote.
4 Veloz gave a zany party for Van and Roz in La Paz.

q/p
5 Paul put a quick quiz on top of the quaint podium.
6 Jacqi may pick a pink pique suit of a unique silk.

key words
7 they quiz pick code next just more bone wove flags
8 name jack flax plug quit zinc wore busy vine third

key phrases
9 to fix it|is to pay|to aid us|or to cut|apt to own
10 is on the|if we did|to be fit|to my pay|due at six

easy
11 Lock may join the squad if we have six big prizes.
12 I am apt to go to the lake dock to sign the forms.

gwam 1' | 1 | 2 | 3 | 4 | 5 | 6 | 7 | 8 | 9 | 10 |

3-14D • 9'
Block Paragraphs

1. Key each ¶ once, using wordwrap (soft returns) if available. The lines you key will be longer than the lines shown if default side margins are used.
2. If time permits, key a 1' writing on one or two of the ¶s.

 Note:

Clearing the screen from time to time between 1' writings avoids confusion when determining *gwam*. Learn how to clear the screen on your software.

Paragraph 1   gwam 1'
When you key lines of drills, strike the return or   10
enter key at the end of each line. That is, use a   20
hard return to space down for a new line.   29

Paragraph 2
When you key copy in this form, though, you do not   10
need to strike return at the end of each line if a   20
machine has wordwrap or a soft return feature.   30

Paragraph 3
But even if your machine returns at line ends for   10
you, you have to strike the return or enter key at   20
the end of a paragraph to leave a line blank.   30

Paragraph 4
Learn now when you do not need to return at ends   10
of lines and when you must do so. Doing this now   20
will assure that your copy will be in proper form.   30

gwam 1' | 1 | 2 | 3 | 4 | 5 | 6 | 7 | 8 | 9 | 10 |

another country, dial the international access code, the country code, the city code, and the local number. For example, to place a call to London, England, you would dial the following sequence of numbers: 011 (international access code) + 44 (country code) + 71 (city code) + seven-digit phone number. Consult the International Calling or similar section of your local telephone directory for country codes. Charges for direct-dial calls begin as soon as the telephone is answered. If you make a direct-dial call and the person you need to speak with is unavailable, you still will be charged for the call.

Person-to-person calls are an expensive type of operator-assisted call where you ask to speak to a particular person. To place a person-to-person call, dial 0 (zero), the area code, and the telephone number of the individual or business you are calling. When you have finished dialing, you will be asked what type of call you wish to place. You will say "person-to-person" and will then be asked the name of the person you are calling. Pronounce the name clearly. Charges for the call begin only after the person you have requested is on the line. If that person is not available, you will not be charged for the call.

The charges for a **collect call** are billed to the telephone number being called, not to the number from which the call was placed. To place a collect call, dial 0 (zero), the area code, and the telephone number. You will be asked what type of call you are placing. Speak clearly into the phone, answering "collect." You will then be asked to give your name. The call will be completed, and the recipient will be asked whether or not the call and the charges will be accepted.

A **conference call** allows three or more persons at different locations to speak together. With some telephone systems, you can use special features to arrange these calls yourself. In many cases, conference calls are set up in advance with a conference operator. Be prepared to give the names, telephone numbers, and locations (cities and states) of the participants as well as the exact time the call is to be placed.

People who travel for business often use a credit card to pay for long-distance charges. The credit card bill provides a convenient summary of the call charges. Another type of card that is often used by travelers is a **prepaid phone card**. This card is purchased in advance and used to pay for a certain number of minutes of phone use.

> **DigiTip**
> For information about using the Internet for conference calls, see page 49.

TOLL-FREE SERVICE

As a convenience to customers who call long-distance, a company may subscribe to toll-free service for callers. This discounted service applies to incoming calls only, and no charge is made to the caller. As with other telephone services, rate plans and regulations for toll-free service plans vary widely. Compare price plans and features from several telephone companies to find the plan that will be most cost-effective for your company.

CONTROLLING TELEPHONE COSTS

Telephone costs for a company can be a significant expense. As a company employee, you will be expected to use the telephone in a cost-effective way. Follow these guidelines to help control telephone costs:

- Use direct-dialing most of the time. Make more specialized, expensive types of calls that require extra assistance only when necessary.

LESSON 3-15

NEW KEYS: backspace, tab, AND quotation mark (")

OBJECTIVES   In this lesson you will:

1. Learn reach technique for the **Backspace** key, **Tab** key, and " (quotation mark).
2. Improve and check keying speed.

3-15A • 7'
Conditioning Practice

Key each line twice SS; then key a 1' writing on line 3; find *gwam*.

```
alphabet  1  Quig just fixed prize vases he won at my key club.
caps lock 2  Find ZIP Codes for the cities in ARIZONA and OHIO.
easy      3  It may be a problem if both girls go to the docks.
gwam 1'   |  1  |  2  |  3  |  4  |  5  |  6  |  7  |  8  |  9  |  10  |
```

3-15B • 7'
New Key: ←Backspace

The BACKSPACE key is used to delete text to the left of the insertion point.

1. Locate the BACKSPACE key on your keyboard.
2. Reach up to the BACKSPACE key with the right little finger (be sure to keep the index finger anchored to the "j" key); depress the BACKSPACE key once for each letter you want deleted; return the finger back to the ; key.

Note:

When you depress and hold down the BACKSPACE key, letters to the left of the insertion point will be deleted continuously until the BACKSPACE key is released.

Backspace key
Right little finger; keep right index finger anchored to **j** key

This symbol means to delete.

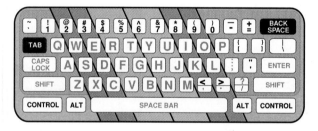

Learn Backspace

1. Key the following.

   The delete

2. Use the BACKSPACE key to make the change shown below.

   The ~~delete~~ backspace

3. Continue keying the sentence as shown below.

   The backspace key can be

4. Use the BACKSPACE key to make the change shown below.

   The backspace key ~~can be~~ is

5. Continue keying the sentence as shown below.

   The backspace key is used to fix

6. Use the BACKSPACE key to make the change shown below.

   The backspace key is used to ~~fix~~ make

7. Continue keying the sentence shown below.

   The backspace key is used to make changes.

USING DIRECTORIES

Many resources are available for you to use when planning a call. You may want to find the telephone number of a business or individual in your local area. You can usually find the number in the white pages of the local directory. In the white pages, individuals and organizations are arranged alphabetically by name. If you are searching for a particular service or product rather than a company, you may find the number in the yellow pages section of the directory. In the yellow pages, information is organized by product or service. You should become familiar with all sections of your local telephone directory.

Local as well as national organizations publish a variety of business and professional directories. National telephone directories can be purchased on CD-ROM. Web sites, such as Switchboard.com and SuperPages.com, provide a variety of data including telephone numbers and services of businesses. You should become familiar with the wide range of information contained in these resources.

If you are unable to locate a telephone number, call the directory assistance operator for help. Dial 411 for a local directory assistance operator. For long-distance directory assistance, dial 1, the area code, and 555-1212.

ACTIVITY 12-7

INTERNET

RESEARCH USING DIRECTORIES

Telephone directories provide a wealth of information for use in planning telephone calls and locating people and services. Practice using your local telephone directory or online directories to find information.

1. Start *Word*. Open the data file *Directory Research*.

2. Use your local telephone directory or online directories to find the information requested. If you need to find an online directory, search the Web using a search term such as *online telephone directory*. Record your answers in the *Word* document.

3. Save the document as *Activity 12-7* in your *DigiTools your name\Chapter 12* folder. Print and close the document.

LONG-DISTANCE SERVICE

Long-distance calls are calls made to numbers outside the service area of your local telephone company. The cost of long-distance service may depend on the time of day the call is placed, length of call, and type of long-distance call.

The consumer chooses a long-distance provider. Long-distance service companies provide a variety of pricing plans. Your local telephone directory may list several long-distance providers (MCI, Sprint, AT&T, etc.) and their numbers for you to contact. You may also visit these companies' Web sites to learn about pricing and special offers and regulations. To place calls efficiently and economically, you must become familiar with the various long-distance services available.

Direct-dial calls, also called station-to-station calls, are long-distance calls placed without assistance from an operator. To make a domestic direct-dial call, dial 1, the area code, and the number you are trying to reach. To make a direct-dial call to a number in

3-15C • 10'
New Key: Tab

The TAB key is used to indent the first line of ¶s. Word processing software has preset tabs called *default* tabs. Usually, the first default tab is set 0.5" to the right of the left margin and is used to indent ¶s (see copy at right).

1. Locate the TAB key on your keyboard (usually at upper left of alphabetic keyboard).
2. Reach up to the TAB key with the left little finger; strike the key firmly and release it quickly. The insertion point will move 0.5" to the right.
3. Key each ¶ once SS. DS between ¶s. As you key, strike the TAB key firmly to indent the first line of each ¶. Use the BACKSPACE key to correct errors as you key.
4. If you complete all ¶s before time is called, rekey them to master TAB key technique.

Tab key *Left little* finger

Tab → The tab key is used to indent blocks of copy such as these.

Tab → It should also be used for tables to arrange data quickly and neatly into columns.

Tab → Learn now to use the tab key by touch; doing so will add to your keying skill.

Tab → Strike the tab key firmly and release it very quickly. Begin the line without a pause.

Tab → If you hold the tab key down, the insertion point will move from tab to tab across the line.

3-15D • 14'
Keyboard Reinforcement

1. Key the lines once SS; DS between 3-line groups.
2. Key the lines again at a faster pace.
3. Key a 1' writing on lines 10–12.

- fingers curved and upright
- forearms parallel to slant of keyboard
- body erect, sitting back in chair
- feet on floor for balance

Reach review (Keep on home keys the fingers not used for reaching.)
1 old led kit six jay oft zap cod big laws five ribs
2 pro quo|is just|my firm|was then|may grow|must try
3 Olga sews aqua and red silk to make six big kites.

Space Bar emphasis (*Think*, *say*, and *key* the words.)
4 en am an by ham fan buy jam pay may form span corn
5 I am|a man|an elm|by any|buy ham|can plan|try them
6 I am to form a plan to buy a firm in the old town.

Shift key emphasis (Reach *up* and reach *down* without moving the hands.)
7 Jan and I are to see Ms. Han. May Lana come, too?
8 Bob Epps lives in Rome; Vic Copa is in Rome, also.
9 Oates and Co. has a branch office in Boise, Idaho.

Easy sentences (*Think*, *say*, and *key* the words at a steady pace.)
10 Eight of the girls may go to the social with them.
11 Corla is to work with us to fix the big dock sign.
12 Keith is to pay the six men for the work they did.

gwam 1' | 1 | 2 | 3 | 4 | 5 | 6 | 7 | 8 | 9 | 10 |

TIME ZONES

Be aware of time zone differences when placing long-distance calls to avoid calling before or after business hours or during lunch. The continental United States and parts of Canada are divided into five standard time zones: Atlantic, Eastern, Central, Mountain, and Pacific. As you move west, each zone is one hour earlier. For example, when it is 1 p.m. in Washington, DC (Eastern zone), it is noon in Dallas (Central zone), 11 a.m. in Denver (Mountain zone), and 10 a.m. in Los Angeles (Pacific zone).

If you are in San Diego and need to speak to a coworker in New York City, you will need to place the call before 2 p.m. Pacific time. Otherwise, the New York office may be closed because it will be 5 p.m. (Eastern time). When making international calls, being aware of the differences in time zones is especially important. Twenty-four time zones are used throughout the world.

A time zone map of the United States is included in most telephone directories. Web sites, such as Maps.com or WorldTimeZone.com, display the current time in several time zones as well as providing a time zone map.

Figure 12-8
This Web site shows the current time in several U.S. time zones.

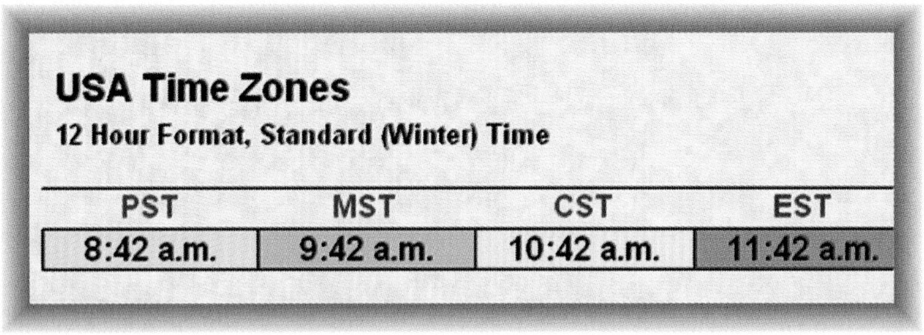

USA Time Zones
12 Hour Format, Standard (Winter) Time

PST	MST	CST	EST
8:42 a.m.	9:42 a.m.	10:42 a.m.	11:42 a.m.

ACTIVITY 12-6 — FIND TIMES IN DIFFERENT TIME ZONES

1. Access the Internet and open your browser. Use a search engine to find at least two sites that provide a time zone map or show times in multiple time zones. Use search terms such as *time zone map* or *time zone converter*. Record the URLs for these two sites. Use one of these sites to help you answer the following questions.

2. You live in Charlotte, North Carolina. You are having problems with your database software and need to call the software company's Technical Support department. The Technical Support department is open from 8:00 a.m. until 5:00 p.m. Pacific time. What is the earliest time you can call technical support from your time zone?

3. You have a friend who is in Florida on vacation. She told you she would call you at 8:00 p.m. your time. What time will it be in Florida when she calls?

4. You need to speak to a coworker in London, England. You want to call sometime between 8 a.m. and 6 p.m. London time. During what hours (in your time zone) can you place the call?

3-15E • 5'
New Key: "

Key each line twice SS (slowly, then faster); DS between 2-line groups.

Note:

On your screen, quotation marks may look different from those shown in these lines.

" Quotation mark Depress left shift and strike " (shift of ') with the *right little* finger

Learn " (quotation mark)

1 ;; "; "; ";" ";" "I believe," she said, "you won."
2 "John Adams," he said, "was the second President."
3 "James Monroe," I said, "was the fifth President."
4 Alison said "attitude" determines your "altitude."

ETHICS

Meeting Commitments

You are working on a special research project with Larry Moore (whose e-mail address is LMoore@hammer.unu.edu). He was supposed to e-mail you a lengthy attachment vital to your completion of the project. He promised to transmit it to you by noon yesterday. At 4:30 p.m., you have not heard from Larry. You are becoming very concerned because the deadline for the project is only one week away. You must study Larry's material carefully before you can complete your part of the project. By your estimate, it will take you approximately nine hours to review Larry's data.

Misleading coworkers about work-related issues, as Larry has done, is both unethical and inconsiderate. Coworkers must be able to depend on you to complete your part of a team project on schedule if the team is to be successful.

For Discussion

1. What reasons might Larry have for failing to send the material on time?

2. What should Larry do if unforeseen circumstances prevent him from meeting a deadline?

Follow Up

Compose and key an e-mail message to Larry that is appropriate for the situation described above. If you do not have e-mail software, key the message in your word processing software. Include the proper heading information for *To:* (Larry's e-mail address) and *Subject*. (The *Date* and *From* information would be added automatically by the e-mail program.) Make the tone of the message positive while stating your concerns and the action that you want Larry to take.

WHAT YOU SHOULD KNOW ABOUT
Voice Mail

Voice mail is a messaging system that uses computers and telephones to record, send, store, and retrieve voice messages. Voice messaging systems are popular because they allow the user to give you information even when you are away from your phone. Each user of a voice messaging system has a voice mailbox. A voice mailbox is a space reserved in a computer to hold recorded voice messages. The caller's message is held in storage until the recipient of the message chooses to access it. Unless a message is deleted, it remains in storage and can be accessed later for reference. Voice mail systems have other features such as the ability to send a recorded message to all users of the system.

When using your voice mail system, prepare a message that delivers appropriate information and instructions to the caller. Include your name, department, and other necessary information. Give instructions as to how to get immediate assistance if the caller cannot wait for you to return the call. Check your voice mail several times a day. Return all calls as soon as possible. Answer your telephone when you are at your desk unless you have visitors or cannot be interrupted for some other important reason. Do not let voice mail routinely answer your phone for you when you are at your desk.

When leaving messages for others on their voice mail systems, give your name, telephone number, company, and a brief reason for your call. Speak slowly and distinctly. Spell out any difficult names (your name, your company's name). Do not communicate bad news or negative statements in the voice message. Wait until you actually speak with the person to give any negative information.

OUTGOING TELEPHONE CALLS

Outgoing telephone calls may be made to a person outside or inside the company. Calls may be local or long-distance. Understanding the procedures for placing outgoing calls will help you be more productive.

Before you make a business call, take time to plan the call. Confirm the name and number of the person you are calling. Identify clearly the main purpose of the call. Outline briefly the points you want to cover during the call. Gather other information or items you need to have available before making the call, such as:

- Dates and times of any meetings or planned events that relate to the call
- Documents that relate to the topic of your conversation
- Questions that you want to ask
- Pen and paper or your computer to take notes during the call

Lesson 3-16

NEW KEYS: apostrophe (') AND hyphen (-)

OBJECTIVES — In this lesson you will:

1. Learn reach technique for ' (apostrophe) and - (hyphen).
2. Improve and check keying speed.

3-16A • 7'
Conditioning Practice

Key each line twice SS; then take a 1' writing on line 3; find *gwam*.

alphabet 1 Nate will vex the judge if he bucks my quiz group.
caps lock 2 STACY works for HPJ, Inc.; SAMANTHA, for JPH Corp.
easy 3 Nancy is to vie with six girls for the city title.
gwam 1' | 1 | 2 | 3 | 4 | 5 | 6 | 7 | 8 | 9 | 10 |

3-16B • 10'
New Keys: ' and –

Key each line twice SS (slowly, then faster); DS between 2-line groups.

Note:
On your screen, apostrophes may look different from those shown in these lines.

' (apostrophe) *right little* finger

- (hyphen) Reach *up* to hyphen with *right little* finger

Learn ' (apostrophe)

1 ;' ;' ;' '; '; I've told you it's hers, haven't I?
2 I'm sure it's Ray's. I'll return it if he's home.
3 I've been told it isn't up to us; it's up to them.

Learn - (hyphen)

4 ;- ;- ;- -; -; -;- -;- Did she say a two-ply tire?
5 I hyphenated self-made, half-pint, and last-ditch.
6 Mary sat on the four-poster bed deep in self-pity.

Combine ' and -

7 ;' ;' ;- ;- ;-' ;-' -'; -'; up-to-date list; x-ray
8 Didn't he say it couldn't be done? I don't agree.
9 I told him the off-the-cuff comment wasn't needed.

Figure 12-7
An Electronic Message Form

Message Form	
To	Mr. Lesinski
Date	8/27/--
Time	1:30 p.m.
From	Ms. Rosanna Robbins
Company	Advanced Realty
Phone	606-555-3478 Ext. 248
Email	rrobbins@advancedrealty.biz
Telephoned ☒	Please Call ☒
Called to See You ☐	Will Call Again ☐
Wants to See You ☐	Returned Your Call ☐
Message	She wants to discuss your meeting scheduled for next week.
By	Marion Witmer

ACTIVITY 12-5

COMPLETE AN ONLINE MESSAGE FORM

During the summer, you work part-time in your father's business. You answered a telephone call from Georgia Steward, of Steward Designs. She is your father's business associate. She called at 4:30 p.m. on the current date. Mrs. Steward asked you to take a message for your father.

Note: This activity instructs you to play a sound file using *Windows Media Player*. As an alternative, your teacher may read information to you.

1. To start *Windows Media Player*, click **Start, (All) Programs, Accessories**. Choose **Entertainment, Windows Media Player**.

2. Choose **Open** from the File menu. Browse to the location where data files are stored. Select the data file *Message* and click **Open**.

3. In a moment you will play the sound file. As you listen to this sound file containing Mrs. Steward's message, take notes about information you will need to complete a message form as shown in Figure 12-7. Click the **Play** button to play the message.

4. Close *Windows Media Player* by clicking the **Close** button in the upper right corner of the window.

5. Start *Word*. Open the data file *Message form*. Enter information in the appropriate form fields. (Click in a gray area and key the data.) Press **Tab** to move to the next field. Click in a checkbox field to check the box.

6. Save the message form as *Activity 12-5* in the *DigiTools your name\Chapter 12* folder. Choose **Word Document (*.doc)** as the document type. Print the message form or send it to your teacher as an e-mail attachment (as directed by your teacher). Close the document.

3-16C • 8'
Speed Check: Sentences

1. Key a 30" writing on each line. Your rate in *gwam* is shown word-for-word above the lines.
2. Key another 30" writing on each line. Try to increase your keying speed.

```
                                                                  gwam  30"
1 When do you think you will go?                                        12
2 Tara just finished taking her exam.                                   14
3 Nancy told the man to fix the car brake.                              16
4 Val could see that he was angry with the boy.                         18
5 Karen may not be able to afford college next year.                    20
6 Jay took three hours to complete the chemistry project.               22
30"  |  2  |  4  |  6  |  8  | 10 | 12 | 14 | 16 | 18 | 20 | 22 |
```

If you finish a line before time is called and start over, your *gwam* is the figure at the end of the line PLUS the figure above or below the point at which you stopped.

3-16D • 18'
Speed Building

1. Key the lines once SS; DS between 2-line groups.
2. Key a 1' writing on each line; find *gwam* on each writing (1' *gwam* = 5-stroke words keyed).

Easy sentences (Key the words at a brisk, steady pace.)

1 Pamela may make a profit off the land by the lake.
2 Eight of the firms may handle the work for Rodney.
3 Vivian may make a map of the city for the six men.
4 Helen held a formal social for eight of the girls.
5 He may work with the men on the city turn signals.
6 The dog and the girl slept in a chair in the hall.
7 Dianna may cycle to the city dock by the big lake.
8 Half of them may be kept busy with the sick girls.

gwam 1' | 1 | 2 | 3 | 4 | 5 | 6 | 7 | 8 | 9 | 10 |

3-16E • 7'
Speed Check: Paragraphs

Key two 1' writings on each ¶; determine *gwam* on each writing.

EASY all letters used

```
           •    2    •    4    •    6    •    8    •
     Keep in home position all of the fingers not
  10    •   12    •   14    •   16    •   18    •
  being used to strike a key.  Do not let them move
  20    •   22    •   24    •   26    •   28    •
  out of position for the next letters in your copy.
           •    2    •    4    •    6    •    8    •
     Prize the control you have over the fingers.
  10    •   12    •   14    •   16    •   18    •
  See how quickly speed goes up when you learn that
  20    •   22    •   24    •   26    •   28    •
  you can make them do just what you expect of them.
```

ACTIVITY 12-4

ROLE-PLAY TELEPHONE CONVERSATIONS

TEAMWORK

Role-playing telephone conversations will help you develop your telephone skills. For the role-playing activities, work with another member of your class. Rotate in different situations between being the caller and being the person answering the telephone.

1. Start *Word*. Open and print the data file *Phone*, which describes several dramatic situations that you will role-play.

2. After you have read each dramatic situation, work with your teammate to prepare a script. Compose the dialog that each person might say in this situation and print a copy for each of you.

3. Practice the telephone conversations. If you have a tape recorder available, record the call. Evaluate yourself and your teammate using the form provided in the data file.

4. Role-play your phone conversations for the class or another team. Have classmates complete an evaluation form for each presentation. Your classmates will also be acting out the situations and you will complete forms to evaluate them.

TAKING MESSAGES

Printed message forms are usually available in the workplace for recording telephone messages. Your company may have software that can be used to complete a message form. Message forms can be created easily in *Microsoft Word*. When you record a message, make sure it is accurate and complete. Verify names and telephone numbers by reading back the information to the caller. Ask for accurate spellings of names if you are in doubt. Record the message carefully. For paper forms, make sure your handwriting is clear and easy to read. Each message should include the following data:

- Date and time of the call
- Name of the caller with the caller's company (Check spellings of any names about which you are uncertain.)
- Caller's telephone number, including area code if it is a long-distance call (Remember to repeat the number for verification.)
- Details of the message
- Your name or initials

Figure 12-7 on page 363 shows an electronic message form created in *Word*. The gray areas are fields in which you can enter information. The gray shading does not print when you print the form. You will use a form like this in Activity 12-5.

LESSON 3-17

SKILLBUILDING

OBJECTIVES — *In this lesson you will:*

1. Improve keying technique on script and rough-draft copy.
2. Improve straight-copy speed and control.

3-17A · 10'
Rough Draft (Edited Copy)

1. Study the proofreaders' marks shown below and in the sentences.
2. Key each sentence DS, making all handwritten editing changes.
3. Key the lines again to improve your editing speed.

∧ = insert
= add space
∽ = transpose
℮ = delete
⌒ = close up
≡ = capitalize

1 A ~~first~~ rough draft is a preliminary or tentative ~~one~~ revision.
2 It is where the ~~creator~~ writer gets his/her thoughts on paper.
3 After the rough draft is created, it will be ~~looked over~~ edited.
4 ~~Reviewing~~ Editing is the step where a ~~persone~~ writer refines the copy.
5 Proofreaders' marks are used to edit the ~~original~~ rough draft ~~copy~~.
6 The editing changes will be then be made to the ~~copy~~ original.
7 After the changes have been made, read the copy again.
8 more changes still may need to be made to the copy.
9 Editing and proofreading does take ~~a lot~~ time and effort.
10 an error-free message ~~copy~~ is worth the trouble, however.

3-17B · 12'
Skill Transfer: Straight Copy to Script and Rough Draft

1. Key each ¶ once SS; DS between ¶s.
2. Key a 1' writing on each ¶; determine *gwam* on each writing; compare the three rates.

Your highest speed should be on ¶ 1 (straight copy); next highest on ¶ 2 (script); lowest on ¶ 3 (rough draft).

3. Key one or two more 1' writings on the two slowest ¶s to improve skill transfer.

Recall
1' *gwam* = total words keyed
A standard word = 5 strokes (characters and spaces).

Straight copy   gwam 1'

    Documents free of errors make a good impression.   10
When a document has no errors, readers can focus on   20
the content. Errors distract readers and may cause   31
them to think less of the message.   38

Script

    Therefore, it is important to proofread the final   10
copy of a document several times to make sure it   20
contains no errors before it leaves your desk.   29
Readers of error-free documents form a positive image   40
of the person who wrote the message.   47

(continued on next page)

- **Assist the caller.** Help the caller as efficiently as you can. Listen attentively to the caller's questions and comments. Make sure that you give accurate information to callers. Avoid passing off a caller to someone else if there is any way that you can be of help yourself.

- **Allow the caller to end the call.** As a general rule, the person who places a call is the one who should end the call and hang up first. If you follow this rule, you avoid making the caller feel as if the conversation has been "cut off" before he or she was ready to hang up. You can ask questions such as, "May I help you with anything else?" to move the call toward conclusion.

Figure 12-6
Have reference materials handy to assist callers.

At times, you must place a caller on hold while you answer another call or find information. Ask the caller if you may place him or her on hold. Check back frequently to reassure the caller that he or she has not been forgotten.

You may need to transfer an incoming call. Calls are usually transferred when the caller has reached a wrong extension, wishes to speak with someone else, or has a request that can be handled more effectively by another person. Always tell the caller why the transfer is necessary. Give the caller the extension number or name of the person to whom the call is being transferred. Then if the call is accidentally disconnected, the caller can reach the appropriate person or extension when he or she calls again.

Rough draft

gwam 1'

When a ~~negative~~ *positive* image of the person who wrote the ~~the~~   10

messge is formed, the message is ~~less~~ *more* likely to succeed.   22

remember, you never get a ~~another~~ *second* chance to make a good *first*   33

impression.   35

3-17C • 13'
Speed Check: Straight Copy

Key two 1' writings on each ¶; determine *gwam* on each writing.

EASY all letters used

gwam 2'

```
       •   2   •   4   •   6   •   8   •
    Time and motion are major items in building         4
  •  10  •  12  •  14  •  16  •  18  •
our keying power.  As we make each move through          9
  •  20  •  22  •  24  •  26  •  28  •
space to a letter or a figure, we use time.  So we      14
  •  30  •  32  •  34  •  36  •  38  •
want to be sure that every move is quick and direct.    20
  •  40  •  42  •  44  •  46  •
We cut time and aid speed in this way.                  24
       •   2   •   4   •   6   •   8   •
    A good way to reduce motion and thus save time      28
  •  10  •  12  •  14  •  16  •  18  •
is just to keep the hands in home position as you       33
  •  20  •  22  •  24  •  26  •  28  •
make the reach to a letter or figure.  Fix your         38
  •  30  •  32  •  34  •  36  •  38  •
gaze on the copy; then, reach to each key with a        43
  •  40  •  42  •  44  •  46  •
direct, low move at your very best speed.               47
```

gwam 2' | 1 | 2 | 3 | 4 | 5 |

DIGITOOLS
DIGITAL WORKBOOK   **CHAPTER 3**

Open the data file *CH03 Workbook*. Complete these exercises in your *DigiTools Digital Workbook* to reinforce and extend your learning for Chapter 3:

- Review Questions
- Vocabulary Reinforcement
- From the Editor's Desk: Creating Possessives
- Math Practice: Converting to Metric Measurements
- Keyboarding: Response Patterns

Your Speaking Skills

When you communicate with others in person, you make them feel welcome by smiling. You show interest and alertness by making eye contact with them. When you communicate by telephone, however, all you have to convey interest, alertness, and courtesy is your voice. Vary the tone of your voice to express feelings and emphasize ideas, but avoid using extremes.

You may speak with people from different parts of your nation or with people from countries all over the world. If you speak too rapidly, the listener may not hear all the information. You must learn to adjust your pace to fit the needs of the listener. Control the volume of your voice so that you are speaking neither too loudly nor too softly. Do not shout or speak so softly that the listener cannot hear what you are saying.

Speaking skills such as word pronunciation, grammar, and vocabulary usage affect the impression you project over the phone. Communication is difficult if the person you are speaking with cannot understand your words. Correct pronunciation of words is essential for understanding. Proper enunciation is also important. When you **enunciate** effectively, you pronounce words clearly and distinctly. For example, you should say "going to" instead of "gonna." State your ideas simply without using highly technical terms or lengthy words. Use proper grammar and avoid use of slang or regional expressions that may not be widely known or understood.

Your Attitude

When you speak to someone over the telephone, your attitude is reflected in your voice. Any boredom, anger, or indifference you are feeling is intensified as your message is sent to the person on the line. On the other hand, a smile and an upbeat, caring attitude are also clearly projected to the person with whom you are speaking. Whether you are the caller or the listener, take part in the call with a sincere, positive attitude.

INCOMING TELEPHONE CALLS

When answering the telephone, you usually do not know who is calling or what the caller wants. You should know how to handle a variety of situations and take care of caller requests, needs, and problems. Every organization wants employees to give a positive, professional impression when communicating over the telephone. Your call may be an initial customer contact—the first time the customer has spoken with someone at your company. How well you handle the call may determine, at least in part, whether the customer will do business with your company. Using proper telephone techniques will help you make a positive impression. Follow these guidelines for handling an incoming call:

- **Answer promptly.** If possible, answer the telephone after the first ring. When you reach for the receiver, also pick up a pen and a notepad or message form. You must be ready to take notes or a message.

- **Identify yourself.** Give your company or department name first and then your name. You may be able to tell that a call is coming from inside your company. In this case, give only your name.

> **DigiTip**
> If you know that it will take several minutes to find information, do not keep the caller waiting. Explain the situation to the caller and offer to make a return call when you have the information.

CHAPTER 4

Keyboarding: Number and Symbol Keys

OBJECTIVES *In this chapter you will:*

1. Review using correct keying position and posture.
2. Learn correct touch-keying for number and symbol keys.
3. Learn to add, subtract, multiply, and divide using the numeric keypad.
4. Build keyboarding skill.
5. Learn how to prevent repetitive stress injuries related to computer use.

Now that you have mastered the touch-system for letter keys, you will learn to use number and symbol keys. Note that number keys appear on both the top row of the keyboard and on the numeric keypad. Many people find that they use both sets of keys, depending on the work they are doing. For example, when keying numbers in a letter or report, using the numbers on the top row is probably more efficient. When using an onscreen calculator or entering numbers in a spreadsheet, however, using the numeric keypad may be more efficient.

WHAT YOU SHOULD KNOW ABOUT
Repetitive Stress Injury

Repetitive stress injury (RSI) is a condition caused by repeated movement of a part of the body. "Tennis elbow" is an RSI that is familiar to many people. Many people may not realize, however, that computers can cause injuries. When using the keyboard, mouse, and digital pen, people repeat the same motions with their hands over and over again. Many people perform thousands of mouse clicks and keystrokes daily. These repeated movements can lead to a form of RSI called **carpal tunnel syndrome (CTS)**.

CTS is a condition that develops gradually. With CTS the wrists, hands, and forearms can become inflamed and painful. CTS symptoms include:

- Numbness in the hand
- Tingling or burning in the hand, wrist, or elbow
- Pain in the forearm, elbow, or shoulder
- Difficulty gripping objects

LESSON 12-2

EFFECTIVE TELEPHONE COMMUNICATIONS

OBJECTIVES In this lesson you will:

1. Learn and apply skills required to make a favorable impression over the telephone.
2. Apply telephone techniques and procedures to handle calls courteously and efficiently.
3. Plan calls efficiently using tools such as published and computerized directories.
4. Complete an electronic message form.
5. Role-play telephone conversations.
6. Learn techniques for controlling telephone costs.
7. Research rates for long-distance service providers.

The most universal tool for voice communications is the telephone. Think of the many calls company personnel receive and place each day. Workers in the company receive and place calls to others both inside and outside the company. They discuss common concerns, place orders, or request information. Messages must be taken and recorded either manually or electronically. Telephone calls are often less time-consuming than a memo, a letter, or even e-mail.

Many workers need to be able to communicate by phone wherever their jobs take them—while traveling or at a work site. Cellular phones allow workers to receive and place calls from many locations. A **cellular phone** is a communications device that uses wireless, radio frequencies to transmit voice and data.

Because the telephone is such an important communications tool, all workers should be able to use proper telephone techniques when answering incoming calls and placing outgoing calls.

MAKING A FAVORABLE FIRST IMPRESSION

Figure 12-5
You represent your organization every time you place or answer a call.

When you handle telephone calls for a company, you are representing that company. The caller's first impression of the organization may be based on how you handle the call. To create a positive image for your organization, you should develop good communication skills. Your voice, pronunciation, grammar, vocabulary, and attitude all contribute to the impression you make when using the telephone.

What You Should Know About

Repetitive Stress Injury

continued

The risk of developing CTS is less for people who pay attention to ergonomics. **Ergonomics** is the science of adjusting the elements of the working environment to the human body. Using a comfortable, well-fitting adjustable chair and placing the keyboard at the right height can improve posture and comfort. For some people, such changes in the ergonomics of a workstation may help relieve the symptoms of CTS. Proper keyboarding techniques, posture and/or muscle-stretching exercises can also be helpful.

Be aware that prolonged handwriting can also cause problems. We all have had cramps in our hands after taking a long test or a long notetaking session. If taken to excess, handwriting with a digital pen can lead to many of the same problems described for keyboard and mouse users.

In the past, RSI usually affected people in their 40s and 50s. However, because of the increasing use of digital devices, such as keyboards, the mouse, cell phones, video games, and remote controls, the hand and wrist are more subject to repetitive stress than at any other time in our history. Increasingly, students as young as 11 and 12 are experiencing adult-like RSI symptoms.

Computer users can reduce the risk of developing RSI/CTS by taking these precautions:

- Arrange the workstation correctly: Position the keyboard directly in front of the chair. Keep the front edge of the keyboard even with the edge of the desk. Place the keyboard at elbow height. Place the monitor about 18 to 24 inches from your eyes with the top edge of the display screen at eye level.
- Use a proper chair and sit correctly. Use a seat that allows you to keep your feet flat on the floor while you are keying. Sit erect and as far back in the seat as possible.
- Use correct arm and wrist positions and movement. Keep your arms near the side of your body in a relaxed position. Your arms should be parallel to the floor and level with the keyboard. Your wrists should be in a flat, neutral position.
- Use proper keyboarding techniques. Keep your fingers curved and upright over the home keys. Keep wrists and forearms from touching or resting on any surface while keying. Strike each key lightly using the fingertip.
- When using a digital pen, hold the pen firmly, but do not grip it too tightly. Do not use excessive pressure when writing. Use good posture when writing as well as when keying.
- Use a variety of input tools. Learn to use the mouse with your left hand or with your right hand. Alternate hands occasionally when using a mouse for long periods.
- Take short rest breaks. Exercise and stretch your neck, shoulders, arms, wrists, and fingers before beginning to work each day and often during the day.

Figure 4-1
Using a variety of input tools can help prevent RSI.

7. On the day before the sale, "red tag sale" labels must be placed on the sale items. This task must be done after the store closes. Add a note to this effect to this task.

- Double-click the task to open the Task window.
- Key a note in the details box.
- Click **Save and Close** on the Standard toolbar.

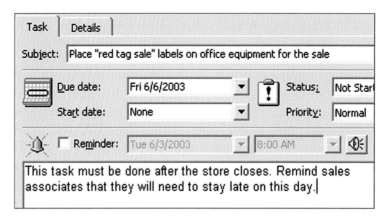

8. Choose **Print Preview** from the File menu to preview the list. To print the tasks list, click **Print** on the Preview bar. Choose **Table Style** under Print Style. Click **OK**. Close *Outlook*.

Self Check Does your final tasks list have three tasks marked as completed?

APPLICATION 12-1 — PERSONAL TASKS LIST

DigiTip
If the Folder List is not displayed, choose **Folder List** from the View menu.

1. Use *Outlook* or *Excel* to create a tasks list. If using *Outlook*, click your folder name under Tasks in the Folder List to store the list in your folder. Delete all the tasks from the office equipment sale project.

2. In the tasks list, include the school assignments or projects you have been assigned for the next several days. Also include an entry for each club meeting, sports activity, or other school-related event you plan to attend during this time period.

3. Assign a date for when you will attend the event or for when the assignment is due. Assign a priority (High, Normal, or Low) to each task or event. Sort the list by date. Print the tasks list.

4. If using *Excel*, save the list as *Application 12-1* in the *DigiTools your name\Chapter 12* folder.

LESSON 4-1

FIGURE KEYS (8, 1, 9, 4, AND 0)

OBJECTIVES *In this lesson you will:*

1. Learn reach technique for **8**, **1**, **9**, **4**, and **0**.
2. Improve skill on script and rough-draft copy.

4-1A • 5'
Conditioning Practice

Key each line twice SS; then key a 1' writing on line 3; determine *gwam*.

alphabet	1	The exquisite prize, a framed clock, was to be given to Jay.
spacing	2	They may try to be at the dorm in time to eat with the team.
easy	3	The maid was with the dog and six girls by the field of hay.
gwam	1'	1 | 2 | 3 | 4 | 5 | 6 | 7 | 8 | 9 | 10 | 11 | 12 |

Note: To determine *gwam*, see 9C directions if necesary.

4-1B • 8'
Learn 8 and 1

Key each line twice SS (slowly, then faster); DS between 2-line groups.

- Reach up without moving hands away from body.
- Reach down without moving hands toward body.
- Use quick-snap keystrokes.

Learn 8

1 k k 8k 8k kk 88 k8k k8k 88k 88k Reach up for 8, 88, and 888.
2 Key the figures 8, 88, and 888. Please open Room 88 or 888.

Learn 1

3 a a 1a 1a aa 11 a1a a1a 11a 11a Reach up for 1, 11, and 111.
4 Add the figures 1, 11, and 111. Only 1 out of 111 finished.

Combine 8 and 1

5 Key 11, 18, 81, and 88. Just 11 of the 18 skiers have left.
6 Reach with the fingers to key 18 and 188 as well as 1 and 8.
7 The stock person counted 11 coats, 18 slacks, and 88 shirts.

CHAPTER 4   Lesson 4-1

UPDATING TASKS IN *OUTLOOK*

An advantage of using *Outlook* to create a tasks list rather than just writing notes on paper is the ability to easily update the tasks list. You can check a task as completed or delete a task from the list. You can change the due date or add notes regarding a task.

To mark a task as complete, you click the checkbox to the left of the task. A line is drawn through the task to show that it has been completed. To delete a task, click the task. Then click the **Delete** button on the toolbar.

ACTIVITY 12-3

UPDATE A TASKS LIST

If you do not have *Outlook*, skip this activity and go to Application 12-1.

1. Start *Outlook*. Click **Tasks** on the Outlook Bar. If the Outlook Bar is not displayed, choose **Outlook Bar** from the View menu.

HELP KEYWORDS
Delete
Delete an item

2. Under Tasks in the Folder List, click your folder name. If the Folder List is not displayed, choose **Folder List** from the View menu.

3. Your manager has decided that the store can use some sale banners that you already have to place in the store windows to promote the sale. You need to delete the task regarding ordering banners from the tasks list. Click the task regarding ordering banners. Click the **Delete** button on the toolbar.

4. The date is now seven days after you began this project. You have completed all the tasks that were a high priority for the first week of the project. You want to keep these tasks displayed on your list until the entire sale project is completed. You want to show, however, that they have been completed. Identify the tasks you were to complete in the first week of the project. Click the checkbox to the left of each of these tasks. Each task you checked will appear with a line drawn through it.

| ☑ | ☑ | ~~Prepare long range plan of things to do~~ |
| ☑ | ☑ | ~~Prepare inventory list of items for sale~~ |

5. Your manager tells you that you need to order "red tag sale" labels to place on the equipment for the sale. These labels should be ordered three weeks before the sale. Add this new task to your tasks list.

DigiTip
See Steps 4–7 of Activity 12-2 to review adding a task to an *Outlook* Tasks list.

6. The date is now about three weeks before the sale. Look at the list of remaining tasks. Identify any tasks that should be done this week. Change the priority of these tasks to High.
- Double-click the task to open the Task window.
- Click the down arrow by **Priority** and choose **High**.
- Click **Save and Close** on the Standard toolbar.

4-1C • 7'
Keyboard Reinforcement

Key each line twice SS (slowly, then faster); DS between 2-line groups.

Figures

1 May 1-8, May 11-18, June 1-8, and June 11-18 are open dates.
2 The quiz on the 18th will be on pages 11 to 18 and 81 to 88.
3 He said only 11 of us got No. 81 right; 88 got No. 81 wrong.

Home/1st

4 ax jab gab call man van back band gala calf cabman avalanche
5 small man|can mask|lava gas|hand vase|lack cash|a small vase
6 Ms. Maas can call a cab, and Jan can flag a small black van.

4-1D • 9'
Learn 9, 4, and 0

Key each line twice SS (slowly, then faster); DS between 2-line groups.

Note:
Use the letter l in line 1. Use the figure 1 in line 2.

Learn 9

1 l l 9l 9l ll 99 l9l l9l 99l 99l Reach up for 9, 99, and 999.
2 The social security number was 919-99-9191, not 191-99-1919.

Learn 4

3 f f 4f 4f ff 44 f4f f4f 44f 44f Reach up for 4, 44, and 444.
4 Add the figures 4, 44, and 444. Please study pages 4 to 44.

Learn 0

5 ; ; 0; 0; ;; 00 ;0; ;0; 00; 00; Reach up for 0, 00, and 000.
6 Snap the finger off the 0. I used 0, 00, and 000 sandpaper.

Combine 9, 4, and 0

7 Flights 904 and 490 left after Flights 409A, 400Z, and 940X.
8 My ZIP Code is 40099, not 44099. Is Tanya's 09094 or 90904?

4-1E • 11'
Speed Check

1. Key a 30" writing on each line. Determine *gwam* on each writing.
2. Key another 30" writing on each line—at a faster pace.

	2	4	6	8	10	12	14	16	18	20	22

1 Suzy may fish off the dock with us.
2 Pay the girls for all the work they did.
3 Quen is due by six and may then fix the sign.
4 Janie is to vie with six girls for the city title.
5 Duane is to go to the lake to fix the auto for the man.

30" | 2 | 4 | 6 | 8 | 10 | 12 | 14 | 16 | 18 | 20 | 22 |

11. Tasks with a high priority display an exclamation point in the priority column. To sort the tasks by priority, click the **Priority** column head.

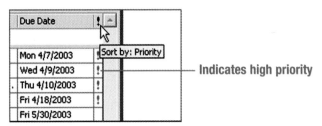
Indicates high priority

12. Sort the tasks by due date. Click the **Due Date** column head until it displays a faint up arrow to sort the list in ascending order.

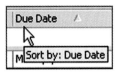

13. Your tasks list should now be sorted by date. It should appear similar to Figure 12-4, but with the dates you determined.

Figure 12-4
Tasks in *Microsoft Outlook*

HELP KEYWORDS

Print
Print a list of messages, contacts, or tasks

14. Choose **Print Preview** from the File menu to preview the list. To print the tasks list, click **Print** on the Preview bar. Choose **Table Style** under Print Style. Click **OK**.

15. Close *Outlook* by choosing **Exit** from the File menu.

Peer Check Compare the dates you determined with those of a classmate who began the project on the same day you did. Discuss any differences and make changes, if needed, after the discussion.

4-1F • 10'
Script and Rough-Draft Copy

1. Key each line once DS (2 hard returns between lines).
2. Key the rough-draft lines again if time permits.

≡ = capitalize
∧ = insert
∼ = transpose
⌿# = delete space
= add space
⌿lc = lowercase
◡ = close up

Script

1 Proofread: Compare copy word for word with the original.
2 Compare all figures digit by digit with your source copy.
3 Be sure to check for spacing and punctuation marks, also.
4 Copy in script or rough draft may not show exact spacing.
5 It is your job to insert correct spacing as you key copy.
6 Soon you will learn how to correct your errors on screen.

Rough draft

7 cap the first word an^d all proper nouns in each every sentence.
8 For example: pablo Mendez is from San juan,Puerto rico.
9 Ami Qwan and her parents will return to Taipie this fall summer.
10 our coffee is comes from Brazil Columbia; tea, fromEngland or china.
11 How many of you have lc Ethnic origins in# a for eign country?
12 Do did you know which of our the states once were part of mexico?

LESSON 4-2 FIGURE KEYS (5, 7, 3, 6, AND 2)

OBJECTIVES — In this lesson you will:
1. Learn reach technique for **5**, **7**, **3**, **6**, and **2**.
2. Improve skill transfer and build speed.

4-2A • 5'
Conditioning Practice

Key each line twice SS; then key a 1' writing on line 3; determine *gwam*.

alphabet 1 Zelda might fix the job growth plans very quickly on Monday.
spacing 2 He will go with me to the city to get the rest of the tapes.
easy 3 The six men with the problems may wish to visit the tax man.
gwam 1' | 1 | 2 | 3 | 4 | 5 | 6 | 7 | 8 | 9 | 10 | 11 | 12 |

4-2B • 5'
Learn 5 and 7

Key each line twice SS (slowly, then faster); DS between 2-line groups.

Learn 5

1 f f 5f 5f ff 55 f5f f5f 55f 55f Reach up for 5, 55, and 555.
2 Reach up to 5 and back to f. Did he say to order 55 or 555?

Learn 7

3 j j 7j 7j jj 77 j7j j7j 77j 77j Reach up for 7, 77, and 777.
4 Key the figures 7, 77, and 777. She checked Rooms 7 and 77.

3. Your folder name will appear under Tasks in the Folder List. Click your folder name. Your folder name will appear on the Folder banner.

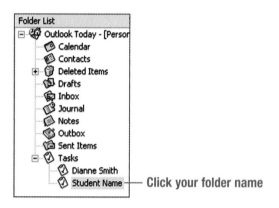
Click your folder name

4. Choose **New** from the File menu, and then click **Task**. The Task window will appear. In the Subject box, type a task name. For example: `Call newspaper and order ads`.

5. Set a priority for the task. Click the down arrow by **Priority** and choose a priority (as you determined earlier) such as **Normal.**

> **HELP KEYWORDS**
>
> **Tasks**
> Set the due date and start date for a task

6. Set a date for the task to be completed using the date you determined. Click the down arrow by **Due Date**. Click the date you want to set as the due date. (If the date is not in the current month, click the right arrow by the month name to move to a later month.)

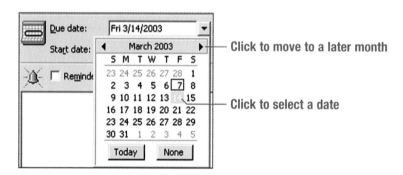
Click to move to a later month
Click to select a date

7. Click **Save and Close** on the Standard toolbar.

8. Repeat Steps 6–9 for each task on the list.

9. View the tasks list. Click **Tasks** under Outlook Shortcuts on the Outlook Bar if the Tasks list is not displayed.

10. When you store a tasks list in a subfolder, the Priority of the task may not display on the Tasks list. To display the Priority field for tasks, choose **Current View** from the View menu. Choose **Customize Current View**. Click the **Fields** button. Select **Priority** under Available Fields and click the **Add** button. Click **OK**. Click **OK** again.

> **HELP KEYWORDS**
>
> **Setting**
> Create a new view or customize an existing one

4-2C · 8'
Figure-Key Mastery

Key each line twice SS (slowly, then faster); DS between 2-line groups.

Straight copy

1 She moved from 819 Briar Lane to 4057 Park Avenue on May 15.
2 The 50-point quiz on May 17 covers pages 88-94, 97, and 100.
3 The meeting will be held in Room 87 on March 19 at 5:40 p.m.

Script

4 The 495 representatives met from 7:00 to 8:40 p.m. on May 1.
5 Social Security Nos. 519-88-7504 and 798-05-4199 were found.
6 My office is at 157 Main, and my home is at 4081 92d Avenue.

Rough draft

7 Runners 180, 90, and 507 were scheduled for August 15.
8 her telephone number was changed to 708-194-5009 on July 1.
9 Review Rules 1-9 on pages 89-90 and rules 15-19 on page 174.

4-2D · 8'
Learn 3, 6, and 2

Key each line twice SS (slowly, then faster); DS between 2-line groups.

Learn 3

1 d d 3d 3d dd 33 d3d d3d 33d 33d Reach up for 3, 33, and 333.
2 Add the figures 3, 33, and 333.   Read pages 3 to 33 tonight.

Learn 6

3 j j 6j 6j jj 66 j6j j6j 66j 66j Reach up for 6, 66, and 666.
4 Key the figures 6, 66, and 666.   Did just 6 of 66 finish it?

Learn 2

5 s s 2s 2s ss 22 s2s s2s 22s 22s Reach up for 2, 22, and 222.
6 Reach up to 2 and back to s.   Ashley reviewed pages 2 to 22.

Combine 3, 6, and 2

7 Only 263 of the 362 flights left on time on Monday, July 26.
8 Read Chapter 26, pages 263 to 326, for the exam on April 23.

3. Consider the list of items on the tasks list you printed. Decide which items should be done first to meet deadlines. Mark these items for High priority. Assign all other items Normal priority. (Write the priority by each item.)

4. Use a calendar to determine the date each task should be completed. Use the time frames indicated on the tasks list. Work backward from the sale date. Keep this work to use in the next activity.

Peer Check Discuss the priority you assigned to each task with a classmate. How do your priorities differ from those assigned by your classmate? Do you want to change any priorities after your discussion?

ACTIVITY 12-2

CREATE AND SORT A TASKS LIST

The steps below instruct you to use *Microsoft Outlook* to create a tasks list. If you do not have *Outlook*, follow the DigiTip at the left to create a list using *Excel*.

DigiTip

If you do not have *Outlook*, create a tasks list in *Excel*. Enter `Equipment Sale Tasks` as the worksheet title. Enter `Due Date`, `Priority`, and `Task` as the column heads. Enter the data for each task in the appropriate columns. Select all the data under the column heads. Click the **Sort Ascending** button on the Standard toolbar to sort the list by due date. Print the list.

1. To start *Microsoft Outlook,* click **Start** on the taskbar. Click **(All) Programs, Microsoft Outlook.**

2. By default, information in a tasks list is saved in the Tasks folder. In a school environment where more than one person uses a computer, you will need to create a subfolder to store your information. To create a Tasks subfolder:
- Click the **Tasks** icon on the Outlook Bar. If the Outlook Bar is not displayed, choose **Outlook Bar** from the View menu.
- Right-click **Tasks** in the Folder banner. Click **New Folder** on the menu that displays.

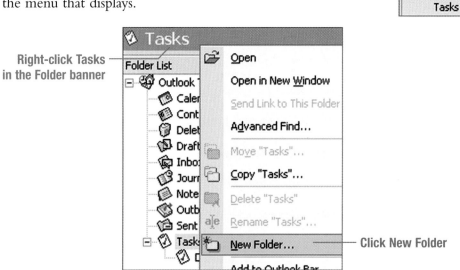

- Key your name in the Name text box. Click **OK**. Click **No** when asked if you want a shortcut folder added to the Outlook Bar.

4-2E • 10'
Skill Transfer

1. Key a 1' writing on each ¶; determine *gwam* on each.
2. Compare rates. On which ¶ did you have highest *gwam*?
3. Key two 1' writings on each of the slower ¶s, trying to equal your highest *gwam* in Step 1.

Note:
Relative speeds on different kinds of copy:
- highest—straight copy
- next highest—script copy
- lowest—statistical copy

To determine *gwam*, use the 1' *gwam* scale for partial lines in ¶s 1 and 2, but count the words in ¶ 3.

| AVG | all letters/figures used | gwam 1' |

You should try now to transfer to other types of copy as much of your straight-copy speed as you can. Handwritten copy and copy in which figures appear tend to slow you down. You can increase speed on these, however, with extra effort. 11 / 23 / 35 / 47

An immediate goal for handwritten copy is at least 90 percent of the straight-copy rate; for copy with figures, at least 75 percent. Try to speed up balanced-hand figures such as 26, 84, and 163. Key harder ones such as 452 and 980 more slowly. 11 / 23 / 35 / 48 / 49

Copy that is written by hand is often not legible, and the spelling of words may be puzzling. So give major attention to unclear words. Question and correct the spacing used with a comma or period. You can do this even as you key. 11 / 23 / 35 / 47

gwam 1' | 1 | 2 | 3 | 4 | 5 | 6 | 7 | 8 | 9 | 10 | 11 | 12 |

4-2F • 14'
Speed Building

1. Key a 1' writing on each ¶; determine *gwam* on each writing.
2. Add 2–4 *gwam* to better rate in Step 1 for a new goal.
3. Key three 1' writings on each ¶ trying to achieve new goal.

| AVG | all letters used | gwam 2' |

When you need to adjust to a new situation in which new people are involved, be quick to recognize that at first it is you who must adapt. This is especially true in an office where the roles of workers have already been established. It is your job to fit into the team structure with harmony. 6 / 12 / 18 / 24 / 30

Learn the rules of the game and who the key players are; then play according to those rules at first. Do not expect to have the rules modified to fit your concept of what the team structure and your role in it should be. Only after you become a valuable member should you suggest major changes. 35 / 41 / 47 / 53 / 59

gwam 2' | 1 | 2 | 3 | 4 | 5 | 6 |

A **personal planner**, also called a day planner or organizer, is a small notebook that contains a calendar and space for recording appointments, listing tasks, writing notes, and keeping a contacts list. **Electronic organizers** may be used in the same way as paper-based systems to keep track of project deadlines, appointments, and work schedules. Rather than being recorded on paper, the information is entered into a computer. Electronic organizers, also called personal information management (PIM) programs or desktop organizers, are available for desktop, laptop, tablet, and handheld computers. *Microsoft Outlook* is an example of personal information management and communications software. *Outlook* helps users manage e-mail messages, contacts, tasks, and appointments.

A reminder file in paper or electronic form that is arranged **chronologically** (in order of time) can also be used to help keep track of tasks and appointments. This file provides a convenient place to keep notes about tasks to be performed on specific dates. Such a file is sometimes called a tickler file.

Figure 12-3
Reminders help ensure that tasks are completed on time.

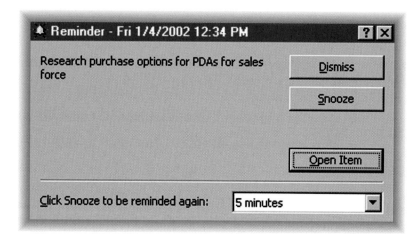

Wall calendars are useful for showing deadlines or stages of a large project. Software designed for managing large or long-term projects is called scheduling or project management software. *Microsoft Project* is an example of this type of software.

ACTIVITY 12-1   PRIORITIZE A TASKS LIST

Your manager has asked you to plan a schedule to prepare for a sale of office equipment. She hands you a rough draft of the inventory list of the products that will be included in the sale. Items marked with an asterisk (*) will have to be ordered from the suppliers so that they arrive in time for the sale.

1. Start *Word*. Open the data file *Task.doc*, which contains a partial list of tasks you must complete before the sale. Print the list.

2. The date of the sale will be two months from the current date. Consult a calendar to determine the date of the sale.

LESSON 4-3

SYMBOL KEYS (/, $, !, %, <, AND >)

OBJECTIVES — *In this lesson you will:*

1. Learn control of **/**, **$**, **!**, **%**, **<**, and **>**.
2. Combine **/**, **$**, **!**, **%**, **<**, and **>** with other keys.

4-3A • 5'
Conditioning Practice

Key each line twice SS; then key a 1' writing on line 3; determine *gwam*.

alphabet 1 Jackie will budget for the most expensive zoology equipment.
figures 2 I had 50 percent of the responses--3,923 of 7,846--by May 1.
easy 3 The official paid the men for the work they did on the dock.
gwam 1' | 1 | 2 | 3 | 4 | 5 | 6 | 7 | 8 | 9 | 10 | 11 | 12 |

4-3B • 15'
Learn /, $, and !

Key each line twice SS (slowly, then faster); DS between 2-line groups.

The / is the shift of the question mark. Strike it with the right little finger.

The $ is the shift of 4. Control it with the left index finger.

The ! is the shift of 1 and is controlled by the left little finger.

SPACING CUE

Do not space between a figure and the / or the $ sign.

Learn / (diagonal or slash) Reach down with the right little finger.

1 ; ; /; /; ;; // ;/; ;/; 2/3   4/5   and/or We keyed 1/2   and 3/4.
2 Space between a whole number and a fraction: 5 2/3, 14 6/9.
3 Do not space before or after the / in a fraction: 2/3, 7/8.

Learn $ (dollar sign) Reach up with the left index finger.

4 f F $f $F fF $$ f$f F$F $4 $4 for $4 Shift for $ and key $4.
5 A period separates dollars and cents: $4.50, $6.25, $19.50.
6 I earned $33.50 on Mon., $23.80 on Tues., and $44.90 on Wed.

Learn ! (exclamation point) Reach up with the left little finger.

7 a A !a !A aA !! a!a A!A 1! 1! I am excited! I won the game!
8 On your mark! Get ready! Get set! Go! Go faster! I won!
9 Great! You made the team! Hurry up! I am late for school!

CHAPTER 4   Lesson 4-3   115

Figure 12-2
You may need to consult with a manager about task priorities.

At times you may need to discuss your priorities with coworkers or a supervisor to be certain that you agree on the order for doing tasks. Once you set your priorities, finish the tasks in priority order. Remain flexible, however, about revising your priorities as circumstances change.

Control Large Projects

Sometimes, getting started on a large project is difficult even though it may be very important. Smaller tasks can be checked off your tasks list with ease; a large task may seem overwhelming. Do not let the size of a project keep you from getting organized and completing the task. Follow these suggestions for handling a large project:

- Break the large project into smaller tasks.
- Determine the steps to be taken in each of the smaller tasks.
- Establish deadlines for each section or smaller task and meet those deadlines.
- Look for ways to improve your procedures and simplify the completion of the project.
- If the large project is one that will be repeated periodically, record procedures and suggestions you want to follow in the future for improvements.

REMINDER SYSTEMS

You may need to keep track of appointments, meetings, travel dates, and deadlines related to your job. Perhaps the most widely used device for keeping track of such items is a calendar. You can use a paper calendar or calendar software.

4-3C · 12'
Learn %, <, and >

Key each line twice SS (slowly, then faster); DS between 2-line groups.

SPACING CUE

Do not space between a figure and the % sign.

The % is the shift of 5. Strike it with the left index finger.

The < is the shift of , and is controlled by the right middle finger.

The > is the shift of . and is controlled by the right ring finger.

Learn % (percent sign) Reach up with the left index finger.
1 f F %f fF % % f%F f%F 5%f 5%f Shift for the % in 5% and 15%.
2 Do not space between a number and %: 5%, 75%, 85%, and 95%.
3 Prices fell 10% on May 1, 15% on June 1, and 20% on July 15.

Learn < ("less than" sign) Reach down with the right middle finger.
4 k K <k <K kK << k<K K<K <, <, <k, <k, <K< 10 < 18; 95 , 120.
5 If a < b, and c < d, and e < f, and a < c and e, then a < d.

Learn > ("greater than" sign) Reach down with the right ring finger.
6 l L >l >L lL >> l>L L>L >. >. >l. >l. >L> 20 > 17; 105 > 98.
7 If b > a, and d > c, and f > e, and c and e > a, then f > a.

4-3D · 10'
Skill Building: Symbols

Key each line twice SS (slowly, then faster); DS between 2-line groups.

Combine /, $, and !
1 Only 2/3 of the class remembered to bring the $5 on Tuesday!
2 I was really excited! I received 1/2 of the $50 door prize!
3 Only 1/10 of the sellers earned more than $100! I felt bad!

Combine % and < >
4 Only 25% of the students got the answer to 5x > 10 but < 20.
5 Yes, 90% of the students scored > 75%, and 10% scored < 75%.
6 Only about 15% of the class understood the < and > concepts!

4-3E · 8'
Speed Building

1. Key three 1' writings on the ¶; determine *gwam* on each writing.
2. Key two 2' writings on the ¶; determine *gwam*.

AVG all letters used   *gwam* 2'

When you key copy that contains both words and numbers,   6
it is best to key numbers using the top row. When the copy   12
consists primarily of figures, however, it may be faster to   18
use the keypad. In any event, keying figures quickly is a   24
major skill to prize. You can expect to key figures often   29
in the future, so learn to key them with very little peeking.   36

gwam 2' | 1 | 2 | 3 | 4 | 5 | 6 |

Figure 12-1
Socialize with coworkers at appropriate times such as lunch or during breaks.

MANAGING YOUR WORK

Using time efficiently requires developing an organized approach to your work. Calendars and time management systems can help you plan your work activities. Planning your daily work activities will help you avoid forgetting tasks that need to be completed. Take five or ten minutes either at the beginning or the close of the workday to plan the coming day's work.

Prepare a tasks list and complete the tasks according to their order of importance or to meet deadlines. Your tasks list can be a simple handwritten or keyed list. If available, you can use personal information management software such as *Microsoft Outlook* to manage appointments and schedule tasks.

Set Priorities

Once you have identified tasks for the day, determine the priority for each task. **Priority** means relative importance. Rank the tasks on your list and complete the most important ones first. To determine the priority of the tasks, ask yourself these questions:

- How much time will the task require?

- By what date (time) should the task be completed?

- Are others involved in completing the task?

- What will happen if this task is not completed on time?

- Do I have all of the information (or materials) I need to complete the task?

LESSON 4-4

SYMBOL KEYS (#, &, +, @, AND ( ))

OBJECTIVES *In this lesson you will:*

1. Learn control of **#**, **&**, **+**, **@**, and **( )**.
2. Combine **#**, **&**, **+**, **@**, and **( )** with other keys.

4-4A • 5'
Conditioning Practice

Key each line twice SS; then key a 1' writing on line 3; determine *gwam*.

alphabet 1 Zack Gappow saved the job requirement list for the six boys.
figures 2 Jay Par's address is 3856 Ash Place, Houston, TX 77007-2491.
easy 3 I may visit the big chapel in the dismal town on the island.

gwam 1' | 1 | 2 | 3 | 4 | 5 | 6 | 7 | 8 | 9 | 10 | 11 | 12 |

4-4B • 15'
Learn #, &, and +

Key each line twice SS (slowly, then faster); DS between 2-line groups.

The # is the shift of 3. The left middle finger controls it.

The & is the shift of 7. Control it with the right index finger.

The + is to the right of the hyphen. Depress the left shift; strike + with the right little finger.

SPACING CUE

- Do not space between # and a figure.
- Space once before and after & used to join names.

Learn # (number/pounds) Reach up with the left middle finger.

1 d d #d #d dd ## d#d d#d 3# 3# Shift for # as you enter #33d.
2 Do not space between a number and #: 3# of #633 at $9.35/#.
3 Jerry recorded Check #38 as #39, #39 as #40, and #40 as #41.

Learn & (ampersand) Reach up with the right index finger.

4 j j &j &j jj && j&j j&j 7& 7& Have you written to Poe & Son?
5 Do not space before or after & in initials, e.g., CG&E, B&O.
6 She will interview with Johnson & Smith and Jones & Beckett.

Learn + ("plus" sign) Reach up with the right little finger.

7 ; + ; + ;+; ;+; +;+ +;+ 7 + 7, a + b + c < a + b + d, 12 + 3
8 If you add 3 + 4 + 5 + 6 + 7, you will get 25 for an answer.
9 If you add 2 + 3 + 4 + 5 + 6, you will get 20 for an answer.

CHAPTER 4   Lesson 4-4

LESSON 12-1   TIME MANAGEMENT

OBJECTIVES   *In this lesson you will:*

1. Learn the importance of time management.
2. Learn effective time management techniques.
3. Prioritize tasks and create a tasks list.

Time management means planning to gain control over how time is spent. Time management is important for success on the job. Managing your time is a process of choosing the most effective way to do your work. You will want to learn how to eliminate time-wasters and handle time work duties efficiently. Analyzing how you spend your time will increase your effectiveness in managing your work. One of the first steps in learning how to use your time is to recognize how it can be wasted.

COMMON TIME-WASTERS

Not all time spent at work is productive. You can waste time without realizing it. Try to eliminate these common time-wasters:

- **Unnecessary telephone conversations.** The telephone can be either a time-saver or a time-waster, depending on how you use it. Limit telephone conversations to work topics. Do not be tempted to discuss other topics or personal concerns during work hours.

- **Frequent interruptions.** Unplanned visits or questions from coworkers, phone calls, and delays in receiving material from others are common interruptions. Try to eliminate interruptions that waste time. Remember, however, that working with coworkers and customers is an important part of many jobs. Not all interruptions are time-wasters.

- **Excessive socializing.** Some socializing will help you maintain good working relations with your coworkers. Too much socializing, however, is misuse of company time. Be careful to limit socializing to your lunch and breaks or other appropriate times.

- **Ineffective communications.** Inaccurate or incomplete messages often require follow-up calls or e-mails that waste time. Be certain the information you give others is specific and accurate. Ask for feedback to be sure that you understand any instructions or information you receive.

- **Disorganization.** Being disorganized can be a major time-waster. Searching for papers, forgetting important deadlines, and shifting unnecessarily from one project to another are all signs of a disorganized person. Take the time to organize your work area and prepare a daily plan for your work. Think through and plan large jobs before starting them.

4-4C • 15'
Learn @, (, and)

Key each line twice SS (slowly, then faster); DS between 2-line groups.

Note:
Use the letter l in lines 4 and 5.

The @ is the shift of 2. Control it with the left ring finger.
 The (is the shift of 9 and is controlled by the right ring finger.
 The) is the shift of 0; use the right little finger to control it.

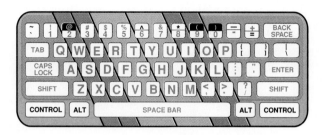

SPACING CUE
Do not space between a left or right parenthesis and the copy enclosed.

Learn @ ("at" sign) Reach up with the left ring finger.
1 s s @s @s ss @@ s@ s@ @ @ The @ is used in e-mail addresses.
2 Change my e-mail address from myers@cs.com to myers@aol.com.
3 I bought 50 shares of F @ $53 1/8 and 100 of USB @ $58 7/16.

Learn ((left parenthesis) Reach up with the right ring finger.
4 l l (l (l ll ((l(l l(l 9(9(Shift for the (as you key (9.
5 As (is the shift of 9, use the l finger to key 9, (, or (9.

Learn) (right parenthesis) Reach up with the right little finger.
6 ; ;);); ;;)) ;); ;); 0) 0) Shift for the) as you key 0).
7 As) is the shift of 0, use the ; finger to key 0,), or 0).

Combine (and)
8 Hints: (1) depress shift; (2) strike key; (3) release both.
9 Tab steps: (1) clear tabs, (2) set stops, and (3) tabulate.

LESSON 4-5 SYMBOL KEYS (=, _, *, \, AND [ ])

OBJECTIVES In this lesson you will:

1. Learn control of =, _, \, *, and [ ].
2. Combine =, _, \, *, and [ ] with other keys.

4-5A • 5'
Conditioning Practice

Key each line twice SS; then key a 1' writing on line 3; determine *gwam*.

alphabet 1 Bobby Klun awarded Jayme sixth place for her very high quiz.
figures   2 The rate on May 14 was 12.57 percent; it was 8.96 on May 30.
easy   3 The haughty man was kept busy with a problem with the docks.
gwam 1' | 1 | 2 | 3 | 4 | 5 | 6 | 7 | 8 | 9 | 10 | 11 | 12 |

CHAPTER 12

Enhancing Workplace Performance

OBJECTIVES *In this chapter you will:*

1. Learn effective time management techniques.
2. Prioritize tasks and create a tasks list.
3. Apply procedures to handle incoming telephone calls effectively.
4. Learn procedures for planning and placing calls.
5. Learn techniques for controlling telephone costs.
6. Learn the purpose of a records management system.
7. Identify the phases of the record life cycle.
8. Describe storage media, equipment, and supplies for records management.
9. Learn and apply basic filing procedures.
10. Learn guidelines for using office copiers effectively.

Your performance in the workplace can be improved greatly by managing time, records, and telephone communications effectively. Planning your work and eliminating time-wasters will give you more time to concentrate on work tasks. Understanding how your company stores its records (information) and how to find information quickly will help you complete tasks or answer customer questions more quickly. Using the telephone effectively will help you communicate information to other employees and make a good impression with customers. In this chapter you will learn time management techniques, how to use the telephone effectively in a business setting, and basic records management procedures.

4-5B • 15'
Learn =, _, and \

Key each line twice SS (slowly, then faster); DS between 2-line groups.

The = is the same key as + and is controlled by the right little finger.

The _ is the shift of the - and is controlled by the right little finger.

The \ is above ENTER. Use the right little finger to control it.

Learn = (equals sign) Reach up with the right little finger.

1 ; ; =; =; ;; == ;= ;= += += The = is used in math equations.
2 Solve the following: 3a = 15, 5b = 30, 3c = 9, and 2d = 16.
3 If a = b + c and c = 5 and a = 9, can you determine what b=?

Learn _ (underline) Reach up with the right little finger.

4 ; ; _; _; ;; __ ;_; ;_; -_ -_ Shift for the _ as you key _-.
5 The _ is used in some Internet locations, e.g., http_data_2.
6 My property has ____ parking spaces and ____ storage bins.

Learn \ (backslash) Reach up with the right little finger.

7 ;; \; \; ;; \\ \;\ \;\ \;\; \;\; Do not shift for the \ key.
8 Use the \ key to map the drive to access \\sps25\deptdir556.
9 Map the drive to \\global128\coxjg$, not \\global127\coxjg$.

4-5C • 15'
Learn *, [, and]

Key each line twice SS (slowly, then faster); DS between 2-line groups.

The * is the shift of 8. Control it with the right middle finger.

The [is to the right of p. Strike it with the right little finger.

The] is to the right of [and also is controlled by the right little finger.

Learn * (asterisk) Reach up with the right middle finger.

1 k k *k *k *k* *k* * She used the * for a single source note.
2 Put an * before (*Gary, *Jan, and *Jay) to show high scores.
3 Asterisks (*) can be used to replace unprintable words ****.

Learn [(left bracket) Reach up with the right little finger.

4 ; ; [; [; [;[[;[[[[[[a [B [c [D [e [F [g [H [i [J [k [L.
5 [m [N [o [P [q [R [s [T [u [V [w [X [y [Z [1 [2 [3 [4 [5 [6.

Learn] (right bracket) Reach up with the right little finger.

6 ; ;];];];]];]]]]] A] b] C] d] E] f] G] h] I] j]]K]l.
7 M] n] O] p] Q] r] S] t] U] v] W] x] Y] z] 7] 8] 9] 10] 11]].

Combine [and]

8 Brackets ([]) are used in algebra: x = [5(a+b)] - [2(d-e)].
9 Use [ ] within quotations to indicate alterations [changes].

Changing and creating pages is lots of fun with *FrontPage* or *Dreamweaver*. *Dreamweaver* has a special tool called the Properties Inspector. This tool palette allows you to quickly make changes to the color scheme. *FrontPage* uses the Standard and Formatting toolbars to change page properties.

Professionals use applications like *Dreamweaver* and *FrontPage* because these programs allow Web designers to view and manage large Web sites. As a Web site grows, it becomes increasingly difficult to keep track of pages and hyperlinks. Web design software helps link pages together and detects broken and missing links. If you want to create and manage large Web sites or just learn an alternate method of creating Web pages, you may want to learn to use an advanced Web design program.

ACTIVITY 11-12 RESEARCH WEB DESIGN PROGRAMS

1. Access the Internet. Use a search engine to find articles or other information about Web design software. If you are interested in a particular program, visit the Web site for the company that sells the program. For instance, visit the Microsoft site (www.microsoft.com) to learn more about *FrontPage*. Visit Macromedia online (www.macromedia.com) to learn more about *Dreamweaver*.

2. Sometimes the company Web sites for Web design software have demonstration or trial versions of the programs available for download. With your teacher's permission, download a demo version and explore the software.

DIGITOOLS
DIGITAL WORKBOOK | **CHAPTER 11**

Open the data file *CH11 Workbook*. Complete these exercises in your *DigiTools Digital Workbook* to reinforce and extend your learning for Chapter 11:

- Review Questions
- Vocabulary Reinforcement
- Math Practice: Calculating Budget Amounts
- From the Editor's Desk: Semicolons and Colons
- Keyboarding Practice: Finger Reaches and Speed Building

LESSON 4-6

NUMERIC KEYPAD KEYS 4/5/6/0

OBJECTIVES — *In this lesson you will:*

1. Learn key techniques for **4**, **5**, **6**, and **0**.
2. Key these home-key numbers with speed and ease.

4-6A • 5'
Numeric Keypad Operating Position

1. Position yourself in front of the keyboard—body erect, both feet on floor.
2. Place this book for easy reading—at right of keyboard or directly behind it.

Input copy at right of (or behind) keypad

Proper position at keypad

4-6B • 5'
Home-Key Position

Curve the fingers of the right hand and place them on the keypad:

- index finger on 4
- middle finger on 5
- ring finger on 6
- thumb on 0

Note:
To use the keypad, the Num (number) Lock must be activated.

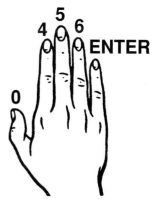

CHAPTER 4   Lesson 4-6   120

FrontPage is part of the professional-level *Microsoft Office* suite. Like other applications, it allows you to view your HTML tags, which are often called code.

Figure 11-18
Viewing the Tags in *FrontPage*

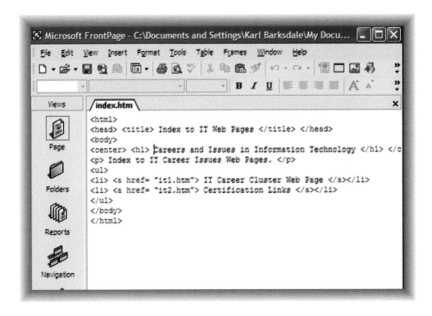

You can also view a Web page as it will look online. This is called the "What You See Is What You Get" or **WYSIWYG** view. Both *FrontPage* and *Dreamweaver* allow users to view pages as they will appear online.

Figure 11-19
With *Dreamweaver*, users can view HTML tags and see the page as it will appear online.

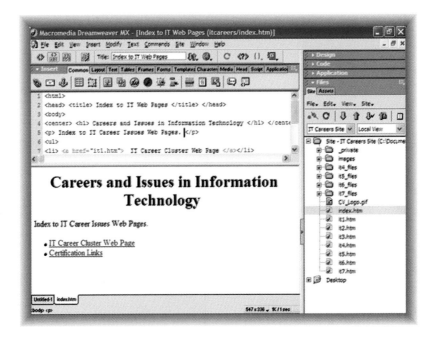

CHAPTER 11   Lesson 11-6   347

4-6C • 40'
New Keys: 4, 5, 6, and 0 (Home Keys)

Use the calculator accessory on your computer to complete the drills at the right.

1. Curve the fingers of your right hand; place them upright on the keypad home keys:
   - index finger on 4
   - middle finger on 5
   - ring finger on 6
   - thumb on 0
2. Key/enter each number: Key the number and enter by pressing the + key with the little finger of the right hand.
3. After entering the last number to be added, press ENTER to get the answer. Verify your answer with the answer shown in blue.
4. Press ESC on the main keyboard to clear the calculator; then enter numbers in the next column.
5. Repeat Steps 2–4 for Drills 1–6 to increase your input rate.

TECHNIQUE CUE

Strike each key with a quick, sharp stroke with the *tip* of the finger; release the key quickly. Keep the fingers curved and upright, the wrist low, relaxed, and steady.
  Strike *0* with the side of the right thumb, similar to the way you strike the Space Bar.

Drill 1

A	B	C	D	E	F
4	5	6	4	5	6
+4	+5	+6	+4	+5	+6
8	10	12	8	10	12

Drill 2

A	B	C	D	E	F
44	55	66	44	55	66
+44	+55	+66	+44	+55	+66
88	110	132	88	110	132

Drill 3

A	B	C	D	E	F
44	45	54	44	55	66
+55	+56	+46	+45	+54	+65
+66	+64	+65	+46	+56	+64
165	165	165	135	165	195

Drill 4

A	B	C	D	E	F
40	50	60	400	500	600
+50	+60	+40	+506	+604	+405
+60	+40	+50	+650	+460	+504
150	150	150	1,556	1,564	1,509

Drill 5

A	B	C	D	E	F
45	404	404	406	450	650
+55	+405	+505	+506	+540	+560
+65	+406	+606	+606	+405	+605
165	1,215	1,515	1,518	1,395	1,815

Drill 6

A	B	C	D	E	F
40	606	444	554	646	456
+50	+505	+445	+555	+656	+654
+60	+404	+446	+556	+666	+504
150	1,515	1,335	1,665	1,968	1,614

2. Create hyperlinks to your *it4.htm*, *it5.htm*, and *it6.htm* pages. This will make these new pages part of your Web site. Enter the following tags below the existing <li> <a href="it2.htm"> Certification Links </a></li> anchor tag.

```
<li> <a href="it4.htm"> Computer Safety and Ergonomics Report </a></li>
<li> <a href="it5.htm"> Injury Reporting (PowerPoint) </a></li>
<li><a href="it6.htm"> RSI Survey Chart </a></li>
```

3. Save the file using the same name. Open the file in your browser. Test the links you have created to your other pages. Close the browser and *Notepad*.

APPLICATION 11-4 — ALTERNATE NAVIGATION LINKS

HELP KEYWORDS

In *Word*
Hyperlink
  Create a hyperlink

1. Use *Notepad* or *Word* and your HTML skills to create an alternate navigation system for your IT Careers Web site. Create text links at the bottom of all of the Web pages in your IT Careers Web site to allow random access navigation.

2. Save each page using the same name. View the pages and test the links in your browser.

APPLICATION 11-5 — CONSISTENT COLOR SCHEME

1. Use *Notepad* or *Word* and your HTML skills to make all of the colors on all of the Web pages in your IT Careers Web site the same. This will give your site a more constant presentation design color scheme.

2. Save each page using the same name. View the pages and test the links in your browser.

LESSON 11-6 — WEB DESIGN APPLICATION PROGRAMS

OBJECTIVES — *In this lesson you will:*

1. Learn about Web design software tools.
2. Research Web design tools online.

Several Web design software tools are on the market. Some of the most popular include *Macromedia Dreamweaver*, *Adobe GoLive*, and *Microsoft FrontPage*. These specially designed applications eliminate the time-consuming task of entering HTML tags. Each is a powerful Web page creation and site management tool.

LESSON 4-7: NUMERIC KEYPAD KEYS 7/8/9

OBJECTIVES *In this lesson you will:*

1. Learn reachstrokes for **7**, **8**, and **9**.
2. Combine the new keys with other keys learned.

4-7A • 5'
Home-Key Review

Review the home keys by calculating totals for the problems at the right.

	A	B	C	D	E	F
	4	44	400	404	440	450
	+5	+55	+500	+505	+550	+560
	+6	+66	+600	+606	+660	+456
	15	165	1,500	1,515	1,650	1,466

4-7B • 45'
New Keys: 7, 8, and 9

Learn reach to 7

1. Locate 7 (above 4) on the numeric keypad.
2. Watch your index finger move up to 7 and back to 4 a few times without striking keys.
3. Practice striking *74* a few times as you watch the finger.
4. With eyes on copy, key/enter the data in Drills 1A and 1B. Do not worry about totals.

Learn reach to 8

1. Learn the middle-finger reach to 8 (above 5) as directed in Steps 1–3 above.
2. With eyes on copy, enter the data in Drills 1C and 1D.

Learn reach to 9

1. Learn the ring-finger reach to 9 (above 6) as directed above.
2. With eyes on copy, enter the data in Drills 1E and 1F.

Drills 2–5

1. Calculate the totals for each problem in Drills 2–5. Check your answers with the problem totals shown.
2. Repeat Drills 2–5 to increase your input speed.

		A	B	C	D	E	F
Drill 1		474	747	585	858	696	969
		+747	+477	+858	+588	+969	+966
		+777	+474	+888	+585	+999	+696
		1,998	1,698	2,331	2,031	2,664	2,631
Drill 2		774	885	996	745	475	754
		+474	+585	+696	+854	+584	+846
		+747	+858	+969	+965	+695	+956
		1,995	2,328	2,661	2,564	1,754	2,556
Drill 3		470	580	690	770	707	407
		+740	+850	+960	+880	+808	+508
		+705	+805	+906	+990	+909	+609
		1,915	2,235	2,556	2,640	2,424	1,524
Drill 4		456	407	508	609	804	905
		+789	+408	+509	+704	+805	+906
		+654	+409	+607	+705	+806	+907
		1,899	1,224	1,624	2,018	2,415	2,718
Drill 5		8	69	4	804	76	86
		+795	+575	+705	+45	+556	+564
		+ 60	+ 4	+ 59	+ 6	+ 5	+504
		863	648	768	855	637	1,154

Figure 11-17
Worksheet and Graph as a Web Page

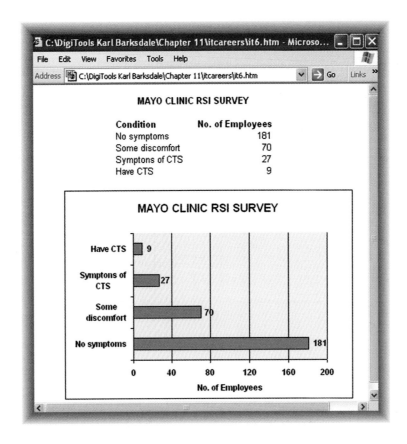

Peer Check Work with a classmate. View your Web page and your classmate's Web page online. Discuss how both pages might be improved. Make changes to your page after your discussion.

LINKING PAGES

Keeping track of pages and linking them properly as a Web site is created is very important. One of the main reasons professionals use applications like *Dreamweaver* and *FrontPage* is their ability to view and manage an entire Web site. As a Web site grows, it becomes increasingly difficult to keep track of all the hyperlinks. The software helps link pages together and detects broken and missing links. Because your Web site has only a few pages, you will be able to link them by entering HTML tags.

ACTIVITY 11-11 — LINK NEW PAGES TO A WEB SITE

1. In *Notepad*, open the file *index.htm* that you created earlier (found in the *DigiTools your name\Chapter 11\itcareers* folder).

LESSON 4-8

NUMERIC KEYPAD KEYS 1/2/3

OBJECTIVES — In this lesson you will:

1. Learn reachstrokes for **1**, **2**, and **3**.
2. Combine the new keys with other keys learned.

4-8A • 5'
Keypad Review

Review the keypad by calculating the totals for the problems at the right.

	A	B	C	D	E	F	G
	45	74	740	996	704	990	477
	+56	+85	+850	+885	+805	+880	+588
	+67	+96	+960	+774	+906	+770	+699
	168	255	2,550	2,655	2,415	2,640	1,764

4-8B • 45'
New Keys: 1, 2, and 3

Learn reach to 1
1. Locate 1 (below 4) on the numeric keypad.
2. Watch your index finger move down to 1 and back to 4 a few times without striking keys.
3. Practice striking *14* a few times as you watch the finger.
4. With eyes on copy, enter the data in Drills 1A and 1B. Do not worry about totals.

Learn reach to 2
1. Learn the middle-finger reach to 2 (below 5) as in Steps 1–3 above.
2. With eyes on copy, enter data in Drills 1C and 1D.

Learn reach to 3
1. Learn the ring-finger reach to 3 (below 6) as directed above.
2. With eyes on copy, enter data in Drills 1E–1G.

Drills 2–4
Calculate totals for each problem and check your answers.

Learn reach to . (decimal point)
1. Learn the ring-finger reach to the decimal point (.) located below the 3.
2. With eyes on copy, calculate the totals for each problem in Drill 5.
3. Repeat Drills 2–5 to increase your input speed.

	A	B	C	D	E	F	G
Drill 1	144	114	525	252	363	636	120
	+141	+414	+252	+552	+363	+366	+285
	+414	+141	+225	+525	+336	+636	+396
	699	669	1,002	1,329	1,062	1,638	801
Drill 2	411	552	663	571	514	481	963
	+144	+255	+366	+482	+425	+672	+852
	+414	+525	+636	+539	+563	+953	+471
	969	1,332	1,665	1,592	1,502	2,106	2,286
Drill 3	471	582	693	303	939	396	417
	+ 41	+802	+963	+220	+822	+285	+508
	+ 14	+825	+936	+101	+717	+174	+639
	526	2,209	2,592	624	2,478	855	1,564
Drill 4	75	128	167	102	853	549	180
	+189	+ 34	+258	+368	+264	+367	+475
	+ 3	+591	+349	+549	+971	+102	+396
	267	753	774	1,019	2,088	1,018	1,051
Drill 5	1.30	2.58	23.87	90.37	16.89	47.01	59.28
	+4.17	+6.90	+14.65	+ 4.25	+ 3.25	+28.36	+ 1.76
	5.47	9.48	38.52	94.62	20.14	75.37	61.04

APPLICATION 11-3 RSI SURVEY WEB PAGE

Excel spreadsheets can be converted into Web pages. First, of course, you must enter data and create the spreadsheet. In this application, you will create a spreadsheet and a bar chart that will reflect the impact repetitive stress injuries have had on a group of employees. Then you will convert the spreadsheet into a Web page.

Consider your parameters for this writing job:

- **Audience:** Other company employees
- **Purpose:** Show RSI data in a chart
- **Parameters and length:** The data and chart should fit on one page.
- **General to specific:** This page contains only specific data.
- **Simplicity:** Format the chart so it is easy to understand.
- **Personality and style:** Use a formal bar chart
- **Professional delivery:** Make your Web page attractive to the reader and easy to use and read.
- **Accessibility:** You will link this page to others in your IT Web site in a later activity.

1. In *Word*, open and read the data file *RSI*.

2. Open *Excel*. Enter MAYO CLINIC RSI SURVEY as the worksheet title. In the worksheet, include the number of employees who:

- Reported no RSI symptoms
- Reported some discomfort or loss of hand function
- Reported symptoms associated with carpal tunnel syndrome
- Met the clinical definition of carpal tunnel syndrome

3. Use the chart feature to create a bar chart that compares these numbers. Format and label the chart to make the data easy to understand. Place the data and the chart on the page in an attractive format.

4. Save the file at *it6* in your *DigiTools your name\Chapter 11* folder. Now save the spreadsheet as an HTML file. (Choose **File, Save As Web Page**.) Save your file as *it6* in your *DigiTools your name\Chapter 11\itcareers* folder. (The .htm will be added automatically.)

5. View your page by opening it in your Web browser. The page might look similar to Figure 11-17 on page 345. Close *Excel* and your browser.

LESSON 4-9

SUBTRACTION AND MULTIPLICATION

OBJECTIVES — *In this lesson you will:*

1. Learn subtraction on numeric keypad.
2. Learn multiplication on numeric keypad.

4-9A · 5'
Keypad Review

Review the keypad by calculating the totals for the problems at the right.

	A	B	C	D	E	F
	17	49	672	513	371	109
	+83	+60	+415	+724	+564	+357
	+52	+93	+808	+690	+289	+620
	152	202	1,895	1,927	1,224	1,086

4-9B · 25'
Subtraction

Learn reach to ⊟ (minus key)

1. Locate – (minus key) on the numeric keypad.
2. Watch your little finger move up to the – (minus key) and back to the + (plus key) a few times without striking keys.
3. Practice striking – (minus key) a few times as you watch the finger.
4. With eyes on copy, key/enter the data in Drills 1–3.
5. Verify your answer with the answer shown in blue.
6. Pres ESC on the main keyboard to clear the calculator; then enter numbers in the next column.

		A	B	C	D	E	F
Drill 1		27	50	893	798	523	401
		−14	−26	−406	−235	−178	−300
		13	24	487	563	345	101
Drill 2		A	B	C	D	E	F
		84	56	996	829	759	83.6
		−17	−38	−476	−514	−420	−41.5
		67	18	520	315	339	42.1
Drill 3		A	B	C	D	E	F
		99	89	505	807	978	63.4
		−16	−10	−264	−234	−220	+37.5
		−23	− 8	− 45	− 65	+461	− 8.9
		−33	−17	− 87	−104	+309	−46.5
		− 9	−24	−156	− 57	−218	+70.1
		18	30	−47	347	1,310	115.6

4-9C · 20'
Multiplication

Learn reach to ✴ (multiplication key)

1. Locate * (multiplication key) on the numeric keypad.
2. Watch your ring finger move up to the * (multiplication key) and back to the 6 key a few times without striking keys.
3. Practice striking * (multiplication key) a few times as you watch the finger.
4. With eyes on copy, key/enter the data in Drills 1–3.
5. Verify your answer with the answer shown in blue.

		A	B	C	D	E
Drill 1		28	54	43	145	68.8
		×13	×60	×89	×271	×19.3
		364	3,240	3,827	39,295	1,327.84
Drill 2		A	B	C	D	E
		603	109	837	468	219
		× 24	× 72	× 55	× 90	× 34
		14,472	7,848	46,035	42,120	7,446

Drill 3

- **A** $3 \times 5 \times 6 = 90$
- **B** $8 \times 7 \times 2 = 112$
- **C** $2 \times 9 \times 4 = 72$
- **D** $4 \times 10 \times 3 = 120$
- **E** $67 \times 13 + 89 = 960$
- **F** $7 \times 70 - 34 = 456$

> **DigiTip**
>
> When you convert your *PowerPoint* slide show to a set of Web pages, some animations and transitions will be lost. However, you should still consider using them, because you may choose to present your slide show in meetings.

> **DigiTip**
>
> As you convert your *PowerPoint* show into HTML files, *PowerPoint* will automatically create a new folder in which to save all of the graphics and design elements. In the case of this *it5.htm* file, the folder will be called *it5_files*.

pages) will allow the reader to see the key points easily and quickly. Placing too many words on any one slide will crowd the slides and make the information less accessible.

Express your personality and style by your choice of words and in the graphics you use to accent your slide. Prepare your slide show for a professional delivery. To review information about creating professional-looking slides, refer to the instructions on slide show design found under the Corporate View intranet. (From the Corporate View Intranet Home page, choose, **Corporate Communications, Style Guide, Presentations**.) Now that you know the general parameters of this writing task, continue with the step-by-step instructions.

1. Access the Corporate View intranet. To review the injury reporting rules, select the **Regular Features** link from the Corporate View Intranet Home page. Choose **Employee Handbook**. Scroll down and select the **Work-Related Injury** link. Read the information on this page.

2. Create a *PowerPoint* show detailing the steps employees should follow in case of workplace injury. Prepare a different *PowerPoint* slide for each step.

3. Save the file at *it5* in your *DigiTools your name\Chapter 11* folder. Now turn your *PowerPoint* document into HTML pages by choosing **File**, **Save As Web Page**. Save your file as *it5* in your *DigiTools your name\Chapter 11\itcareers* folder. (The .htm extension will be added automatically.)

4. View your HTML pages by opening them in your Web browser. (Choose **Web Page Preview** from the File menu.) Make changes to your slides, if needed, in *PowerPoint* and save them again. Your slide show will display in the browser with links at the side and buttons at the bottom for navigation as shown in Figure 11-16.

Figure 11-16
PowerPoint Show as Web Pages

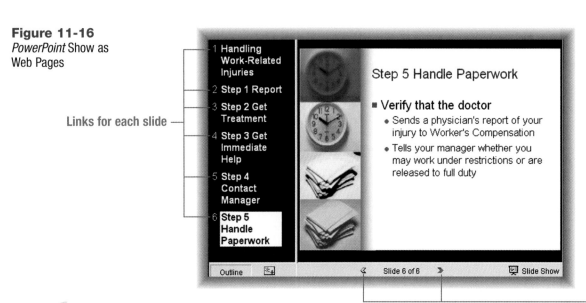

Peer Check Work with a classmate. View your slides and your classmate's slides online. Discuss how both pages might be improved. Make changes to your slides after your discussion.

LESSON 4-10

DIVISION AND MATH CALCULATIONS

OBJECTIVES *In this lesson you will:*

1. Learn division on numeric keypad.
2. Learn to complete math calculations on numeric keypad.

4-10A • 5'
Keypad Review

Review the keypad by calculating the totals for the problems at the right.

A	B	C	D	E	F
20	92	872	613		
+65	−43	−115	+716	438	704.9
+39	+20	+178	−690	× 4.8	× 5.03
124	69	935	639	2,102.4	3,545.65

4-10B • 25'
Division
Learn reach to / (division key)

1. Locate / (division key) on the numeric keypad.
2. Watch your middle finger move up to the / (division key) and back to the 5 key a few times without striking keys.
3. Practice striking / (division key) a few times as you watch the finger.
4. With eyes on copy, enter the data in Drills 1–3.
5. Verify your answer with the answer shown in blue.

	A	B	C	D
Drill 1	51.17 6/307	42 10/420	179 5/895	106 7/742
Drill 2	32.25 12/387	75.52 66/4,984	229.13 32/7,332	96.42 52/5,014
Drill 3	44.20 6.9/305	7.94 47.6/378	98.27 12.7/1,248	173.37 31.2/5,409

4-10C • 20'
Math Calculations

Use the numeric keypad to solve the six math problems shown at the right. Compare your answers to those of a classmate.

1. Michael opened a checking account with $100. He wrote checks for $12.88, $15.67, $8.37, and $5.25. He made one deposit of $26.80 and had a service charge of $1.75. What is his current balance?

2. Marci purchased six tickets for the Utah Jazz basketball game. Four of the tickets cost $29.50; the other two cost $35.00. There was a service charge of $1.50 for each ticket. How much did Marci pay for the six tickets?

3. Four friends went out for dinner. The cost of the dinner came to $47.88. They decided to leave a 15 percent tip and to split the cost of the dinner equally among them. How much did each person have to pay?

4. Jay filled his car up with gas. The odometer reading was 45,688 miles. Jay drove to New York to see a Yankees game. When he got there he filled the car up again. It took 15.7 gallons. The odometer now read 45,933. How many miles per gallon did Jay get on the trip to New York?

(continued on next page)

WHAT YOU SHOULD KNOW ABOUT
Work-Related Injuries

People are injured on the job from time to time. Although the overall number of work-related injuries has declined steadily over the past four decades, injuries still happen. Protecting your health and safety on the job should be an important goal for both you and your employer.

Knowing your employer's policy related to workplace injuries is important. You should report an injury at work to your supervisor as soon as it happens. If the injury is serious, go immediately to the nearest medical facility for treatment. If the injury is life-threatening, call for an ambulance or have someone drive you to the nearest emergency room.

After the emergency has passed, you will need to do some paperwork to protect your rights under the Worker's Compensation regulations. Follow the advice of your physician and physical therapist. Keep accurate records of all your medical visits and what you are instructed to do by your healthcare providers. When your physician says you are able, you can return to work.

DigiTip
Worker's Compensation is a government-mandated insurance program paid for by employers and workers. Funds are used to help workers who suffer injury or disease related to their job. To learn more, visit the U.S. Department of Labor Web site at www.dol.gov.

CONVERTING OTHER FILES INTO WEB PAGES

You can convert *PowerPoint* or *Excel* files into Web pages as easily as you converted a *Word* document. Simply use the Save As Web Page command as you do to convert *Word* documents.

The same eight technical writing principles that you applied to your Web page on computer safety also apply to creating Web pages from *PowerPoint* files. *PowerPoint* shows, by definition, are linear structures. Your audience will move step-by-step through a series of slides to learn more about a topic.

APPLICATION 11-2 — WORKPLACE INJURY WEB PAGE

In this application, you will create a *PowerPoint* show and convert it to Web pages. Your audience will be other company employees. Your purpose will be to explain how to report workplace injuries to an employer. In terms of parameters and length, your slide show should contain six to nine slides. Write your slide show to move from the general topic to specific steps. Your first slide should introduce the major topic or theme of the slide show. The remaining slides should list the specific steps employees must follow to report an injury according to the rules provided by the Human Resources Department of the company.

Remember to keep your slides simple and easy to understand. Make each slide accessible to the reader. This means making slides that (after they are converted to Web

5. Mary bowled six games this week. Her scores were 138, 151, 198, 147, 156, and 173. What was her average for those six games?

6. There are 800 points available in the Chemistry class. Roberto wants an A in the class. To get an A, he needs to achieve 95 percent or better. What is the minimum number of points he will need to earn the A?

4-10D • 15'
Speed Building

1. Key three 1' writings on the ¶; determine *gwam* on each writing.
2. Key two 2' writings on the ¶s; determine *gwam*.

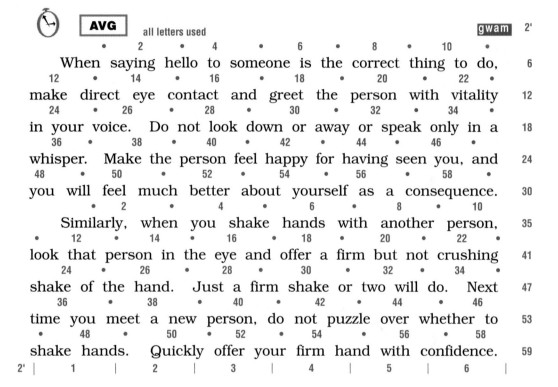

When saying hello to someone is the correct thing to do, make direct eye contact and greet the person with vitality in your voice. Do not look down or away or speak only in a whisper. Make the person feel happy for having seen you, and you will feel much better about yourself as a consequence.

Similarly, when you shake hands with another person, look that person in the eye and offer a firm but not crushing shake of the hand. Just a firm shake or two will do. Next time you meet a new person, do not puzzle over whether to shake hands. Quickly offer your firm hand with confidence.

DIGITOOLS
DIGITAL WORKBOOK — CHAPTER 4

Open the data file *CH04 Workbook*. Complete these exercises in your *DigiTools Digital Workbook* to reinforce and extend your learning for Chapter 4:

- Review Questions
- Symbol Review
- From the Editor's Desk: Numbers in Text
- Math Practice: Calculating Mark-up and Mark-down Amounts
- Keyboarding: Tab Technique

CHAPTER 4   Lesson 4-10

DigiTip

When people open Web pages, they decide within a few seconds whether or not the information pertains to them. Because of hyperlinking, users can quickly move on to another page. You must catch the reader's interest quickly and show clearly the topic of the Web page in the first few lines of text. Otherwise, the visitor may not continue reading.

2. Access the Internet. Use a search engine to find additional information on the topic. Take notes on what you find, keeping in mind your purpose and what your audience needs to know about the topic. Keep a record of the sources you use so you can cite and list the references appropriately.

3. Open *Word* and outline the information you found before you begin writing. Remember the general to specific rule. First, note the main points of the information in your outline. Then, enter more specific ideas and statements below the main points as subtopics.

4. Begin writing your Web page in *Word*. Apply the simplicity rule by making your writing easy to understand. Use a main title to help readers identify the topic quickly when they access the page. Use side headings to identify parts of the document as you would in a report. Explain complicated terms and make each statement clear and easy to follow. Use bulleted or numbered lists, if appropriate. Remember the personality and style rule. Web pages should never be boring. Use an interesting writing style. Let your personality come through in your writing. Try to hold your readers' interest, even when writing about a serious topic such as computer injuries.

5. Proofread and edit your document. Make sure you have met your parameters and length requirement of 500-1,000 words. Add or cut as needed, to meet the length requirement.

DigiTip

In *Word*, choose **Word Count** from the Tools menu to see the number of words in your document.

6. Apply the next online writing principle, professional delivery. Make your writing attractive by formatting your document in an appealing and interesting way. You can use a variety of formatting elements as learned in Chapter 5. For example, you can format the title and side headings in larger fonts or different font styles. You may wish to add a graphic to the page to make it more interesting or to illustrate a point.

7. The final technical writing principle to consider is accessibility. In a later lesson, you will link this Web page to your *itcareers* Web site for easy access. For now, convert your *Word* document into an HTML page. Choose **Save As Web Page** from the File menu. Save your file as *it4* in your *DigiTools your name\Chapter 11\itcareers* folder. Remember, in the Save as type box, choose the **Web Page (*.htm, *.html)** option.

DigiTip

A folder called *it4_files* will be created automatically if you use graphics in your Web page.

8. Open and view your Web page in your Web browser. Make adjustments, if needed, and save the file again. Close the browser and *Word*.

Peer Check Work with a classmate. View your Web page and your classmate's page online. Discuss how both pages might be improved. Make changes to your Web page after your discussion.

CHAPTER 5

Word Processing

OBJECTIVES *In this chapter you will:*

1. Create memos, letters, reports, and tables.
2. Edit and format text.
3. Cut, copy, and paste text.
4. Preview and print documents.
5. Select page orientation and alignment options.
6. Check documents for spelling and grammar.
7. Proofread documents and use proofreaders' marks.
8. Create documents with headers.
9. Insert and format graphics in a document.
10. Learn about services and DigiTools that speed mail processing.

In Chapter 1, you learned some basic commands and features of *Microsoft Word*. In this chapter, you will learn many word processing commands that will help you create documents such as memos, letters, and reports. At the same time, you will continue to develop your keyboarding skill.

LESSON 5-1 MEMOS

OBJECTIVES *In this lesson you will:*

1. Learn the purpose and format of a memo.
2. Select, format, and edit text.
3. Apply character formats.
4. Use the Insert Date and Undo commands.
5. Set tabs.
6. Create and print memos.

REVIEW

The features and commands you learned in Chapter 1 are listed below. Review these commands now, if needed, so you will be ready to learn new commands in this lesson.

INFORMATION DESIGN

Information design deals with the creation of a Web site's message. The words you write and the way you organize your text affects how successful your design efforts will be. Consider these eight technical writing principles when writing Web page content.

- **Audience.** Know toward whom the message is directed. (What group of users/readers do you want to reach?)

- **Purpose.** Clearly understand the precise goals and objectives of the message. (What do you want to accomplish with the message?)

- **Parameters and length.** Know what length the document should be and your completion deadlines.

- **General to specific.** Move from general information at the beginning of the Web page to more specific information later in the document.

- **Simplicity.** State your message in a way that is easy to understand.

- **Personality and style.** Project an appropriate mood, tone, flair, voice, or attitude in your writing. (Should the message be serious or fun, formal or informal?)

- **Professional delivery.** Make your Web page attractive to the reader and easy to use and read.

- **Accessibility.** Make the most important information the most accessible or easy to find and read. Also, make information accessible by hyperlinking to it.

APPLICATION 11-1 COMPUTER SAFETY AND ERGONOMICS WEB PAGE

Computer-related injuries have become a difficult problem for many office professional and IT workers. You have been asked to prepare a Web page report on computer safety and ergonomics.

In the preparation of this Web page, you will apply all eight of the information design principles. The starting point for creating any Web page is to research your topic thoroughly. As you research and take notes, keep in mind three of the eight technical writing principles: audience, purpose, parameters and length.

Your purpose is to create a Web page that will educate IT employees about computer safety and ergonomics. In this case, your audience will be employees of the company in which you work. Before you start researching your topic, consider your parameters and length. In this case, your Web page must be between 500 and 1,000 words. Now that you know the general parameters of this writing task, continue with the step-by-step instructions.

1. Research the assigned topic *Computer Safety and Ergonomics*. Refer to these parts of your textbook for information about the topic:

Chapter 4, What You Should Know About... Repetitive Stress Injury, page 108
Appendix B, Using Your Equipment and Media Properly and Safely, page App-11

Command	Page Reference
Start *Word*	Activity 1-4, page 16
Close *Word*	Activity 1-4, page 16
Enter text	Activity 1-5, page 18
Navigate in a document	Activity 1-5, page 18
Save a document	Activity 1-6, page 19
Open a document	Activity 1-7, page 20
Close a document	Activity 1-7, page 20
Preview and print a document	Activity 1-8, page 22
Access Help	Activity 1-9, page 23

MEMO FORMAT

A **memo**, more formally called a memorandum, is an informal document. A memo is most often used to communicate within an organization, such as a company or a club. The memo is often printed on inexpensive paper. The content is usually not confidential. **Confidential** means private or secret.

Many people now use an e-mail to send a short message rather than a memo. A memo is still useful, however, for longer, informal messages.

ACTIVITY 5-1 — RESEARCH MEMO FORMAT

Corporate View provides information about memos and a sample memo format on the intranet. Access the intranet to learn about the proper format for a memo.

1. Open your browser and access the Corporate View intranet. Use the bookmark on your Favorites list to access the Style Guide page. (If you do not have a bookmark, choose **Corporate Communications, Style Guide** from the Corporate View Intranet Home page.) Click the **Memos** link.

2. Read the information about memos and the sample memo.

3. What heading information is included in a memo?

4. What side margins should you use for a memo?

5. How should the paragraphs of a memo be spaced?

6. Why are titles often omitted in the heading information on a memo?

7. When might sending an e-mail rather than a memo be appropriate? When might sending a memo rather than an e-mail be appropriate?

8. Close your browser.

LESSON 11-5

INFORMATION DESIGN FOR WEB PAGES

OBJECTIVES — *In this lesson you will:*

1. Use *Word* to create Web pages.
2. Learn and apply eight technical writing principles for creating Web pages.
3. Convert *Word*, *PowerPoint*, and *Excel* files to Web pages.
4. Link Web pages to create a Web site.

MICROSOFT WORD AND HTML

DigiTip
Many programs have HTML capabilities to make it more convenient to create Web pages. For example, you can turn almost any *Word*, *PowerPoint*, or *Excel* file into an HTML Web page.

Word's **Save as Web Page** command allows users to save documents created in *Word* as Web pages. When a document is saved with this special command, *Word* creates HTML tags in the document. However, *Word* adds a lot of extra tags that you would not use if you entered the tags. Some people do not like using *Word* to create Web pages for this reason. Using *Word* is convenient for many users, however, and can save a lot of time when creating Web pages because the HTML tags do not have to be entered manually. In Activity 11-10, you will use *Word* to create a Web page.

ACTIVITY 11-10 — CONVERT A *WORD* DOCUMENT TO A WEB PAGE

1. Start *Word*. Open the data file *aboutcv*. This document contains information about Corporate View that you will convert to a Web page.

2. Insert a graphic at the top of the page at the left margin. Use the data file *CV_Banner.gif*. (See Lesson 5-5 in Chapter 5 to review inserting and working with graphics.)

DigiTip
When you have graphics in a *Word* document that you convert to HTML, a special folder will be created that will include your graphics and other files that are created automatically. In the case of this *aboutcv.htm* file, the folder will be called *aboutcv_files*.

3. Convert your *Word* document into an HTML page. Choose **Save As Web Page** from the File menu.

4. Create a new subfolder in your *DigiTools your name\Chapter 11* folder named *corpview*. Save your file as *aboutcv* in your *DigiTools your name\Chapter 11\corpview* folder. In the Save as type box choose the **Web Page (*.htm,*.html)** option. (The .htm will be added automatically.)

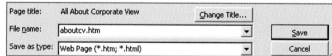

5. To open and view your *aboutcv.htm* document in your Web browser, choose **Web Page Preview** from the File menu. Close your browser. Close the file in *Word*.

LINE SPACING

HELP KEYWORDS

Line spacing
Adjust line or paragraph spacing

Most documents you create will use single or double line spacing. *Word's* default line spacing is single. With **single spacing**, no blank lines are placed between lines of text. **Double spacing** leaves a blank line between keyed lines.

Line spacing can be changed using the Paragraph command on the Format menu. It can also be changed using the Line Spacing button on the Formatting toolbar.

DigiTip If the Line Spacing button is not displayed, click the chevrons at the right of the toolbar (>>) to display additional formatting options.

SELECTING TEXT

HELP KEYWORDS

Select text
Select text and graphics

You can make format changes to text you have entered. You must first select the text you want to change. Selected text is highlighted in black. You can select text using the mouse, the keyboard, or a digital pen. To deselect text, click or tap anywhere outside of the selected text.

Figure 5-1
Selected text appears highlighted in black.

> Text is keyed in the application window. To change or edit text, you must move the insertion point around within the document. **You can move to different parts of the document by using the mouse or the keyboard.**

To select text with the mouse or digital pen:

To Select:	Do This:
Any amount of text	Click or tap at the beginning of the text and drag over the text.
A word	Double-click or tap the word.
A line	Click or tap in the area left of the line.
Multiple lines	Drag in the area left of the lines (selection bar).
A paragraph	Double-click or tap in the selection bar next to the paragraph.

To select text with the keyboard:

To Select:	Do This:
One character to left or right	**Shift+Left** or **Right Arrow**
To the end of a line	**Shift+End**
To the beginning of line	**Shift+Home**

Figure 11-15
Images in Different Sizes

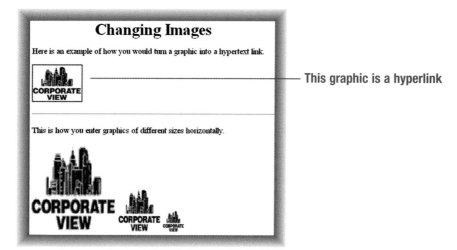

ETHICS

Graphics from the Web

At work one day, you noticed that a coworker, Jeremy, was using his computer to download copyrighted clip art. You observed him pasting the clip art into the department's online newsletter and posting it to the company's Web site. When you asked Jeremy about his activities, he shrugged and said that no one should mind because the chances of his getting caught were very slim. "Besides," he said, "I'm not really hurting anyone."

Capturing copyrighted pictures off the Internet and using them without permission is unethical and, in many cases, illegal. Copyrighted images cannot be used without the permission of the owner. If you wish to use copyrighted graphics, you must ask permission and receive a formal written reply. This request is commonly made via e-mail.

Clip art collections are available on the Web that you may use legally and without cost. Other images can be purchased inexpensively and used without copyright infringement.

For Discussion

What should you do regarding the situation with Jeremy?

- Let the matter drop and ignore his illegal actions.
- Inform your supervisor about Jeremy's use of the Internet.
- Send an e-mail message to Jeremy condemning him for this actions.
- Talk with Jeremy again and list reasons why you think his behavior is inappropriate.

Follow Up

Use your Web search tools to locate free clip art. Search using the term *free clip art*. Add one or two Web sites that are good sources of free clip art to your Favorites list for future reference.

SHOW/HIDE CODES COMMAND

HELP KEYWORDS
Show
  Show or hide formatting marks

Each time you press **Enter**, *Word* inserts a paragraph mark (¶) and starts a new paragraph. Thus, a **paragraph** may consist of a single line or several lines of text or other characters followed by a paragraph mark (¶). Being able to see where you have pressed **Enter**, **Space**, or **Tab** is sometimes helpful.

The **Show/Hide** command shows (or hides) nonprinting characters:

Paragraph marker ¶
Space ·
Tab →

Click the Show/Hide button on the Standard toolbar to use this command.

ACTIVITY 5-2 UPDATE A MEMO

1. Start *Word*. Open the data file *Cardinal Group*. Print the document. Note on the memo how the format of this memo is different from the sample memo you viewed. Return to the Style Guide intranet page, if needed, to compare the documents.

2. The first difference you may note is that the information following the heading words is not aligned properly. Show the non-printing characters in the document by clicking the **Show/Hide** button. This will help you see where **Tab** has been used. Position the insertion point and insert tabs as needed to align the heading words.

3. Another difference you may note is that the line spacing is not correct. Review the information on page 129 about selecting text. Select the four heading lines (**To, From, Date,** and **Subject**). Click the down arrow on the Line Spacing button and select **2**.

4. The paragraphs also are not spaced correctly. Select the four paragraphs and the blank lines between them. Set the line spacing to **1**.

5. The memo should now be in the correct format. You may have noticed some errors in the text of the document. You will correct those later. Save the memo as *Activity 5-2* in your *DigiTools your name\Chapter 5* folder. Close the document.

CHARACTER FORMATS

HELP KEYWORDS
Font
  Apply bold
  Apply italic
  Change the size of text

Character formats are commands that change the normal appearance of text. Bold, underline, and italic, for example, are character formats. Character formats can be applied to letters, numbers, and punctuation marks.

This text is in bold. *This text is in italic.* <u>This text is underlined.</u>

4. Start *Notepad*. Enter the following tags in *Notepad*.

```
<html>
<head> <title> Inserting Images </title> </head>
<body>
<center> <h1> Changing Images </h1> </center>
<p> Here is an example of how you would turn a graphic into a hypertext link. </p>
<p> <a href="http://www.corpview.com"> <img src="CV_Logo.gif"></a></p>
<hr>
<p> This is how you enter graphics of different sizes horizontally. </p>
<img src="CV_Logo.gif" height=200 width=200>
<img src="CV_Logo.gif" height=100 width=100>
<img src="CV_Logo.gif" height=50 width=50>
<hr>
<p> This is how you enter graphics of different sizes vertically. </p>
<p> <img src="CV_Logo.gif" height=200 width=200></p>
<p> <img src="CV_Logo.gif" height=100 width=100></p>
<p> <img src="CV_Logo.gif" height=50 width=50></p>
<hr>
<p> This is how you enter graphics and move them to the left and right of a page. </p>
<p> <img src="CV_Logo.gif" height=50 width=50 align=right></p>
<p> <img src="CV_Logo.gif" height=50 width=50 align=left></p>
</body>
</html>
```

5. Save your file as *it3.htm* in your *DigiTools your name\Chapter 11\itcareers* folder. View your page in your browser. The top portion of the page is shown in Figure 11-15 on page 338. Close *Notepad* and your browser.

DigiTip You can use a graphic as a background. Try this on your *index.htm* page. Enter the following in the body tag: <body background="CV_Logo.gif">. View your page and see how it looks. It is probably a bit messy! Return to your page in *Notepad* and delete the new tag. However, you may find another image that would serve as a great texture background for your Web pages. And now you know how to do it!

The font style and font size are also character formats. The **font style** determines the look or artistic style of the letters. The font size determines how large the letters appear.

This is Arial font style in font size 12.

This is Times New Roman font style in font size 14.

This is Comic Sans font style in font size 12.

The Formatting toolbar provides a quick way to apply character formats. You can apply a character format before or after keying text. To apply a format after keying text, select the text. Then click a button or choose an option on the toolbar. To apply formats as you key, click a button or choose an option on the toolbar. Key the text. Then click the button again to turn off the format.

Figure 5-2
Character Formats on the Toolbar

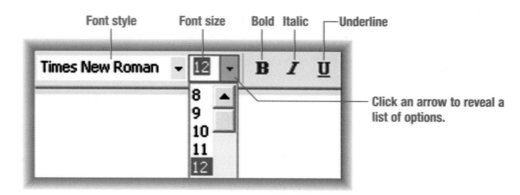

TABS

HELP KEYWORDS
Tab
  Set tab stops

Tab stops allow you to line up text at a certain place. Text or numbers can be aligned at the left, right, or center. Numbers can also be aligned at a decimal point. Tab settings appear on the ruler.

Figure 5-3
Types of Tab Stops

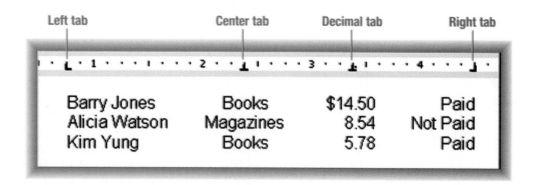

Word sets a left tab every one-half inch from the left margin by default. You can change these tab stops. Tabs stops can be changed using the Tabs command on the Format menu or using the ruler. You will practice setting tabs in Activity 5-3.

GRAPHICS

Hypertext links made the World Wide Web a powerful tool for people to use. Graphics, on the other hand, make the Web fun for people to use. Adding graphics to a Web page is not difficult, as you will see.

When you insert a graphic, you specify its size in the HTML tag. The picture size is measured in pixels. A **pixel** is a tiny dot or square of color, the basic element that makes up an image displayed on a computer screen. The term pixel is short for *picture element*. You can actually view the tiny dots that are pixels in your computer monitor with a magnifying glass. The table below shows two tags that are used to insert graphics.

LINKING HTML TAGS

Tags	Function
<img src="filename.gif" height=100 width=100 align=center>	Adds a picture or image to a Web page, sets its height and width at 100 pixels, and aligns the picture to the center.
<a href="http://www.corpview.com"> <img src="filename.gif"></a>	Turns an image into a hypertext link.
<hr>	Creates a graphical line called a horizontal rule in a Web page.

ACTIVITY 11-9 — INSERT GRAPHICS IN A WEB PAGE

1. Access the Corporate View intranet. Select the **Regular Features** link from the Corporate View Intranet Home page followed by **Intranet FAQs**. Scroll down to Section 5 and select the **Copying and Pasting Information from the Web and the Intranet** link. Click on the **Copy Pictures or Graphics** link and read how to capture pictures online.

2. Scroll to the top of the Web page that you are reading and right-click on the logo in the top, left-hand corner of the page.

Right-click on the graphic in the top corner of the page

3. Choose **Save Picture As** from the menu that appears. (**Note:** This wording may be different for your specific Web browser such as: **Save Image As** or **Download Image to Disk**.) Save it with the same name, *CV_Logo.gif*, in your *DigiTools your name\Chapter 11\itcareers* folder.

> **DigiTip**
>
> When entering the names of graphics in HTML, you must be precise. For example, in the name *CV_Logo.gif*, the letters CV and L are uppercase letters. The underscore symbol (_) is used because spaces are not allowed in Web filenames. The underscore key is located between the 0 and = keys on the top row of the keyboard. In your speech software you can simply say *underscore*.

ACTIVITY 5-3

EDIT A MEMO

HELP KEYWORDS

Undo
Undo mistakes

DigiTip
The Undo button reverses the last command or action. The Redo button restores the action or command changed by Undo. Click the down arrow on either button to see a list of actions to undo or redo.

1. Start *Word*. Open *Activity 5-2* from your *DigiTools your name\Chapter 5* folder.

2. Practice changing character formats. Choose **Select All** from the Edit menu to highlight the entire document. Click the down arrow by the font style name and choose a different font such as **Arial**. Click anywhere in the document to deselect it.

3. You decide you don't like the look of this font. Click the **Undo** button on the Standard toolbar to reverse the change.

4. Select the entire document again. Click the down arrow by the font size number and choose **11**. Click anywhere in the document to deselect it.

5. Select the document title, *MEMO*. Click the **Bold** button on the Formatting toolbar to bold the word.

6. Select the **TO** heading. Hold down the **Ctrl** key and select the **FROM** heading. Continue to hold down **Ctrl** and select the **DATE** and **SUBJECT** headings. All four headings should be selected (but not the words after the headings). Click the **Bold** button to format the headings.

7. Practice your navigation and correction skills. Read the body of the memo. You will find several words that should be capitalized. Navigate to each error and correct it.

DigiTip To delete text to the left of the insertion point, press the **Backspace** key. To delete text to the right of the insertion point, press the **Delete** key.

HELP KEYWORDS

Insert date
Insert current date or time

8. You notice that the date of the memo is not current. You can use the **Insert Date** feature to insert the current date in a document. Select the date. Choose **Date and Time** from the Insert menu. The date in a memo should show the month spelled out and followed by the day and year in numbers, such as *January 23, 2003*. Select this form of the current date from the list in the Date and Time dialog box. Click **OK**.

9. The name of a book, *Sarah's Home*, appears twice in this memo. Highlight both instances of this book name. (Hold down **Ctrl** while you select the second one.) Click the **Italic** button on the Formatting toolbar to format the text.

10. You decide to add some information to the end of the memo about books the club might read in the future. Move the bottom of the document. Leave one blank line, and then key the text below.

```
A few books I am considering for the future are shown
below. The one I like best is underlined.
```

11. Press **Enter** twice to begin a new paragraph. Set tabs to position the title, author, and price of each book you will list. Choose **Tabs** from the Format menu.
    - In the Tab stop position box, enter **1**. Choose **Left** under alignment. Click **Set**.
    - In the Tab stop position box, enter **3**. Choose **Center** under alignment. Click **Set**.
    - In the Tab stop position box, enter **5**. Choose **Decimal** under alignment. Click **Set**. Click **OK**.

ACTIVITY 11-8 — USE HEXADECIMAL VALUES FOR COLOR

1. Start *Notepad*. Open the file *it1.htm* that you created earlier.

2. Change the color scheme using hexadecimal values as shown here in bold.

```
<html>
<head> <title> Information Technology Careers </title>
</head>
<body bgcolor=000000 text=ffffff>
<center> <h1> Careers in Information Technology </h1>
</center>
<p> <font face=helvetica size=4 color=ff6600> Enter
your description of the function of IT and possible IT
opportunities between these tags. </font> </p>
<ol>
```

3. Save your file using the same name. Open the file in your Web browser. It should look similar to Figure 11-14.

Figure 11-14
Hexadecimal Color Scheme

4. This page is easier to read; however, you may still not like its appearance. Experiment with hexadecimal colors of your own. What Web-safe color combinations can you invent? You can also find references that show hexadecimal values for many colors by searching the Web. Use the search term *hexadecimal values*.

5. Save and view your page to see the colors you have created or selected. Close the browser and *Notepad*.

Figure 5-4
The Tabs Dialog Box

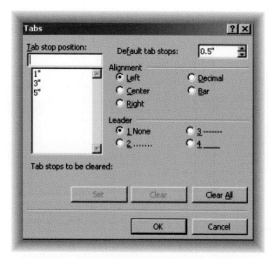

12. Key the book information shown below using the tabs you set. Apply italic to the names of the books. Also apply underline to the book you like best. To apply underline, select the text and click the **Underline** button on the Formatting toolbar.

```
The Adventures of Tom Sawyer    Mark Twain        $5.95
A Tale of Two Cities            Charles Dickens    4.95
The Mistletoe Mystery           Caroline Keene     4.99
```

13. After reading the memo, you decide that the second paragraph would sound better at the end of the memo. Select the second paragraph. Click and drag it to place it at the bottom of the document. Adjust the spacing, if needed, so you have one blank line between the paragraphs. (Click the **Show/Hide** button to view the paragraph marks.)

14. Save the memo under a new name as *Activity 5-3* in your *DigiTools your name\Chapter 5* folder. Print preview the document. Make adjustments to the format if needed. Print the document. Close the document.

APPLICATION 5-1   CREATE A MEMO

1. Start *Word*. Key the text on page 134 in proper memo format. Apply bold and italic as shown. Use the Insert Date command for the date. Proofread the memo carefully and correct all errors.

2. Save the memo as *Application 5-1* in your *DigiTools your name\Chapter 5* folder. Print preview the document. Make corrections if needed. Print the document. Close the document.

Peer Check Discuss your memo with a classmate. Make corrections, if needed, using feedback from the discussion.

Figure 11-13
The color scheme used on this Web page makes the text hard to read.

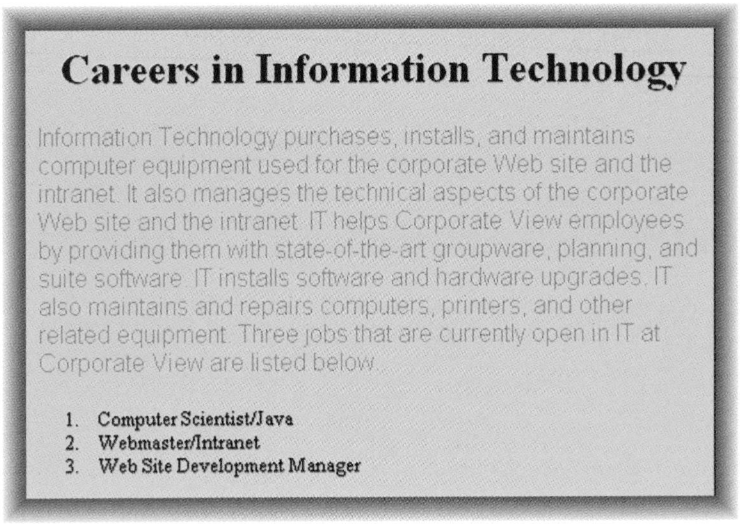

Careers in Information Technology

Information Technology purchases, installs, and maintains computer equipment used for the corporate Web site and the intranet. It also manages the technical aspects of the corporate Web site and the intranet. IT helps Corporate View employees by providing them with state-of-the-art groupware, planning, and suite software. IT installs software and hardware upgrades. IT also maintains and repairs computers, printers, and other related equipment. Three jobs that are currently open in IT at Corporate View are listed below.

1. Computer Scientist/Java
2. Webmaster/Intranet
3. Web Site Development Manager

3. As you can see, the use of color affects how easy text is to read on a Web page. You will change these colors in a later activity. For now, close *Notepad* and your browser.

HEXADECIMAL VALUES

You can use hexadecimal values in your tags to define colors more precisely than using words such as *yellow* or *blue*. **Hexadecimal values** are a base 16 system that use the numbers 0 to 9 and the letters A, B, C, D, E, and F. With just six pairs of hexadecimal values, you can create colors that can be displayed easily by any Web browser. These six pairs of hexadecimal values are:

```
00   33   66   99   cc   ff
```

Choose any three pairs of these hexadecimal numbers and you will create Web safe colors. Examples are shown in the following table.

SAMPLE HEXADECIMAL COLOR VALUES

Value	Color
000000	black
0000FF	blue
00FF00	green
CCFFFF	light blue
FFFFCC	light yellow
FF6600	orange
99FF99	pastel green
FF0000	red
FFFFFF	white
FFFF00	yellow

1" or 2" top margin

MEMO
　　　　DS　↓ Align heading information　　　　　　　　　　1" or default side margins
TO:　　　　All Sunwood Employees
　　　　　　　　　　　DS
FROM:　　Julie Patel, Human Relations
　　　　　　　　　　　DS
DATE:　　 Current date
　　　　　　　　　　DS
SUBJECT:　Eric Kershaw Update
　　　　　　　　　　DS

We were notified by Eric Kershaw's family that he was admitted into the hospital this past weekend. They expect that he will be there for another ten days. Visits and phone calls are limited, but cards and notes are welcome.

A book, *Discovering Our National Parks*, is being sent to Eric from the Sunwood staff. Stop by our office before Wednesday if you wish to sign the card. If you would like to send your own card to Eric, send it to:

Eric Kershaw
County General Hospital
Room 401
Atlanta, GA 38209-4751

At least 1" bottom margin

Memo

CUT, COPY, AND PASTE COMMANDS

HELP KEYWORDS

Copy text
Move or copy text and graphics

The **Cut** command deletes selected text or objects, such as pictures, from a document. The text or object is placed on the clipboard (a temporary storage area) and can be pasted into a document.

The **Copy** command creates a copy of selected text or objects. The original text or object remains in the document. The text or object is placed on the clipboard and can be pasted into a document.

CHAPTER 5　　Lesson 5-1　　　　　　　　　　　　　　　　　　　　　　　134

ATTRIBUTES

Your existing Web pages are somewhat plain, but you can add life to them quickly by altering a few attributes and values. **Attributes** are detailed instructions in a tag that provide information about what the tag is supposed to do. For example, by adding the *bgcolor* and *text* attributes to the <body> tag, the colors of the words and the background of the page can be changed dramatically. The choice of colors is defined by the values that are chosen. **Values** are options for attributes such as numbers or color choices. In this example, a yellow Web page will have bright blue text.

```
<body bgcolor=yellow text=blue>
```

Admittedly, these are awful color choices, but they give you an idea of how you can use color. It will be up to you to select more complementary colors in a later activity.

HTML TAGS WITH ATTRIBUTES

Tags and Attributes	Function
<body bgcolor=yellow text=blue>	Changes the background and text colors found on a Web page.
<font face=helvetica size=4 color=orange>	Changes the style, size, and color of a font or words on the page.

ACTIVITY 11-7 — USE ATTRIBUTES AND VALUES

DigiTip

When you change text in the body tag <body text=blue>, all of the text in the page will turn blue unless you use the <font color=> </font> tags to precisely change a specific word, sentence, paragraph, or other selection of text. The <font> tag overrides the text changes made in the <body> tag.

1. Start *Notepad* and open your *it1.htm* file. Add the tags, values, and attributes shown in bold below to your file. Make sure you place them precisely in the same position seen here:

```
</title> </head>
<body bgcolor=yellow text=blue>
<center> <h1> Careers in Information Technology </h1>
</center>
<p> <font face=helvetica size=4 color=orange> Enter your
description of the function of IT and possible IT
opportunities between these tags. </font> </p>
```

2. Save your file using the same name. Open the *it1.htm* file in your Web browser. It should look similar to Figure 11-13 on page 334.

DigiTip

When you cut or copy text, it is stored temporarily on the clipboard. You can paste this text into any *Office* document. For example, you can copy some text from a *Word* document and paste it into an *Excel* spreadsheet.

The **Paste** command places cut or copied text or objects in a document at the cursor location. You can paste text or objects to another location in the same document. You can also paste text or objects to another location in a different document or program.

The Cut, Copy, and Paste commands can be accessed from the Edit menu or from buttons on the Standard toolbar.

Figure 5-5
Cut, Copy, and Paste Buttons

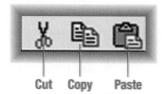

ACTIVITY 5-4 — E-MAIL MESSAGE

1. Start your e-mail program. Write an e-mail to a friend who asked about Eric Kershaw earlier today. Use the e-mail address of a classmate in the **TO** box. Enter an appropriate subject.

2. Tell your friend that Eric is ill and in the hospital. Then give the address where your friend can send a card to Eric.
- Open the memo you created in Application 5-1. Select the address lines and click **Copy** on the Standard toolbar.
- Switch to your e-mail message window. Position the insertion point and click **Paste** on the toolbar.

3. Send the e-mail message. Print a copy of the message from your *Sent* folder. Close your e-mail program.

LESSON 5-2 — LETTERS

OBJECTIVES

In this lesson you will:

1. Learn the purpose and format of a letter.
2. Use the Spelling and Grammar feature.
3. Align text vertically and horizontally.
4. Insert page breaks.

LETTER FORMAT

A **letter** is a formal type of correspondence. A letter is most often used to communicate with individuals outside the company or organization. A letter may be sent to someone within the company if a formal correspondence is appropriate. The content of a letter may or may not be confidential.

2. Save this new file as *index.htm* in your *DigiTools your name\Chapter 11\itcareers* folder. View your index file which should be similar to Figure 11-12.

Figure 11-12
Hierarchical Index Navigation Page

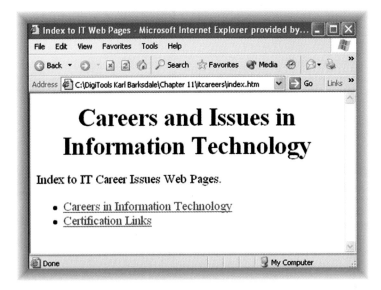

3. Click on either of your hypertext links to jump to the page you have listed in your index. Click the **Back** button to return to your *index.htm* page.

4. Close *Notepad* and your browser.

Self Check Do all of your links work? If not, compare your HTML document with the tags shown in the text to see how they differ.

LESSON 11-4: ADDING COLOR AND GRAPHICS TO WEB PAGES

OBJECTIVES *In this lesson you will:*

1. Use attributes and values in tags to enhance a Web page.
2. Use hexadecimal values for color in a Web page.
3. Use tags to insert graphics in a Web page.

Many Web sites are almost works of art using color, graphics, and layout design to hold the viewer's attention. Exciting color schemes, fonts, graphics, and multimedia enhance Web sites. A **color scheme** is a set of color choices that can be applied to an entire Web site. If you wish to be a Web designer, you should consider taking as many art classes as you can. This training will help you design Web pages that look attractive.

Letters are also written by individuals. People often write to businesses, organizations, or other individuals. This type of letter is called a personal-business letter.

Business letters are usually printed on letterhead stationery. **Letterhead stationery** is good quality paper that has company information printed on it. The information usually includes the company name, address, and telephone number. Some type of graphic and an e-mail or Web address might also be included.

Personal business letters are often printed on plain paper. When a personal-business letter is printed on plain paper, the writer's return address should be keyed on the two lines above the date. Some people create a personal letterhead design using text and graphics in their *Word* document. This letterhead design prints at the top of the letter page.

ACTIVITY 5-5 RESEARCH LETTER FORMAT

Corporate View provides information about letters and a sample letter format on the intranet. Access the intranet to learn about the proper format for a letter.

DigiTip
You can also access the style guide by choosing **Corporate Communications, Style Guide** from the Corporate View Intranet Home page.

1. Open your browser and access the Corporate View intranet. Use the bookmark on your Favorites list to access the Style Guide page. Click the **Letters** link.

2. Read the information about letters and the sample letter.

3. What parts are included in the sample letter?

4. What side margins should you use for a letter?

5. How should the paragraphs of a letter be spaced?

6. In block letter style, where do all lines begin?

7. In open punctuation style, what punctuation is placed after the salutation? after the complimentary close (closing)?

8. Why would you want to use the "you" approach when writing a letter?

9. What information should be included in a second-page heading when two pages are required for a letter?

10. Close your browser.

ALIGNING TEXT VERTICALLY AND HORIZONTALLY

Letters may be printed on letterhead paper with the date positioned about one inch below the letterhead. Another acceptable format is to center the letter vertically on the page. This format works well for short letters and letters printed on plain paper.

The Center Page command is used to center text vertically on a page. This command can be accessed by choosing Page Setup from the File menu.

Text can be aligned horizontally on the page in different ways. Text can be aligned at the left, center, or right as shown in the example on page 137. Text can also be **justified**. This means that the lines of a paragraph end evenly at the right edge.

4. Print your Web page. In your browser, choose **Print** from the file menu. Select a printer, if needed, and click **Print**. Examine the printed page. Note that it contains information such as the Web page name (as shown in the title bar), the page number, the path, and the current date.

5. Close your browser and the *Notepad* file.

Self Check Do the links on your Web page take you to the other Web sites? If not, open your Web page in *Notepad*. Try to determine how your HTML document differs from the tags shown in the text.

INDEX PAGE

An **index page** is the first or default page that a Web browser looks for when visiting a new Web site. Traditionally, these pages have been called *index.htm*. These pages can also be called *default.htm* or carry other file extensions depending on the multimedia effects used on the page. Index pages are important. For example, suppose you want to visit the site www.corpview.com. The actual page that will be viewed will be an index page such as www.corpview.com/index.htm. If your *itcareers* folder was on the site, you could launch it by entering www.corpview.com/itcareers/. The www.corpview.com/itcareers/index.htm page would be opened automatically.

You have created two Web pages, but you still do not have a Web site. All the pages found on a Web site must be linked together with some sort of navigation system. To create an interactive navigation system, you will need to use the anchor tag again. This time instead of linking to other Web sites around the world, you will create an index page and link to Web pages on your own site.

ACTIVITY 11-6 — CREATE AN INDEX PAGE AND LINKS

1. Open *Notepad* and create a new Web page by entering the tags shown below. This page will create a navigation index for all of your IT pages.

```
<html>
<head> <title> Index to IT Web Pages </title> </head>
<body>
<center> <h1> Careers and Issues in Information Technology
</h1> </center>
<p> Index to IT Career Issues Web Pages. </p>
<ul>
<li> <a href="it1.htm"> Careers in Information Technology
</a></li>
<li> <a href="it2.htm"> Certification Links </a></li>
</ul>
</body>
</html>
```

```
This text is left aligned.
                This text is center aligned.
                            This text is right aligned.
This text uses justify alignment.  This alignment causes
lines to end evenly at the right margin.  The problem with
this setting is that it can leave too much white space
between words.
```

SPELLING AND GRAMMAR

A letter, memo, or report that has spelling or grammar errors will make a poor impression on the reader. Keying errors, such as those in dates, names, or amounts of money, may also cause serious problems or misunderstandings.

You can use *Word's* Spelling and Grammar Checker to help you locate and correct errors. *Word* checks spelling and grammar automatically as you key. Wavy red underlines show possible spelling errors. Wavy green underlines show possible grammatical errors. When you right-click a marked word, a list of possible correct choices appears. You can click one of the choices to insert it in the document.

Figure 5-6
Word marks possible spelling errors as you key.

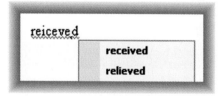

You can check spelling and grammar for an entire document by choosing Spelling and Grammar from the Tools menu. You can also access this feature by clicking the Spelling and Grammar button on the Standard toolbar.

DigiTip If you do not see wavy red lines under misspelled words, choose **Options** from the Tools menu. On the Spelling and Grammar tab, choose **Check spelling as you type**.

ACTIVITY 5-6 — FORMAT AND CHECK A LETTER

1. Start *Word*. Open the data file *Special Sale*.

2. Center the text of the letter vertically on the page. Choose **Page Setup** from the File menu. Click the **Layout** tab. In the Vertical alignment box, choose **Center**. Click **OK**.

ACTIVITY 11-5

INSERT ANCHOR TAGS

1. Start *Notepad* and enter the following tags to create a new Web page.

```
<html>
<head> <title> Certification Links </title> </head>
<body>
<center> <h1> Certification in Information Technology </h1>
</center>
<p> Information about IT training and certification
programs can be found at the following Web sites: </p>
<ul>
<li> <a href="http://www.microsoft.com.com"> Microsoft
</a></li>
<li> <a href="http://www.sun.com"> Sun Microsystems
</a></li>
<li> <a href="http://www.cisco.com"> Cisco </a></li>
<li> <a href="http://www.adobe.com"> Adobe </a></li>
<li> <a href="http://www.macromedia.com"> Macromedia
</a></li>
<li> <a href="http://www.oracle.com"> Oracle </a></li>
</ul>
</body>
</html>
```

2. Save your file as *it2.htm* in your *DigiTools your name\Chapter 11\itcareers* folder. Then view your HTML page in your Web browser. It should look similar to Figure 11-11.

Figure 11-11
Web Page with Hyperlinks

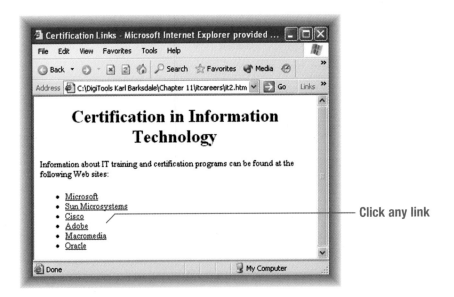

Click any link

3. If you are connected live to the Internet, you can click on any one of the links to go to the welcome pages of these companies. Each company offers IT certification programs.

Figure 5-7
Page Setup Dialog Box

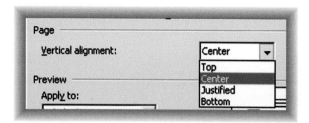

HELP KEYWORDS

Grammar
Check grammar

3. Click the **Spelling and Grammar** button on the Standard toolbar. *Word* will highlight possible errors in spelling or grammar. For each instance, click an appropriate option:
   - If the word is correct, click **Ignore Once** or **Ignore All**.
   - If the word is not correct, key the correct word or choose from the list of suggested words. Then click **Change**.
   - If you want to add this word to *Word's* dictionary, click **Add to Dictionary**.
   - Click **Change All** to make the change every time the word appears in the document.

If a statistics box appears when the check is complete, click **OK** to close the box.

Figure 5-8
Spelling and Grammar Dialog Box

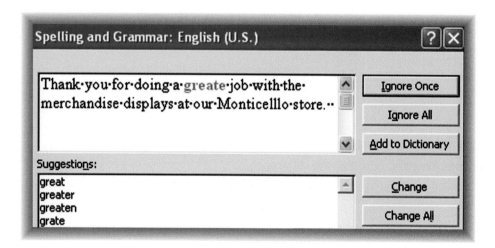

4. Proofread your document carefully to find any errors the Checker may have missed. Save the letter under a new name, *Activity 5-6*, in your *DigiTools your name\Chapter 5* folder. Print the document. Close the document.

APPLICATION 5-2 — PERSONAL-BUSINESS LETTER

1. Start *Word*. Key the letter shown on page 139 in block style with open punctuation. Place the writer's return address on the two lines above the date. Center the letter vertically on the page. Use Insert Date for the date.

2. Check the letter for spelling errors. Save the letter as *Application 5-2* in your *DigiTools your name\Chapter 5* folder. Print the letter and close the document.

10. View your file. It should look similar to Figure 11-10. Close your browser.

Figure 11-10
IT Careers Web Page

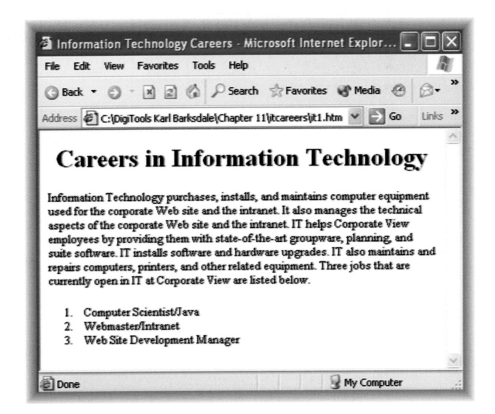

Self Check Compare your Web page to Figure 11-10. If they differ significantly, examine your HTML tags to see if you can determine the cause of the difference.

ANCHOR TAGS

One reason the World Wide Web has become so huge and so widely used is that Web designers can link to pages on many other Web sites. The sites may be located nearby or around the world. An **anchor tag** allows users to link Web pages. For example, the anchor tag shown below creates a hypertext link that jumps to the Corporate View welcome page.

```
<A HREF="http://www.corpview.com"> Corporate View Web
Page </A>
```

Center vertically on page

Return address → 610 Grand Avenue
Laramie, WY 82070-1423
Date → February 20, 20—
                    QS

Letter address → Elegant Treasures   1" side margins
388 Stonegate Dr.   or default
Longview, TX 75601-0132
                    DS
Salutation → Dear Armani Dealer

Body → The Giuseppe Armani figurines in your window are very attractive. I noticed them last week. Do you have other figurines?

A friend gave me a pamphlet showing three Armani millennium sculptures: **Stardust** (Years 1-999), **Silver Moon** (Years 1000-1999), and **Comet** (Year 2000 and beyond). I want to buy all three sculptures. Do you have them in stock or could you order them? If not, could you refer me to a nearby dealer?

I look forward to adding these beautiful pieces to my collection.

Complimentary Close → Sincerely
                    QS

Writer → Cynthia A. Maustin

At least 1" bottom margin

Personal-Business Letter in Block Format

DigiTip

When working with HTML Web page folders, it is best not to include spaces in the names of your folders. Therefore, *IT Careers* should be written without a space as *ITCareers*. Using a mixture of uppercase and lowercase letters is acceptable. Subfolders are usually left in lowercase letters, however, to make Web addresses easier for Web site visitors to input, as in *itcareers*.

8. Create a new subfolder called *itcareers* in the *DigiTools your name\Chapter 11* folder to save your HTML files. Click **Save As** from the **File** menu.

- In the Save As dialog box, navigate to the *DigiTools your name\Chapter 12* folder. Click the **New Folder** button.

- Enter a name for the new folder *itcareers* and press **Enter**. Double-click the folder name to move the name to the Save in box.

- Key a name for the file *it1.htm* in the File name box. The Save as type and Encoding options should be as shown here. Click **Save**. (**Note:** If you are using a program other than *Notepad*, be sure to select the text format option when saving the file.)

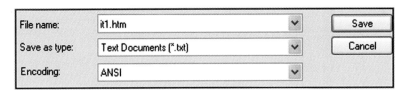

9. To view your HTML page, start your Web browser. Open the *it1.htm* file. Instructions for *Internet Explorer* are shown below:

***Internet Explorer* Instructions**
- Choose **File**, **Open**.
- Choose the **Browse** button. Search for the *DigiTools your name\Chapter 11\itcareers* folder.
- Choose your *it1.htm* HTML file, and choose **Open**.
- Choose **OK**.

Figure 11-9
Browse to locate a file.

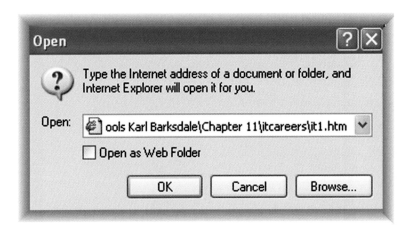

DigiTip Web files can be saved with either an .html or .htm extension. With most Microsoft products, the filename extension is usually truncated, or shortened, to .htm.

APPLICATION 5-3 BUSINESS LETTER

1. Start *Word*. Key the letter in block style with open punctuation. Assume that the writer's return address will be shown in a letterhead. (No return address will be keyed above the date.) Make corrections as indicated by the proofreaders' marks. Center the letter vertically on the page. Use Insert Date for the date.

2. Check the letter for spelling and grammar errors. Save the letter as *Application 5-3* in your *DigiTools your name\Chapter 5* folder. Print the letter and close the document.

Current date

Mr. Jon A. Richardson
283 Mt. Pleasant Dr.
Oklahoma City, OK 73110-6661

Dear Mr. Richardson

Thank you for your kind letter. I'm glad you ~~liked~~ enjoyed my presentation at the convention last week.

Using the internet in my class has made learning history ~~a lot~~ more interesting for my students. Students were bored with just reading about history from a book. Now, students read and view information about people and events from history the on Internet. As a result, students are more interested and learn more.

One Internet site that Students often use is PBS's "The american Explorer." The URL is www.pbs.org/wgbhy/amex/whoweare.html. This site contains stories of people and events, ~~which~~ that shaped our country. I hope your students will enjoy using this site.

Sincerely

Martin G. Anderson, Professor

Self Check Compare the format of the letter with the model document on page 139. Make changes to the format, if needed.

ACTIVITY 11-4

CREATE A WEB PAGE

1. To find the information you will need for the Web page, access the Corporate View intranet. Select the **Information Technology** link from the main Corporate View Intranet Home page. Select and read **About Information Technology**.

2. Now learn about three IT jobs at Corporate View. Select the **Human Resources & Management** link from the navigation bar. Choose **Current Job Openings at TeleView**, then **Information Technology**. Read about the IT jobs that are currently open. Note the names of the three jobs for later use. Close the Corporate View intranet.

3. Open *Notepad* by choosing **Start (All) Programs, Accessories, Notepad**.

4. Enter these starting tags exactly as shown in the text. Be careful not to leave out a single angle bracket, letter, or slash.

```
<html>
<head> <title> </title> </head>
<body>
<center> <h1> </h1> </center>
<p> </p>
<ol>
<li> </li>
<li> </li>
<li> </li>
</ol>
</body>
</html>
```

5. Enter a title for the page between the <title> </title> tags as shown:

```
<head> <title> Information Technology Careers</title>
</head>
```

6. Enter a heading for the information between the <h1> </h1> tags as shown:

```
<center> <h1> Careers in Information Technology </h1>
</center>
```

7. Enter your own brief description of IT between the <p> </p> tags. End the paragraph with a sentence to introduce the three IT jobs that are now open at Corporate View. Enter three jobs in IT between the list tags (<li> </li>) as shown below.

```
<li> List one job in IT between these tags. </li>
<li> List another job in IT between these tags. </li>
<li> List yet another job in IT between these tags.
</li>
```

DigiTip

If you choose to use your voice to enter tags, the / is called a *slash*. The < and > are spoken as *less than* and *greater than*. You can also say *open angle* and *close angle*. If you are using your handwriting software, use the On-screen Standard and Shifted keyboards to enter </>.

DigiTip

In *Notepad*, turn on word wrap to make working with text easier. Choose **Word Wrap** from the Format menu.

CHANGE CASE COMMAND

The **Change Case** command allows you to format selected text for small or capital letters. For example, you can change lowercase (small) letters to capital letters. You can change uppercase (capital letters) to lowercase. Other options are also available. This feature is helpful when working with titles and envelope addresses. The Change Case command is found on the Format menu.

Figure 5-9
Change Case Dialog Box

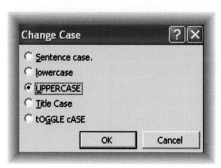

ENVELOPES

Word's **Envelopes and Labels** feature allows you to create and print envelopes for letters. You can add information to your envelope such as a return address or a POSTNET (Postal Numeric Encoding Technique) bar code. POSTNET codes can be read by the U.S. Postal Service's automated mail-handling equipment and help your mail be processed faster. The POSTNET bar code contains U.S. ZIP Code information for the delivery address.

You can format the text on an envelope and choose from various sizes of envelopes. You can also add a graphic, such as a company logo, to an envelope. When you create an envelope, you can print the envelope without saving it. If you want to save the envelope for editing or later use, you can attach it to a document. When you save the document, the envelope is also saved.

The U.S. Postal Service recommends using a certain address style. This style helps ensure that addresses can be read automatically by DigiTools such as scanners and sorters. The recommended address style for the delivery address includes block format, all capital letters, and no punctuation except in the ZIP code. The return address is shown in traditional style using initial capitals and punctuation.

Figure 5-10
An envelope using the U.S. Postal Service recommended address format.

DigiTip In *Notepad*, choose **Find** from the Edit menu. Enter the opening tag you want to find such as *<center>* in the Find dialog box. Click **Find Next**. If the tag is not in the document, a "Cannot find" message will display. If the tag is in the document, the insertion point will move to its location.

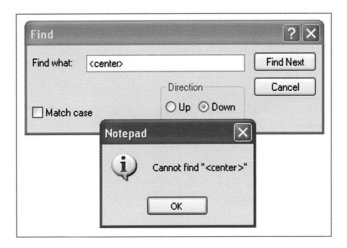

5. Close *Notepad* and the Corporate View intranet.

CREATING SIMPLE WEB PAGES

OBJECTIVES *In this lesson you will:*

1. Create a simple Web page using HTML tags.
2. Create a Web page with hyperlinks to other sites.
3. Print a Web page that you have created.
4. Create an index page and link to other Web pages on the same site.

CREATING A WEB PAGE USING NOTEPAD

Creating a Web page in a text editor such as *Notepad* is quite simple. First, plan the information you want to include on the page. What title should appear on the title bar? What headings should appear on the page? What text paragraphs or lists should be used? Once you have identified this information, open the program. Enter the text and the HTML tags to tell a browser how to display the page. Proofread the information carefully. Save the page in a folder created just for the pages of this Web site.

Start this lesson by visiting the Corporate View Web site to gather information about the Information Technology department. You will use this information to prepare a simple Web page.

ACTIVITY 5-7

CREATE AN ENVELOPE

HELP KEYWORDS

Envelope
Create and print a single envelope

1. Start *Word*. Open the letter file you created earlier, *Application 5-3*.

2. Click **Tools** on the menu bar. Point to **Letters and Mailings** and then click **Envelopes and Labels**. Click the **Envelopes** tab. The letter address from your envelope will appear in the Delivery address box.

3. Click in the **Return address** box, and key your name on the first line. Key your street address on the second line. Key your city, state, and ZIP code on the third line.

(To omit a return address, you could select the Omit checkbox.)

DigiTip

Always use the two-letter state abbreviations in addresses on letters and envelopes. A list of two-letter state abbreviations can be found in Appendix A, on page App-3.

4. Click **Options**. The Options dialog box opens. Here you can select an envelope size and other options. You will use the default envelope size, size 10. This is a standard size envelope used for business letters. (To choose another envelope size, you would click the down arrow for Envelope size and select an option.) Click the **Delivery point bar code** option. Click **OK** to close the Options box.

5. Click **Add to Document**. This will attach the envelope to the current document.

6. Select the delivery address. Click **Format** on the menu bar. Choose **Change Case**. Choose **Uppercase** and click **OK**. Delete all punctuation from the delivery address except the hyphen in the ZIP code.

HELP KEYWORDS

Sentence case
Capitalize text

7. Be sure that envelopes are loaded into the printer or print the envelope on regular paper as directed by your teacher. To print the envelope, place the insertion point in the envelope. Choose **Print** from the File menu. Choose the **Current page** option under Print range. Click **OK**.

8. Save the document as *Activity 5-7* in your *DigiTools your name\Chapter 5* folder. Close the document.

WHAT YOU SHOULD KNOW ABOUT
Postage Meters

In Lesson 5-2, you learned how to use a software DigiTool to add a POSTNET bar code to speed processing of outgoing mail. A postage meter is a hardware DigiTool that also speeds mail processing. A **postage meter** is a machine that prints postage in the amount needed. The meter prints the postage either directly onto the envelope or onto a label that you apply to the envelope or package. For example, you can use the numeric keys on the postage meter to set it to print postage for a letter weighing one ounce. You can easily reset it to print postage for a letter weighing three ounces.

The postage meter prints the date as well as the postage amount. Always be sure the correct date is set on the meter. Some meters can also be set to print a business slogan or advertisement next to the postmark. Because metered mail is already dated and postmarked, it can be processed faster than stamped mail.

BASIC HTML TAGS

Sets of HTML tags	What They Do
`<html> </html>`	Instructs the Web browser of the beginning and ending points of an HTML file.
`<head> </head>`	Includes information about a Web page. However, this information is not displayed to the user.
`<title> </title>`	Identifies the title of the Web page in the title bar of the Web browser.
`<body> </body>`	Marks the main portion of a Web page, the part that is displayed in the browser's main window.
`<center> </center>`	Marks text that should be centered in a Web browser's window.
`<h1> </h1>` `<h3> </h3>` `<h6> </h6>`	Marks various levels of headings in Web page documents. The numbers indicate the level of importance for marked headings with 1 being the most prominent (larger text) and 6 being the least prominent (smaller text).
`<ul></ul>`	Starts and ends a bulleted or an "unordered" list.
`<ol></ol>`	Starts and ends an "ordered" or numbered list.
`<li> </li>`	Marks items to be listed.
`<p> </p>`	Creates a paragraph break or double space.
`<a href="http://www.corpview.com"> </a>`	The anchor tag tells the browser the location of a Web page or resource. The anchor tag is the key to Web navigation and interaction. This tag allows movement from page to page.

ACTIVITY 11-3 LOCATE HTML TAGS

1. Review the table, *Basic HTML Tags*, before continuing.

2. Access the Corporate View intranet. Select the **Regular Features** link from the Corporate View Intranet Home page followed by the **Intranet FAQs** link. Scroll down to part 5 and select the **Copying and Pasting Information from the Web and the Intranet** link.

3. Choose **Source** on the View menu to display the source code in *Notepad*.

4. Scan the Web page file and examine the tags. Can you find any tags from the *Basic HTML Tags* table that are *not* used in this Web page? Note which tags are not used on the Web page. See the DigiTip that follows on page 326 to learn a quick way to look for tags.

What You Should Know About Postage Meters
continued

Figure 5-11
An electronic postage meter prints the postmark and the postage on mail.

For some postage meters, you take the meter with you to the post office to buy postage. A postal worker resets the meter for the amount of postage purchased. As you use the postage meter, the meter setting decreases, showing you how much postage remains. Do not let the postage get too low before buying more. The meter locks when the postage runs out.

Several companies offer postage meters that allow you to purchase postage online. The user connects the meter, which contains a modem, to a standard phone line. Using a keypad, the user indicates the amount of postage to be purchased. The meter is updated and the user is billed for the cost of the postage.

LESSON 5-3 TABLES

OBJECTIVES
In this lesson you will:

1. Create tables.
2. Edit and navigate tables.
3. Format tables and add shading.
4. Merge and split cells in a table.
5. Copy a table into a memo.

A **table** is a document that shows data arranged in a grid of columns and rows. The data may be text or numbers. A table may be a complete document by itself or it may be placed in a memo, letter, or report.

ACTIVITY 11-2   VIEW HTML TAGS

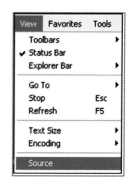

DigiTip

Different browsers use slightly different words for the Source command, so you may need to hunt for the correct option. In *Netscape,* choose **View, Page Source**.

1. Access the Corporate View intranet. When the Corporate View Intranet Home page appears, choose **Source** from the View menu in *Internet Explorer.*

2. When the source code for the Web page appears in *Notepad,* as seen in Figure 11-8, scroll down and look at the HTML tags that create the page. Does any of this make sense to you? Well, it soon will. In the next activity you will unravel the mystery of the tags. Close *Notepad* and your browser.

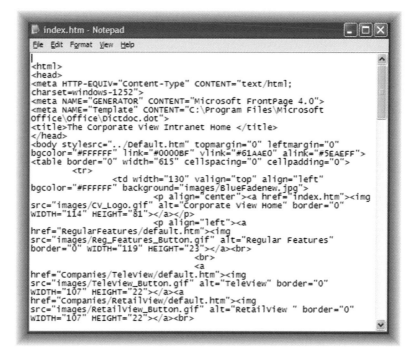

Figure 11-8
Look for the HTML tags enclosed in angle brackets.

HOW HTML WORKS

Web pages can be displayed by a variety of Web browsers, such as: *Internet Explorer, Mozilla, Opera,* or *Netscape.* Web content can be displayed on computers, handheld devices, Tablet PCs, even cell phones and pagers.

HTML uses tags to describe Web pages to browsers. HTML tags pass instructions to the Web browser on how to display pages. Tags can appear in uppercase <CENTER> </CENTER> or lowercase <center> </center> letters. HTML tags usually appear in pairs containing an open tag and a close tag. A close tag has a slash (/) in it. An open tag starts a process and a close tag stops the process. The tags tell the browser what to do and then disappear from view. For example, this string of tags, *<center> <b> DigiTools! </b> </center>,* will center and bold the word *DigiTools!*

HELP KEYWORDS

Table
About Tables

A **column** is a vertical list of data in a table. Columns are identified with letters (A, B, C) from left to right. A **row** is a horizontal list of data in a table. Rows are identified with numbers (1, 2, 3) from top to bottom. A **cell** is one block of data in a table. A cell is identified using a column letter and a row number (A1, B4, C6). In Figure 5-12, the column headings are in Row 1. The name *Avery, Thomas* is in Row 2. The names are in Column A. The regions are in Column B, and the sales figures are in Column C. The shaded cell that contains *Southeast* is Cell B4.

A table may be a simple listing of data or it may contain a title and column headings. The title may be placed on the first row of the table grid or above the table grid. In both cases, the title is keyed in all caps and centered horizontally over the table. Column headings are often centered within the columns. The title and the column headings may be placed in bold print for emphasis. In Figure 5-12, the column heading are center aligned and in bold. The data in Columns A and B are left aligned. The data in Column C are right aligned.

Figure 5-12
Table Elements

TOP SALES REPRESENTATIVES

Name	Region	Sales ($)
Avery, Thomas	Northeast	100,500
Brewer, Lorraine	Plains	100,250
Dolton, Celeste	Southeast	98,500
Heil, Clinton	Northwest	95,650
Packard, Hillary	Southwest	94,800
Stevens, Richard	Central	92,500

CREATING A TABLE

HELP KEYWORDS

Table
Create a table

You can create tables using the Insert command on the Table menu. You can create a table quickly using the Table button on the Standard toolbar. Either method produces the same results. Read about creating tables below. Later in this lesson, you will use the Table button to create tables.

To create a table using the Insert Table button:

- Click the **Insert Table** button on the Standard toolbar. A drop-down grid displays.

- Click the left mouse button and drag the pointer across and down to highlight the number of rows and columns you want in the table. The table displays when you release the left mouse button.

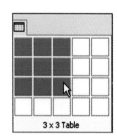

DigiTip

When using speech recognition, say the names of most links to choose them. A text-only navigation system ensures navigation for voice users. For example, say "Corporate Communications" in voice command mode to access that link.

9. Exit the database seminar and return to the **Regular Features** section. Scroll down to the bottom of the page. The random access navigation structure is repeated in a text only format similar to Figure 11-7. This is an alternate access system. If the hyperlinked graphics do not load properly, visitors can still find a way around the Web site easily. Click on the **Intranet Home** link.

Figure 11-7
A Text-Only Navigation System

10. Close the Corporate View intranet and your browser.

LESSON 11-2 HTML TAGS

OBJECTIVES *In this lesson you will:*

1. Learn how HTML tags are used in Web documents.
2. View the source code for an HTML document.
3. Find and compare HTML tags in a document.

In Lesson 11-1 you learned about three possible designs for Web page navigation. In this lesson, you will learn about the technical "magic" underneath the Web site's design.

HTML TAGS

HTML (hypertext markup language) is a document description language. It defines how an HTML document—commonly called a Web page—should look. Web browsers download HTML files containing text characters and symbols. These characters and symbols are called *tags*. HTML tags tell Web browsers how a Web page should appear on your screen. Tags are marked with angle brackets, like this: <tag>.

In the next activity, you will see how many tags it takes to create even a simple Web page. These tags may seem confusing at first. If you become a Web page designer, however, using them will become second nature to you.

MOVING WITHIN A TABLE

When a table is created, the insertion point is in cell A1. To move within a table, use the **Tab** key or simply click or tap within a cell. Refer to this table as you learn to enter text in a table:

Press	Movement
Tab	To move to the next cell. If the insertion point is in the last cell, pressing **Tab** will add a new row to the end of the table. If you add a new row by mistake, click the **Undo** button to remove it.
Shift+Tab	To move to the previous cell.
Enter	To increase the height of the row. If you press **Enter** by mistake, press **Backspace** to delete the line.

ACTIVITY 5-8 CREATE A TABLE

1. Start *Word* and open a new blank document.

2. Key a title for a table: COLLEGE SPORTS PROGRAM. Apply bold to the title. Select the title and click the **Center** button on the Formatting toolbar to center the title over the table.

3. Press **Enter** two times to leave one blank line after the title.

4. Click the **Table** button on the Standard toolbar and drag to display a 4 × 3 table. Release the mouse button. If the table grid that appears in your document is not what you wanted, click the **Undo** button to delete it and try again.

5. Key the information below in the table. Highlight the column heads in Row 1 and apply bold and center alignment.

COLLEGE SPORTS PROGRAM

Fall Events	Winter Events	Spring Events
Football	Basketball	Golf
Soccer	Gymnastics	Baseball
Volleyball	Swimming	Softball

6. Center the document vertically on the page. (**Review:** Choose **Page Setup** from the File menu. On the Layout tab, select **Center** from the drop-down list for Vertical alignment.)

7. Check the table for spelling errors. Save the table as *Activity 5-8* in your *DigiTools your name\Chapter 5* folder. Print the table and close the document.

Self Check Does your table appear similar to the one in the textbook? If not, what is different?

3. Notice that a new navigation system appears to the left side of the page. This is called the *navigation bar.* Notice that all of the Corporate View departments are listed. You can go to each department with one simple click. This is the distinguishing feature of a random access navigation system.

Figure 11-5
The Corporate View intranet uses a mixed navigation structure.

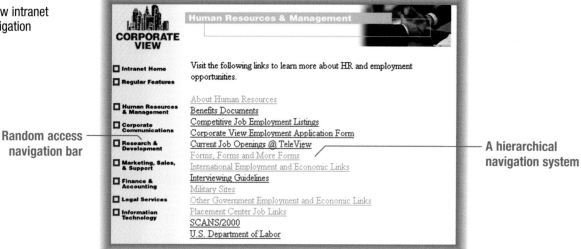

Random access navigation bar

A hierarchical navigation system

4. Notice the list of links in the center of the screen in Figure 11-5. These links are organized in a hierarchical fashion. They can only be accessed from this page. To move down one level of the hierarchy, click the **Corporate View Employment Application Form** link.

5. Notice that the hierarchical list disappears. The way to move back up the hierarchy is to click the Back button on the browser toolbar. Click the **Back** button now.

6. Move down two levels of the hierarchy. Start by choosing the **Benefits Documents** link. Then choose the **Dental Insurance Comparison Chart** link. Move back up the hierarchy by choosing the **Back** button twice.

7. Visit a section of the Corporate View Web site with a linear design. Choose the **Regular Features** link, followed by **Employee Training and Evaluations**, and then **Understanding Databases**. Choose **Click here** to open the seminar.

8. Choose the link for the first topic, **Database Elements**. This training seminar requires step-by-step instruction, so a linear design is used. Notice the Back/Next options on each page as seen in Figure 11-6. Also notice that the Web designer lets you know what page you are on and the total number of pages. Click the **Next** and **Back** links several times to view different pages of the Database seminar.

Figure 11-6
The Understanding Databases seminar pages use a linear navigation design.

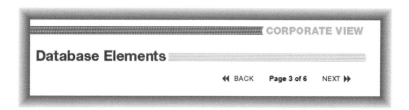

MODIFYING A TABLE

After you have created a table, you may find that you need a different number of rows or columns to add or remove data. You can insert and delete rows and columns in a table. You might want to change the width of a column so the table does not have so much blank space between columns. You can do so by clicking and dragging the grid line that separates the columns.

Changing Rows and Columns

Read about inserting and deleting rows and columns below. You will practice these procedures in Activity 5-9.

To insert rows or columns in a table, place the insertion point where you want the new row or column. Choose **Insert** from the Table menu. Choose **Rows Above** or **Rows Below** to insert rows. Choose **Columns to the Right** (or **Left**) to insert columns.

> **HELP KEYWORDS**
> **Table**
> Add a cell, row, or column to a table

Figure 5-13
Insert a Row

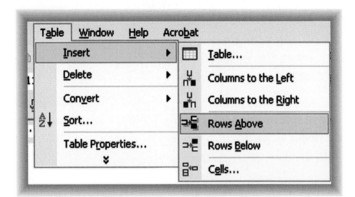

To delete rows or columns from a table, place the insertion point in the row or column or select several rows or columns. Choose **Delete** from the Table menu. Then choose **Rows** or **Columns**. (Choose **Table** to delete the entire table.)

Figure 5-14
Delete Columns in a Table

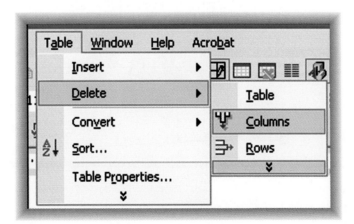

LINEAR DESIGN

A **linear design** allows users to move forward or back through Web pages in a controlled sequence. (See Figure 11-3.) Only two alternatives can be chosen from each page—moving forward or moving back along the straight line. This structure is used for step-by-step instructions or information that should be presented in a particular order—as in a slide show.

Figure 11-3
Linear Navigation Design

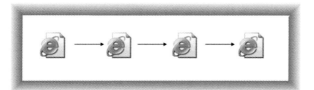

ACTIVITY 11-1

EXPLORE WEB SITE DESIGN

Because each Web design structure has advantages and disadvantages, the Corporate View intranet Web site mixes all three of these designs. In this activity, you will see the three design structures described above.

1. Access the Corporate View intranet. On the Corporate View Intranet Home page, you will see, in the center of the screen, a random access navigation design. Notice how each major department (or subsection) is represented in a circle around a graphic. (See Figure 11-4.)

Figure 11-4
The Corporate View Home page uses a random access structure.

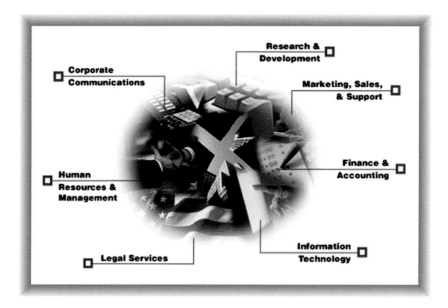

2. In this random navigation structure, all employees need do is to click on their department. They are taken directly to the information they need. Click the **Human Resources & Management** link now.

HELP KEYWORDS

Table
Resize all or part of a table

Adjusting Column Widths

Tables extend from margin to margin when they are created. Some tables, however, would be more attractive and easier to read if the columns were narrower. Column widths can be changed manually using a mouse or digital pen or automatically using AutoFit. The **AutoFit** feature adjusts column widths to the length of the data in the columns. Using a mouse or digital pen, you can adjust the widths as you like. Once you change the width of a table, you will need to center it horizontally.

To adjust column widths using a mouse or pen, point to the gridline between the columns in the table. The pointer will change to a double arrow. Drag the gridline to the left to make the column narrower or to the right to make the column wider.

Figure 5-15
Drag a column gridline to resize the column.

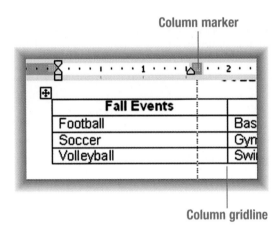

Adjust the column widths attractively. Leave approximately 0.5" to .75" between the longest line and the border. Use the horizontal ruler as a guide.

DigiTip You can display the width of the columns by pointing to a column marker on the ruler, holding down the **Alt** key, and clicking the left mouse button.

ACTIVITY 5-9 MODIFY A TABLE

1. Start *Word*. Open the table, *Activity 5-8*, that you created earlier.

2. Click in Cell A2 (which contains *Football*). Insert a new row above this cell: Choose **Insert** from the Table menu. Choose **Rows Above**. Enter data in the new Row 2:

| Wrestling | Weight Training | Track |

3. Click in any cell of the table. Choose **AutoFit** from the Table menu. Choose **AutoFit Contents**. The column widths will change.

4. To insert a new column, click in Column C. Choose **Insert** from the Table menu. Choose **Columns to the Right**. A new Column D will be inserted.

LESSON 11-1

WEB SITE INTERACTION DESIGN

OBJECTIVES *In this lesson you will:*

1. Learn about three organizational designs for Web sites.
2. Explore Web site designs using the Corporate View Web site.
3. Practice navigating a Web site using hyperlinks and a navigation bar.

A goal of Web site designers is to make it easy for Web users to find the information they are looking for. Generally, designers use three types of structures to organize information: random access design, hierarchical design, and linear design.[2]

RANDOM ACCESS DESIGN

A **random access design** allows users to move to each major subsection of a Web site from any other subsection. No single subsection stands out as more important than the others. Jumping from one subsection to another simply requires selecting a single link. A random access design only works well when the site has a limited number of hyperlinks. Usually, eight or fewer subsections would be used with this design. (See Figure 11-1.)

HIERARCHICAL DESIGN

A **hierarchical design** allows users to move through levels of Web pages. As seen in Figure 11-2, this design looks like an upside down plant, with the root at the top and branches going out to important subsections of information below. This is a great structure for large amounts of information with links leading to more detailed resources further along a logical path of related Web pages.

Figure 11-1
Random Access Navigation Design

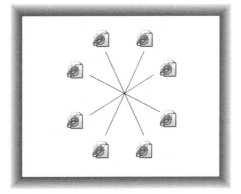

Figure 11-2
Hierarchical Navigation Design

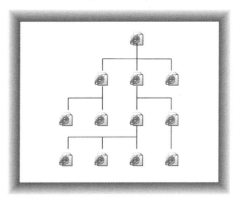

---

[2] Stubbs, Barksdale, Crispin, *Web Page Design.* Cincinnati: South-Western Educational Publishing, 2000, pp. 67–69.

5. Resize column D. Click and drag the right gridline of Column D until it is about the same size as the other columns. Enter the data below in Column D.

Summer Events
Volleyball
Golf
Baseball
Swimming

6. Click and drag to resize all the columns to have about 0.5" of space between words in the columns. All column heads should fit on one line.

7. Notice that the table is not centered horizontally on the page. When you change column widths or add or delete columns, you should center the table horizontally. Choose **Table Properties** from the Table menu. Click the **Table** tab. Choose **Center** under Alignment and click **OK**.

8. Save your table as *Activity 5-9 Table 1* in your *DigiTools your name\Chapter 5* folder. Print the table and keep the document open.

9. Click in cell A1 of the table, *Activity 5-9 Table 1*. To delete Column A, choose **Delete** from the Table menu. Choose **Columns**. Notice that the table position is adjusted automatically so the table is still centered horizontally.

10. Save your table as *Activity 5-9 Table 2* in your *DigiTools your name\Chapter 5* folder. Print the table and close the document.

> **HELP KEYWORDS**
>
> **Table**
> Position a table on a page

Peer Check Compare your table with a classmate's table. How do they differ? Make changes to the table, if needed, after your discussion.

APPLICATION 5-4

VERIFY ZIP CODES

The company you work for often sends volume mailings using standard U.S. mail. Periodically, an automated check of the mailing list for correct ZIP codes is done. During the last check, the computer identified ten addresses that may have incorrect ZIP codes.

1. Start *Word*. Open the data file *ZIP codes*, which contains a list of ten addresses.

2. Create a two-column table. Place the addresses in column A and the ZIP codes in column B. (Cut and paste the data into the table grid.) Use appropriate column heads. Enter UPDATED ADDRESSES as the table title. Format the data and size the columns for an attractive table. Center the table vertically and horizontally on the page.

CHAPTER 11

HTML and Web Site Design

OBJECTIVES *In this chapter you will:*

1. Learn about three levels of Web site design.
2. Explore hierarchical, linear, and random access Web site navigation structures.
3. Create Web pages using HTML tags.
4. Save, display, and print Web pages.
5. Create hyperlinks and link pages to create a Web site.
6. Apply HTML attributes and values.
7. Insert graphics into Web pages.
8. Convert *Word, Excel,* and *PowerPoint* documents into Web pages.
9. Apply the principles of interaction, presentation, and information design.
10. Learn about Web design tools such as *FrontPage* and *Dreamweaver*.

To the casual observer, **Web sites** are simple creations. They are just collections of Web pages about a topic or theme. Most Web sites serve a single purpose. For example, the Corporate View Web site serves the digital communications needs of the employees of the company.

Web sites can be created with simple text editors like *Notepad*. They can also be created with applications such as *Microsoft Word, PowerPoint,* and *Excel.* Web sites can be complex creations. Web site designers often use powerful software such as *Dreamweaver, GoLive,* or *FrontPage* to create Web sites. These sites often combine text, graphics, animation, and sound. Today's complex Web sites must be planned and designed. There are three levels of Web site design.[1]

- **Interaction:** The way visitors navigate, find, and interact with the information they are looking for on the Web site.
- **Presentation:** The look and feel of the Web site. This includes color schemes, fonts, use of graphics, and the arrangement of all the various elements on the site's pages.
- **Information:** The message of a Web site.

---

[1] Stubbs, Barksdale, and Crispin, *Web Page Design*. Cincinnati: South-Western Educational Publishing, 2000, p. 46.

3. Find and record the correct ZIP code for each address on the list. Use a printed ZIP code directory or access the U.S. Postal Service Web site at www.usps.gov to find current ZIP code information. Update any ZIP codes in the table that are incorrect with the correct information.

4. Proofread and spell-check the table. Correct all errors. Save the table as *Application 5-4* in the *DigiTools your name\Chapter 5* folder. Print and close the document.

Peer Check Compare your table with a classmate's table. Discuss any differences. Make changes, if needed, based on your discussion.

SELECTING IN A TABLE

If you want to apply formats, such as bold, or merge or split columns, you must first select part of the table. Read the table below that tells how to select various parts of a table.

To select	Move the insertion point
Entire table	Over the table and click the table move handle
Column	To the top of the column until a solid down arrow appears, then click the left mouse button
Row	To the left area just outside the table until an open arrow appears; then click the left mouse button

Use Show/Hide to display the table markers. These markers are useful when editing tables.

Figure 5-16
Table Markers

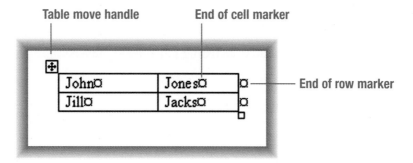

MERGING AND SPLITTING CELLS

HELP KEYWORDS

Table
Merge cells into one cell
Split cells into multiple cells

Cells in a table can be **merged** (joined) or **split** (divided) to better fit your data. Merging cells is useful when you want to put the title of your table in the table grid. Splitting cells is useful when you want to divide only some of the cells in a column. To merge cells, select the cells and click the **Merge Cells** button on the Tables and Borders toolbar. To split cells, select the cells and click the **Split Cells** button on the Tables and Borders toolbar.

6. Develop the detailed contents of the presentation. Use information found in articles from your research and in the *Flextime* file to provide supporting details. Anticipate questions and write sample questions and answers.

7. Create a *PowerPoint* slide show to use as visual aids for the presentation. Use a digital camera or scanner, if available, to capture pictures related to your topic for your slides.

8. Decide who will present each part of the presentation and practice with your team.

9. Deliver the presentation to your class or to another team as directed by your teacher. Include time for a question-and-answer session.

10. Ask your listeners to complete an evaluation form such as the one you developed earlier. Review the evaluation forms and write a summary of how the listeners rated your presentation. Note what you need to improve in future presentations. Save the self-evaluation as *Evaluation 3* in the *DigiTools your name\Chapter 10* folder.

DIGITOOLS
DIGITAL WORKBOOK — CHAPTER 10

Open the data file *CH10 Workbook*. Complete these exercises in your *DigiTools Digital Workbook* to reinforce and extend your learning for Chapter 10:

- Review Questions
- Vocabulary Reinforcement
- Math Practice: Calculating Currency Conversions
- From the Editor's Desk: Confusing Word Usage
- Keyboarding Practice: Reach Technique and Speed Building

Figure 5-17
Tables and Borders Toolbar

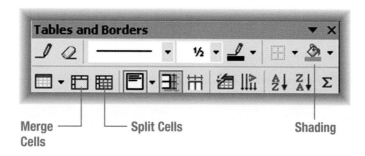

SHADING CELLS

HELP KEYWORDS

Table
Add shading, color, or graphic fills

Shading is a colored background that can be added to cells. Shading is useful for highlighting information or showing groups of data. To shade cells, select the cells. Then click the down arrow on the Shading button on the Tables and Borders toolbar. (Locate the Shading button in Figure 5-17.) Select a color for the shading. Be careful to select a shading color that allows text in the cells to be read easily. Do not make the shading too dark.

Figure 5-18
Shading Options for Cells

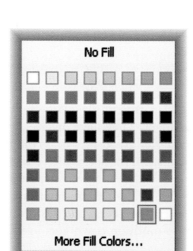

ACTIVITY 5-10 — CHANGE TABLE STRUCTURE

1. Start *Word* and open a new blank document.

2. Create a table with one column and six rows. Key the data below in the table.

Effect
Superscript
Subscript
Strikethrough
Shadow
Small Caps

3. Add supporting details for each main topic. Add additional comments in the Notes pane to help you in delivering the presentation. Write questions listeners may ask and write sample answers.

4. Choose a design template and color scheme for your slides. Review the text to appear on each slide. Make sure it is concise and includes only the important points. Add graphics and motion or sound to enhance your slides. Use a digital camera or scanner, if available, to capture pictures related to your topic.

5. Practice your presentation in front of a mirror. If possible, videotape your practice session. Focus on using your visuals correctly so they help to emphasize the key points in your presentation. Notice your posture and gestures, and concentrate on using a pleasant voice.

6. Deliver the presentation to your class or to a group of classmates as directed by your teacher. Include time for a question-and-answer session.

7. Ask your listeners to complete an evaluation form such as the one you developed earlier. Review the evaluation forms and write a summary of how the listeners rated your presentation. Note what you need to improve in future presentations. Save the self-evaluation as *Evaluation 2* in the *DigiTools your name\Chapter 10* folder.

APPLICATION 10-2

FLEXTIME TEAM PRESENTATION

TEAMWORK

INTERNET

You work in the Accounting Department for a large insurance company. Approximately 25 workers are in your department. They have diverse ages and lifestyles. During the past year, you worked with several other employees in writing a proposed plan for employee flextime options. Recently, management decided to test the flextime schedule with a two-month pilot program for employees in your department only. If the pilot program is successful, the company will offer the flextime options to most employees.

You and a team of coworkers have been asked to prepare and deliver a ten-minute presentation. The purpose is to motivate coworkers in your department to participate in the flextime pilot program.

1. Work with a team of three classmates to complete this assignment.

2. Use the Internet or other reference sources to find articles or reports about flextime. Each team member should find one or two articles (different from those found by other team members). Print or copy the articles, if possible. If not, read them online.

3. For each flextime article, key the name of the article and complete source information. Then compose and key a summary of the main points you learned from reading the article. Discuss the articles you found with your teammates.

4. Open and print the data file *Flextime*, which contains specific details about the flextime program at your company.

5. Plan the presentation. Create storyboard worksheets or develop an outline of the main points of the presentation.

3. Select all the cells in the table by clicking the **Table Move** handle. Click the **Split Cells** button on the Tables and Borders toolbar. Enter **2** for the Number of columns. Enter **6** for the Number of rows. Do not select **Merge cells before split**. Click **OK**.

> **HELP KEYWORDS**
>
> Font
> Apply strikethrough formatting

4. Enter the data below in Column B. Resize the column so all the text in each cell fits on one line. To make the various font effects, key the text without any special steps. (Key **82** for 8^2.) Select the character or words to have the special effect. Choose **Font** from the Format menu. Click the appropriate option under Effects such as **Strikethrough** or **Superscript**. Click **OK**.

Example
Add 8^2 + 10^3.
H_2O is water. CO_2 is carbon dioxide.
~~Strikethrough places a line through text~~.
This is shadow text.
THIS IS SMALL CAPS.

5. Insert a new row above Row 1. Select the cells in Row 1. Click the **Merge** button on the Tables and Borders toolbar. Key a table title, FONT EFFECTS, in Row 1. Bold the title and use center alignment.

6. Center and bold the column headings. Select Row 2. Click the **Shading** button on the Tables and Borders toolbar. Choose a light green shading. Center the table vertically and horizontally on the page.

7. Spell-check the document and correct all errors. Save the table as *Activity 5-10* in the *DigiTools your name\Chapter 5* folder. Print and close the document.

APPLICATION 5-5 EXPORTS TABLE

1. Start *Word* and open a new blank document. Create a table as shown on page 152. Merge and split cells as needed to create the table structure shown.

2. Apply light green shading to the first and last rows. Center-align the column headings. Right-align the numbers in Rows 4–9. Adjust column widths for an attractive table. Center the table vertically and horizontally on the page.

3. Spell-check and proofread the table. Correct all errors. Save the table as *Application 5-5* in the *DigiTools your name\Chapter 5* folder. Print and close the document.

ACTIVITY 10-14 — CREATE AN EVALUATION FORM

In this activity, you will create a presentation evaluation form. You will ask your classmates to use this form in evaluating your presentation.

1. Review the information from this chapter about effective speaking and presentations. Think about the information that a listener might want to comment on. Also think about the information you would find useful for improving future presentations.

2. In a *Word* document, create a list of evaluation points or questions to be included on the evaluation form.

3. Create an evaluation form using a format that will be easy to use for both you and the listeners. A sample evaluation form is shown in Figure 10-18. Include an appropriate title for the form and blanks to record the speaker name, topic, and date of the presentation. Also include a place for the listener to make comments on the form.

4. Print the form. Save the document as *Evaluation Form* in the *DigiTools your name\Chapter 10* folder.

ACTIVITY 10-15 — DELIVER A PRESENTATION

TEAMWORK

In this activity, you will deliver the presentation you created about employee benefits at Corporate View. You will also evaluate your performance and that of your classmates.

1. Work with three classmates to complete this activity. Take turns with the other students in your group so that all students complete Steps 2 and 3.

2. Deliver your presentation about Corporate View employee benefits to the group. Include a question-and-answer session at the end of your presentation. Ask fellow students to evaluate your presentation using the form you created in Activity 10-14.

3. Review the evaluations of your presentation prepared by your classmates. Then write a brief self-evaluation of your presentation. Comment on your overall performance, areas where you performed particularly well, and areas where you will try to improve for future presentations. Save the self-evaluation as *Evaluation 1* in the *DigiTools your name\Chapter 10* folder.

APPLICATION 10-1 — INDIVIDUAL PRESENTATION

In this application, you will plan, create, and deliver a five-minute presentation to inform and educate your classmates on a topic of your choice.

1. Choose a hobby, sport, or activity that you really enjoy and about which you are knowledgeable. Plan a presentation about this topic.

2. Using *PowerPoint*, create an outline to include each main idea in your topic. Enter the purpose of the presentation, listener interests, and listener advantages in the Notes for Slide 1.

INTERNATIONAL EXPORTS				
Exports	2001		2002	
Goods and Services	$ Millions	% of Total	$ Millions	% of Total
Agriculture	3,798	8.8	4,783	9.5
Mining	23,587	54.6	25,261	50.4
Manufacturing	11,582	26.8	15,438	30.8
Other Goods	410	1.0	518	1.0
Services	3,791	8.8	4,155	8.3
Total	43,168	100.0	50,155	100.0

Self Check Does your table appear similar to the one in the textbook? If not, how do they differ?

WHAT YOU SHOULD KNOW ABOUT
Private Mail Delivery Service

Almost all areas are served by express transportation companies. These companies deliver documents, packages, and freight shipments worldwide. Check under *Delivery Service* in the yellow pages telephone directory, or search the Internet to find companies in your area. Ask about services and fees to help you choose the delivery company that best meets your needs.

Figure 5-19
FedEx® is the world's largest express transportation company.

Ask your audience to evaluate the presentation also by providing an evaluation form. Evaluation forms are valuable tools that can help you improve your communication skills. To be effective, though, the evaluation form has to gather appropriate information from your listeners. Be specific about the feedback you want from the audience. Figure 10-18 shows a sample evaluation form. For example, ask listeners the following questions:

- Have you convinced them to take a course of action?
- Have they learned something from your presentation?
- Was the length of your presentation appropriate?
- Could they relate the content of your presentation to their personal experiences?

You will get the most accurate feedback if you ask your listeners to complete the evaluation immediately following your presentation. Their reactions are fresh and their comments will be more specific.

Learn from the evaluation comments and use the information constructively to improve the content and/or the delivery of your message. Each time you speak before a group, you will grow in confidence and ability. If you have the opportunity to give the same presentation again, you can refine and improve it using the feedback you receive.

Figure 10-18
Evaluation forms provide valuable feedback.

PRESENTATION EVALUATION FORM

Topic: _____ Speaker: _____
Date: _____

Please check one of the choices at the right for each of the statements below. Place your completed form on the table by the door on your way out. Thank you for your feedback.

	Very Much	Some-what	Not At All
The coverage of the topic met my expectations.	☐	☐	☐
The topic was of interest to me.	☐	☐	☐
The length of the presentation was appropriate.	☐	☐	☐
The presenter addressed the topic effectively.	☐	☐	☐
The concepts were presented clearly.	☐	☐	☐
The presenter related the information to my experiences.	☐	☐	☐
The information presented will be useful to me.	☐	☐	☐
The visuals used were helpful and appropriate.	☐	☐	☐

Comments:

What You Should Know About

Private Mail Delivery Service

continued

You must prepare a delivery form to accompany the package that includes information such as:

- Your name, address, and phone number
- The recipient's name, address, and phone number
- The type of delivery service and the weight of the package
- The payment method or account number

Many delivery companies have Web sites that allow you to track packages that have been sent using the delivery company. The Web sites also include information such as the services provided, rates, and pickup locations.

APPLICATION 5-6

RESEARCH MAILING COSTS

TEAMWORK

INTERNET

You have been asked to compare rates for the U.S. Postal Service and two private mail delivery services for mailing several types of items. Work with two classmates to complete this activity.

1. Identify two private mail delivery companies that serve your area. Search the Internet to find Web sites for the U.S. Postal Service and the two private companies.

2. Start *Word*. Open and read the data file *Mail Costs*. This file contains a simulated e-mail message to you from your manager.

3. Research the cost of mailing the items listed in the e-mail message using the U.S. Postal Service and the two private delivery services. The items will be mailed from your school location or the nearest drop-off location. Create a table that shows the costs for each service for each item. A sample of how the first part of the table might look is shown below.

MAILING COSTS COMPARISON			
Item	U.S. Postal Service	FedEx®	Airborne Express
Package to Chicago, IL 10 pounds, delivery in 10 days or less	$9.10	$6.52	$5.24
Letter to Tampa, FL 6 ounces, delivery next day	$13.65	$14.65	$12.38

ACTIVITY 10-13 — PREPARE QUESTIONS AND ANSWERS

In Activity 10-15, you will deliver the presentation you have prepared about employee benefits at Corporate View. In this activity, anticipate questions that listeners may ask and write answers to the questions.

1. Start *PowerPoint*. Open the presentation file *CV Benefits9* that you created earlier. Review each slide in the presentation.

2. Start *Word*. Write at least seven questions that listeners may ask about employee benefits after hearing your presentation. Under each question, write an answer to the question. Remember that listeners may ask for information you have already covered. They might not remember what you said, or they might not have understood your earlier comments.

Example

Question: Are part-time employees who work 20 hours per week eligible for sick leave?
Answer: No. Employees must work at least 30 hours per week to be eligible for sick leave.

3. Refer to the Employee Handbook on the Corporate View intranet to help you write answers to questions listeners may have that are not covered on your slides. (The Handbook is found in the Regular Features section of the intranet.)

4. Save the *Word* document as *Q and A* in the *DigiTools your name\Chapter 10* folder. Print the document.

Peer Check Ask a classmate to read your questions and answers and give you feedback. Are the answers clear? Should other information be included? Update your document, if needed, after your discussion.

Present Closing Remarks

Following the question-and-answer session, you have one last chance to get your point across. Your closing remarks should be a concise review of the major points in your presentation. Be careful, however, to word your closing so that you do not repeat exactly what you have already said. Restate the specific points. Then close the presentation, thanking your audience for listening.

EVALUATING YOUR PRESENTATION

After the presentation is completed, evaluate yourself. Consider the strong points of your presentation and what seemed to be effective. Think about what you could do to improve it. Did you forget to mention something? Could you have used more visuals, or did you use too many? What would you do differently the next time?

4. Write a short memo to your supervisor, Jeremy Waters, from your team to report your findings. Use the current date and an appropriate subject line.
- Recommend the mail service that your team thinks should be used for routine mailings. Recommend the service that should be used for items that must have guaranteed overnight or second-day delivery.
- Determine whether each service has a system available for customers to track packages that have been mailed. Include this information in your memo.

5. Write a short paragraph to introduce the table you created and include it in your memo. Place the table a DS below the last paragraph. Center the table horizontally.

6. Check the document for spelling and grammar errors. Save the document as *Application 5-6* in your *DigiTools your name\Chapter 5* folder. Print the memo and close the document.

LESSON 5-4: REPORTS

OBJECTIVES *In this lesson you will:*

1. Create reports.
2. Format and edit reports.
3. Use internal citations and a reference page.
4. Create headers, footers, page numbers, and lists.
5. Create a report title page.

A **report** is a document that provides information (facts, opinions, ideas, or recommendations) on a certain topic. Short reports are often prepared without covers or binders. If they have more than one page, the pages may be fastened together in the upper left corner with a staple or paper clip. Such reports are called **unbound reports**.

Refer to Figure 5-20 on page 155 as you read about reports. A report begins with a title that describes the report. The title may require one line or two lines. The paragraphs, or body, of the report present the information. The body of a report is double spaced and the paragraphs are indented 0.5" from the left margin. Side headings, which describe parts of the report, may be used to guide readers. If the report requires more than one page, the second and following pages are numbered.

A report may contain quoted or paraphrased material. References, called **textual citations**, are used to give credit for quoted or paraphrased material. They are keyed in parentheses in the body of the report. Textual citations include the name(s) of the author(s), year of publication, and page number(s) of the reference material. A **Reference page** or section at the end of the report gives complete information for each reference used in the report. This list may also be called *Bibliography* or *Works Cited*.

Figure 10-17
Plan time in your presentation for questions and answers.

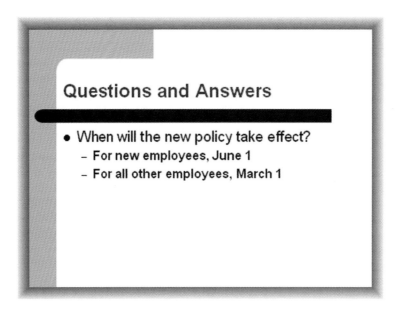

Restate the Question

When you receive a question from the audience, the entire group needs to hear the question. Generally, the person asking directs the question to the speaker, and the entire audience does not hear the question. Restate the question for everyone to hear. Doing so gives you time to think about the answer and enables you to confirm to the listener that you understand his or her question.

Respond to Questions

Respond to all questions in a courteous and sincere manner. Keep your answers brief to maintain the exchange between you and the audience. Direct your answer to the entire group, not just the person who asked the question. Maintain eye contact to keep the audience focused on the discussion. When appropriate, provide supporting details or evidence to back up your answer.

Be honest if you do not know the answer to a question. If appropriate, offer to find the answer and communicate it to the individual later. Do not become frustrated if the question relates to information already covered in your presentation. Provide the details or explain the point again briefly, perhaps using different words. Offer to provide more details on an individual basis later if appropriate.

Sometimes one individual will ask more than one question and begin to dominate (take over) the question-and-answer session. If this happens, break eye contact with the individual before he or she has an opportunity to ask another question. Upon giving an answer to that individual's question, establish eye contact with another person and ask, "Does anyone else have a question to ask?"

In a team presentation, the team members should decide in advance how they will handle questions from the audience. For example, they may decide that one team member will direct the questions to the appropriate person for an answer. All presenters should be included in the question-and-answer session.

Figure 5-20
An Unbound Report

All caps and center align title

Indent ¶s .5" and DS

1" side margins

Side Heading

SS and indent long quotes

Numbered list

2" top margin on page 1

TRENDS FOR BUSINESS DRESS
QS

   Casual dress in the workplace has become widely accepted. According to a national study conducted by Schoenholtz & Associates in 1995, a majority of the companies surveyed allowed employees to dress casually one day a week, usually Fridays (Tartt, 1995, 23). The trend continued to climb as shown by the 1997 survey by Schoenholtz & Associates. Fifty-eight percent of office workers surveyed were allowed to dress casually for work every day. About 92 percent of the offices allowed employees to dress casually occasionally (Sutphin, 2000, 10).

<u>Decline in Trend</u>

   The trend to dress casually that started in the early 1990s may be shifting, states Susan Monaghan (2000, 34):

   Although a large number of companies are allowing casual attire every day or only on Fridays, a current survey revealed a decline of 10 percent in 1999 when compared to the same survey conducted in 1998. Some experts predict the new trend for business dress codes will be a dress up day every week.

   What accounts for this decline in companies permitting casual dress? Several reasons may include:

1. Confusion of what business casual is. Employees may slip into dressing too casually (work jeans, faded tee-shirts, old sneakers, and improperly fitting clothing).

2. Casual dress does not portray the desired image of the company.

3. Employees are realizing that promotion decisions are affected by a professional appearance.

<u>Guidelines for Business Dress</u>

   Companies may hire image consultants to teach employees what is appropriate business casual dress. They help plan the best business attire to project the corporate image. Erica Gilreath (2000), the

At least 1" bottom margin

What You Should Know About Body Language
continued

Tapping fingers on the desk or fidgeting while talking or listening indicates a lack of concentration or nervousness. Constantly glancing out at the corridor or at a clock while speaking with a colleague can convey a lack of interest and may reduce the effectiveness of what you are saying.

When talking with coworkers or participating in meetings, observe the body language of the speaker and those around you. Paying attention to body language as well as words will help you understand better what others are trying to communicate. When communicating in person with coworkers or giving a formal presentation, be aware of your body language. Use your body language to reinforce the message you want to communicate.

USING VISUALS EFFECTIVELY

If you've prepared well, you have some great visuals to help you communicate your ideas. These visuals, however, are not the key to your presentation. You are the key element. Begin by drawing the listeners' attention to yourself. Then, when appropriate, direct their attention to the visuals to make your message more powerful.

Stand to the left or right of the display of your visuals. Be sure your listeners can see you and the visuals. As you display your visuals, look at your listeners, not at your visuals. Continue to maintain eye contact. Pause to allow listeners time to view and think about your visuals. For example, look for nods of agreement or expressions of disagreement. If appropriate, ask for their feedback before continuing.

QUESTIONS AND CLOSING REMARKS

Allow your audience time to ask questions. Question-and-answer sessions are valuable to you as well as the audience. They provide you an opportunity to hear from your listeners as they share what they are thinking.

Anticipate Listener Questions

Many speakers become anxious about receiving questions from an audience. They are afraid they will not be able to answer the questions or that they will lose control of the situation. As with the overall presentation, the key to feeling calm and confident when receiving questions is preparation. Anticipate what your audience will ask you following your presentation. If you have prepared well, you have probably addressed many of their concerns in the content of your presentation. However, you may get questions about content you have already covered. This means that the listener either did not understand or did not retain that information.

Perhaps when you give your audience the opportunity to ask questions, no one will do so. This situation happens frequently. Just in case no one in the audience asks a question, have some questions ready. You may even want to include a slide that contains one or two questions. This will help to fill the time you have allotted for questions and may spark some questions from the audience.

Figure 5-20
An Unbound Report (*cont.*)

Page number in header

1" top margin on page 2
1" side margins

Bulleted list

2" top margin on Reference page

SS references use .5" hanging indent

References listed in alphabetical order by author's last names

---

2

author of *Casual Dress,* a guidebook on business casual, provides excellent advice on how to dress casually for business success. She presents the following advice:

- Do not wear any clothing that is designed for recreational or sports activities, such as cargo pants or pants with elastic waist.
- Press pants, skirts, shirts, and other clothing. Wrinkled clothing does not present a professional image.
- Do not wear sneakers.
- Be sure clothing fits properly. Avoid baggy clothes or clothes that are too tight.

In summary, conscientious employees need to plan their dress carefully. If business casual is appropriate, it's best to consult the experts on business casual to ensure a professional image.

3

REFERENCES
QS

Gilreath, Erica. "Dressing Casually with Power." http://www.dresscasual.com (23 March 2001).

Monaghan, Susan. "Business Dress Codes May Be Shifting." *Business Executive,* April 2000, 34-35.

Sutphin, Rachel. "Your Business Wardrobe Decisions Are Important Decisions." *Business Management Journal,* January 2000, 10-12.

Tartt, Kelsey. "Companies Support Business Casual Dress." *Management Success,* June 1995, 23-25.

- Control your posture and gestures. Do not constantly pace back and forth, shuffle your feet, or shift weight from one leg to the other. Stand with your feet slightly apart and firmly planted, moving or shifting your weight only occasionally. Leave your hands at your sides until you use them for natural gestures that enhance your words. Your listeners will be watching your body language as they listen to you speak. Make sure your posture and gestures are not communicating something different than the message you want them to receive.

Figure 10-16
Gestures can enhance or detract from a speaker's words.

WHAT YOU SHOULD KNOW ABOUT
Body Language

More than words are a part of communicating when talking with coworkers in person or giving a presentation. You also communicate to others through your body language. **Body language** is posture, body movements, gestures, and facial expressions that serve as nonverbal communication. Understanding nonverbal cues can help you communicate your messages more effectively.

Facial expressions are important nonverbal cues. A smile can convey understanding or support for what is being said. A frown, on the other hand, may indicate lack of understanding or disagreement. A smile at an inappropriate time, however, may convey smugness or insensitivity. Raised eyebrows may convey surprise or disapproval.

Eye movement can signal your attentiveness and openness. Making eye contact frequently shows your interest in what is being said and that you are being honest or open when speaking. Letting your eyes roam around and seldom making eye contact can signal that you are not paying attention or that you are being evasive.

Posture and gestures are important elements of body language. Good posture projects confidence and being at ease. Slouching or stooping indicates an indifferent attitude or lack of self-confidence. Leaning closer or nodding conveys interest. Leaning or turning your body away conveys discomfort or disagreement with what is being discussed. Crossed arms show a defensive or unwelcome attitude. Sitting calmly with hands folded in your lap conveys an openness to listen. Placing your hand to your cheek generally indicates that you are evaluating or considering, while placing your hand over your mouth generally indicates your disapproval.

MARGIN SETTINGS

Margins are the blank space between the edge of the paper and the printed text. The default settings are 1.25" side margins and 1" top and bottom margins. You will need to change margins when you key reports. You can change margins using the Page Setup dialog box. To open this box, choose **Page Setup** from the File menu. On the Margins tab, enter the setting you want for margins and click **OK**.

Figure 5-21
Page Setup Dialog Box

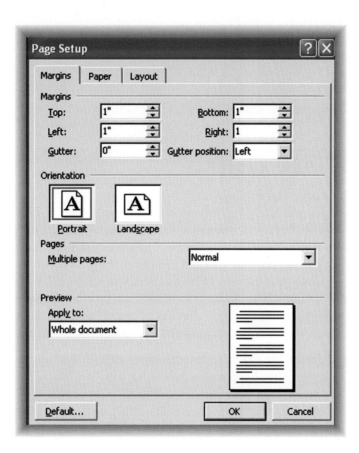

NUMBERED AND BULLETED LISTS

Lists are used to emphasize information in a report or make it easier to read and understand. When the items in the list require a certain order, a numbered list is used. When the order of the items is not important, a bulleted list is used. See Figure 5-20 on pages 155 and 156 for examples of bulleted and numbered lists.

Single-space a list when each item in the list is short and fits on one line. For lists with longer items, single-space the items and double-space between the items. To create a list, key the items. Select the items and click the **Bullets** or **Numbering** button on the Formatting toolbar.

Numbering button — Bullets button

ACTIVITY 10-12   PREPARE OPENING REMARKS

In Activity 10-15, you will deliver the presentation you have prepared about employee benefits at Corporate View. In this activity, plan the opening remarks you will make in the presentation.

1. Start *PowerPoint*. Open the presentation file *CV Benefits8* that you created earlier.

2. On the Notes pane for Slide 1, write the comments you will make at the beginning of the presentation. Pretend you work in the Human Resources Department at Corporate View. Introduce yourself, welcome the new employees, and tell them the purpose of your presentation.

3. Save the presentation file as *CV Benefits9* in the *DigiTools your name\Chapter 10* folder.

4. Print the Notes for Slide 1. Close the file and *PowerPoint*.

COMMUNICATING WITH YOUR AUDIENCE

To deliver your message effectively, you must communicate with your audience. Your listeners will likely be interested in your presentation if you appear relaxed and comfortable. Being nervous is normal. Naturally, you want to do well, and you may experience some nervousness. Take deep breaths, concentrate on talking slowly, and think about what you're going to say next. You may think that you are talking too slowly, but that is generally not the case. The listeners will understand that you are giving thought to what you are about to say. Maintain a positive self-image and an upbeat attitude. Remind yourself that you are well prepared and can deliver the message effectively.

Follow these guidelines to help you deliver an effective presentation:

- Maintain eye contact. Making eye contact helps you involve your listeners and know whether your audience is following you. You can judge their reactions to what you are saying, and your listeners will feel more involved and important.

- Avoid using non-words—verbal sounds that have no meaning. Often we do not realize we say non-words such as "uhh," and "ah." Review your videotape or have a friend help you identify non-words that you use. Also, avoid using real words in an unmeaningful way. Do not begin many sentences with *okay, you know,* or *well*.

- Speak with enthusiasm and conviction. If you believe in what you are saying, let your listeners know. Show a sincere interest in helping your listeners meet their needs. Describe experiences or examples with which your listeners can identify. Speak with a pleasant voice that is neither too high nor too low. Your voice should sound relaxed and have an even tone.

- Keep the audience focused. If they seem confused or distracted, back up and rephrase your point. If you sense that you are losing their attention, try to focus again on listener interests and advantages.

ACTIVITY 5-11

CREATE A SIMPLE REPORT

1. Start *Word* and open a new blank document.
2. Click **Page Setup** on the File menu. Enter **1"** for the top, bottom, left, and right margins. Click **OK**. (Refer to Figure 5-21.)
3. Click the **Line Spacing** button and choose **2**. Press **Enter** three times to leave an approximate 2" top margin.
4. Key the report shown below. Key the title in all caps and use center alignment. Press **Enter** twice. Press **Tab** to move to the default .5" tab stop and begin keying the first paragraph.
5. When you press **Enter** after the first paragraph, *Word* will automatically indent to the first Tab stop. Press **Backspace** to move to the left margin to key the side heading.
6. Key the items in the list. Select the items and change to single spacing. Select the items and click the **Numbering** button on the Formatting toolbar.
7. Check the spelling and format of your document. Correct all errors. Save the document as *Activity 5-11* in the *DigiTools your name\Chapter 5* folder. Print and close the document.

HELP KEYWORDS
Margins
Change page margins

HELP KEYWORDS
Number
Add bullets or numbering

```
                    PROCESSING INCOMING MAIL

     Mail for various individuals and departments is
all mixed together when it is delivered to a company.
Most companies want all mail sorted quickly so that it
can be handled promptly.

Prioritizing Mail

     Incoming mail should be prioritized (sorted in
order of importance) to ensure proper handling. The
following guidelines are usually satisfactory for
sorting mail, moving from the top to the bottom of the
stack:

          1. Urgent messages that require prompt attention
          2. Personal and confidential letters
          3. Business letters, memos, or other
             correspondence of special importance
          4. Letters containing checks or money orders
          5. Letters containing orders
```

(continued on next page)

PREPARING TO PRESENT

Review each of your visuals and the notes you have created to accompany them. Rehearse out loud exactly what you plan to say. Make sure you state each idea from the listeners' point of view. Also be sure to provide listener advantages.

For team presentations, each presenter must know the content he or she will present. Although each member should use his or her own style of speaking, the overall theme should be reflected by everyone. All team members should be present, even if some of them do not present. Most importantly, the team should practice the presentation as a group. You may wish to videotape your presentation so you can evaluate and critique yourself.

The Meeting Room

There are many factors to consider regarding the meeting room. Make sure the seating arrangement is appropriate. Arrive early to set up equipment and support materials. Practice ahead of time with the specific equipment to be used in the presentation, such as a computer or projector. Test the audio equipment and know who to call to get help with the equipment.

Check the lighting in the room and determine the best light level to use for the presentation. Even though your visuals may look good in a dark room, you want enough light so your audience can clearly see you. Check the room temperature. Remember that bodies heat up a room. Setting the temperature to about 68 degrees will usually provide a comfortable environment when the room is filled with people.

Your Appearance

Your appearance can affect how your audience receives your message. Dressing appropriately can help you gain their respect and hold their attention. When inappropriate, your appearance can distract your listeners. Good grooming is important. Be neat and clean in your appearance. Get a good night's rest before the presentation so you can look and be alert. For formal presentations, business suits are appropriate. For informal presentations, your attire can be more casual depending on the audience. Dress comfortably but conservatively.

OPENING REMARKS

In a small- or large-group presentation, another person may introduce you to your audience. You should also introduce yourself. In your opening remarks, be sure to state your purpose. Your opening remarks help to set the tone for your presentation. You may choose to use a visual for this introduction.

Figure 10-15
Use a visual to introduce a key opening remark.

```
            6. Letters containing bills, invoices, or other
               requests for payment
            7. Advertisements, newspapers, and magazines
            8. Packages

       Safety Precautions

            Office workers should take care to protect
       themselves against dangerous substances that might be
       present in envelopes or packages received via mail.
       Wearing gloves and a face mask can provide some
       protection from airborne substances that might be
       dangerous. When handling mail, avoid touching your
       face and mouth to help prevent the transfer of germs.
       Wash your hands with disinfectant soap after handling
       mail. The United States Postal Service recommends that
       you not handle a piece of mail that you suspect is
       dangerous.
```

ETHICS

Protecting Confidential Information

Confidential information is information that is private or secret. Giving out confidential information from your workplace is unethical. Release of such information could cause harm to the business or its employees, clients, or customers.

You may come in contact with confidential information as you handle mail, prepare documents, or access employee or customer records. The following guidelines will help you keep business information confidential:

- Know what information you should and should not give to visitors or callers. When in doubt, ask your supervisor or manager.

- Place confidential mail in a folder or in a secure location where it will be seen only by the intended recipient. Do not send confidential information by fax or e-mail. Use overnight mail services if speed is a consideration.

- Take precautions to keep others from reading confidential information from your computer screen. Place confidential mail or papers in a drawer when you leave your desk—even for a few moments. Shred confidential documents rather than placing them in your wastebasket.

- Reduce electronic information loss. Use password log-on and log-off procedures. Change your password frequently. Be alert in removing printouts from the printer when you finish the print job, particularly if the printer is shared with others. Make backup copies of confidential files and place them in a secure location.

Questions for Discussion

1. Give some examples of types of company information that might be confidential.

2. Why do you think confidential information should not be sent by e-mail?

3. Start *PowerPoint*. Choose **New** from the File menu. In the Task pane, choose **From AutoContent Wizard**. On the AutoContent Wizard dialog box, click **Next**.

4. Click the **General** button and then click **Training**. Click **Next**.

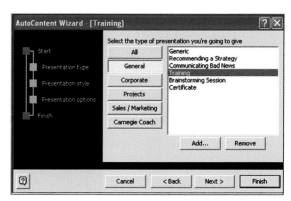

5. For type of output, choose **On-screen presentation**. Click **Next**.

6. For the presentation title, key `Phone Script Training`. In the Footer box, key `Copyright 20-- Corporate View`. Click **Next**. Click **Finish**. The slide show created by the Wizard appears. Play the show to review the slides and animation effects used.

7. Edit the content of the slides. Include all the main points about creating phone scripts from the printed intranet pages. Add subpoints to the slides. Insert more slides or rearrange slides if needed. Add graphics where appropriate. For example, you might add clip art of a telephone to the title slide.

8. Save the file as *Phone Training* in your *DigiTools your name\Chapter 10* folder. Close the file and *PowerPoint*.

Peer Check Ask a classmate to review your slides and offer suggestions for improvement. Update the slides after your discussion.

LESSON 10-3 DELIVERING THE PRESENTATION

OBJECTIVES *In this lesson you will:*

1. Practice, deliver, and evaluate presentations.
2. Learn how personal dress and the meeting room conditions affect presentation success.
3. Apply proper techniques for communicating with an audience.
4. Use visuals effectively.
5. Conduct question-and-answer sessions.

Now that you have prepared your message, you are to put your ideas in motion. The more experienced you become in speaking to others, the less practice you will need. If you are new at making presentations, you will definitely want to practice. You can practice before friends or colleagues, or you can rehearse the presentation on your own.

HEADERS AND FOOTERS

A **header** is information placed at the top of each page in a document. A **footer** is information placed at the bottom of each page in a document. You can use the Header feature to place page numbers in a report. Read about creating a header below. You will practice creating a header in Activity 5-12.

To create a header for page numbers:

- Click **Header and Footer** on the View menu. A grid area for the header information and the Header and Footer toolbar will display.

Figure 5-22
Header and Footer Toolbar

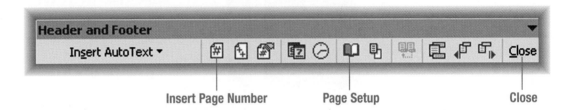

- Press the **Tab** key twice to move to the right margin position in the header grid area.
- Click the **Insert Page Number** button to insert a page number. This feature will automatically number each page correctly as it is added to your report.
- Click the **Page Setup** button. Choose **Different First Page** and click **OK**. This will prevent the header (the page number) from printing on the first page.
- Click **Close** on the toolbar. Use Print Preview to view the header.

INDENTATION OPTIONS

Word allows you to set options that will automatically indent text in paragraphs. You can use the Indentation Left and Right options to indent a paragraph from the left or right margin. The Indentation Left option is useful for indenting long quotes in a report.

Two types of Special Indentation options are used in creating reports. The **First line** option allows you to tell *Word* to indent the first line of each paragraph. This option is useful for keying the body of a report. The **Hanging** option sets the first line of a paragraph to begin at the left margin and all other lines to be indented. This option is used in creating reference entries. To access the indentation options, choose **Paragraph** from the Format menu.

Figure 5-23
The Paragraph Dialog Box

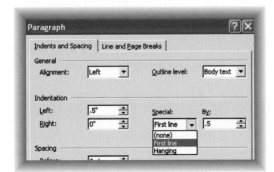

14. Notice that Notes has a red check mark beside it on slides that have notes. Because notes are usually for the presenter's use, you may want to remove notes from slides before saving them as a Web page.

Figure 10-14
Presentation Web Page

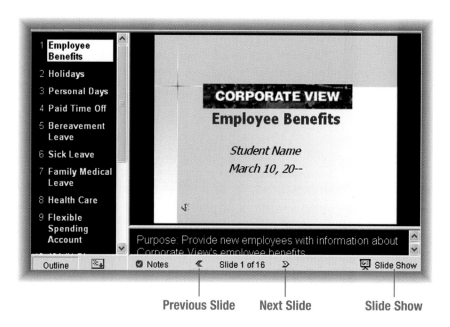

15. Move to Slide 1. Click **Slide Show** on the bottom right of the screen to play the show. Click the mouse or press the **Right Arrow** key to move to the next bullet point or slide. Press the **Escape** key if you want to exit before viewing all slides. The sounds you inserted should play on the first and last slides.

16. Close your browser.

Self Check Does your Web page appear similar to Figure 10-14?

ACTIVITY 10-11 USE A PRESENTATION WIZARD

HELP KEYWORDS

Wizard
Create a presentation using suggested content

PowerPoint provides an AutoContent Wizard to help you create certain types of presentations more easily. You choose a few options about the type of presentation. Then the Wizard creates a slide show, complete with a design template and sample content.

You have been asked to help train Corporate View employees to write effective phone scripts. A phone script might be used when making sales calls or taking a survey. In this activity, you will use the AutoContent Wizard to create your presentation.

1. Access the Corporate View intranet. Choose the **Corporate Communications** link followed by **Style Guide**. Choose **Phone Scripts**.

2. Print pages 1 and 2 of the Web page. Choose **Print** from the File menu. Under Print Range, select **Pages**. Enter 1 - 2 in the text box. Click **OK**. Close your browser. Read the information about phone scripts.

PAGE BREAKS

As you key text in a document, *Word* automatically begins a new page when the first one is filled. These page breaks are called soft page breaks because they will shift automatically if you later key more text in the document. When you create reports, however, you may need to use manual or hard page breaks.

A hard page break is entered by the user and will remain in place unless you move it. When you create reports, you must use page breaks that make the report easy to read. For example, keep a side heading with at least one line of the text that follows it. When dividing a page in the middle of a paragraph, leave at least two lines of the paragraph at the bottom of the first page and carry at least two lines to the second page.

To insert a hard page break, hold down the **Ctrl** key and press **Enter**. Use **Backspace** to delete a hard page break.

ACTIVITY 5-12 — CREATE A REPORT WITH REFERENCES

1. Start *Word* and open a new blank document.

2. Click **Page Setup** on the File menu. Enter **1"** for the top, bottom, left, and right margins. Click **OK**.

3. Click the **Line Spacing** button and choose **2**. Press **Enter** three times to leave an approximate 2" top margin.

4. Key the title in all caps and use center alignment. Press **Enter** twice after the title.

5. Set the indentation for the report body. Choose **Paragraph** from the Format menu. Click the down arrow under **Special**. Choose **First line**. Enter **.5"** in the By text box. Click **OK**.

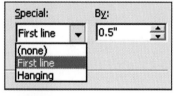

HELP KEYWORDS
Number
Add bullets or numbering

6. Key the report shown on pages 162–164. Key the items in the bulleted list. Select the items and change to single spacing. DS between the items. Select the items and then click the **Bullets** button on the Formatting toolbar.

7. Insert a hard page break (**Ctrl+Enter**) after the last paragraph to start a new page for the references. Press **Enter** three times to leave about a 2" top margin on the References page. Key REFERENCES in all caps and center-align it. Press **Enter** twice.

8. Change the line spacing to 1. Set a paragraph indent for the references. Choose **Paragraph** from the Format menu. Click the down arrow under Special. Choose **Hanging**. Enter **.5"** in the By text box. Click **OK**. Key the references as shown. DS between references.

9. Examine how the page breaks between pages 1 and 2. Enter a hard page break if needed.

5. Click in the text box and key

```
Contact:
    Nigel Stewart, Human Resources
    Phone: 555-0145
```

6. Select the text and change the font size to 28.

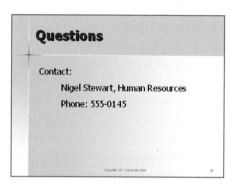

Note: If you were not able to record a sound file in Activity 10-4, skip Steps 7 and 8; go to Step 9.

7. Move to Slide 1. Choose **Movies and Sounds** from the Insert menu. Choose **Sound from File**. Navigate to your *DigiTools your name\Chapter 10* folder. Select the *Corporate View1* file you recorded earlier. Click **OK**. Choose **Yes** to have the sound play automatically. Drag the Sound icon to the bottom left corner of the screen.

8. Play the show to hear your sound file played automatically. Press **Escape** on Slide 2.

9. Move to the last slide in the show. Choose **Movies and Sounds** from the Insert menu. Choose **Sound from File**. Navigate to your data files. Select the *Benefits1* file. Click **OK**. Choose **Yes** to have the sound play automatically. Drag the Sound icon to the bottom left corner of the screen. Play the show to hear your sound file played automatically.

10. Save the file again using the same name.

11. Now save the file for use on the intranet. Choose **Save as Web Page** from the File menu. Save the file as *Employee_Benefits* in your *DigiTools your name\Chapter 10* folder. **Web Page (*.htm, *html)** should appear in the Save as type box. Click **Save**. Close *PowerPoint*.

12. Open your browser. Choose **Open** from the File menu. Click the **Browse** button and navigate to your *DigiTools your name\Chapter 10* folder. Select *Employee_Benefits*. Click **Open**. Click **OK**.

13. Notice that the titles of your slides appear at the left of the window. Slide 1 is displayed at the right. You can navigate by clicking a slide title or by using the **Previous Slide** and **Next Slide** buttons at the bottom of the screen. (See Figure 10-14.) Practice using these buttons or clicking a slide title to move from slide to slide.

HELP KEYWORDS

Sound
Add music or sound effects

10. Create a header for the page number:
- Click **Header and Footer** on the View menu.

- Press the **Tab** key twice to move to the right margin position in the header grid area.

- Click the **Insert Page Number** button to insert a page number.

- Click the **Page Setup** button.

- Choose **Different First Page** and click **OK**.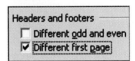

- Click **Close** on the toolbar.

11. Check the spelling and format of your document. Correct all errors. Save the document as *Activity 5-12* in the *DigiTools your name\Chapter 5* folder. Print and close the document.

```
                TRENDS FOR BUSINESS DRESS

     Casual dress in the workplace has become widely
accepted. According to a national study conducted by
Schoenholtz & Associates in 1995, a majority of the
companies surveyed allowed employees to dress casually
one day a week, usually Fridays (Tartt, 1995, 23). The
trend continued to climb as shown by the 1997 survey
by Schoenholtz & Associates. Fifty-eight percent of
office workers surveyed were allowed to dress casually
for work every day. About 92 percent of the offices
allowed employees to dress casually occasionally
(Sutphin, 2000, 10).

Decline in Trend

     The trend to dress casually that started in the
early 1990s may be shifting, states Susan Monaghan
(2000, 34):

          Although a large number of companies are
     allowing casual attire every day or only on
     Fridays, a current survey revealed a decline of
     10 percent in 1999 when compared to the same
     survey conducted in 1998. Some experts predict
     the new trend for business dress codes will be a
     dress up day every week.
```

(continued on next page)

ACTIVITY 10-9 — ADD MOTION

In this activity, you will add slide transitions and motion effects for bulleted lists.

1. Start *PowerPoint*. Open the file *CV Benefits 6* that you created earlier. Save the file using the new name *CV Benefits 7*.

2. Choose **Animation Schemes** from the Slide Show menu. Choose the **Wipe** animation scheme from the Subtle category. Click **Apply to All Slides**.

3. Move to Slide 1. Choose **No Animation** from the animation schemes list. Play the show to see the animation effects. Click the mouse or press **Space Bar** or **Right Arrow** to make each bullet point appear.

4. Save the show again using the same name.

Self Check Do all your slides have animation except Slide 1? If not, try Steps 2 and 3 again.

ACTIVITY 10-10 — ADD SOUND

New employees at a company are often presented with a lot of new information in a short time. The new Corporate View employees may want to review the information you present about employee benefits at a later time. In this activity, you will add a slide to the end of the show and add sound to two slides. You will save the file as a Web page that can be placed on the Corporate View intranet. Employees can access the show on the intranet to review it later.

1. Start *PowerPoint*. Open the file *CV Benefits 7* that you created earlier. Save the file using the new name *CV Benefits 8*.

2. Move to the last slide in the presentation. Choose **New Slide** from the Insert menu. Choose a **Title Only** slide from the layouts available in the Task pane.

3. Enter `Questions` for the slide title.

4. Click the **Text Box** button on the Drawing toolbar. Draw a text box under the title box on the slide.

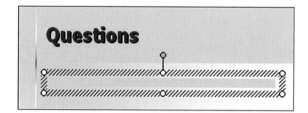

What accounts for this decline in companies permitting casual dress? Several reasons may include:

1. Confusion of what business casual is. Employees may slip into dressing too casually (work jeans, faded tee-shirts, old sneakers, and improperly fitting clothing).

2. Casual dress does not portray the desired image of the company.

3. Employees are realizing that promotion decisions are affected by a professional appearance.

Guidelines for Business Dress

Companies may hire image consultants to teach employees what is appropriate business casual dress. They help plan the best business attire to project the corporate image. Erica Gilreath (2000), the author of *Casual Dress*, a guidebook on business casual, provides excellent advice on how to dress casually for business success. She presents the following advice:

- Do not wear any clothing that is designed for recreational or sports activities, such as cargo pants or pants with elastic waist.

- Press pants, skirts, shirts, and other clothing. Wrinkled clothing does not present a professional image.

- Do not wear sneakers.

- Be sure clothing fits properly. Avoid baggy clothes or clothes that are too tight.

In summary, conscientious employees need to plan their dress carefully. If business casual is appropriate, it's best to consult the experts on business casual to ensure a professional image.

(continued on next page)

To format a header or footer, you must access the slide master. Any information placed on the slide master will appear on all slides except the title slide. To access the slide master, choose **Master** from the View menu. Then choose **Slide Master**. On the master, select the header or footer placeholder you want to make changes to. You can resize the placeholder or move it. You can change the text and the font style or size for a header or footer.

ACTIVITY 10-8

ADD A FOOTER

> **HELP KEYWORDS**
> **Footer**
> Add headers and footers

In this activity, you will add a footer to your Corporate View slides.

1. Start *PowerPoint*. Open the file *CV Benefits5* that you created earlier. Save the file using the new name *CV Benefits6*.

2. Choose **Header or Footer** from the View menu. On the Slide tab, select the **Footer** checkbox. Key `Copyright 20-- Corporate View` in the footer text box. (Use the current year.)

3. Select the **Slide number** checkbox. Select the **Don't show on title slide** checkbox. Deselect (uncheck) the **Date and time** checkbox. Click the **Apply to All** button.

4. Play the show to view the footer. Note that the slide number changes automatically for each slide. Note also that the footer does not appear on the title slide.

5. Save the file using the same name. Print Slide 2. Choose **Print** from the File menu. Select **Slides** under Print range. Key 2 in the Slides text box. Under Print what, select **Slides**. Under Color\grayscale, choose **Grayscale**. Click **OK**. Note that the footer is printed at the bottom of the slide.

ADDING MOTION AND SOUND

You can add motion to an electronic slide presentation in a variety of ways. Bullet lines can be set to appear on screen one at a time at your command or after a certain period of time (usually seconds) that you specify. Using this technique is effective when you want to emphasize each point as you present it.

Transitions in an electronic slide show are the motions used to move from one slide to the next. A variety of transition styles are available. To animate bullet points or apply a transition in *PowerPoint*, choose **Animation Schemes** from the Slide Show menu. Choose an animation scheme from the Subtle, Moderate, or Exciting category. Because you do not want the motion to distract listeners from your message, choices from the Subtle category are usually most appropriate.

> **HELP KEYWORDS**
> **Animation**
> Animate text and objects

You can place links in your slides to video or audio you want to play during the presentation. You might include an audio file of a company executive speaking, or you might play a video that shows how to operate a piece of equipment. Be sure the audio or video is appropriate for the content of the message and that it is fairly short so you do not lose the attention of the audience.

REFERENCES

Gilreath, Erica. "Dressing Casually with Power." http://www.dresscasual.com (23 March 2001).

Monaghan, Susan. "Business Dress Codes May Be Shifting." *Business Executive*, April 2000, 34-35.

Sutphin, Rachel. "Your Business Wardrobe Decisions Are Important Decisions." *Business Management Journal*, January 2000, 10-12.

Tartt, Kelsey. "Companies Support Business Casual Dress." *Management Success*, June 1995, 23-25.

Self Check Compare your report to report illustrations on pages 155 and 156. Make changes, if needed, to correct the report format.

APPLICATION 5-7 — TEAM REPORT

In Exercise 1-2 for Unit 1, you worked in a team to find information about a school, university, college, or technical training center that provides educational programs related to careers in the Agriculture and Natural Resources cluster. Your team was asked to write a short description of each school and what it offers to students in each of the areas being researched. Work with the same teammates again. Use the information you found earlier to create a short report.

1. Start *Word* and open a new blank document. Key an appropriate title for the report. Use the name of the various schools you describe as side headings. Use internal citations and a References page to list the Web sites or other sources where you found information.

2. Format the report in unbound style. Create a header for page numbers. Create a References page. Insert hard page breaks as needed.

3. Spell-check and proofread your report. Correct all errors. Save the report as *Application 5-7* in the *DigiTools your name\Chapter 5* folder. Print and close the document.

What You Should Know About

Digital Cameras and Scanners

continued

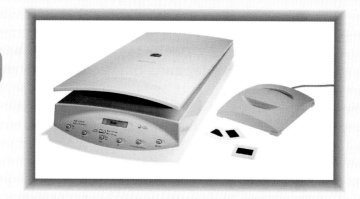

Figure 10-12
A scanner can be used to convert photographs to a digital image.

Once you have created an image using a digital camera or a scanner, you can modify the image using a graphics program. *Adobe Photoshop* is a popular program for editing pictures. You may want to change the file format of your image depending on how you plan to use it. File formats such as TIFF or BMP work well for documents that will be printed. The JPG file format works well for images that will be viewed online, in a Web page for instance.

HEADERS AND FOOTERS

A header is information that appears at the top of slides in a presentation. A footer is information that appears at the bottom of slides in a presentation. In a footer, for example, you might want to include the slide number, the company name, a copyright notice, or the date.

To create a footer or header, choose **Header or Footer** from the View menu. On the Slide tab, select options and key text to appear in a footer on your slides. On the Notes and Handouts tab, select options and key text to appear in a footer or header on your notes or handouts.

Figure 10-13
Header and Footer Dialog Box

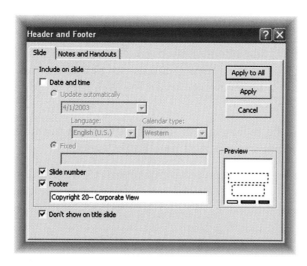

LESSON 5-5

GRAPHICS

OBJECTIVES *In this lesson you will:*

1. Insert graphics in a document.
2. Size and move graphics.
3. Copy and paste graphics.
4. Set page orientation.

Some documents, such as reports, flyers, or invitations, are more interesting when they contain graphics. A **graphic** is an image such as a photo, a drawing, or a chart. A colorful photo or drawing can catch the reader's eye and help gain the reader's attention. Other graphics, such as bar graphs or pie charts, may help readers understand information more easily than a table of numbers.

Graphics can be placed in the document using the Copy and Paste commands or the Insert Picture command. After an image has been placed in a document, you can move and resize the image.

CLIP ART

HELP KEYWORDS

Clip Art
Find a clip

Microsoft Office provides a collection of images called **Clip Art** that you can use in documents. The clips are organized into collections (Animals, Plants, Special Occasions, etc.) to help you locate a particular kind of clip more easily. You can use the Clip Organizer to browse and view images in the collection as shown in Figure 5-24.

Figure 5-24
Microsoft Clip Organizer

WHAT YOU SHOULD KNOW ABOUT
Digital Cameras and Scanners

A **digital camera** is a device that creates photographs in an electronic format. These photographs can be added to slides to enhance a presentation. For example, suppose you are creating a presentation for a sports booster group. You want to convince them to raise money to buy new equipment for your sports team. Action pictures of the team playing hard would be perfect for your slides.

The features of digital cameras vary from basic to more advanced models. Consult the user's manual to learn the specific features and operating instructions for a particular camera. Many cameras have auto modes or settings that make using them easy.

Figure 10-11
A digital camera can be used to take photos for a presentation.

Resolution and compression are two settings that you need to be familiar with when using digital cameras. The resolution is the number of pixels the camera uses when creating an image. Images with high resolution show more detail. However, they also require more space. This means that your camera's memory card cannot hold as many high-resolution pictures as it can low-resolution pictures.

Compression settings also affect the quality of your pictures. Using a low compression setting means you will have better quality pictures but large file sizes. Using a higher compression setting generally gives you a lower-quality picture but a smaller file size. You may have to decide on a compromise between image quality and manageable file sizes.

A **scanner** is a device that makes an electronic image of a document or other object, such as the cover of a book, a photograph, or a slide. Flatbed scanners have a glass plate and a cover where you place objects to be scanned, similar to a photocopier. With other scanners, you feed items (sheets of paper or photographs) through the scanning device.

The features of scanners vary, depending on the model. Many scanners allow you to choose options such as:

- Scanning to a black-and-white, grayscale, or color image
- Cropping or sizing an image
- Rotating an image
- Reducing or enlarging an image
- Scanning the text of a document to a text file that can be edited
- Changing the print quality or image resolution

Many scanners allow you to choose from several file formats when you save the scanned image.

DRAWING TOOLBAR

Commands and options for working with images are found on the Drawing toolbar. The Insert Clip Art button allows you to search for clip art to place in a document or to open the Clip Organizer to browse for images. The Insert Picture button allows you to insert images from other sources.

To display the Drawing toolbar, choose **Toolbars** from the View menu. Then choose **Drawing**.

Figure 5-25
Drawing Toolbar

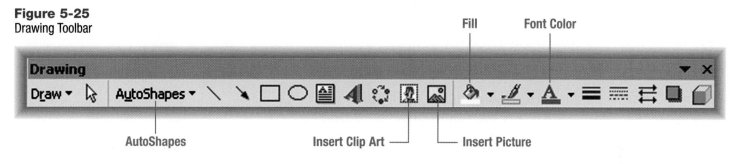

RESIZING AND MOVING IMAGES

To change the size of an image in a document, first click on the image to select it. The image will display with handles along its border. To size the image, point to a handle. When the cursor displays as a double arrow, click and drag the image. Drag a corner handle to keep the image in its original shape (square, rectangle) as it is sized. Drag a handle along the top or bottom border to make the image longer. Drag a handle along a side border to make the image wider.

HELP KEYWORDS
Clip Art
Position graphics and text

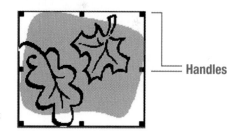

When some types of images are placed in a document, they have the *In line with text* format by default. To make moving an image easier, change the format. Right-click on the image and choose **Format Picture**. Click the **Layout** tab on the Format Picture dialog box. Choose **Tight** under Wrapping style. Click **OK**.

Figure 5-26
Format Picture Dialog Box

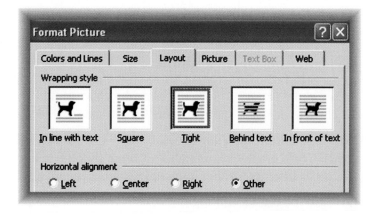

9. Move to the slide about health care. Use AutoShapes to add and group graphics on this slide.
- Click **AutoShapes** on the Drawing toolbar. Choose **Basic Shapes**. Choose a square.
- Click and drag on the slide to draw a rectangle. Move the rectangle to the bottom right portion of the slide.
- Click the rectangle to select it. Click the **Fill Button** on the Drawing toolbar. Choose a black fill color.

10. Now add a second shape.
- Click **AutoShapes** on the Drawing toolbar. Choose **Basic Shapes**. Choose a cross.
- Click and drag on the slide to draw a cross. Move the cross to center it on top of the rectangle.
- Click the cross to select it. Click the **Fill Button** on the Drawing toolbar. Choose a red fill color.

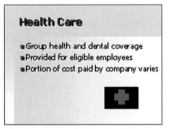

11. Group the two AutoShapes.
- Click the rectangle to select it.
- Hold down the **Shift** key and click the cross. Both objects should now be selected.
- Choose **Group** from the Draw menu on the Drawing toolbar.

The two objects can now be moved, resized, copied, or deleted as if they were one object.

12. Move to the slide about paid time off. Change the slide layout. Choose Slide Layout from the Format menu. Choose the **Title, Text, and Content** layout from the Task pane. Insert clip art that relates to a vacation or travel. Choose **Picture** from the Insert menu. Choose **Clip Art**. In the search box, key `travel`. Click on the globe and baggage image to place it in your slide.

13. Move to the slide about stock options. Change the slide layout to **Title and Text over Contents**. You will insert a chart showing Corporate View stock prices in this slide.

14. Open *Excel*. Open the data file *Stock Chart*. Click the chart to select it. Click **Copy** on the Standard toolbar. Move to *PowerPoint* and click in the stock options slide. Click **Paste** on the Standard toolbar. Resize the chart to fill the bottom half of the slide. Move the chart to center it approximately on the slide. Close *Excel*.

15. Choose from the View menu. Look over your slides. Five slides should contain graphics. Add graphics to at least two more slides. Use clip art, draw AutoShapes, or search the Internet for free (noncopyrighted) clip art or photos.

16. Save the file again using the same name. Close the file.

To move the image, move the cursor over the image until the cursor becomes a four-sided arrow. Click and drag the image to the desired location. Any text in your document will wrap around the image.

Figure 5-27
Text can be wrapped around an image.

PAGE ORIENTATION

Page orientation is the direction in which text is placed on the page relative to the shape of the page. A standard size piece of paper used for documents is 8.5" by 11". The short side of the paper is placed at the top and text runs down the page. This layout is called portrait orientation.

Sometimes you may have a table or an image that is too wide to fit on the page in portrait orientation. You can change the layout so the long side of the paper is placed at the top. This is called landscape orientation. This layout is often used for flyers or charts.

To change the page orientation, choose **Page Setup** from the File menu. Choose the desired orientation and click **OK**.

Portrait Orientation

Landscape Orientation

ACTIVITY 5-13 — INSERT CLIP ART

HELP KEYWORDS
Orientation
Select page orientation

1. Start *Word*. Open the data file *Valentine*.

2. Change the page orientation. Choose **Page Setup** from the File menu. Click the **Margins** tab. Choose **Landscape** under Orientation. Click **OK**.

3. To insert clip art, click the **Insert Clip Art** button on the Drawing toolbar. In the Task Pane, enter `valentine` or `hearts` or `party` in the Search text box. Click **Search**.

4. In the Slide pane, click in the title text box. Key `Lodge Discounts` for the title of the slide. Click in the text box on the left side of the slide. Enter these points:

- `Corporate View Lodge offers 20% discount`
- `Boulder location only`
- `All employees are eligible`

5. Now copy and paste a picture of the Corporate View Lodge into your slide.
- Access the Corporate View intranet.
- Choose the following links: **Regular Features, Corporate View Archive, Corporate Communications Documents, CVTravel Pictures.**
- Click the first image of a Corporate View Lodge, *Lodeg1.jpg*.
- Right-click the image and choose **Copy** from the menu.
- Move to *PowerPoint*. Click on the Lodge Discounts slide. Click the **Paste** button on the Standard toolbar.
- Close the Corporate View intranet.

> **HELP KEYWORDS**
> Move slide
> Change slide order

6. After giving the presentation more thought, you decide to present the slides in the order shown below. To rearrange slides, choose **Slide Sorter** from the View menu. Click and drag the slides to place them in the new order.

1. Title slide
2. Holidays
3. Personal Days
4. Paid Time Off (PTO)
5. Bereavement Leave
6. Sick Leave
7. Family Medical Leave
8. Health Care
9. Flexible Spending Account
10. 401(k) Plan
11. Stock Option Plan
12. Overtime
13. Jury Duty
14. Training and Dues
15. Lodge Discounts

7. Move to Slide 1. Edit the presentation title to read *Employee Benefits.* (Delete Corporate View.) Click outside the title text box. Choose **Picture** from the Insert menu. Choose **From File**. Navigate to where your data files are stored. Choose *CV_Banner.gif*. Click **Insert**.

8. Click and drag the graphic to place it over the words *Employee Benefits*. Click a corner handle on the graphic and drag to resize it. Make the graphic about the same width as the title words *Employee Benefits*. Move the graphic to center it approximately over the title.

Click and drag a handle to resize a graphic

DigiTip

To browse for images, click the **Clip Organizer** option near the bottom of the Task Pane.

HELP KEYWORDS

AutoShape
 Add a shape

4. When the search results display, point to the image you want to use. Click the arrow button on the image and choose **Insert** to place it in the document. (You may choose a different image than the one shown here.)

5. Change the layout of the image. Right-click on the image and choose **Format Picture**. Click the **Layout** tab on the Format Picture dialog box. Choose **Tight** under Wrapping style. Click **OK**. (See Figure 5-26 on page 166.)

6. Add heart shapes to the flyer. Click the **AutoShapes** button on the Drawing toolbar. Choose the **Basic Shapes** category. Click on the **heart** shape. A drawing canvas frame appears in the document with a message "Create your drawing here." Click and drag down and to the right in the drawing canvas to create the heart. (If a drawing canvas does not appear, draw the shape directly in the document. The option for a drawing canvas could be turned off on your system.)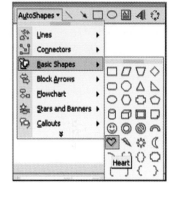

7. Click the heart to select it. Click the **Fill** button on the Drawing toolbar and choose a red color for the heart.

8. Click and drag the heart to move it off the drawing canvas to the place where you want it to appear. Click in the now empty drawing canvas and press **Delete** to delete the drawing canvas frame. If you want to use more than one heart image, copy the image. Click the image to select it. Hold down the **Ctrl** key and drag to the right until a duplicate image appears. Release the mouse. Move the image or size it as desired.

9. Change the font size, font style, and the horizontal alignment to make an attractive flyer for the party. For example, you might want to make the most important words, *Valentine Party,* font size 72. Make the other lines smaller font sizes. If you want to change the color of text, select the text. Click the down arrow on the **Font Color** button on the Drawing toolbar and choose a color. You might want to make *Valentine Party* red, for example.

10. Move and size the images and adjust the font sizes and alignments as you desire to create an attractive flyer. Proofread and correct all errors. Save the flyer as *Activity 5-13* in the *DigiTools your name\Chapter 5* folder. Print and close the document.

The order of slides in a presentation can be changed. To rearrange slides, choose **Slide Sorter** from the View menu. Click and drag the slides to place them in the order you want.

Figure 10-10
Slide Sorter View

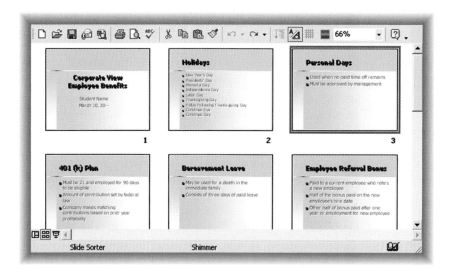

GRAPHICS

Graphics, such as photos, clip art, or charts, can be used to enhance your slides. You can cut and paste graphics from another program or Web site. You can use the Insert Picture command to add a picture stored on your hard drive. AutoShapes can be selected from the AutoShapes options on the Drawing toolbar. Charts can be created in *PowerPoint* or copied from another program such as *Excel*.

Once a graphic has been inserted, you can resize, move, copy, cut, paste, or delete the graphic. Many of the options for working with graphics in *PowerPoint* are the same as the ones you learned in *Word*.

ACTIVITY 10-7 — UPDATE A PRESENTATION

In this activity, you will delete, insert, and rearrange slides. You will also add graphics to enhance your presentation.

1. Start *PowerPoint*. Open the file *CV Benefits4* that you created earlier. Save the file using the new name *CV Benefits5*.

2. You have decided to delete the slide about the employee referral bonus benefit. Move to this slide. Choose **Delete Slide** from the Edit menu.

3. You need to insert a new slide about employee discounts at the Corporate View Lodge. Move to the Personal Days slides. Choose **New Slide** from the Insert menu. In the Task pane, scroll down to see the choices under Text and Content Layouts. Click the **Title, Text, and Content** option.

ACTIVITY 5-14: INSERT IMAGES IN A REPORT

1. Start *Word*. Open the data file *TravelView History*. Format the information as an unbound report. Add the report title HISTORY OF TRAVELVIEW.

2. Add these two paragraphs to the end of the report. Make corrections as shown by the proofreaders' marks.

   Because of our ^positive experience with the lodge, we decided ⟨ot⟩ purchase a ^small travel agency. ≡a company our size has extensive travel demands. We determine^d that we could easily support our own agency ^and ~~of~~ reduce our travel costs. We've been able to cut Corporate ≡view travel expense^s by almost a third and still remain profitable⊙ We also give ourselves great service because we are ⟍our main client.

   In only two short years, TravelView has grown into a profitable division of Corporate View. TravelView contributes 5 percent of the total earnings for Corporate View as shown in the pie chart below.

3. Insert the Corporate View Travel logo into the document. Click the **Insert Picture** button on the Drawing toolbar. Navigate to the folder where data files are stored. Choose the file *CVTravel.bmp*. Click **Insert**. Change the layout of image to **Tight**. (See Step 5 of Activity 5-13.) Move the image to position it somewhere near the top of the report, such as centered over the title.

4. Access the Corporate View intranet. On the Corporate View Intranet Home page, choose **Regular Features**. Choose **Corporate View Archive**, then **Corporate Communications Documents**, and then **CVTravel Pictures**.

5. Click on the first image, *Lodge*, to enlarge it. Right-click on the enlarged image and choose **Copy**. Return to your *Word* document. Click in the document at the bottom of the page. Click **Paste**. Change the layout of the image to **Tight**. (See Step 5 of Activity 5-13.) Move the image to position it near the right margin in the third paragraph of the report that describes the lodge. Size the image to make it about 25 percent smaller. Close the Corporate View intranet.

ACTIVITY 10-6 — EDIT AND FORMAT TEXT

In this activity, you will examine the text you have used in your outline and make it appropriate for your slides.

1. Start *PowerPoint*. Open the file *CV Benefits3* that you created earlier. Save the file using the new name *CV Benefits4*.

2. Close the Task pane if it is open by clicking its **Close** button. Look at each of your slides carefully in the Slides pane. (Click the **Next Slide** button or the **Previous Slide** button on the Slides pane scroll bar to move to a different slide.)

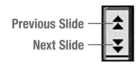

Previous Slide
Next Slide

3. Is each item in the bulleted lists on your slides clear and concise? To change the wording of text, click in the text on the Slides pane or click on the slide outline. Delete the unwanted text and key the new text. Review every slide. If a slide contains too much information, delete some text. Show only the main points on the slide. If there is other information you want to discuss, key a reminder in the Notes pane for that slide.

4. Move to Slide 1. In the Slide pane, click in the box that says "Click to add subtitle." Key your name followed by a comma and `Presenter`. Press **Enter** and key the date for the presentation. Use a date one week from today.

5. You can format text on a slide using options such as alignment, bold, font size, or font style similar to how you format text in *Word*. Practice some of these formatting options.
- Click in the presentation title. Click the **Center** button on the Formatting toolbar.
- Select your name and the date. Click the **Center** button on the Formatting toolbar.
- Select your name and the date. Click the **Italic** button on the Formatting toolbar.
- Select your name and the date. Choose **36** in the font size list from the Formatting toolbar.

HELP KEYWORDS

Format
Make text bold, italic, or underline

6. Play your show to see how it looks so far. Move to Slide 1. Choose **View Show** from the Slide Show menu. Click the mouse or press the **Right Arrow** key to move to the next slide. Press the **Escape** key if you want to exit before viewing all slides.

7. Save the file again using the same name. Close the presentation.

INSERTING, DELETING, AND ARRANGING SLIDES

You can change the structure of a *PowerPoint* presentation by deleting and inserting slides. To delete a slide, move to the slide you want to delete. Choose **Delete Slide** from the Edit menu.

To insert a slide, choose **New Slide** from the Insert menu. Several types of slides are available. Some slides are set up to contain only text. Others have place holders for graphics. You can choose the slide layout that best fits your data.

6. Open *Excel*. (Click **Start, (All) Programs, Microsoft Excel**.) Open the data file *TravelView Chart*. This *Excel* file contains the chart mentioned in the last paragraph of your report.

7. Click just inside the chart border to select it (but not on a chart element). Click the **Copy** button on the Formatting toolbar. Switch to your *Word* document. Go to the end of the document and press **Enter** twice after the last paragraph. Click the **Paste** button on the Formatting toolbar. The chart will appear in your report. Close *Excel*.

8. In your *Word* document, click the chart to select it. Change the layout of the image (chart) to **Tight**. Choose **Center** under Alignment in the Format Picture dialog box. (See Step 5 of Activity 5-13.)

9. Insert a hard page break to place the last paragraph and the pie chart on page 2. Create a header to place the page number on page 2 of your report.

10. Proofread and spell-check the document. Correct all errors. Save the report as *Activity 5-14* in the *DigiTools your name\Chapter 5* folder. Print and close the document.

Peer Check Compare your report with a classmate's report. Discuss any differences. Make changes, if needed, after your discussion.

CLICK AND TYPE

HELP KEYWORDS
Click and Type
About Click and Type

Click and Type is a feature that automatically applies paragraph formatting using the position of the insertion point. When you double-click anywhere in a blank *Word* document, such as the middle of a blank page, *Word* will assign the paragraph formatting that most closely matches the insertion point position. For example, when you double-click near the horizontal center of the page, Center alignment will be applied. When you double-click near the right margin area, right alignment will be applied. This feature can be useful when creating flyers or title pages for reports.

ACTIVITY 5-15   CREATE A TITLE PAGE

1. Access the Corporate View intranet. Access the Style Guide using the bookmark on your Favorites list or choose **Corporate Communications, Style Guide**. Choose **Unbound Report Example**. Study the example of a report title page. Note that it contains a company logo graphic, the title of the report, the date, and other information.

2. Start *Word* and open a new blank document. Prepare a title page for the report you completed in Activity 5-14. Use **Click and Type** to position your cursor about 3" from the top of the page and near the horizontal center. Key the name of the report. Insert the CVTravel logo above the report title. (Data file *CVTravel.bmp*.)

4. When you point to a template, the template name will display. Locate the Shimmer template. Click the **Shimmer** template to apply it to your slides.

5. The default color scheme for this template uses white text on a dark blue background. Some of the graphics you will add to the slides later might look better on a light background. To choose a different color scheme, click **Color Scheme**s at the top of the Task pane. Click the color scheme that has a light blue background to apply it to your slides.

Save the file again using the same name. Close the file and *PowerPoint*.

Self Check Does your title slide look like Figure 10-9? If not, try choosing the template design and color scheme again.

Figure 10-9
Slide with Design Template Applied

TEXT

Limit text on visuals. If the text on a visual is crowded, it results in a confusing appearance. If you have too much text, the audience becomes involved in reading the content of the visual instead of listening to what you are saying. Use bulleted lists to help the audience follow the presentation. Make the wording on all visuals parallel in verb tense. Use strong, active verbs. Whenever possible, make the points short and concise.

Limit the use of different text styles, sizes, and colors on slides to avoid a confusing appearance. For example, you might choose to use Arial, 24 point, blue text for main points or headings and Times Roman, 18 point, black text for subheads or supporting details. Lowercase text is easier to read than all caps. As a general rule, use all caps sparingly. If you use a design template in *PowerPoint*, appropriate text sizes and styles are part of the template design.

3. Click near the middle of the page and type the name of this class, such as *Digital Communication Tools*. Key your teacher's name a DS below the class name.

4. Click about 2" from the bottom of the page and near the center. Key your name. Key the date a DS below your name. Use Print Preview to see if the title page is formatted attractively. Make adjustments if needed.

5. Proofread and correct errors. Save the document as *Activity 5-15* in the *DigiTools your name\Chapter 5* folder. Print and close the document.

Self Check Does the title begin about 3" from the top of the page with the graphic above it? Is the class name placed at about the vertical center of the page? Are your name and the date placed about 2" from the bottom of the page? Are all lines centered horizontally?

APPLICATION 5-8   EVENT FLYER

1. Start *Word* and open a new blank document.

2. Choose a school event that will occur in the next few weeks. The event might be a concert, party, club meeting, sports match or game, debate, or academic competition.

3. Design a flyer to announce this event. Include the name of the event, where and when it will take place, who is involved or invited to attend, the price of admission (or no admission charge), and other relevant information. Use landscape orientation for the flyer. Insert clip art, AutoShapes, or other images to make the flyer interesting. Use varying font sizes and colors to create an attractive flyer.

4. Proofread your flyer and correct all errors. Save the flyer as *Application 5-8* in the *DigiTools your name\Chapter 5* folder. Print and close the document.

DIGITOOLS
DIGITAL WORKBOOK   CHAPTER 5

Open the data file *CH05 Workbook*. Complete these exercises in your *DigiTools Digital Workbook* to reinforce and extend your learning for Chapter 5:

- Review Questions
- Vocabulary Reinforcement
- Math Practice: Calculating Mailing Volumes
- From the Editor's Desk: Subject and Verb Agreement
- Keyboarding Practice: Opposite Hand Reaches and Speed Building

Keep the design of your visuals simple. Remember that the purpose of the visual is to help you hold your listeners' attention and to help your listeners remember your message. Do not make your visuals too complicated or difficult to read or understand.

When you create slides, you should be concerned about the overall theme and color choices used. The text size and font style should be chosen carefully. You should limit the number of words used on each slide for easy readability.

COLOR CHOICES

The colors you choose to use in visual aids such as transparencies or multimedia slide shows will affect the way your audience reacts to your presentation. Use color effectively in your slides or transparencies. Just because color is available does not mean many colors should be used. In fact, limiting the number of colors used is often best.

Consider the generally accepted associations of colors. For example, in business, red usually relates to cost and green usually relates to profit. In general, red draws the most attention and evokes excitement. Blues and greens are relaxing. Earth tones can be soothing, but can also be dull or lack impact. Borders are effective for adding and using color wisely. They help to guide the viewer's attention and give the visual a professional look.

The colors you choose will depend on the media you are developing for the presentation. Overhead transparencies are most effective with dark text on a light background. Computer slides and 35mm slides can be effective with a variety of background colors and textures. Choose your background color, text colors, and image colors to create a pleasing effect. Color should not distract your audience from the message. Presentation software programs such as *PowerPoint* have design templates you can use to give your slides a professional look. Coordinating colors for background, text, bullets, and other design elements are part of the template design.

ACTIVITY 10-5 — CHOOSE COLORS AND DESIGN

In this activity, you will learn more about using color on slides. Then you will choose a theme and colors for a slide show.

1. Access the CorporateView intranet. Choose the **Corporate Communications** and **Style Guide** links. Choose **Presentations**. Read the information provided under the **Color Choices** link. Close the CorporateView intranet.

2. Start PowerPoint. Open the file *CV Benefits2* that you created earlier. Save the file using the new name *CV Benefits3*.

3. Choose **Slide Design** from the Format menu. Sample design templates appear in the Task pane. These templates use predefined colors, font styles and sizes, and other design elements. Use the Task pane scroll bar to scroll down and view several of the templates. Click on any template to see it displayed in the Slides pane. Click on several different templates to see how they look with your title slide.

HELP KEYWORDS
Design template
Apply a design template

CHAPTER 6

Handwriting Recognition

OBJECTIVES *In this chapter you will:*

1. Explore mobile computing.
2. Learn to use Microsoft handwriting recognition.
3. Control the Language bar and Writing Pad or Input Panel.
4. Adjust automatic recognition.
5. Write with printing and cursive styles.
6. Develop input speed using two-line writing.
7. Edit and correct handwriting errors.
8. Practice entering letters and commands with the on-screen keyboard.
9. Create documents using handwriting recognition.

MOBILE COMPUTING

Many people who could benefit from using a computer cannot work at a desktop computer all day. Doctors, police officers, and lawyers might all benefit from using mobile computers. Manufacturing workers, students, and salespersons also need to work on the go. A **mobile computer** is one that is small enough to carry with you easily as you attend class or do your work. Tablet PCs and personal digital assistants are two types of mobile computers.

Figure 6-1
A mobile computer helps a student be more productive.

CHAPTER 6   172

When preparing a team presentation, selecting a leader is helpful. The leader can help keep the team focused on the project. Each team member should have assigned tasks to complete. These tasks can vary. All team members may not be involved in delivering the presentation. For example, one or more team members may research and develop content for the presentation; another may create the visuals. One or more members may deliver the presentation.

Once the content and visuals have been developed and organized, the team should review the entire presentation to make sure all the elements flow together well. The team can then set time allowances and choose the content to be presented by each person. The team should practice delivering the presentation as a group if more than one person will present.

LESSON 10-2 CREATING VISUALS

OBJECTIVES *In this lesson you will:*

1. Choose a design and color scheme for slides.
2. Format text on slides.
3. Insert and delete slides.
4. Change the order of slides.
5. Insert, size, and format graphics on slides.
6. Apply transition effects to slides.
7. Insert sound files in slides.
8. Save a presentation as a Web page.

Visuals are an important part of a presentation. They add interest and help listeners remember the information you present. You learned about several types of visuals in Lesson 10-1. In this lesson, you will create slides in *PowerPoint* to use as visuals for a presentation.

DESIGN STRATEGIES

Limit your design to one main idea per visual, and use plenty of white space. **White space** is the blank area on a document or slide. Make the orientation of your visuals consistent throughout the presentation. Visuals in landscape orientation are wider than they are tall. Visuals in portrait orientation are taller than they are wide. Transparencies and slides (both individual slides and slides in an electronic multimedia presentation) are usually created in landscape orientation.

Tablet PCs

A **Tablet PC** is a mobile computer with the power and capabilities of many desktop computers. At first glance, a Tablet PC looks much like a typical laptop computer. Unlike a laptop, however, the user can input text by writing on the computer screen. In some models, the computer/screen can be separated from the keyboard. In other models, the screen can be rotated and folded on top of the keyboard to make writing more convenient. A typical size for a Tablet PC is 8" by 10.5". Speech recognition can also be used for input with Tablet PCs.

Users can carry a Tablet PC with them as they attend classes, go to meetings, or work away from their desks. Text and other data is input by writing or tapping on the screen with a digital pen. When desired, a keyboard can be attached to the computer. Users can input data using a digital pen or a traditional keyboard depending on which is most convenient in the situation.

> **DigiTip**
> Once you get used to using your digital pen, it can prove much faster than a mouse when doing the same sorts of tasks.

Personal Digital Assistants

A **personal digital assistant (PDA)** is a handheld mobile computing device. PDAs are small, easy to carry, and lightweight. A PDA is less powerful than a PC. The power of PDAs continues to increase, however. With some PDAs, users can input data with a digital pen. A PDA can help users perform tasks such as these:

- Automatically track telephone calls
- Access and send e-mail messages and instant messages
- Work with programs such as word processors or spreadsheets
- Do currency conversion calculations
- Upload data to or download data from desktop or laptop computers
- Record notes of telephone calls
- Recognize schedule conflicts and set reminders for appointments

Some PDAs are known as Pocket PCs or Palm Pilots. These two are competitive versions of palm-sized computing devices. PDA features are now being built into digital phones, combining the best of both technologies.

Figure 6-2
A PDA phone combines the features of a PDA with a telephone.

5. Your computer should have a microphone attached. The microphone may have an on/off switch. If so, turn on the microphone. Click the **Record** button and read the sentence below. Then click the **Stop** button.

```
Corporate View provides a variety of benefits for
employees. To learn more, click the Next button.
```

6. Click the **Play** button to hear your recording. You may need to adjust your microphone or change your computer's microphone volume settings. Ask your teacher for help if you think you need to make these adjustments. Read and record the sentence two or three times until you are happy with the recording. Before you record each time, choose **New** from the File menu. Choose **No** when asked if you want to save changes.

7. When you are happy with your recording, choose **Save As** from the file menu. Save the file as *Corporate View1* in the *DigiTools your name\Chapter 10* folder. The **Sounds (*.wav)** option should be selected in the Save as type box. Close *Sound Recorder*.

HANDOUTS

Because the audience will remember only a small portion of your presentation, provide handouts for your listeners. **Handouts** are printed documents used to summarize information or provide more details. Handouts can be used later for reference and as reminders of the key points in your presentation. The handouts may include graphics as well as text. Handouts can be generated easily from a slide show you create with *PowerPoint*. Throughout the presentation, the handout can be used for note taking.

If you have several handouts to distribute during a presentation, consider printing them on paper of different colors. The colored pages will make it easy for you and your listeners to distinguish between the handouts. Consider using a color copier to reproduce color handouts that will add impact to your presentation.

TEAM PRESENTATIONS

Workers are often involved in giving team presentations. An advantage of working as a team is that the knowledge and skills of all team members can be used to create an effective presentation. A disadvantage of working as a team is that extra care must be taken to make the content and visuals consistent throughout the presentation.

Figure 10-8
Individuals must work together to prepare a team presentation.

OFFICE XP HANDWRITING RECOGNITION

Even if you don't have a Tablet PC or PDA, you can use handwriting recognition with a desktop PC. Since May 2001, handwriting recognition has been available to all *Microsoft Office XP* users on desktop and laptop computers. To use handwriting on full sized PCs, you need an electronic writing tablet, such as the one in Figure 6-3. An **electronic writing tablet** is a detachable piece of hardware that allows users to input data with a digital pen.

Figure 6-3
An electronic writing tablet allows PC users to use handwriting recognition.

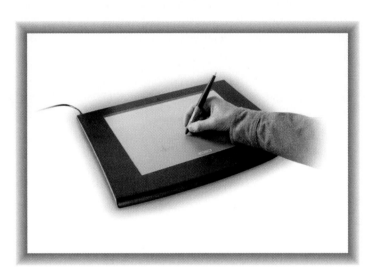

LESSON 6-1: CONTROL THE LANGUAGE BAR OR INPUT PANEL

OBJECTIVES *In this lesson you will:*

1. Control the Language bar and Writing Pad or Input Panel.
2. Set the Automatic Recognition option.

HANDWRITING TOOLS

You can use handwriting recognition in *Word, Excel, Outlook, PowerPoint,* and other *Office XP* applications. On a Tablet PC, speech and handwriting tools are installed at the factory before you purchase the computer. When using a desktop PC, however, you will need to make sure that both your speech and handwriting recognition tools have been installed properly. See the DigiTip below for more information.

DigiTip Learn how to install handwriting in *Office XP* using the Help feature. In *Word*, choose **Help** on the menu bar. Choose **Microsoft Word Help**. On the Index tab, enter the keywords `install handwriting`. Choose **install handwriting recognition** from the topics list.

Using Audio Effectively

From soft music playing as your audience gathers to resounding applause for a job well done, sounds can enhance your presentation. Sounds used in presentations are called audio. **Audio** means related to the reproduction of sound. Audio collections are readily available and contain a wide variety of music, sound effects, and common phrases. Remember that copyright restrictions apply to music files as well as art files. Use copyrighted files only with permission. If you require specific music or text, you may wish to record your own audio or have it prepared professionally.

If you use audio in a presentation, make sure it is appropriate. The audio should not be overbearing or distracting. Sound can be used effectively to introduce a topic, build excitement, or add special effects. If you decide to use sound, be sure you are not competing with it when you are talking.

ACTIVITY 10-4 — PLAY AND RECORD AUDIO

In this activity, you will use *Windows Sound Recorder* to play audio files. You will also record a sound file.

1. Choose **Start, (All) Programs, Accessories, Entertainment, Sound Recorder**.

2. Choose **Open** from the File menu. Navigate to your data files and choose *30Tacket*. Click the **Play** button to play this music file. This upbeat music might be used as background when you want to talk about fast-paced action. For example, employees might be rushing orders to meet a tight deadline.

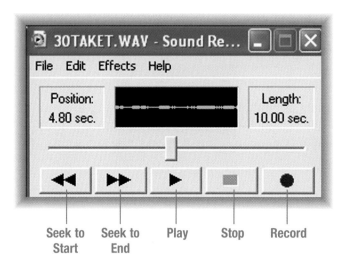

Figure 10-7
Windows Sound Recorder

3. Open the data file *Benefits1*. Play the file. This voice file might be used in a presentation that users access on an intranet or Web page. It might be indicated by a Help or question mark graphic.

4. Now practice recording a sound file that you might use in a presentation. Choose **New** from the File menu.

The tools used for handwriting recognition in *Office XP* and on a Tablet PC work in very similar ways. However, the tools look a bit different. Some of the buttons and screens have different names. As you complete the first two activities in this chapter, choose either the *Office XP* or the *Tablet PC* instructions depending on the DigiTools you are using. In later activities, differences will be noted where they occur.

ACTIVITY 6-1 — OPEN THE WRITING PAD OR INPUT PANEL

Office XP

With *Microsoft Office XP*, you will use the Language bar as shown in Figure 6-4. The **Language bar** is a toolbar that contains speech and handwriting recognition commands and options.

1. Open *Word*. You should see the Language bar. If you don't, choose **Speech** from the Tools menu.

2. When the microphone has been turned off, the Language bar will appear in its collapsed state, as shown in Figure 6-4. If the microphone is on, the Language bar will be expanded and have additional buttons. You will not use speech recognition at this time. Tap the **Microphone** button to turn speech recognition off and return to a collapsed Language bar if it is currently expanded.

HELP KEYWORDS

Language
Show or hide the Language bar

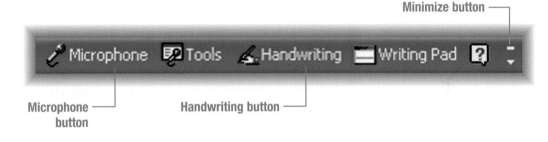

Figure 6-4
The Language bar provides access to handwriting recognition commands and options.

3. Tap the **Minimize** button, shown in Figure 6-4, to send the Language bar to the taskbar. When the Language bar is on the taskbar, it will look similar to Figure 6-5. Choose the **Restore** button to restore (maximize) the Language bar.

Figure 6-5
The minimized Language bar appears on the taskbar.

DigiTip

If you are using an older Windows operating system, your minimized Language bar will look like the image shown below. Choose the icon and then select **Show the Language bar** to restore the Language bar.

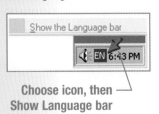

Choose icon, then Show Language bar

Figure 10-5
This visual introduces the topic and the speaker.

Selecting the Media

Media are materials or means used to communicate. The media you choose for your visuals will depend on your budget and the equipment you have available. Your audience and whether the presentation is formal or informal will also influence the media you use.

- Flip charts are very effective for small, informal groups. They are inexpensive and easy to create and use. For one-to-one communications, desktop easels can be effective.
- Overhead transparencies are effective for small or large groups and for formal and informal presentations. Overhead transparencies are inexpensive to create. They can be created in black and white or in color.
- Electronic multimedia presentations are effective for both large and small groups. With presentation software packages available, such as *Microsoft PowerPoint*, you can create professional presentations quickly and easily. You will create a presentation using *PowerPoint* in a later lesson.
- Traditional slides displayed through a slide projector can be effective visual aids. You can have individual slides created commercially at photo centers. The slides can be produced from photos or illustrations. They can also be developed from the slides you create using presentation software.
- A poster can be used to reinforce a main idea or theme in your presentation. Display the poster to restate key points, introduce a new product, or review a visual element.

Figure 10-6
Overhead transparencies are effective for a variety of presentations.

4. Tap the **Handwriting** button on the Language bar. Choose **Writing Pad**. The Writing Pad will open as shown in Figure 6-6. You will learn to enter text in the Writing Pad in a later activity. For now, close the Writing Pad by tapping the **Close** button in the upper right corner of the Writing Pad.

Figure 6-6
The Office XP Writing Pad

Tablet PC

1. If you do not see the Input Panel after you start your Tablet PC, tap the **Tablet PC Input Panel** button that can be found on the taskbar near the Start button.

2. Your Tablet PC input panel may have been minimized. If so, tap the **Show Pen Input Area** button as marked in Figure 6-7.

> **HELP KEYWORDS**
> In Input Panel, choose **Help** from the Tools menu
> **Input Panel**
> Understanding Input Panel

Figure 6-7
Tap the **Show Pen Input Area** button to maximize the Input Panel.

3. With the Tablet PC Input Panel maximized, tap the **Writing Pad** tab as seen in Figure 6-8. You will learn to use the Writing Pad in a later activity.

Figure 6-8
The Writing Pad in Input Panel

> **DigiTip**
> To open the Input Panel quickly, make several quick strokes back and forth across the bottom of your Tablet PC screen (without lifting the pen).

4. To hide the Input Panel completely, tap the **Close** button as marked in Figure 6-8. Repeat steps 1-3 of this exercise to practice accessing the Writing Pad. Then close the Input Panel.

10. Print the Notes for slides that have information in the Notes pane. Choose **Print** from the File menu. Under Print what, choose **Notes Pages**. Under Color/grayscale, choose **Grayscale**. Under Print Range, choose **Slides**. Key the numbers of the slides that have notes in the Slide box. Separate the slide numbers with commas. Click **OK**.

11. Close the presentation file and *PowerPoint*.

Self Check Do you have 15 slides in your presentation? Did you add details to all slides except the title slide?

ETHICS

Honesty in Presentations

When you give a business presentation, you want to present your company in a positive light. You want to inform or educate your listeners or persuade them to take a course of action. To achieve these goals, you may be tempted to "stretch" the truth a bit. For example, your insurance provider might sell insurance to your company at a lower cost if company employees had only one or two fewer work accidents last year. In another example, the people you want to invest in your company might be persuaded more easily if the company's sales were only a little higher. In these instances, you may be tempted to "overlook" a couple of accidents or overstate the sales figures. Do not give in to the temptation to alter data you use in a presentation. Giving false information in a presentation is dishonest and unethical.

When preparing a presentation, always check your data carefully for accuracy. Verify that the sources of information you use are reputable. Sometimes you may have to include information that is unfavorable to your workgroup or the company. In that case, try to include positive points that will show how a problem is being corrected or a change in policy will give more positive results in the future.

Follow-Up

1. Have you heard messages from businesses that you thought stretched the truth or gave false information? The message could be in a television commercial, a printed ad, or from a telephone salesperson. Describe the message.

2. What impression do you have of a business that gives information you think is probably false? Would you want to do business with such a company?

VISUALS AND AUDIO

Images are very powerful. A **visual aid** is an image that helps the listener understand or draws the listener's attention. A visual aid can be a picture, chart, poster, graphic, slide, or object that the presenter displays. Studies show that we remember about 10 percent of what we hear in a presentation and about 20 percent of what we see. However, we remember about 50 percent of what we both see and hear. Visuals are very effective in helping listeners remember your presentation.

Plan a visual for each main idea in your presentation. The first visual should introduce the topic and set the tone for the presentation. Other visuals should carry out the theme of the presentation. You will practice creating visual aids in Lesson 10-2.

HANDWRITING OPTIONS

DigiTip

Tablet PC users: Close the Text Preview panel if it is displayed. Tap the down arrow on the Tools menu. Remove the check next to **Text Preview.**

Microsoft handwriting recognition has an important setting called Automatic Recognition. The **Automatic Recognition** option allows you to automatically insert handwritten text into an application after a short pause. If you turn Automatic Recognition off, the system will wait until you decide when the text is ready to be inserted in your document. Turning off this option will allow you to erase mistakes as you practice your handwriting skills. In Activity 6-2, you will turn the Automatic Recognition option off.

After you become comfortable with your handwriting tools, you can return to the Handwriting Options dialog box and turn the Automatic Recognition option on again. You can adjust the speed of the recognition action by moving the Recognition Delay slider to a faster or slower setting. The Handwriting Options dialog box also allows you to change the color and width of your pen.

ACTIVITY 6-2

TURN OFF AUTOMATIC RECOGNITION

Office

1. Open *Word* and open the Writing Pad as explained in Activity 6-1.

2. Tap the down arrow on the Writing Pad menu. (See Figure 6-6.) Choose **Options**. Choose the **Common** tab. Remove the check mark by Automatic recognition. Tap **OK**.

HELP KEYWORDS

Automatic recognition
Change handwriting recognition options

Figure 6-9
Handwriting Options Dialog Box

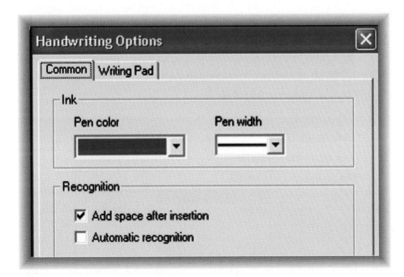

3. Close the Writing Pad and *Word*.

2. Think about each benefit on your outline. Is there a particular listener interest or advantage that relates to this benefit? If yes, note it on your printed outline.

3. Once again, think about each employee benefit on your outline. Is there an objection or concern the listener may have related to this benefit? If yes, note it on your printed outline. Then note any ideas you have about how to counter this objection or address this concern.

Example: Holidays
Listener objection: A particular holiday I want to observe is not listed as a Corporate View holiday.
Counter: You can use your one floating holiday for this day.

4. Close the Corporate View intranet. Open the presentation outline you created earlier, *CV Benefits1*. Save the file under the new name *CV Benefits2*.

5. Click the **Outline** tab in the Outline and Slides pane to display your outline, if needed. Click in the outline on Slide 3, Personal Days. Press **Enter**. Click the **Increase Indent** button or press **Tab** to make a lower outline level for Slide 3. Enter the two or more details you noted about personal days. Use short phrases or clauses rather than complete sentences. Capitalize the first word and do not place a period at the end of the subpoint.

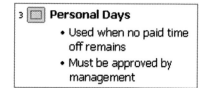

DigiTip
You can resize the panes in the *PowerPoint* window by clicking and dragging a pane border. You may want to make the Outline pane wider to make keying your outline easier.

HELP KEYWORDS
Check
  Check spelling

6. Repeat this process of adding subpoints to your outline for all the remaining slides. Some slides may need four or five subpoints; others will only need two.

7. Click on Slide 2 in your outline. Then click in the Notes pane. Enter the listener objection and counter as shown in the example in Step 3. Click on other slides and enter listener interests, advantages, or objections you noted as you read about each benefit.

8. Spell-check the slides. Click the **Spelling** button on the Standard toolbar. Choose the appropriate options from the dialog box. Click **OK** when the spell check is complete.

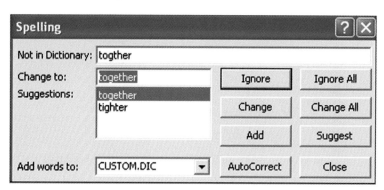

9. Save the file again using the same name. Print the outline for all slides.

> **HELP KEYWORDS**
>
> In Input Panel, choose **Help** from the Tools menu
> **Handwriting recognition**
> Customizing handwriting recognition

Figure 6-10
Handwriting Options Dialog Box

Tablet PC

1. Open *Word* and open the Writing Pad as explained in Activity 6-1.

2. Tap the down arrow on the Tools menu. (See Figure 6-8.) Choose **Options**. Choose the **Writing Pad** tab. Remove the check mark by the option that reads: *Automatically insert text into the active program after a pause.* Tap **OK**.

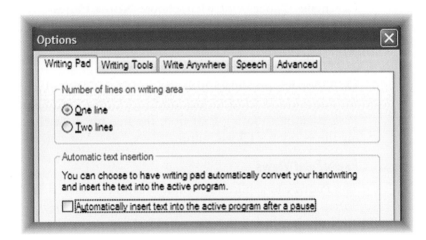

3. Close the Input Panel and *Word*.

LESSON 6-2: INPUT WITH WRITING PAD

OBJECTIVES *In this lesson you will:*

1. Write using block printing style.
2. Correct handwriting errors.
3. Create a summary of an article using handwriting recognition skills.

WRITING STYLE

With handwriting technology, penmanship matters! Skills you may not have thought about since elementary school will quickly be brought to mind. The key to using handwriting recognition successfully is writing clearly. Writing Pad can understand several styles of handwriting. Vertical printing, slanted printing, vertical cursive, or slanted cursive writing are all recognized.

To begin, try using your usual handwriting style. Later you can practice block-style printing and cursive styles to see if they work better for you.

10. You will add more details for the slides later. Save the file again using the same name.

11. Print the outline. Choose **Print** from the File menu. Under Print what, choose **Outline View**. Under Color/grayscale, choose **Grayscale**. Be sure that **All** is selected under Print Range. Click **OK**.

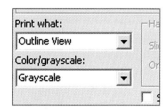

12. Print the notes for Slide 1. Choose **Print** from the File menu. Under Print what, choose **Notes Pages**. Under Color/grayscale, choose **Grayscale**. Under Print Range, choose **Slides**. Enter 1 in the slides text box. Click **OK**.

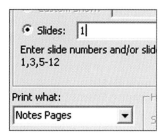

13. Close the presentation file by choosing **Close** from the File menu or clicking the window's **Close** button. Exit *PowerPoint* by choosing **Exit** from the File menu or clicking the window's **Close** button.

PowerPoint Views

The *PowerPoint* application window can be displayed in three views: Normal, Slide Sorter, and Slide Show. Normal view is the main view used to create and edit slides. Normal view has the Outline and Slides pane on the left, the Slide Pane on the right, and the Notes pane below the Slides pane. You used Normal view in Activity 10-2.

Slider Sorter view shows all the slides in your presentation. This view makes it easy to rearrange or delete slides. Slide Show view shows the slides using the full screen in the way they will appear when you present them. To change views, you can choose a view from the View menu or click a View button.

> **HELP KEYWORDS**
> View
> About *PowerPoint* views

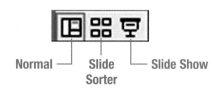

Normal — Slide Sorter — Slide Show

Listener Advantages and Objections

Once you have developed your ideas and your listener's interests, list the advantages for your listeners if they accept your ideas. If possible, prioritize these advantages in order of importance to the listeners.

Consider all the objections your listeners may have regarding your ideas. **Objections** are reasons to disapprove or reject ideas. When you anticipate the listeners' objections, you can decide how to address them. You may be able to offer solutions or alternatives to what the listeners see as a problem. Your goal is to address the objections so the listeners no longer see them as a problem.

ACTIVITY 10-3   **ADD DETAILS**

You need to learn more about Corporate View's employee benefits so you can add details to your outline.

1. Access the Corporate View intranet. Choose the **Regular Features** and **Employee Handbook** links. Choose the **Compensation and Employee Benefits** link. Refer to the outline you printed. For each benefit on your outline, choose its link on the intranet page. Read about the benefit. Make notes on your outline of two or more important details you should include about this benefit in your presentation.

ACTIVITY 6-3

WRITE IN WRITING PAD

1. Open *Word* and Writing Pad as explained in Lesson 6-1.

2. In *Office XP*, tap the **Text** button.

   In Tablet PC Input Panel, tap the down arrow on the Send button. Choose **Send as Text**.

3. Before you begin writing, tap your digital pen at the position in *Word* where you want the text to appear. Position your pen to write on the line provided in Writing Pad. Using your best penmanship, write this sample sentence: `This is a test!`

DigiTip
Tap the **Expand** button to show all the tools that are available in *Office XP* Writing Pad.

Figure 6-11
Write clearly in the Writing Pad.

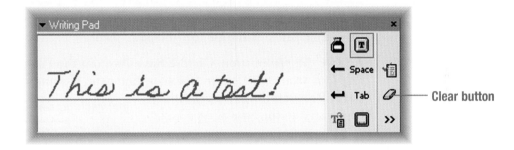

Clear button

4. In *Office XP*, choose the **Recognize Now** button.

   In Tablet PC Input Panel, choose the **Send** button.

5. Tap the **Enter** button twice to create a double space before you try another sentence. You can write sentences in several parts, tapping **Recognize Now** or **Send** after each phrase. For example, write `Benjamin Franklin lived.`

6. Tap the **Recognize Now** or **Send** button. Finish the sentence by writing `in colonial times.` Tap the **Recognize Now** or **Send** button.

7. If your penmanship is lacking, you can use the **Clear** button in *Office XP* to erase any poorly written words in Writing Pad. (See Figure 6-11.)

   On the Tablet PC, tap the **Del** button to erase mistakes. To practice, write `Thomas Jefferson.` Erase *Thomas Jefferson*.

8. Write `Abraham Lincoln.` Erase *Abraham Lincoln*.

9. Write `Franklin Roosevelt.` Send this text to your document. (Choose the **Recognize Now** button in *Office XP*. In Tablet PC Input Panel, choose the **Send** button.)

10. Close the *Word* document without saving.

DigiTip
In Input Panel, quickly swish your pen horizontally across an error two or three times (without lifting the pen) to erase it.

Self Check Did you use the Clear or Del buttons to erase poorly written text before sending it to the document?

4. Click on Slide 1 in the outline to display it again. Click in the Notes pane. Key the purpose of the presentation and press **Enter**.

```
Purpose: Provide new employees with information about
Corporate View's employee benefits
```

5. Key the overall listener interest for the presentation and press **Enter**. You may note more specific listener interests on some slides later.

```
Listener interest: New employees will want to know
what benefits are available
```

6. Key the overall listener advantage for the presentation and press **Enter**. You may note more specific listener advantages on some slides later.

```
Listener advantage: Corporate View provides services
and benefits that employees would otherwise have to
pay for and manage on their own
```

7. Save the presentation. Choose **Save As** from the File menu. Navigate to your *DigiTools your name\Chapter 10* folder. Enter the filename `CV Benefits1` and click **Save**. **Presentation (*.ppt)** should be selected in the Save as type box.

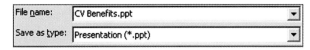

8. Click in the outline on Slide 2. Enter the name of the first benefit you want to discuss, `Holidays`. Press **Enter**. Click the **Increase Indent** button on the Formatting toolbar.

Enter the name of the first holiday, `New Year's Day`. Press **Enter**. Key the names of the other holidays shown below. Press **Enter** after each one.

```
Presidents' Day
Memorial Day
Independence Day
Labor Day
Thanksgiving Day
Friday Following Thanksgiving Day
Christmas Eve
Christmas Day
```

9. Click the **Decrease Indent** button on the Formatting toolbar to create Slide 3.

Key the name of the next benefit, `Personal Days`. Press **Enter**. Continue keying to create a slide for each benefit you will discuss in your presentation as listed below.

```
 4. 401(k) Plan                 10. Jury Duty
 5. Bereavement Leave           11. Overtime
 6. Employee Referral Bonus     12. Paid Time Off (PTO)
 7. Family Medical Leave        13. Sick Leave
 8. Flexible Spending Account   14. Stock Option Plan
 9. Health Care                 15. Training and Dues
```

WRITING SKILLS

Block printing is an effective writing style for many users. Study the writing styles shown below.

Figure 6-12
Block-Style Printed Letters

Lowercase Letters

a b c d e f g h i j k l m n o p q r s t u v w x y z

Capital Letters

A B C D E F G H I J K L M N O P Q R S T U V W X Y Z

In the next activity, you will practice every letter of the alphabet using three nonsense sentences. For now, try block-style printing. Wait to practice cursive letters in the next lesson.

ACTIVITY 6-4

WRITE USING BLOCK PRINTING

DigiTip
You can correct errors by selecting the mistaken word, tapping a button or symbol, and choosing the correct word from a list.

In *Office XP*, tap the **Correction** button. If the Correction button is not visible, tap or click the down arrow in the bottom right corner of the Language bar and place a check mark next to **Correction**.

In Input Panel, tap the little green symbol on the edge of the word to see the list of possible corrections.

1. Open *Word* and Writing Pad. Tap the **Text** button in *Office XP* or choose the **Send as Text** option in the Input Panel.

2. Practice writing the following nonsense sentence using block style printed letters. Leave a space between words. Write punctuation marks normally. Use a dot (.) for a period and a short, slightly curved, downward stroke for a comma (,). Write the sentence in parts. Then tap the **Send** or **Recognize Now** button to post your sentence phrase by phrase. If you make a mistake, erase the phrase and rewrite it before you send it to *Word*.

```
My great aunt had a bat, a cat, a dog, an elk, a fox, a
kangaroo, a turtle, a monkey, and a zebra in her yard.
```

3. To correct an error, use your pen to highlight the error in the *Word* window by tapping and dragging over the error. Move to the Writing Pad and write the word again. Tap **Recognize Now** or **Send** to update the *Word* document. Practice correcting errors in your document.

Detailed Notes

After you have organized the main points, write more detailed notes for your presentation. The notes can be just words or phrases, or they can be complete sentences. Whenever possible, provide evidence or details that support your ideas. For example, you can state facts or offer statistics to back up your points. You can use examples to help listeners understand. You can relate a situation to personal experiences of your listeners or experiences of your own. When you use *PowerPoint*, you can write your notes in the Notes pane.

ACTIVITY 10-2 CREATE AN OUTLINE

In this activity, you will create an outline of the main points you want to cover in the Corporate View Employee Benefits presentation. Because you know that you want to create slides for the presentation using *PowerPoint*, you will also use *PowerPoint* to outline the presentation.

HELP KEYWORDS
Presentation
Create a new presentation using blank slides

1. To start *PowerPoint*, click **Start**. Choose **(All) Programs, Microsoft PowerPoint**.

2. In *PowerPoint*, you will use the Outline pane to enter the main points of your presentation. The overall purpose, listener interest, and listener advantage will be recorded in the Notes pane. The Notes pane is for the presenter's use. It does not display when you play a slide show. Locate the Outline pane and the Notes pane in Figure 10-4.

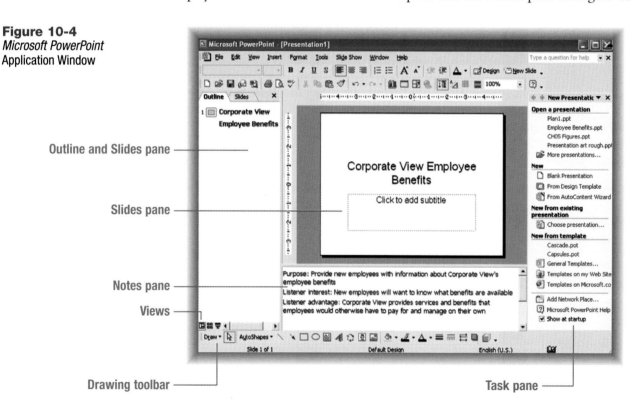

Figure 10-4
Microsoft PowerPoint Application Window

3. Click the **Outline** tab in the Outline and Slides pane. Key the name for the presentation and press **Enter**.

```
Corporate View Employee Benefits
```

4. Continue to practice your writing. Remember to tap **Enter** to leave a blank line between sentences before continuing to the next sentence. Write the two sentences below.

```
The Queen of England dropped a large orange pie last
night on the rug at the hotel.

He put his umbrella on the vine above the xylophone in
the back of the wagon.
```

5. Select all of the sentences by tapping and dragging over all of them. Press either the **Delete** or **Backspace** key to delete all of your practice writing.

Self Check Did you remember to use block-style printed letters as you wrote the sentences?

APPLICATION 6-1   ARTICLE SUMMARY

1. Open *Word* and Writing Pad. Tap the **Text** button in *Office XP* or choose the **Send as Text** option in the Input Panel.

2. Open the data file *History of Computers*. Print this file, which contains an article on the history of computers.

3. Read the article. As you read, use your digital pen to record important points from the article in a *Word* document. Write in complete sentences.

4. Go to the top of the document. Enter a title, HISTORY OF COMPUTERS SUMMARY, for your document. Center the title. Write an appropriate short opening paragraph to introduce your list of computer history facts. Format the points you recorded in a bulleted list below the introductory paragraph.

> **DigiTip** If you have trouble writing in all caps, write the title in lowercase. Then use the **Change Case** command to make the title all caps.

5. Spell-check and proofread the document. Correct all errors. Save the document as *Application 6-1* in your *DigiTools your name\Chapter 6* folder. Print and close the document.

Peer Check Compare your summary with a classmate's summary. Discuss how both summaries could be improved.

THE MESSAGE

The **message** of a presentation is the main points or ideas you want to communicate. The message of your presentation must accomplish the purpose of the presentation. The message must also relate to your listeners' needs or interests. With the purpose clearly in mind, create a list of major points or ideas you must include to accomplish your purpose.

For a work presentation, you will usually have at least some basic information about the topic of the presentation. However, you may need to find more information by searching the Internet, reading books or magazine articles, or talking with someone knowledgeable about the subject.

Organization of Ideas

After you have identified the ideas or main points to be included in the presentation, create a storyboard or an outline. A **storyboard** contains the main points and basic organization for a presentation.

To create a storyboard, fill out a form for each idea or main point in your presentation. (See Figure 10-3.) Consider how you can relate each major point or idea to the interests of your audience. Once all your ideas are written down, you can improve the flow by rearranging the pages. Organize the information in a logical way that your listener will understand.

Figure 10-3
A storyboard worksheet helps you organize your presentation.

STORYBOARD WORKSHEET	
Purpose	Motivate and influence sales staff to increase sales during the fall campaign
Main Idea	Commission and bonus opportunities will increase
Support for Idea	Commissions on sale items raised from 10% to 15%; $500 bonus for top ten total sales
Listener Interest	Commissions and bonuses that may be earned during the campaign
Listener Advantage	More income for the staff member
Listener Objection	Large number of clients to be handled during the sale. Counter: The extra effort required will be rewarded with higher income
Visual Element	Growing dollar sign

An alternative to completing storyboard forms is to create an outline. List the main topics (or ideas) and then supporting points for each main topic. Once you key your ideas, you can edit and rearrange them quickly and easily. Be sure to think about listener interests, support for your ideas, and listener advantages for each idea.

LESSON 6-3

IMPROVE WRITING SKILLS

OBJECTIVES *In this lesson you will:*

1. Practice writing using two lines for automatic input.
2. Write using cursive style.
3. Use digital ink to write a signature.
4. Create a personal-business letter using handwriting recognition skills.

CURSIVE STYLE WRITING

Cursive is a style of writing in which the letters in each word are joined. Because you do not lift your pen between letters, many people find using cursive style quicker than printing block letters. Review the style for cursive letters in Figure 6-13.

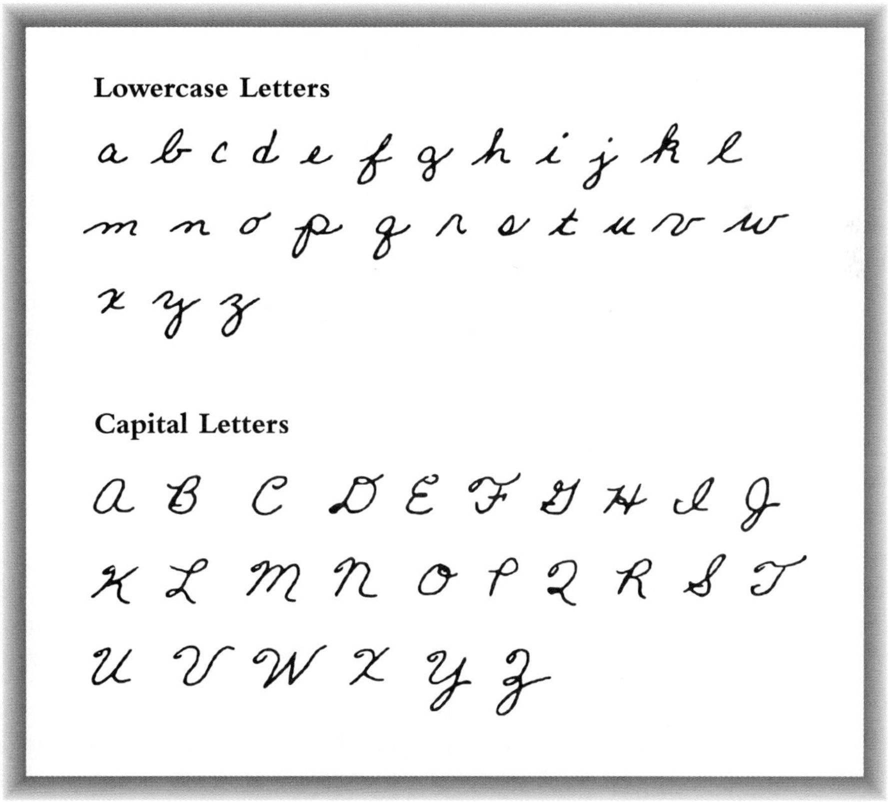

Figure 6-13
Cursive Style Letters

Lowercase Letters

a b c d e f g h i j k l
m n o p q r s t u v w
x y z

Capital Letters

A B C D E F G H I J
K L M N O P Q R S T
U V W X Y Z

Most people find it best to write slightly slanted cursive letters. You can also use a combination of printed and cursive characters. For example, a printed capital **F** or **Q** might be easier to write than their traditional cursive forms.

CHAPTER 6　　Lesson 6-3　　182

Developing a listener profile helps you identify the interests of your listeners. At first this may seem difficult, but if you try to think as they think, you should be able to do so. For example, if your audience is a particular department in your company, you should learn all you can about what the department does and what its concerns are. You may be able to find information on the company intranet, as shown in Figure 10-2. You could also talk with someone who works in the department.

Figure 10-2
Information about each department may be provided on a company intranet.

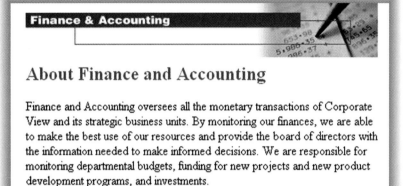

ACTIVITY 10-1 PROFILE LISTENERS

A new group of employees will soon begin work at Corporate View. You have been asked to provide information about Corporate View's employee benefits at an orientation meeting for these employees. You need to profile your listeners before you begin preparing the presentation.

The new employees will be working in the Information Technology; Finance & Accounting; and Marketing, Sales, and Support departments. They are men and women of various ages. All have college degrees. Some have only limited experience in their chosen careers. Others have been working in their career areas for many years.

1. Start *Word* and open a new blank document.

2. Key this heading at the top of the document:

    LISTENER PROFILE
    Corporate View Benefits Presentation

3. Key a bulleted list of points that describe your listeners. For example, you might begin the list by noting that all listeners are new employees at Corporate View. What else do you know or can you guess about the listeners?

4. Access the Corporate View intranet. On the Corporate View Intranet Home page, choose the **Information Technology** link. Read the About Information Technology section. Add a couple of new points to your listener profile using what you have learned about this department. Repeat this process for the Finance & Accounting and the Marketing, Sales, and Support departments.

5. Close the Corporate View intranet. Save your listener profile as *Activity 10-1* in the *DigiTools your name\Chapter 10* folder. Print the document.

TWO-LINE INPUT METHOD

The Writing Pad provides an option that allows you to display two lines for text in the Writing Pad. This option provides the advantage of keeping text in the Writing Pad until you signal that it should be sent to *Word*. However, the option allows you to write long sections without having to move your pen to tap the Recognize Now or Send button.

When using the Writing Pad with two lines displayed, write until the first line is filled. You can pause to look for mistakes if you sense you may have written a word poorly. You can use the Clear or Del button as you learned earlier. If you think the text on the first line is fine, start writing on the second line. When you begin to write on the second line, text on the first line will be sent to *Word*. When you fill the second line, continue writing on the first line. Text on the second line will be sent to *Word*. This method allows you write without interrupting your train of thought to tap command buttons.

ACTIVITY 6-5 — WRITE USING TWO LINES

1. Open *Word* and Writing Pad.
2. Set the number of lines for writing to **2**.

In *Office XP*, tap the down arrow to open the Writing Pad menu. Choose **Options**. Tap the **Writing Pad** tab. In the Number of lines box, enter **2**. Tap **OK**.

Figure 6-14
Increase the Number of lines to **2** in *Office XP*.

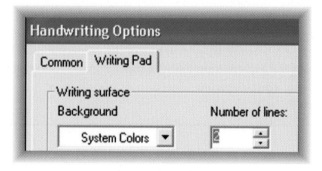

In Tablet PC Input Panel, tap the down arrow to open the Tools menu. Choose **Options**. Choose **Two lines** in the Number of lines on writing area section. Tap **OK**.

Figure 6-15
Choose **Two lines** in Tablet PC Input Panel.

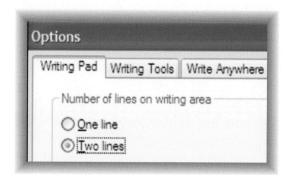

Figure 10-1
Presentations are an effective way to communicate with a large group.

PURPOSE OF THE PRESENTATION

When you give a presentation, it will likely be for one of two purposes. You will either want to motivate and influence your listeners or to inform and educate them. The message of your presentation will include the main ideas and supporting details you want to present. When you are speaking to motivate or influence, your message needs to be persuasive. Your purpose is to get your listeners to take a course of action. When you are speaking to inform, your message should be clear and concise. Your purpose is to communicate the information so your listeners can understand and use it. Identifying the overall purpose of the presentation and the specific goals you want to accomplish is the first step in preparing a presentation.

LISTENER PROFILE

Regardless of the purpose of your presentation, in order to hold your audience's attention, your message must be important to them. Developing a profile of your listeners and learning what is important to them is the next step in preparing a presentation. A **profile** is a description or mental picture of a person or group. You must determine your listeners' interests or needs so you can relate your message to something they want to hear about.

To determine what is important to your listeners, you must first describe them in as much detail as possible. Put yourself in their shoes. Write down everything you know about them:

- What do the listeners like or dislike?
- What do the listeners need?
- What is the expertise of the listeners?
- What biases do the listeners have?
- What responsibilities do the listeners have?
- Are the listeners decision-makers?

3. Now practice writing using two lines and cursive style. On the first line of Writing Pad, write `This is a sample of`.

Do not tap the **Recognize Now** or **Send** button. Write the following words (including the period) on the second line:

`writing with a digital pen.`

4. Continue practicing this double line entry method by completing the paragraph below that you have started. You have text on line 2. Begin writing on line 1. Tap the **Enter** button twice to leave a blank line between sentences.

`My great aunt had a bat, a cat, a dog, an elk, a fox, a kangaroo, a turtle, a monkey, and a zebra in her yard.`

`The Queen of England dropped a large orange pie last night on the rug at the hotel.`

`He put his umbrella on a vine above the xylophone in the back of the wagon.`

5. Close the document without saving.

Self Check Did you remember to use cursive letters as you wrote the sentences? Which style of writing gives you more accurate results, block print or cursive?

DIGITAL INK

In previous activities, you have chosen to have your handwriting converted to typed characters. When using *Word*, you can also choose the option to leave text in handwritten form. The **Ink** option allows you to insert text into a document in handwritten form. Once the handwriting is inserted in the document, you can format it. For example, you can change the font size and color.

ACTIVITY 6-6   WRITE WITH INK

1. Open *Word* and Writing Pad.

2. Select the **Ink** option.

In *Office XP*, click the **Ink** button on the Writing Pad toolbar.

In Tablet PC Input Panel, click the down arrow on the Send button. Choose **Send as Ink.**

3. In the Writing Pad, write your signature. Tap the **Recognize Now** or the **Send** button.

CHAPTER 10

Presentations

OBJECTIVES *In this chapter you will:*

1. Plan, research, and write oral presentations.
2. Prepare visuals for oral presentations.
3. Deliver oral presentations individually and as part of a team.
4. Evaluate oral presentations.

You will need to express yourself clearly in your oral communications at work so others will understand you. This is true whether you are talking with a coworker, addressing meeting participants, or giving a formal presentation. In most cases, you will speak either to motivate and influence or to inform and educate. You will also need to listen effectively so you can give appropriate feedback, answer questions, or carry out instructions. This chapter focuses on oral presentation skills and their importance to your success at work.

LESSON 10-1: PLANNING AND PREPARING A PRESENTATION

OBJECTIVES *In this lesson you will:*

1. Identify the purpose of your message.
2. Profile your listeners.
3. Address the interests of your listeners.
4. Develop ideas for your message and organize them in an outline.
5. Plan visuals and handouts.
6. Play and record audio.
7. Learn strategies for organizing team presentations.

Regardless of your job or career, you will need to express yourself clearly to others at work. Even though presentations may not be a part of your daily work, occasionally you may need to present information to others, formally or informally. The situation may require you to speak to a small group of your peers or to a large audience. Regardless of the size of your audience, you must keep your goals and your listeners' interests in mind as you develop your presentation.

4. Use **Ink** to write five other names of friends and family members. Tap the **Enter** button after each name to send it to *Word* and start a new line.

5. Practice formatting the names. Select the names and change the font size or style. Underline one name. Apply bold to one name.

6. Close the file without saving the document.

APPLICATION 6-2 — PERSONAL-BUSINESS LETTER

1. Open *Word* and Writing Pad.

2. Use your handwriting recognition skills to write a personal-business letter. Review the format for a personal-business letter on page App-6 in Appendix A.

3. Use your return address on the two lines above the date. Use the Text option to send words to *Word* for all but your signature. Use the Ink option to place your signature above your typed name. Increase the font size of the signature if needed, so that it looks realistic and as if you had signed the letter with a pen.

4. Spell-check and proofread the letter carefully. Correct all errors. Save the letter as *Application 6-2* in your *DigiTools your name\Chapter 6* folder. Print the letter and close the document.

```
Current Date | Mr. Andrew Chaney | 324 Brookside Ave. NW |
Salem, OR 97034-9008 | Dear Mr. Chaney
```

Thank you for speaking to our Music Club at (your school name). It was great learning more about the "Masters" from you. I particularly enjoyed learning more about the German composers. It is amazing that so many great classical musicians are from Germany.

Your insights into what it takes to make it as a professional musician were also enlightening for our members. Those of us who want to become professional musicians know we have to dedicate ourselves to that goal if we are going to be successful.

Sincerely

Your Signature

Your Typed Name
Music Club President

Self Check Compare the format of your letter with the personal-business letter shown on page App-6 in Appendix A. Make corrections to the format if needed.

DIGITOOLS
DIGITAL WORKBOOK — CHAPTER 9

Open the data file *CH09 Workbook*. Complete these exercises in your *DigiTools Digital Workbook* to reinforce and extend your learning for Chapter 9:

- Review Questions
- Vocabulary Reinforcement
- Math Practice: Calculating Selling Prices and Gross Profit Amounts
- From the Editor's Desk: Comma Usage Part 2
- Keyboarding Practice: High-Frequency Words and Speed Building

ETHICS

Loyalty

Managers expect employees to be loyal to the company and to their work group or department. Being **loyal** means showing commitment or support for the efforts of your company and workgroup. A loyal employee does not make unfavorable remarks about the company or workgroup outside the group or take other actions that may harm the company. For example, several members of a department may have different ideas about plans for a new project. Each employee may offer suggestions and criticisms of proposed plans. Once the manager or the group has made a decision about how to proceed, however, all members of the team are expected to do their best to make the plan successful.

Managers expect employees to be honest and behave in an ethical manner. Although employees are expected to be loyal to their companies, no employer should expect employees to engage in illegal or immoral behavior. Legislation, called **whistle-blower laws**, has been passed to protect the rights of employees who report the wrongdoing or illegal acts of their employers.

For Discussion

1. Have you heard workers speak in a negative away about the plans or policies of their employers? What impression would you have of a worker who did this?

2. Do you know a worker who thinks he or she has been treated unfairly after complaining about a company action or policy? Do you think this worker is justified in thinking this way?

Follow-Up

INTER N E T

Search the Internet to find articles or other information about whistle-blower laws. Read the articles and note the important points. Be prepared to discuss the articles with your classmates.

LESSON 6-4: INPUT WITH THE ON-SCREEN KEYBOARD

OBJECTIVES

In this lesson you will:

1. Use the on-screen keypad.
2. Create an e-mail message using handwriting recognition skills.

For creating most messages in *Word* or other programs, using the Writing Pad can be fast and efficient. Sometimes, however, you may find entering text with the on-screen keyboard more accurate. E-mail and Web addresses are examples of data you may want to enter using the keyboard.

Figure 9-14
The Report Wizard makes creating a report easy.

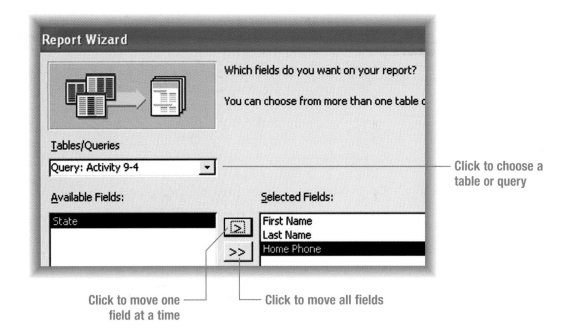

4. You do not want any grouping levels for the report; click **Next**.

5. Sort the records in the report. Click the down arrow by the first sort box and choose **Last Name**. Leave the sort order as **Ascending**. Click **Next**.

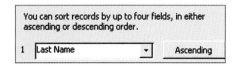

6. For report layout, choose **Tabular** and **Portrait**. Click **Next**.

7. For the report style, choose **Corporate**. Click **Next**.

8. Name the report *Kentucky Customers*. Select the **Preview the report** option. Click **Finish**. Your report displays in the Print Preview window.

9. Click the **Print** button on the Standard toolbar to print the report. Notice that the title appears at the top of the page. The data is formatted neatly in columns with field names as the column heads. The current date and the page number print at the bottom of the page.

10. The report is already saved and can be used again later. Click the **Close** button to close the report.

APPLICATION 9-4　SUPPLIERS REPORT

1. Use the Report Wizard to create a report to show data from the Suppliers table.

2. Include fields that contain the company name, phone number, and URL. Sort the data in the report by the company name. Name the report *Suppliers*. Print the report.

You can also use the on-screen keyboard to do quick keystroke operations just as you can with a regular keyboard. You can tap **Alt** and an underlined letter found on a menu option to open it. For example, tapping **Alt** followed by the **F** key opens the File menu. You can tap **Ctrl** and then a keyboard letter to give basic formatting commands. For example, **Ctrl** followed by **c** is the command for **Copy**. **Ctrl** followed by **v** is the command for **Paste**.

In the next activity, you will practice using the on-screen keyboard to enter text and commands.

ACTIVITY 6-7 USE THE ON-SCREEN KEYBOARD

1. Start *Word*.

2. Follow the steps below for your system to access the on-screen keyboard.

In *Office XP*, open *Word* and the Language bar. Tap **Handwriting** on the Language bar and choose **On-screen Standard Keyboard** from the menu.

Figure 6-16
The on-screen keyboard is useful for giving commands.

In the Tablet PC Input Panel, tap the **Keyboard** tab.

Figure 6-17
On-screen Keyboard in Input Panel

— Keypad tab

3. Tap the letters to spell the following URLs. Tap the **Shift** button first to reveal capital letters or to enter special characters such as ~, !, @, #, $, %, ^, &, *, (,), _, and +. Tap the **Space Bar** to create a space.

```
Corporate View's Web site = www.corpview.com
Microsoft's Web site = www.microsoft.com
My e-mail is mkim@corpview.com
```

APPLICATION 9-3 SUPPLIERS FORM

1. Start *Access*. Open the *Customers* database. Open the Suppliers table you created in Application 9-1.

2. Use AutoForm to create a form based on the Suppliers table. Save the form and name it *Suppliers*.

3. Create a new record and enter the data below using the form.

Company name	`Natorp's Garden Store`
Address	`4400 Reading Rd., Cincinnati, OH 45229`
Phone	`513-242-3743`
URL	`www.natorp.com`
Products	`trees, shrubs, plants, flowers, fertizler, garden ornaments, edging stones, tools`

4. Print the new record from the form.

REPORTS

A **report** is a database object used to display data in specified ways. Reports can contain data from tables or queries. The data displayed can be summarized and formatted in attractive ways. *Access* provides several ways to create reports. As with forms, you can create an AutoReport with data from an open table or query. You can use the Report Wizard to create reports or mailing labels. You can also create or change the layout of reports in Design view. In this lesson, you will create a report using the Report Wizard.

ACTIVITY 9-6 CREATE A REPORT

1. Start *Access*. Open the *Customers* database.

2. Click **Reports** in the Objects bar on the Database window. Double-click **Create report by using wizard**.

3. A Report Wizard dialog box will appear. See Figure 9-14 on page 281. Cick the arrow for Tables/Queries and select **Query: Activity 9-4**. Click the right arrow four times to place all the fields under Selected Fields. Click **Next** to move to the next step.

HELP KEYWORDS

Shortcut
Keyboard shortcuts

4. Select (tap and drag over) the line *My e-mail is mkim@corpview.com*. To cut the line, tap **Ctrl** followed by the letter **x**.

5. Tap your pen at the top of the page to move the insertion point. Paste the line you cut by tapping **Ctrl** followed by the letter **v**.

6. Use your on-screen keyboard to save the document in your *DigiTools your name\Chapter 6* folder. To open the Save As dialog box, tap the **F12** function key on the top row. (In Input Panel, tap the **Func** key to reveal the function keys.) Enter a document name, *Activity 6-7*, and tap **Save**. Close the document.

APPLICATION 6-3 — E-MAIL MESSAGE

Use your e-mail software, Writing Pad, and the on-screen keyboard to create an e-mail message to a friend. If your e-mail program will not accept text from *Office XP's* handwriting program, create the message for your e-mail in *Word*. Then copy and paste the message into your e-mail message window.

1. Start your e-mail program and create a new message. Use the e-mail address of a classmate or friend in the **To** box. Enter Shakespeare in the **Subject** box. Key these paragraphs for the e-mail message.

```
Last week you mentioned that you will be taking an
Introduction to Shakespeare course next semester. Here are
four Internet locations dealing with Shakespeare that you
may find helpful.

http://www.shakespeare-globe.org/Default.htm
http://www.wfu.edu/-tedforri/shakesp.htm
http://www.jetlink.net/-massij/shakes/
http://www.albemarle-london.com/map-globe.html

As I come across other resources, I will forward them to
you. Have a good vacation.
```

2. Send your message. Print a copy of the message from your *Sent* folder. Close your e-mail program.

7. Enter a record for a new job using the data below. Press the **Tab** key to move to the next field or **Shift+Tab** to move to the previous field. Use the current year for the dates.

Customer #	76444
Customer Name	Mr. Les McTighe
Address	501 Keslinger St., Aurora, IN 47001
Home Phone	(812) 555-7834
Job Estimate	$3,300.00
Start Date	4-28
Target Completion Date	5-12
Referral	No
Project Leader	Roberts

8. When you press **Tab** after entering data in the last field, the record is saved and a new blank record appears. Close the form by clicking the **Close** button in the form window or selecting **Close** from the File menu.

9. After closing the form, the Activity 9-5 table is still open. You will need to update it to make the new record appear in the table. Press **Shift+F9** to update the table. Notice that the new record appears at the bottom of the table. Close the table.

10. Practice opening a form you have created. In the Database window, click **Forms** on the Objects bar. Select the name of the form and click **Open** on the Database window toolbar.

HELP KEYWORDS
Record
Delete a record

11. You can print all the records as shown in the form or just selected records. Choose **Print** from the File menu. Under Print Range, choose **Selected Record(s).** Click **OK**. The record displayed in the form will print.

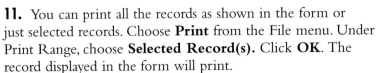

12. Mr. Goldstein has decided to move to another home. He has cancelled the landscaping project planned for his current home. You need to delete his record from the database. Use the navigation buttons or the Find feature to locate the record for Mr. Goldstein, Customer # 35801.

DigiTip
To delete a record in datasheet view, select the record. Click the **Delete Record** button.

13. When the record with Customer # 35801 is displayed in the form, click the **Delete Record** button on the Standard toolbar. Click **Yes** at the prompt asking if you want to delete the record.

14. Close the form and the table. Exit *Access*.

Self Check Open the Activity 9-5 table to confirm that the Customer # 35801 record has been deleted from the table.

DIGITOOLS
DIGITAL WORKBOOK — CHAPTER 6

Open the data file *CH06 Workbook*. Complete these exercises in your *DigiTools Digital Workbook* to reinforce and extend your learning for Chapter 6:

- Review Questions
- Vocabulary Reinforcement
- Math Practice: Calculating Sales Trends
- From the Editor's Desk: Run-on Sentences
- Keyboarding Practice: Letter Location and Speed Building

The fastest and easiest way to create a form for a table is to create an AutoForm. The **AutoForm** command creates a form that includes all the fields in the current table. An AutoForm is created using the New Object, AutoForm command. A form you have created can be saved and used again. You will practice creating and saving an AutoForm in Activity 9-5. If you want to create custom forms, you can do so in Design view.

ACTIVITY 9-5

CREATE AN AUTOFORM

HELP KEYWORDS

Form
Create a form

1. Start *Access*. Open the *Customers* database.

2. Make a copy of the Activity 9-3 table. Name it Activity 9-5. (See Step 3 in Activity 9-2 to review copying a table.) Open the Activity 9-5 table.

3. Click the arrow on the **New Object** button on the Table Datasheet toolbar. Select **AutoForm**.

4. The new form appears with data for Record 1 displayed as shown in Figure 9-13.

Figure 9-13
A Database Form

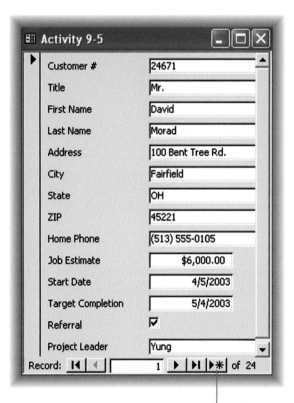

New Record button

DigiTip

Use the navigation buttons at the bottom of a form window to move through the records.

5. Click the **Save** button on the Standard toolbar. Use *Activity 9-5* for the form name. Click **OK**.

6. Click the **New Record** button to create a blank form in which to enter data.

CHAPTER 7

Speech Recognition

OBJECTIVES *In this chapter you will:*

1. Position the microphone properly and adjust settings.
2. Use the Language bar and complete initial training.
3. Choose your user profile.
4. Speak voice commands, punctuation, and words clearly.
5. Navigate a document with voice commands.
6. Scratch and correct errors.
7. Train unique words, names, and terms.
8. Format text using voice commands.
9. Create documents using voice input.

USES OF SPEECH RECOGNITION

Computer scientists have long dreamed of computers that can understand human speech. Once considered science fiction, that dream has become a reality. Today, **speech recognition** programs allow people to input data and commands to computers using spoken words. The dictated words can be input at speeds around 120 to 170 words per minute.

Speech recognition can be used with computerized devices of various sizes and types. Car owners can talk to the instrument panels on their vehicles. Individuals can access numbers on their cell phones with voice commands. Businesspeople can create e-mails, letters, and reports or enter data in a spreadsheet by dictating. Children can interact with computerized toys by speaking. Students can complete reports by dictating to their Tablet, portable, or desktop PCs. Several programs, such as those listed below, can be used for speech recognition:

- *Dragon NaturallySpeaking*
- *IBM ViaVoice*
- *Microsoft Office*
- *Windows XP Tablet PC Edition*

ETHICS

Protecting Personal Data

Companies often have electronic records that contain personal data about their customers. Many people enter their personal data when using the Internet to purchase products, register for prizes or free services, or take surveys. How the company may use the personal data you give them varies. Records are stored in electronic format so they may be accessed quickly. Having records in electronic format also means that they may be shared easily.

Some companies only use the data for the direct purpose stated when you enter the data. Other companies may use your personal data to send you advertisements for their products in the form of regular (paper) mail, e-mail, or phone calls. Some companies may give or sell your personal data to other companies.

Many people think selling customer data without the customer's consent is unethical. Before entering your personal data on a Web site, look for links such as *Privacy Policy* or *Privacy Statement*. A privacy policy tells you how your data will be used by the company.

For Discussion

Do you think it is ethical or unethical for companies to sell customer data? Why?

LESSON 9-4: FORMS AND REPORTS

OBJECTIVES — *In this lesson you will:*

1. Create and save a form.
2. Enter data and print data using a form.
3. Delete a record.
4. Create and print a report.

Data can be entered into a database using a table in datasheet view. You have entered data this way in earlier activities. You have also used a table to view and print your data. Using a table is not always the best way to accomplish these tasks. In this lesson, you will learn to use forms to enter data and reports to display data.

FORMS

When a table is large with many records and fields, entering data using a table can become difficult. A different database object, called a form, can make entering data easier. A **form** is a database object that is primarily used to enter data in a database. A form can be designed to display the field names and values for one record at a time, making data entry easier. Forms also make it easier to modify and view data.

Figure 7-1
Speech recognition is an effective input method for many people.

DigiTip
A tutorial for *Dragon NaturallySpeaking* can be found on the Instructor's Resource CD for this text.

Speech programs have many similarities, but they also have differences. In the exercises that follow, differences in instructions for *Office* and the Tablet PC speech programs are noted as needed. Follow the instructions for your DigiTool.

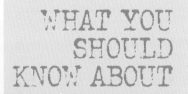

WHAT YOU SHOULD KNOW ABOUT

Protecting Your Voice

The consistent use of speech recognition can reduce your risk of developing carpal tunnel syndrome (CTS) in your wrists and repetitive stress injuries (RSI) in your hands, arms, or shoulders caused by excessive typing and mouse clicking.

Using speech recognition is not without its own risks. Dictating can place an added strain on your vocal cords. Keep a water bottle handy and take sips frequently as you dictate. Just as you should take frequent breaks from prolonged typing, take frequent breaks from dictating, too.

MICROPHONE POSITION

HELP KEYWORDS

Speech
Position the speech recognition microphone

Placing your microphone in the proper position is important. Follow this "rule of thumb": Place the microphone a thumb's width away from the side of your mouth. If your microphone's arm is too long to stay at the side of your mouth, place the microphone even with, or slightly below, your lower lip. In this position, the microphone should not pick up the sounds of your breathing. Breathing sounds from either your nose or your mouth can cause words to appear on the screen by mistake. If breathing errors occur, try moving your microphone slightly further away from your mouth and nose.

7. Note that the datasheet displays the first and last names and home phone numbers of all customers who live in Kentucky. The records are sorted by last name. This is the information you asked for in the query.

8. Print the query datasheet. (Follow the same steps used for printing a table datasheet.)

9. Close the query. (Click the **Close** button on the query window or choose **Close** from the File menu.)

10. Practice opening a query. Click **Queries** on the Objects bar. Select the name of the query, **Activity 9-4**. Click **Open** on the Database window toolbar.

11. Close the query and exit *Access*.

Self Check Does your query datasheet look like the one in Figure 9-12? If not, examine your query in Design view and compare it to the illustration in Step 4. Make changes if needed.

APPLICATION 9-2 — PROJECTS BY LEADER QUERY

The four project leaders who work for Evergreen Landscaping are responsible for completing the jobs on time and in a manner that is satisfactory to the customers. Each leader is responsible for a large amount of revenue for the company. Create a query to answer the question: For which projects is each project leader responsible and what are the start dates and job estimates for these projects?

1. Start *Access*. Open the *Customers* database.

2. Think about the question you need to answer with a query. What fields will you need to include? Does it make sense to sort the date by one of the fields?

3. Create the query. Use the Activity 9-3 table for the data. Save the query as Application 9-2. View the query datasheet to see the results. Print the query datasheet.

4. Use the query results to create a worksheet in *Excel*. Make the Project Leader names the column heads. Enter the data for Job Estimate amounts for each project leader under his or her name. Total the Job Estimate amounts for each project leader. Create a pie chart that shows the percentage of the Job Estimate amounts for each project leader compared to the total.

5. Save the worksheet as *Chart 9-2* in the *DigiTools your name\Chapter 9* folder. Print the worksheet with the chart.

Peer Check Compare your database query design with a classmate's query. How do they differ? Make changes, if needed, after your discussion.

Figure 7-2
Place your microphone properly as shown in the Microsoft Microphone Wizard.

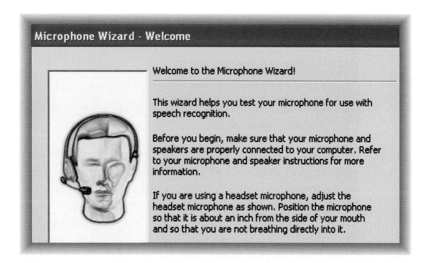

Place your microphone in the same position each time you use your software. Avoid touching the microphone with your hands or your mouth. This can cause unnecessary errors. If you can, obtain your own personal headset to use every day.

INITIAL TRAINING

OBJECTIVES In this lesson you will:

1. Open the Language bar.
2. Complete initial speech training.
3. Choose your user profile.

The **Language bar** is the control center for your speech program. This bar provides access to speech commands and options. On a Tablet PC, speech commands and options are also available on the Tablet PC Input Panel. To use the program effectively, you must create and use your own personal speech user profile. This is accomplished during the enrollment or "training" process. This **user profile** is your voice database that contains information about the way you speak and the unique words you use.

You should *always* use your own personal user file. Be careful not to choose the speech profile of another person. If you do, your accuracy will drop drastically and you may add incorrect words to another person's speech user file.

After your first training session, the program may recognize only 80 to 90 percent of your words correctly. You may want to do additional training sessions later to improve your accuracy. When you want to do more training, choose Training from the Tools menu on the Language bar.

3. In the Show Tables dialog box, click the **Tables** tab. Select the table you want to use, **Activity 9-3**. Click **Add**. Click **Close** to close the dialog box.

> **HELP KEYWORDS**
>
> **Query**
> About designing a query

4. Add fields to the Field row in the design grid to ask your question. For example, suppose you want to find the first and last names and home phone numbers of all customers who live in Kentucky. You would also like the records sorted by last name.

- Click in the first column in the cell by Field. A down arrow appears. Click the down arrow and choose **First Name**. (This tells the query you want to know the first name of the customers.)
- Click in the second column in the Field row. Click the down arrow and choose **Last Name**.
- Under the Last Name, click in the **Sort** cell. Click the down arrow and choose **Ascending**. (This tells the query to sort the records in ascending order by last name.)
- In the next column, choose **Home Phone** from the field names list.
- In the next column, choose **State** from the field names list. In the criteria row for this column, enter KY. (This tells the query to only show records with *KY* in the State field.) The Show box should be checked in each column where you entered a field name.

5. Click the **Save** button on the Standard toolbar. The Save As dialog box displays. Name the query Activity 9-4 in the Query Name text box. Click **OK**.

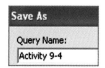

6. To view the query results (the answer to your question), click the **Datasheet View** button. The datasheet displaying the query results should look like Figure 9-12.

Figure 9-12
Query Results Datasheet

Lesson 9-3

ACTIVITY 7-1

CREATE A PROFILE AND TRAIN

The steps in this activity work best for users with a single-user login for *Windows XP*. If your computer has *Office* and *Windows 2000* or an earlier operating system, these steps should also work well for you. If you have both *Windows XP* and *Office*, your teacher may instruct you to create a new user in *Windows XP* rather than following the steps in this activity. These users will automatically be asked to create a new user profile the first time they start speech recognition.

HELP KEYWORDS

Speech
Create and use speech recognition profiles

1. Start *Word*. If the Language bar does not appear, choose **Speech** from the Tools menu.

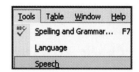

2. Set up your profile.

DigiTip

The instructions in this chapter direct you to give commands by clicking the mouse. You may also use your digital pen and tap buttons, menus, or options to give commands. In a later lesson, you will learn to use speech recognition tools to give commands.

Office

Click **Tools** on the Language bar. Choose **Options**. This will open the Speech Properties dialog box.

Tablet PC

Click **Tools** on the Language bar. Choose **Options**. This will open the Speech Input Settings dialog box. Click **Advanced Speech** near the bottom of the dialog box. This will open the Speech Properties dialog box.

3. Click the **New** button. Follow the instructions on the screen to complete your first training session. This will take 10 to 15 minutes. After you complete the training, click **Finish.** Click **OK** to close the Speech Properties dialog box.

Figure 7-3
The Speech Properties Dialog Box

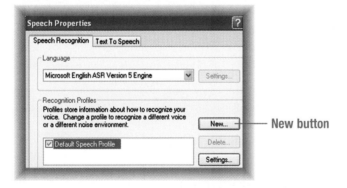

4. To select your username, choose **Current User** from the Tools menu on the Language bar. A check mark should appear in front of your name.

Figure 7-4
Choose your user profile.

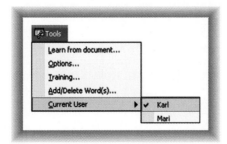

8. To remove the filter, click **Remove Filter/Sort** from the Records menu.

9. Create a filter to display only records for projects assigned to Waters. Click a cell in the Project Leader field that contains the value *Waters*. Click the **Filter by Selection** button.

10. Sort the records by Start Date in ascending order. Hide the Home Phone, Referral, Title, and First Name columns. Print the table using Landscape orientation.

11. Close the table without saving changes. Exit *Access*.

Self Check Did you answer the questions about the table correctly? In Step 3, records for IN appear at the top of the table. Ten customers are in Kentucky. In Step 4, the latest project start date is September 13. Seven projects will start in May. In Step 7, 11 records are displayed. Yung, Waters, and Roberts have projects in Ohio.

USING QUERIES

A **query** is a question that you ask a database. For your *Customers* database for example, you want to ask: What records have *Roberts* in the Project Leader field and *OH* in the State field? A query can give you an answer to this type of question.

Several types of queries can be created. In this lesson, you will learn about the most commonly used type of query, a select query. A select query finds and lists the records that meet criteria you set (the question you ask).

To create a select query, you must open the database in which you want to create the query. You create a new query (ask your question) by choosing a table in which to look for records and setting the criteria. A query can also look for records in several tables or other queries. You will practice creating a query in Activity 9-4.

ACTIVITY 9-4   CREATE A QUERY

1. Start *Access*. Open the *Customers* database.

2. In the Database window, click **Queries** on the Objects bar. Select **Create query in Design view** and click **Open**.

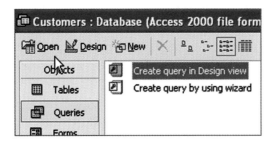

WHAT YOU SHOULD KNOW ABOUT
Readjusting Your Microphone Settings

Your speech profile automatically remembers your audio settings from session to session. However, you should adjust the microphone settings if one of the following occurs:

- Your background noise or acoustic conditions change.
- There is a noticeable decline in your recognition accuracy.
- Another person has used the speech system before you on the same computer.

To adjust the microphone, access the Speech Properties dialog box as explained in Activity 7-1. Click the **Configure Microphone** button. Read and follow the instructions.

DigiTip
Tablet PC users can also choose **Microphone Adjustment** from the Speech Tools menu on the Input Panel.

LESSON 7-2: DICTATING VOICE COMMANDS AND TEXT

OBJECTIVES — *In this lesson you will:*

1. Speak voice commands.
2. Dictate text.

Your speech recognition software listens for three types of input: voice commands, words, and punctuation. First, you will focus on commands, which will allow you to use your voice to accomplish everything you can do with a mouse or a digital pen. Then you will practice entering text by voice.

VOICE COMMANDS

Follow these guidelines when saying commands:

- Pause or hesitate briefly before and after each command. Say <pause> `Voice Command` <pause>
- When there are several words in the command, say the command as a phrase, not as separate words. Do not stop in the middle of a multiword command. Say <pause> `Voice Command` <pause>, not `Voice` <pause> `Command`.
- Clearly say each word in the command. Do not rush or slur commands. Say <pause> `Voice Command` <pause>, not `VoiComan`.
- Do not shout or whisper commands. Use a normal tone of voice. Say <pause> `Voice Command`<pause>, not `VOICE COMMAND!` or <softly> `voice command`.

To sort dates and times from earlier to later, use ascending order. Use descending order to sort from later to earlier. When you sort a field in ascending order, any records in that field which are blank are listed first.

You can reverse the sort and place the records back in the order they were in before the sort. To remove a sort, choose **Remove Filter/Sort** on the Records menu.

FILTERING RECORDS

Sometimes seeing only a part of the records in a table datasheet can be helpful. In your *Customers* database, for example, you might want to see the records for only the customers who live in Ohio. You might also want to see only the records for jobs for a certain project leader. A filter allows you to do this. A **filter** displays a subset of records selected according to criteria you give.

To create a filter in a table datasheet, find one instance of the value you want to display. (For example, if you want to display all records for a certain project leader, find that project leader.) Click in the cell that contains this value. Click the **Filter by Selection** button on the Table Datasheet toolbar. See Figure 9-11.

To remove the filter and display all the records, choose **Remove Filter/Sort** on the Records menu. A filter cannot be saved. If you apply a filter and close the table, the full table will display when it is opened again.

ACTIVITY 9-3

SORT AND FILTER RECORDS

1. Start *Access*. Open the *Customers* database. (Choose **Open** from the File menu. Navigate to the *DigiTools your name\Chapter 9* folder. Select the filename *Customers*. Click **Open**.)

2. Make a copy of the Activity 9-2 table. Name it Activity 9-3. (See Step 3 in Activity 9-2 to review copying a table.) Open the Activity 9-3 table.

HELP KEYWORDS
Sort
Sort records

3. To sort the records in ascending order by the State field, click in a cell in the State field. Click the **Sort Ascending** button on the Table Datasheet toolbar.
- Records for customers in which state appear at the top of the table?
- How many customers are located in Kentucky?

4. To sort the records in descending order by the Start Date field, click in a cell in the Start Date field. Click the **Sort Descending** button on the Table Datasheet toolbar.
- What is the latest start date for a project?
- How many projects will start in May?

5. To sort the records in ascending order by the Customer # field, click in a cell in the Customer # field. Click the **Sort Ascending** button on the Table Datasheet toolbar.

6. Click the **Save** button. Print the sorted table.

HELP KEYWORDS
Filter
Create a filter

7. Create a filter to display only records for customers in Ohio. Click a cell in the State field that contains the value *OH*. Click the **Filter by Selection** button.
- How many records are displayed in the datasheet?
- Which project leaders have been assigned jobs in Ohio?

ACTIVITY 7-2

PRACTICE VOICE COMMANDS

DigiTip

Tablet PC users, if you do not see the Input Panel after you start your Tablet PC, tap the **Tablet PC Input Panel** button. It can be found on the taskbar near the Start button.

Figure 7-5
The Language Bar

1. Open *Word* and display the Language bar by choosing **Tools, Speech**. Tablet PC users, open the Input Panel and choose **Tools, Speech.**

2. Turn on your microphone by clicking the **Microphone** button on the Language toolbar.

You can click the Dictation and Voice Command buttons on the Language bar to switch between dictating text and giving commands. You can also say `Dictation` or `Voice Command` to switch input modes. **Dictation** mode allows you to input words and punctuation. **Voice Command** mode allows you to input commands.

Tablet PC users can also use the Dictation and Command buttons on the Input Panel to switch between dictating text and giving commands. To display the speech tools, click the **Tools** button. Choose **Speech** from the menu. Speech tools will display as shown in Figure 7-6.

Figure 7-6
Speech Commands in Input Panel

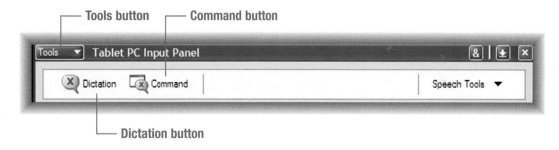

HELP KEYWORDS

Speech
Use speech recognition

3. Practice switching between Voice Command and Dictation modes by saying the following commands several times: `Dictation` <pause> `Voice Command`. Notice that the button selected on the Language bar or Input Panel changes each time you speak.

4. To turn the microphone off, say `Microphone`. The Language bar returns to a collapsed state and the Dictation and Voice Commands buttons are not displayed on the Language bar. *Not listening* is displayed on the Input Panel.

5. Turn on the microphone. Activate Voice Command mode by saying `Voice Command`. Say the following commands to open and close various menus on the menu bar. Turn your microphone off after practicing these commands.

```
File <pause> (Note that the File menu is open.) Cancel/Escape
Edit <pause> (Note that the Edit menu is open.) Cancel/Escape
View <pause> (Note that the View menu is open.) Cancel/Escape/Escape
```

3. Each team member, open your *Customers* database and create a new table named Suppliers. Follow your plan for field names and types. Save the table structure. Enter your data into the table. Save and print the table.

Peer Check Compare your database table plan with another team's plan. How do they differ? How could each one be made better?

LESSON 9-3: ARRANGING AND FINDING DATA

OBJECTIVES *In this lesson you will:*

1. Sort data in a database table.
2. Filter data in a database table.
3. Create and run a query.

Access provides several tools you can use to sort or display data in various ways. In this lesson, you will learn to sort records, filter records, and query records.

SORTING RECORDS

Records are often entered into a database table in the order the information or source documents are received. This is not always the most useful order for viewing the data. In your *Customers* database, for example, you might want to see the records in order by the start date or by the project leader. The **Sort** command allows you to arrange records in ascending or descending order using the data in a specific field.

To sort records in a table datasheet, click in a cell in the field you want to use to sort the records (such as Start Date). Click the **Sort Ascending** or **Sort Descending** buttons on the Table Datasheet toolbar. The records will be sorted in that order.

Figure 9-11
Table Datasheet Toolbar

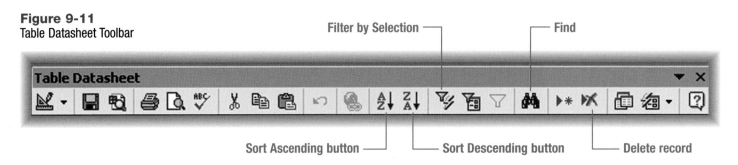

> **DigiTip**
> Pause or hesitate briefly before and after each command. Say <pause> `Voice Command` <pause>.

6. Use voice commands to display and hide a toolbar. Turn your microphone on. Say these four commands: `Voice Command, View, Toolbars, Drawing`. The Drawing toolbar is displayed. Repeat the commands to hide this toolbar. Turn your microphone off.

7. Click the **Minimize** button on the Language bar to send it to the taskbar. (See Figure 7-5.) Close the document without saving and close *Word*.

TEXT

When you input by speaking, pay attention to your enunciation. **Enunciation** means speaking words clearly and distinctly. Do not worry about correcting mistakes at this early stage. Everyone makes mistakes in the beginning. You will learn to correct errors in later lessons. Until then, practice speaking clearly and smoothly with pauses before commands. Remember these guidelines:

- Continue talking until you reach the end of the sentence, thought, or idea.
- Do not break words into syllables. For instance, say *popcorn,* not *pop corn.*
- Do not run words together or drop letter sounds out. For example, say *candy bar,* not *canybr.*
- Do not shout and do not whisper. Do not let your speaking volume trail off toward the end of a sentence. Say *New Paragraph,* not *NewParagra…*
- Do not speak too fast or too slowly. Speak at a speed that is just right for you.

ACTIVITY 7-3 — DICTATE TEXT

> **HELP KEYWORDS**
> Speech
> How to talk to a computer

1. Open *Word* to a new blank document and display the Language bar. Choose your user profile. Tablet PC users, display the Input Panel's Speech bar.

2. Turn on your Microphone. Click the **Dictation** button.

3. Dictate the following sentences. Say `period` when you come to a period at the end of a sentence. (**Note**: It sometimes helps to pause briefly before saying punctuation marks like *period.*) Turn off your microphone after dictating the paragraph.

```
Talk to your computer. Speak normally and say each
word clearly. Don't shout and don't whisper. Don't
stop in the middle of a sentence. Practice speaking
clearly.
```

> **DigiTip**
> Tablet PC users, you can use the Text Preview option to view text before sending it to *Word*. However, most Tablet PC users find this an unnecessary step. To view or close the Text Preview panel, choose **Tools, Text Preview** on the Input Panel.

4. How did you do? Remember to speak each word clearly. Next, you will clear all the text from your document so you can try again. Turn on your microphone. Say these five commands: `Voice Command, Edit, Select All, Edit, Cut`.

5. Switch to Dictation mode. Dictate the paragraph from step 3 again. Remember to say the periods.

Figure 9-10
Use the Find feature to quickly locate a record.

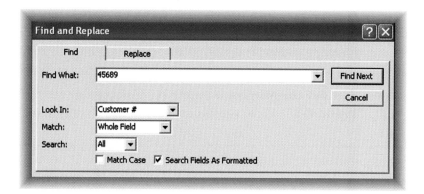

12. Enter the project leader names in the Project Leader fields for these records:

Project Leader	Customer #s
Perez	45689, 35228, 54333, 42631, 46782
Roberts	25788, 43215, 45678, 56495, 43221, 53251
Waters	54303, 32679, 34788, 35801, 54291, 54214, 46655
Yung	24671, 54666, 56789, 29807, 54102, 35789

13. When you have entered all the project leader names, unhide the columns. Choose **Unhide Columns** from the Format menu. Click the box beside the column name for each hidden column to unhide it. Click **Close**.

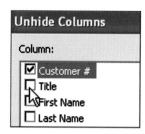

14. Click the **Save** button on the Standard toolbar. Print the table. Choose **Close** from the File menu to close the database table. Choose **Exit** from the File menu to close *Access*.

HELP KEYWORDS
Hide
Show or hide columns in a datasheet

APPLICATION 9-1

SUPPLIERS DATABASE TABLE

TEAMWORK

Work with a classmate to complete this application.

Pretend that the Evergreen Landscaping business is located in your area. You have a database table to store information about customers. Create a database table to store information about businesses from which you will buy supplies. These supplies would include items such as trees, shrubs, plants, flowers, fertilizer, mulch, garden ornaments, edging stones or timbers, tools, and other items needed to do the landscaping jobs.

INTERNET

1. As a team, use a Yellow Pages phone directory for your area or search the Internet to find businesses from which you can buy supplies. Record the company name, full address, phone number, and Web site URL (if available) for each company. Note the type of materials you could purchase from this company. Find information for at least six companies. If you live in a small town or rural area, you may have to buy from suppliers in other cities.

DigiTip
Create a field with the Memo data type to enter data of varying lengths such as the list of items you could purchase from a store. Use the Hyperlink data type for a field to hold URLs.

2. As a team, plan a database table to hold the information you found. Look at the information you have and decide what fields you will need to store the data. Make a list of field names and the data type you will use for each field.

6. How did you do this time? Clear your screen before moving to the next dialogue. Delete all the information with a new set of voice commands. Say these four commands: `Voice Command, Edit, Select All, Backspace`.

Note: You will no longer be reminded to turn your microphone on and off during an activity. Turn off your microphone when you will not be dictating. Turn it on when you are ready to dictate again.

7. Switch to Dictation mode. Dictate the paragraph below. Remember to say `period` at the end of each sentence.

```
Don't break any speed records when you speak to your
computer. Speaking fast will cause you to run words
together. Speaking slowly can also cause problems. If
you speak very fast or very slowly the system will
have trouble understanding you. Talk at a speed that
is just right for you.
```

8. In Voice Command mode, say `Enter` twice to insert a couple of blank lines after the paragraph. Dictate using the following "starter sentences." Starter sentences are designed to begin a thought and allow you to finish it. Read the sentences silently and think about what you will say before you begin dictating. Dictate the text shown and then finish the sentence with your own words.

```
This is a test of the emergency alert system. In a
real emergency you should...

All people are created equal. This means...
```

9. Practice copying and pasting text using voice commands. Switch to Voice Command mode. Select all the text in the document. (Say `Edit, Select All`.) Copy the text by saying `Copy`. Go to the end of the document by saying `Move to End of Document`. Paste a copy of your text into your document by saying `Paste`.

10. Save your document as *Activity 7-3* in your *DigiTools your name\Chapter 7* folder. Use Voice Command mode to close your document and *Word*. In Voice Command mode, say `File, Exit`.

Self Check Did you turn off your microphone each time you paused from dictating for more than a few seconds?

6. Change the field size for the Customer # field so you won't accidentally enter too many numbers in this field.
- Click the **Design View** button to switch to Design view.
- Click the **Customer #** field name. In the Field Properties window, change the field size to **5**. See Figure 9-8.
- Click the **Save** button to save the new structure.
- Click the **Datasheet View** button to view the table datasheet.

7. Enter the data from the source documents for 21 records. Click in the blank cell under the Customer # for record 3 and begin entering data. Press **Tab** to move to a new cell. Do not enter the name of the person who referred our company to the customer. Simply check the Referral box if a referral was made. Use the current year when entering dates.

> **HELP KEYWORDS**
> Table
> About changing a field's data type

8. After thinking about your database design, you decide that the Start Date and Target Completion fields should be Date/Time fields rather than Text fields. Change the data type for these fields.
- Click the **Design View** button to switch to Design view. Click in the data type cell for the Start Date field. Click the down arrow and choose the **Date/Time** data type.
- Click the **Save** button to save the new structure.
- Click the **Datasheet View** button to view the table datasheet.
- Repeat the process for the Target Completion field.

> **HELP KEYWORDS**
> Table
> Add a field to a table

9. The company manager has now assigned a project leader to each landscaping job. You will add a field named *Project Leader* and enter the data for this field.
- Click the **Design View** button to switch to Design view.
- Click in a blank row on the Field Grid below Referral. Enter the name of the field, `Project Leader`. Choose **Text** for the data type.
- Click the **Save** button to save the new structure.
- Click the **Datasheet View** button to view the table datasheet.

10. To make entering the project leader information easier, hide several of the columns in the datasheet so you can see both the Customer # and Project Leader fields without scrolling. Point to the Title column head until the cursor becomes a dark down arrow. Click on the column head and drag right to select several columns. Choose **Hide Columns** from the Format menu.

> **HELP KEYWORDS**
> Find
> Find or replace a value in a field

11. Use the Find command to find the first record you need to update. Choose **Find** from the Edit menu. In the Find What box, enter the customer # for the first record you want to update: `45689`. See Figure 9-10 on page 271. Choose the **Customer #** field name from the Look In list if it is not already displayed. Click **Find Next**. The cursor will move to the record with Customer # 45689. Click **Cancel** to close the dialog box. Strike **Tab** to move to the Project Leader field and enter the name: `Perez`.

WHAT YOU SHOULD KNOW ABOUT

The Origins of Speech Recognition

Early research into speech recognition started in the late 1950s at International Business Machines (IBM). In the early 1960s, computers were beginning to recognize the human voice. Early speech systems, called **discrete speech**, forced speakers to pause before saying each word. Discrete speech is still used by individuals with respiratory handicaps, but the system proved too cumbersome for the average user.

At long last, in April 1997, Dragon Systems released the first large-vocabulary, continuous speech recognition (CSR) system for the average person. This program was called *NaturallySpeaking*. *Dragon NaturallySpeaking* allowed users to speak normally and to have their words entered as text accurately at speeds between 120 and 170 words per minute.

Shortly thereafter, IBM also released its first continuous speech recognition system, which today is known as *ViaVoice*. Several years later, in May 2001, Microsoft released its first continuous speech recognition program with its *Office XP* software. Microsoft's speech recognition was enhanced and built into the first Tablet PCs, released in November 2002.

LESSON 7-3: NAVIGATION AND PUNCTUATION

OBJECTIVES
In this lesson you will:

1. Navigate in a document using commands in Dictation and Voice Command modes.
2. Dictate punctuation marks.
3. Improve speech recognition accuracy by completing additional training.

You can speak various commands, even when in Dictation mode, that allow you to navigate through a document. Punctuation marks can also be entered in Dictation mode. You will practice navigation and using punctuation marks in this lesson.

NAVIGATION

The New Paragraph and New Line commands are essential to writing and formatting documents. Saying these commands is like pressing the **Enter** key on your keyboard. The **New Line** command causes the insertion point to move down to the next line, as when you press the **Enter** key once. The **New Paragraph** command causes the insertion point to move down two lines to create a double space. To make your dictating easier, you can say either of these commands while in Dictation mode when using *Office XP*.

Other navigation voice commands that you will find useful are listed below.

DigiTip
Tablet PC users can use these navigation commands also: Go to Bottom, Go to Top, Next Line, Previous Line.

- Move to Beginning of Document
- Move to End of Document
- Move to End of Line
- Move to Beginning of Line
- Move to Next Line
- Move to Previous Line

ACTIVITY 9-2

MODIFY A TABLE AND ENTER RECORDS

When an employee talks with a customer about a landscape job, he or she completes a form with the customer's contact information. The form also includes basic information about the job. In this activity, you will use these forms as source documents for database records.

In a real business situation, you would open and update the table you created earlier as often as needed, keeping the same name for the table. For the exercises in this chapter, however, you will make a copy of the table before changing it. This allows you and your teacher to see your work for different exercises.

1. Start *Word*. Open and print the data file *Evergreen*. This file contains the source documents you will use to enter data in the table. (The forms are shown several per page to save printing resources.) Set the source documents aside for later use. Close *Word*.

2. Start *Access*. Open the *Customers* database. (Choose **Open** from the File menu. Navigate to the *DigiTools your name\Chapter 9* folder. Select the filename, *Customers*. Click **Open**.)

3. Follow these steps to make a copy of the Activity 9-1 table.
   - Click the **Activity 9-1** table to select it.
   - Click the **Copy** button on the Standard toolbar.
   - Click the **Paste** button on the Standard toolbar. Name the copy of the table Activity 9-2. **Structure and Data** should be selected under Paste Options. Click **OK**.

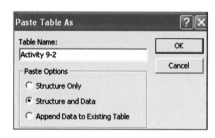

4. Open the Activity 9-2 table. Select the table name and click **Open** on the Database window toolbar.

HELP KEYWORDS

Table
Resize a column or row

DigiTip

Another way to resize a column is to click the border of the column and drag it to the desired width.

5. Some of the columns in the datasheet are not wide enough to display all the data. Some columns are wider than needed to display the data. Change the size of all columns. Double-click the right edge of each column heading to size it to fit the longest data in the column.

Figure 9-9
Resize Column Width

Notice how the cursor appears when you point to the right edge of the column head

Sometimes people cut off the initial or final sounds of a command. For instance, some people say *...ew Paragraph* or *New paragra...* instead of *New Paragraph*. Remember not to pause between words in a command. For example, say <pause> `New Paragraph` <pause>, not `New` <pause> `Paragraph`.

PUNCTUATION

You must say punctuation marks to insert them in your document. Pausing briefly before and after you say punctuation marks can be helpful. Pausing briefly can help you breathe normally, relax, and speak calmly. Some punctuation mark commands you can say in Dictation mode are shown below.

- Backslash (\)
- Close Quote (")
- Colon (:)
- Comma (,)
- Dash (– or —)
- Exclamation Point/Mark (!)
- Open Quote (")
- Period (.)
- Question Mark (?)
- Semicolon (;)
- Slash (/)

ACTIVITY 7-4 — NAVIGATE AND INSERT PUNCTUATION

1. Open *Word* to a new blank document and display the Language bar. Turn on your microphone. Tablet PC users, display the Input Panel's Speech bar.

2. Say or click **Dictation**. Practice the New Line and New Paragraph commands by saying each sentence below clearly.

```
This is a complete sentence. <pause> New Line
Speak clearly without stopping. <pause> New Line
Speak with a normal voice. <pause> New Paragraph
```

3. Continue to practice dictating with the following sentences.

```
Think about your audience. <pause> New Paragraph
Evaluate your purpose in writing. <pause> New Paragraph
Use an appropriate writing style or personality. <pause>
New Paragraph
Don't make your documents too long. <pause> New Paragraph
```

4. Practice moving around your document by saying the commands below (in Voice Command mode).

```
Move to Previous Line          Move to End of Line
Move to Next Line              Move to Beginning of
Move to End of Document          Line
Move to Beginning of
   Document
```

CHAPTER 7   Lesson 7-3

To delete a field in a table, open the table in Design view. Select the table name and click **Design** on the Database window toolbar. Click the field selector box to the left of the Field Name to select the field you want to delete. Press the **Delete** key. Choose **Yes** at the prompt asking if you want to delete the field. Click the **Save** button to save the new structure. Click the **Datasheet View** button to return to the datasheet.

Figure 9-6
Select a field and press **Delete** to delete it.

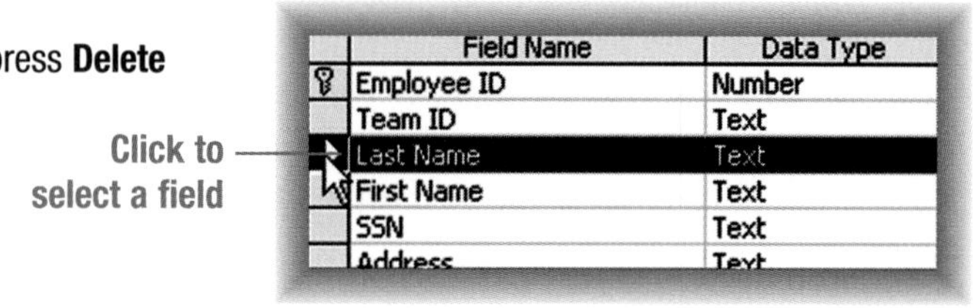

Click to select a field

CHANGING FIELDS

Fields can be changed by giving them a different data type or by setting options for the field properties. To change a field data type, open the table in Design view. Click in the data type cell for the field you want to change. Click the down arrow and choose a different data type. Click the **Save** button to save the new structure. Click the **Datasheet View** button to view the table datasheet.

Figure 9-7
Select a Data Type

Click the arrow and choose a data type

Field properties can be set to make entering data easier or to help prevent errors when entering data. For example, all customer numbers in the *Customers* database you created have five characters. By default, text fields allow 255 characters in the field. By changing the field size to five characters, you prevent a user from accidentally entering an incorrect customer number with six characters.

To change the properties for a field, open the table in Design view. Click the field name of the field you want to change. The field's properties will display in the Field Properties window. Make changes, such as to the field size, as desired. Click the **Save** button to save the new structure. Click the **Datasheet View** button to view the table datasheet.

Figure 9-8
The Field Size property can be changed.

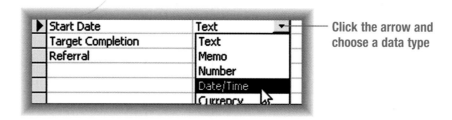

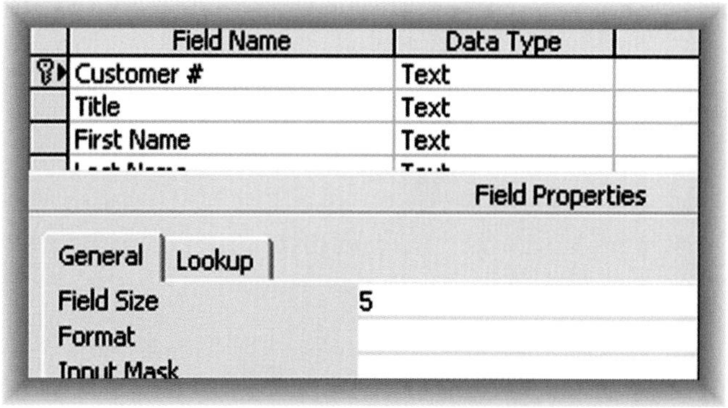

5. Move to the end of the document. Switch to Dictation mode. Dictate the following sentences to practice inserting punctuation. Use the New Paragraph command at the end of each paragraph.

```
Is that your final answer?  Congratulations!  You have
just won the grand prize.
I know what you want: a dog, a cat, and a tropical
fish.
Don't speak very fast; don't speak very slowly. Speak
at just the right speed for you.
I think this is going rather well—better than
expected!
Thomas said, "Using speech recognition software is
easy."
The document is located in the Thomas\letters folder.
```

6. Switch to Voice Command mode. Say `File, Save As` to open the Save As dialog box. Save the file as *Activity 7-4* in your *DigiTools your name\Chapter 7* folder. When you have navigated to the correct folder and entered the filename, say `Save` to close the dialog box and save the file. Close the document and exit *Word*.

Self Check Did you remember to pause briefly before saying punctuation?

ACTIVITY 7-5   COMPLETE ADDITIONAL TRAINING

In a later lesson, you will learn how to correct mistakes made when dictating. However, the best way to avoid having to correct mistakes is not to make them. This requires improving your software's recognition accuracy. In this activity, you will read an additional training story so your system will understand your voice better.

Office XP

1. Display the Language bar. Make sure that the Speech Tools option has been selected. Click the **Options** down arrow. Place a check mark by the **Speech Tools** option if it is not checked.

HELP KEYWORDS

Table
Move between records or fields

15. Practice using the navigation buttons to move in the datasheet. Review Figure 9-2 on page 264.
- Click the **First Record** button to move to the first record. What is the first name of the customer in Record 1?
- Click the **Last Record** button. What data is in the State field for this record?
- Click the **Previous Record** button. What data is in the City field for this record?

16. Choose **Close** from the File menu to close the database table. Choose **Exit** from the File menu to close *Access*.

Self Check Did you find these answers for Step 14? First record, customer first name: David. Last record, data in State field: KY. Record 2, data in City field: Madison.

LESSON 9-2 CHANGING TABLE STRUCTURE AND FIELDS

OBJECTIVES *In this lesson you will:*

1. Modify a database table structure.
2. Change field data types and properties.
3. Enter data from source documents.

CHANGING TABLES

After a database table has been created, you may find that additional fields are needed in the table. You can easily add or delete fields to a database table.

To add a field to a table, open the table in Design view. Select the table name and click **Design** on the Database window toolbar. Click in a blank row on the Field Grid. Enter the name of the field under Field Name and choose a data type. Click the **Save** button to save the new structure. Click the **Datasheet View** button to view the table datasheet.

Figure 9-5
Open a Table in Design View

Figure 7-7
The Options Menu

2. Verify that your speech profile has been selected. Click **Tools** on the Language bar. Choose **Current User**. A check mark should appear next to your name. Choose your user profile if necessary.

3. Click the **Tools** button on the Language bar and choose **Training**. Choose a new training story and read it aloud as directed. The training exercise may take 10 to 15 minutes. You do not need to say punctuation marks during the training. Click **Finish** after your profile is updated.

Tablet PC

1. Display the Speech tools on the Input Panel. (Choose **Speech** from the Tools menu.)

2. Click the **Speech Tools** button on the Speech bar and choose **Voice Training**. Choose a new training story and read it aloud as directed. The training exercise may take 10 to 15 minutes. You do not need to say punctuation marks during the training. Click **Finish** after your profile is updated.

LESSON 7-4 CORRECTING ERRORS

OBJECTIVES *In this lesson you will:*

1. Correct errors using the Scratch That command.
2. Correct errors using the Correction button.
3. Correct errors by substituting words.

SCRATCH THAT COMMAND

The **Scratch That** command erases the last word or continuous spoken phrase. This is one of the quickest and most commonly used methods to erase mistakes immediately. You can say the Scratch That command several times in succession to remove multiple phrases or mistakes. Remember to pause before you say Scratch That.

10. Choose **Save As** from the File menu. Enter `Activity 9-1` as the name for the table. Under As, accept the default, **Table**. Click **OK**. Choose **Datasheet View** from the View Menu.

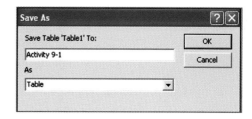

> **HELP KEYWORDS**
> Table
> Add or edit data

11. Enter the data for three records in the table. (You will enter additional records later.) Press **Tab** to move to a new field. Click the checkbox in the Referral field to indicate *Yes*. This means the customer was referred by another customer. Leave the box empty to indicate *No*. Use the current year in the dates. In the Job Estimate field, do not key the dollar sign ($); it will be added automatically.

Customer #	24671	25788	29807
Title	Mr.	Mrs.	Miss
First Name	David	Dineen	Jenna
Last Name	Morad	Ebert	Erickson
Address	100 Bent Tree Rd.	27 Main St.	34 Emerson St.
City	Fairfield	Madison	Covington
State	OH	IN	KY
ZIP	45221	47250	41011
Home Phone	(513) 555-0105	(812) 555-3999	(606) 555-3201
Job Estimate	$6,000.00	$2,000.00	$2,500.00
Start Date	4-5	5-3	5-15
Target Completion	5-4	6-2	6-15
Referral	Yes	Yes	No

12. Click the **Save** button on the Standard toolbar. After you enter data in a cell on the datasheet and press **Tab** to move to another cell, the data is automatically saved. However, clicking the **Save** button before closing a table is still a good idea. This will save any changes you may have made to the datasheet, such as a change in the width of a column.

13. Choose **Page Setup** from the File menu. Click the **Page** tab. Select **Landscape** under Orientation. Click OK.

14. Choose **Print** from the File menu. Click **OK**. Two pages will be required for the table. Review the printed table. Notice that not all the data displays in some of the columns. You will practice changing column widths in the next activity.

> **HELP KEYWORDS**
> Table
> Print a database object

ACTIVITY 7-6

USE THE SCRATCH THAT COMMAND

1. Open *Word* to a new blank document and display the Language bar. Tablet PC users, display the Input Panel's Speech bar.

2. Turn on your Microphone and select Dictation mode. Say the following:

```
Computer (Voice Command) Scratch That (Dictation)
```
```
I want to buy a computer (Voice Command)  Scratch That (Dictation)
```
```
Automobile (Voice Command) Scratch That (Dictation)
```
```
I want a new automobile (Voice Command) Scratch That (Dictation)
```
```
Airplane (Voice Command) Scratch That
```
```
I want to fly an airplane (Voice Command) Scratch That (Dictation)
```

3. You can erase multiple phrases by using the Scratch That command several times in a row. Dictate the text below. Then delete phrases using the Scratch That command.

```
In the meantime, Michael, let's start working on that
business plan. We must have it ready for the meeting on
Thursday. (Voice Command) Scratch That
```

(Continue saying `Scratch That` until the paragraph disappears.)

4. Close the document without saving.

> **HELP KEYWORDS**
> Speech
>   Correct speech recognition errors
>   Tips

> **DigiTip**
> Tablet PC users can restore the last phrase by saying `Undo That.`

CORRECTION COMMAND

You can correct errors in documents you have dictated in the usual ways. You can use the **Backspace** or **Delete** keys to delete text. You can select a word and key, handwrite, or speak a replacement. If you correct errors by choosing replacement words from a list, however, you will help your software remember the way you speak. The likelihood of making the same mistakes again will decrease.

Use your digital pen or mouse in combination with your voice to correct mistakes. You can select from a list of possible alternatives for a word you have highlighted. If no correct replacement is listed, you can choose to delete the error and reenter the word.

Some of the tricky words you will need to correct are called homonyms. **Homonyms** are words that sound the same but have different meanings. For example, *there*, *their*, and *they're*; *for*, *four*, and *fore*; and *by*, *buy*, and *bye*. Other similar sounding word pairs can also cause trouble, such as *all ready* and *already*; *advise* and *advice*; *affect* and *effect*. You will practice correcting some of these words in this lesson.

HELP KEYWORDS

Table
Save a database object

Figure 9-3
The Database Window

5. Click the **Create** button. The Database window displays. Click **Tables** in the Objects bar to select it. Then double-click **Create table in Design view**.

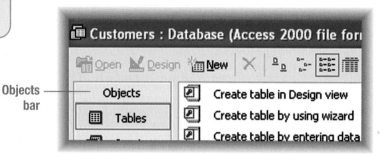

HELP KEYWORDS

Tables
About Tables
Table Design View

Figure 9-4
Enter field names and data types in the Field Grid.

6. The Table Design view window displays. In the top portion, called the *Field Grid*, you will enter a name and choose a data type for each field. You can also enter a description for the field if desired. In the lower portion of the window, called *Field Properties*, you can set options for the fields.

	Field Name	Data Type
🔑	Customer #	Text
	Title	Text
	First Name	Text
	Last Name	Text
	Address	Text
	City	Text
	State	Text
	ZIP	Text
	Home Phone	Text
	Job Estimate	Currency
	Start Date	Text
	Target Completion	Text
	Referral	Yes/No

7. Under Field Name, enter `Customer #`. Press **Tab**. *Text* will automatically appear. Press **Tab** to accept that data type. (To choose a different data type, click the down arrow in the cell and choose a data type.)

8. Enter the other field names and data types as shown in Figure 9-4. Do not change any of the default options under Field Properties at this time.

HELP KEYWORDS

Tables
About primary keys

9. A **primary key** is a field that uniquely identifies a record. Set the Customer # field as the primary key for the table. Right-click on **Customer #** and choose **Primary Key.**

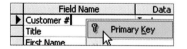

CHAPTER 9   Lesson 9-1   265

ACTIVITY 7-7 CORRECT ERRORS

Office XP

1. Open *Word* to a new blank document and display the Language bar.

2. You may need to add the Correction command button to your Language bar. To do so, click the **Options** button on the Language bar. Choose **Correction**. The Correction button now appears on the Language bar.

3. Dictate the sentences below. Remember to dictate the punctuation marks. Use the New Paragraph command after each paragraph.

```
There are many words that I don't understand.
After all, I am just a computer!

Sometimes I confuse the words an, and, and end.
What do you expect?  Perfection?
```

4. Proofread your sentences carefully, checking for mistakes. Highlight an incorrect word with your digital pen or mouse.

5. Click the **Correction** button on the Language bar or say `Correction`. If the correct word appears in the numbered list, click the correct word. You can also choose the correct word by saying `Select` and the number of the item. For example, `Select 2`. If the correct word does not appear in the list, delete the word. Key or speak the correct word.

DigiTip
Sometimes a single word you speak is recognized incorrectly as two words. You can select two or three words at once and use the Correction command.

Figure 7-8
Choose the correct word from the list.

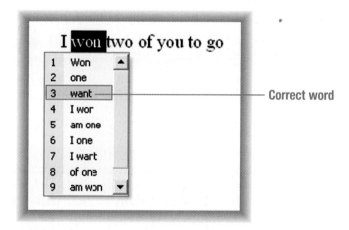

6. Practice correcting other words. Close the document without saving.

Tablet PC

1. Open *Word* to a new blank document. Display the Language bar. Open the Tablet PC Input Panel's Speech bar.

2. Open the Text Preview window to make correcting errors easier. In Input Panel, click the down arrow on the Tools button to display the menu. Choose **Text Preview**.

Data Type	Description
Text	For text characters or numbers that do not require calculations
Memo	For lengthy notes that can contain text and numbers
Number	For numbers (other than currency) to be used in calculations
Date/Time	For dates and times
Currency	For dollar values
AutoNumber	Consecutive numbers assigned automatically by *Access*
Yes/No	For data that can only be Yes or No
OLE Object	For holding objects such as an *Excel* spreadsheet (used with more advanced features)
Hyperlink	For hyperlinks to other locations (uses URLs or paths)
LookUp Wizard	For choosing a value from another table or list

NAVIGATING IN A DATABASE TABLE

Once a database table has been created, you can move around the table and enter data just as you would in a *Word* table. The **Tab, Arrows, End**, and **Home** keys move your insertion point to the next fields or up and down one record (row).

Navigation buttons at the bottom of the table are also used to move around in the table. The buttons allow you to move to the next record, the previous record, the first or last record, or to a new row for a blank record. The total number of records in the table is also displayed. Scroll bars allow you to display different parts of the table if it does not all display on the screen.

Figure 9-2
Navigation Buttons

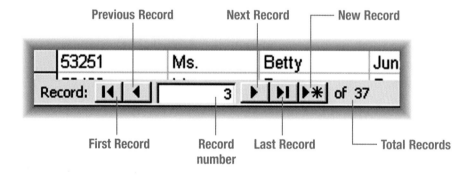

ACTIVITY 9-1

CREATE A DATABASE FILE AND TABLE

1. To start *Access*, click the **Start** button. Choose **(All) Programs, Microsoft Access**.

2. In the Task Pane under New, click **Blank Database**.

3. Specify a location for the database in the Save in box. Save the database in the *DigiTools your name\Chapter 9* folder.

4. Key a filename. Use the name *Customers* for your database. Indicate the Save as Type as **Microsoft Access Database**.

3. Dictate the sentences below. Remember to dictate the punctuation marks. Use the New Line command after each line. Do not tap **Send Text**.

```
There are many words that I don't understand.
After all, I am just a computer!

Sometimes I confuse the words an, and, and end.
What do you expect?  Perfection?
```

4. Proofread your sentences carefully, checking for mistakes. Highlight an incorrect word.

5. Tap the green symbol that appears by the word to display a list of alternatives. You can also say `Correction`. If the correct word appears in the list, click or tap the correct word. You can also choose the correct word by saying `Select` and the number of the item, for example, `Select 2`. If the correct word does not appear in the list, choose **Rewrite/Respeak** and write or say the correct word. Enter (key, speak, or handwrite) the correct word.

> **DigiTip**
> Sometimes a single word you speak is recognized incorrectly as two words. You can select two or three words at once and use the Correction command.

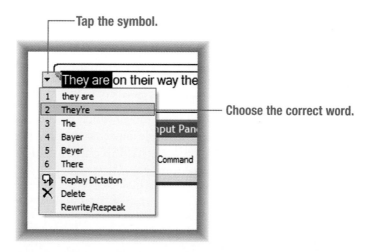

Figure 7-9
Choose the correct word from the list.

6. When all errors have been corrected, close the *Word* document without saving.

ACTIVITY 7-8 CORRECT HOMONYMS

1. Open *Word* to a new blank document and display the Language bar. Tablet PC users, display the Input Panel's Speech bar.

2. Dictate the sentences below that contain some homonyms.

```
They're on their way there.
He went to the store to buy some books after he said
bye to his family.
Enough already!  Are you all ready to go or not?
My advice to you is to advise your brother carefully
on the issue.
After the law went into effect, it affected everyone
profoundly.
```

LESSON 9-1   CREATING A DATABASE

OBJECTIVES   *In this lesson you will:*

1. Create a database file.
2. Create a database table and choose data types.
3. Enter data in a database table.
4. Save and print a database table.

To create a database, you first open and save a new database file. Once you have created a database file, you can create objects, such as tables, to store your data.

DATABASE TABLES

A **table** is the main data storage container in a database. All databases contain at least one table. A table stores data about a single subject, such as products or customers. Data in a table is stored in records. A **record** is composed of fields and contains all the data about one particular person, company, product, or other item in a database. A **field** is part of a record and contains a single piece of data for the subject of the record. For example, in a customer record, one field might contain the customer's last name. Fields are displayed in a table in columns.

Figure 9-1
A Database Table

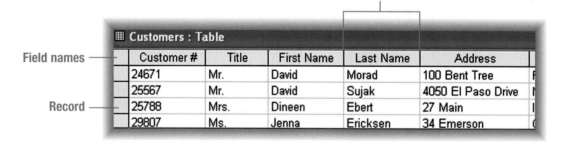

DATA TYPES

HELP KEYWORDS
Field
Field data types available in *Access*

When you create a database table, you must indicate the data type for each field you plan to include. The **data type** identifies the kind of information to be placed in a field. You can use several different types of data:

CHAPTER 9   Lesson 9-1   263

3. Correct all of the errors in your sentences as you learned to do in the previous activity. Save your work as *Activity 7-8* in your *DigiTools your name\Chapter 7* folder. Print and close the document.

Peer Check Are all the homonyms and other confusing words in your document correct?

SUBSTITUTING WORDS

At times, you may wish to substitute one word for another. This substitution technique is one of the most popular ways to correct words you may have misspoken.

To substitute words, first switch to Voice Command mode. Say `Correct <word or phrase>` or `Select <word or phrase>`. Switch to Dictation mode, and say the new word or phrase. Practice substituting words in the next activity.

ACTIVITY 7-9 — SUBSTITUTE WORDS

1. Open *Word* to a new blank document and display the Language bar. Tablet PC users, display the Input Panel's Speech bar.

2. In Dictation mode, say the following sentence.

`Benjamin Franklin discovered electricity.`

3. Choose *Benjamin Franklin*. Replace the name by saying *Thomas Jefferson* in Dictation mode. Choose *discovered* and replace it by saying *invented*. Choose *electricity* and replace it by saying *a swivel chair*. Your sentence should now read:

Thomas Jefferson invented a swivel chair.

4. Say the following sentence:

`Benjamin Franklin wrote a very famous book.`

5. Using your voice, replace *Benjamin Franklin* with *Thomas Jefferson*. Then replace *a very famous book* with *the Virginia Constitution*. The sentence should read as follows:

Thomas Jefferson wrote the Virginia Constitution.

6. Close the document without saving.

> **DigiTip**
> Pause or hesitate briefly before and after each command. Say <pause> `Voice Command` <pause>. Say <pause> `Dictation` <pause>.

CHAPTER 9

Databases

OBJECTIVES *In this chapter you will:*

1. Learn about database objects and field types.
2. Create a database.
3. Plan, create, and modify database tables.
4. Enter data and navigate in a database.
5. Save and print database elements.
6. Find records and delete records.
7. Filter and sort data.
8. Create queries, forms, and reports.

A **database** is a collection of related information. Businesses use databases to manage many types of data related to customers, vendors, employees, products, and other topics. Databases contain objects such as tables, forms, queries, and reports. Using database objects, businesses can find, sort, and isolate data quickly. A database can be used to answer questions such as:

- Which customer accounts are overdue?
- Which customers have ordered a particular product?
- How many of our customers are in a particular city or state?
- Which employees were hired in a particular year?
- Which products had the highest sales volume in a particular period?

In this chapter, you will learn to create a database file, enter data, and use database objects to manage data and answer questions.

ETHICS
Avoiding Gossip

Informal communications are common in an organization. The informal network by which employees communicate in an unofficial manner is sometimes called the "grapevine." All informal communication is not undesirable. Employees are naturally interested in the plans and events that affect the company and its employees. Consider the following example.

Kim, an employee at an insurance company, heard his department manager mention to another worker that he was planning to work on budget cuts all afternoon. A few minutes later, Kim walked by the manager's office as he was using speech recognition to prepare a document. Kim overheard "... and Gloria House, Fred Cates, and Katrina Yetz will probably be the ones."

When Kim talked with Gloria, Fred, and Katrina at the afternoon break, he took it upon himself to break the "bad news" that they would probably soon be a victim of budget cuts.

Unfortunately, rumors and **gossip**, which are incomplete and/or false statements about individuals or situations, are also often spread by informal communications. Rumors and gossip may be harmful to the company or its employees. Use discretion in your informal communications with others. Avoid discussing company plans or events that you do not know are correct or that may be confidential at the present time. Do not discuss personal issues or affairs of fellow employees.

Follow Up

1. Did Kim demonstrate ethical behavior by telling his coworkers the information he overheard? Why or why not?

2. Can we tell from what Kim overheard that Gloria, Fred, and Katrina will be downsized (laid off)? What are some other possible actions or events for which Gloria, Fred, and Katrina "will probably be the ones."

LESSON 7-5: ADDING WORDS TO THE DICTIONARY

OBJECTIVES *In this lesson you will:*

1. Add words to the speech program's dictionary.
2. Practice saying unique words.

Although your speech recognition program can recognize thousands of words, you may need to add a few words. For example, you will need to add certain names and unique expressions that you use. The speech program provides a simple way to accomplish this task.

4. Accept the data range that you selected earlier. **Columns** should be selected for Series in. Click **Next**.

5. Enter `January Sales` for the chart title. Click **Next**.

6. Select **As object in** for the chart location. The sheet name, 2004, will show in the text box. Click **Finish**.

7. Click on the chart legend and move it closer to the pie graphic. Move the chart so it is approximately centered under the worksheet data.

8. Select the text of the chart title, January Sales. Change the font size to 16 points. (Click the arrow by Font Size on the Formatting toolbar and choose **16**.)

9. Add data labels to the pieces of the pie. Right-click on the pie graphic. Choose **Format Data Series**. Click the **Data Labels** tab. Select **Percentage** and click **OK**.

10. Save again using the same name. Open the data file *Sales Memo*. Insert your name and the current date where indicated. Edit the last paragraph to provide the data indicated from your worksheet.

11. In your worksheet, click the chart to select it. Click **Copy** on the Standard toolbar. In your memo, place the insertion point below the last paragraph. Click **Paste** on the Standard toolbar. Center-align the table and resize it, if needed, for attractive placement.

12. Save the memo as *Activity 8-15 Memo* in your *DigiTools your name\Chapter 8* folder. Print the memo. Close the memo and the workbook.

Peer Check Compare your completed memo with a classmate's memo. Did you both choose the same product as the top seller? If not, reexamine the worksheet to determine the correct product.

DIGITOOLS
DIGITAL WORKBOOK — CHAPTER 8

Open the data file *CH08 Workbook*. Complete these exercises in your *DigiTools Digital Workbook* to reinforce and extend your learning for Chapter 8:

- Review Questions
- Vocabulary Reinforcement
- Math Practice: Calculating Volume Discounts
- From the Editor's Desk: Comma Usage Part 1
- Keyboarding Practice: Number Reaches and Speed Building

ACTIVITY 7-10 — TRAIN UNIQUE WORDS

HELP KEYWORDS

Speech
Add to or delete from the speech recognition dictionary

1. Open *Word* to a new blank document and display the Language bar. Tablet PC users, display the Input Panel.

2. Click **Tools** on the Language bar. Choose **Add/Delete Word(s)**.

3. In the *Word* entry box, key the word or phrase you want to add (for example, *trilithiun crystals*) as shown in Figure 7-10.

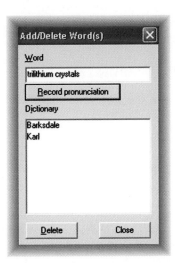

Figure 7-10
Record new and unique words.

4. Click the **Record Pronunciation** button and say the new word or phrase clearly. The new word or phrase will be added to your Dictionary. Choose **Close** to return to your document.

5. In *Star Trek* movies and television shows, the characters use many unique words and names. Train these interplanetary terms:

```
Cardassian
Romulan
Klingon
Captain Picard
```

6. Try saying the following sentence to test your accuracy. If you make a mistake, correct your errors in the same way you learned how to do in the previous lesson.

```
The Klingon took the trilithium crystals from the
Cardassian leader, Captain Picard. He smuggled them
across the Romulan neutral zone.
```

7. Train (add to the Dictionary) the names of five friends or family members. Dictate the names of the five friends and family members to check your accuracy. If you make a mistake, correct these names as you would any other words.

8. Close the document without saving.

MODIFYING CHARTS

HELP KEYWORDS

Chart
Change values in a chart

After a chart has been created, you can modify it in several ways. For example, you may wish to change the color used for an element of the chart, change the title, or even change the chart type. If the data in the worksheet is changed, the chart will be updated automatically.

To change or format a chart element, click in the chart to select it. Then click on the element you want to change or format. Make the appropriate changes or choices in the dialog box.

ACTIVITY 8-14 MODIFY A COLUMN CHART

1. Start *Excel*. Open *Activity 8-13* that you created earlier.

2. Save the file as *Activity 8-14* in the *DigiTools your name\Chapter 8* folder.

3. The January sales for Product 3 should be 250. Click in Cell B5 and change the number to 250. Note that the chart is updated automatically to show the new data.

4. Change the Value axis title from *Number* to *Quantity*. Click on the chart to select it. Choose **Chart Options** from the Chart menu. Key the new name for the Value axis title `Quantity`. Click **OK**.

5. To change the color of the plot area, right-click in the plot area. Choose **Format Plot Area** from the menu. On the Format Plot Area dialog box, click a light color such as light yellow. Click **OK**.

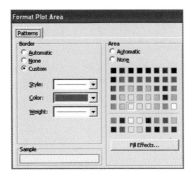

6. Save the file again using the same name. Print and close the worksheet.

ACTIVITY 8-15 CREATE A PIE CHART

A **pie chart** (so called because the graph wedges look like pieces of a pie) is a display of how a part contributes to the whole. The whole circle represents 100 percent and each wedge represents a portion of the whole. Each wedge should be identified with an appropriate label, color, or pattern. For example, you might want to compare sales of all products for a given month. You will create this type of chart in this activity.

1. Start *Excel*. Open *Activity 8-9* that you created earlier. Save the file as *Activity 8-15* in the *DigiTools your name\Chapter 8* folder.

2. Create a pie chart that compares January sales for all six products. Select Cells A2:B8. Click the **Chart Wizard** button on the Standard toolbar.

3. Under Chart type, choose **Pie**. Choose the first option under Chart sub-type. Click **Next**.

LESSON 7-6 CHANGING CASE

OBJECTIVES *In this lesson you will:*

1. Change case using the Capitalize command.
2. Change case using the Change Case dialog box.

When using your speech software, you must apply the rules of capitalization. Your speech software will normally capitalize the first word of each sentence, but it may miss a few capitalized words in your document. You should make corrections as needed so your document is capitalized properly. Dictating your sentences first and then making corrections for capitalization works best.

Words entirely in **UPPERCASE** (in all capital letters) are often used in titles. To change a word to all UPPERCASE letters, you can use voice commands to open the Change Case dialog box. There you may select UPPERCASE as an option.

Sometimes words are capitalized that should not be. You can select and fix such mistakes using voice commands. To change a word to **lowercase** (all small letters), you can use voice commands to open the Change Case dialog box. There you may choose lowercase as an option.

Figure 7-11
Change Case Dialog Box

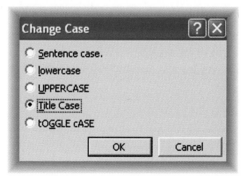

ACTIVITY 7-11   PRACTICE CHANGING CASE

1. Open *Word* and display the Language bar. Tablet PC users, display the Input Panel's Speech bar. Turn on your microphone and dictate the following sentence:

   `Mary attended the marine wildlife institute.`

2. Using your voice, a digital pen, or the mouse, select the words **marine wildlife institute**.

Office XP

Say `Cap That` or `Capitalize`. The first letter of each word should now be capitalized.

6. Enter `Unit Sales` for the Chart title. Enter `Product` for the Category axis title. Enter `Number` for the Value axis title. Click **Next**.

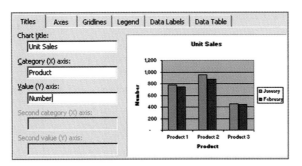

7. Select **As object in** for the chart location. The sheet name, 2004, will show in the text box. Click **Finish**.

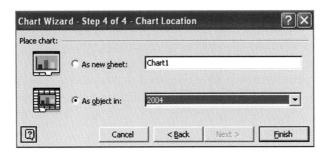

8. Your chart should look like the one in Figure 8-20. The chart can be moved or resized like any graphic object. Click the chart border and drag to move the chart. Click and drag a chart handle to resize the chart. Experiment with changing the size of the chart. Move the chart under the data in the worksheet.

Figure 8-20
Unit Sales Column Chart

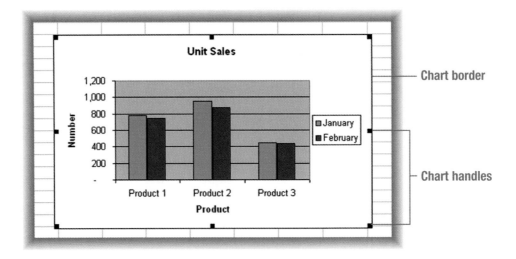

9. Use Print Preview to view the worksheet data and the chart. Move the chart again, if needed, to center it approximately under the worksheet data. Print the worksheet.

10. Save the file again using the same name. Close the file.

Tablet PC

Switch to Voice Command mode. Say `Cap That` or `Capitalize`. The first letter of each word should now be capitalized. (You must use Voice Command mode each time you use the Capitalize command. You will not be reminded to switch modes.)

3. Say the following title:

`The alpine country club official membership list`

4. Select the title and say `Cap That` or `Capitalize`.

5. Say the following title:

`The National Legal Review`

6. Select the title and say `Cap That` or `Capitalize`.

7. Say the title: `Treasure Island`.

8. Select the words *Treasure Island*. Switch to Voice Command mode. Say `Format, Change Case` to open the Change Case dialog box. Say `Uppercase`. Say OK.

9. Say the following newspaper headline:

`Japan attacked Pearl Harbor early this morning`

10. Select the newspaper headline from 1941. Change the entire title to uppercase using the commands you practiced in step 8.

11. Select the title *TREASURE ISLAND*.

12. Switch to Voice Command mode. Say `Format, Change Case` to open the Change Case dialog box. Say `Title Case`. Say OK. The title should now read: *Treasure Island*.

13. Select the newspaper headline:

JAPAN ATTACKED PEARL HARBOR EARLY THIS MORNING

14. Use voice commands to open the Change Case dialog box. Then say: `lowercase`. Using voice commands, capitalize *Japan* and *Pearl Harbor*. Add a period to end the sentence so it appears: *Japan attacked Pearl Harbor early this morning.*

15. Close the document without saving.

> **DigiTip**
> Tablet PC users can also say `All Caps That` or `Uppercase That` to change words to uppercase.

> **DigiTip**
> Tablet PC users can say `No Caps That` or `Lowercase That` to change words to lowercase.

APPLICATION 7-1

PERSONAL-BUSINESS LETTER

1. Use your dictation skills to create the personal-business letter shown on page 210. Use block format and open punctuation.

2. Proofread and correct all errors. Save the letter as *Application 7-1* in the *DigiTools your name\Chapter 7* folder. Print the letter.

> **DigiTip**
> See page App-6 in Appendix A to review block letter format.

CREATING CHARTS

To create a chart, first you must enter the chart data in a worksheet. Arrange the data logically in rows and columns. The data for the chart that appears in Figure 8-18 is shown in Figure 8-19.

Figure 8-19
Chart Data

	Diversity	Team Building	Communications
2004	160	105	100
2005	190	95	150

To create a chart:

- Select the data you want to include in the chart.
- Click the **Chart Wizard** button on the Standard toolbar.
- Follow the instructions in the Chart Wizard.

ACTIVITY 8-13 — CREATE A COLUMN CHART

HELP KEYWORDS
Chart
  Create a chart

A **column chart** is used to show a comparison of values for two or more items. For example, you might want to compare sales of three products for two or three months. You will create this type of chart in this activity.

1. Start *Excel*. Open *Activity 8-9* that you created earlier.

2. Save the file as *Activity 8-13* in the *DigiTools your name\Chapter 8* folder.

3. Create a column chart that compares January and February sales for Products 1, 2, and 3. Select Cells A2:C5. Click the **Chart Wizard** button on the Standard toolbar.

4. Under Chart type, choose **Column**. Choose the first option under Chart sub-type. Click **Next**.

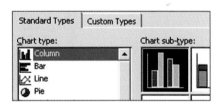

5. Accept the data range that you selected earlier. **Columns** should be selected for Series in. Click **Next**.

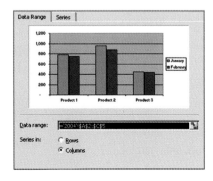

1245 Park Avenue
New York, NY 10128-2231
April 1, 20—

Mrs. Tara Cruz
4221 Beekman St.
New York, NY 10038-8326

Dear Mrs. Cruz

Thank you for helping chaperone the fifth grade class on their field trip to Broadway. When I visited Mrs. Kensington classroom yesterday, the children were still excited about having attended the play. Their thank-you note is enclosed.

Because of parents like you, educational experiences outside the classroom are possible. These experiences bring to live what the students learn in school. I'm glad our children have this enrichment.

Thank you again for accepting the challenge of watching over the fifth graders on their exciting trip to Broadway. I know the task wasn't easy, but I felt it was well worth our time.

Sincerely

Marsha Rhodes
Parent volunteer

Enclosure

Peer Check Compare your letter with a classmate's letter. Discuss any differences and make corrections if needed.

LESSON 8-5 CHARTS

OBJECTIVES   *In this lesson you will:*

1. Create a chart.
2. Modify a chart.
3. Preview and print a chart.

Sometimes data from a worksheet is easier to understand when it is displayed in a chart. Charts can make comparing data or remembering data easier for readers. Charts can be created in many different styles. The common ones include column charts, bar charts, line charts, and pie charts. A column chart is shown in Figure 8-18.

Charts are created in two ways. An embedded chart appears on the same sheet as the data. This lets the reader see both the data and the chart at the same time. Charts can also appear on a separate sheet in the workbook called a chart sheet. Chart sheets are useful when you want to prepare charts for part of a presentation.

Figure 8-18
Column Chart

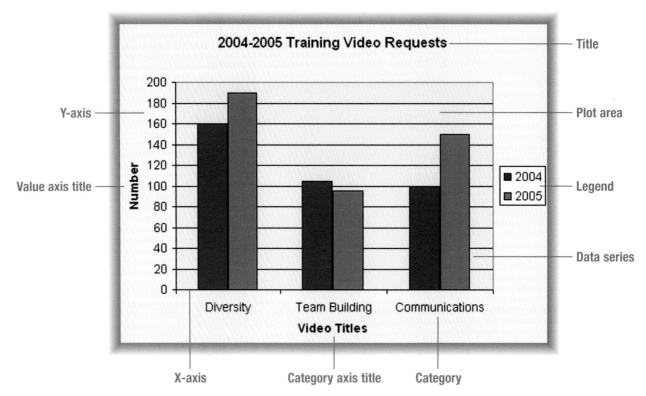

CHAPTER 8   Lesson 8-5   257

LESSON 7-7

FORMATTING TEXT WITH VOICE COMMANDS

OBJECTIVES *In this lesson you will:*

1. Apply bold, italic, and underline formats.
2. Align text.
3. Change line spacing.

You can use voice commands to format text. Text formats, such as bold, italic, and underline, are frequently used to make text stand out. You can also change the horizontal alignment of text and the amount of space between lines. In this lesson, you will practice several formatting commands. Once you learn how to choose menu items and open dialog boxes using voice commands, you can easily transfer this learning to using voice for other commands. As you continue to use speech input in the future, experiment with giving commands by voice.

CHARACTER FORMATS

DigiTip
Make sure that your Formatting toolbar is visible. The Formatting toolbar can be opened by saying `View, Toolbars, Formatting` in Voice Command mode.

Bold text is used to display important text or titles clearly. <u>Underline</u> format is used less frequently because underlined words can be confused with hyperlinks. Nevertheless, underline is still used in printed documents. *Italic* format is used for titles of books, newspapers, magazines, and other printed materials.

The following commands can be used in Voice Command mode to format text:

- Bold
- Italic
- Underline
- UnBold That
- UnItalic That
- UnUnderline That

ACTIVITY 7-12   EMPHASIZE TEXT

1. Open *Word* and display the Language bar. Tablet PC users, display the Input Panel's Speech bar. Dictate the following sentences:

```
The biology teacher was speaking about pollution.
According to U.S. News, we spend billions each year on
sports activities.
There is nothing you can do about it!
```

APPLICATION 8-2 — REUNION BUDGET WORKSHEET

Apply the skills you have learned to create the worksheet shown below.

1. Enter the data shown in the worksheet. (You will format the data in a later step.) Use Fill Series to enter the dates March through August.

	A	B	C	D	E	F	G	H	I
1	CLASS OF 20-- REUNION BUDGET								
2									
3		February	March	April	May	June	July	August	Total
4	Revenue								
5	$30/per person	$ 240.00	$ 120.00	$ 240.00	$ 450.00	$ 450.00	$ 750.00	$ 750.00	
6	Expenses								
7	Food	$ 100.00						$1,850.00	
8	DJ			$ 100.00				$ 200.00	
9	Tickets					$ 25.00			
10	Decorations					$ 75.00	$ 75.00		
11	Door Prizes						$ 100.00		
12	Flyers			$ 40.00		$ 40.00			
13	Paper	$ 10.00				$ 25.00			
14	Labels	$ 30.00							
15	Name Badges							$ 30.00	
16	Postage		$ 30.00			$ 30.00			
17	Total Expenses								
18	Cash on Hand								
19									

2. Enter formulas to calculate the totals in Column I and in Row 17. Leave Cell I18 blank.

3. To find the cash on hand for February, enter a formula to subtract the total expenses for February from the revenue for February.

4. The cash on hand for March equals the February cash on hand plus the March revenue minus the March expenses. Enter a formula to calculate this amount. The cash on hand amounts for April through August are calculated in a similar way. Use Fill to copy the March formula to April through August.

5. Format the sheet approximately as shown using currency style, bold, fill color, and borders. Center the column heads. Use 12 point font size for the worksheet title. Merge and center the title over the worksheet data. Change the row height for Rows 6 and 18 to 22.

6. Change the page orientation to Landscape and center the data vertically and horizontally on the page. Place the filename in a footer. Print the worksheet.

7. Save the worksheet as *Application 8-2* in the *DigiTools your name\Chapter 8* folder. Close the worksheet.

HELP KEYWORDS

Speech
Format text by using speech recognition

2. Select the entire first sentence. Switch to Voice Command mode and say `Bold`. Select the second sentence and say `Italic`. Select the third sentence and say `Underline`.

3. Select all three sentences and return them back to normal text. Say the commands: `UnBold That, UnItalic That, UnUnderline That`.

You can also remove formatting with these commands: `Format, Styles and Formatting, Clear Formatting`.

4. Another way to change the font, font style, and the size of text is to use the Font dialog box. (See Figure 7-12.) To apply bold, select all three sentences and say: `Format, Font, Font Style, Bold, OK`. To remove the bold, say `Format, Font, Font Style, Regular, OK`.

Figure 7-12
The Font Dialog Box

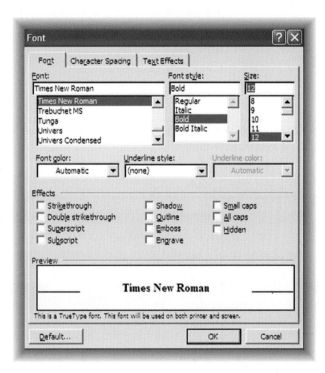

5. Select the lines again. Use the Font dialog box to change the font to Verdana, Georgia, or some other font. Change the font size to 14 points and italicize all of the text.

6. Close the document without saving.

Self Check When using dialog boxes, did you speak each command in order just as you would click options with a mouse?

Headers and footers are selected or entered on the Header/Footer tab of the Page Setup dialog box. To select a header or footer, choose **Header and Footer** from the View menu. Click the arrow for Header or Footer and choose an option. Click **OK**.

To create a custom header or footer, click the **Custom Header** or **Custom Footer** button on the Header/Footer tab. Enter data for each section. Click **OK**.

Figure 8-17
Header/Footer Options

ACTIVITY 8-12 SELECT HEADER AND FOOTER

1. Start *Excel*. Open *Activity 8-10* that you created earlier.

2. Save the worksheet as *Activity 8-12* in the *DigiTools your name\Chapter 8* folder.

3. Choose **Header and Footer** from the View menu. Click the arrow for Header and choose **November, Confidential, Page 1**. Click the arrow for Footer and choose the filename **Activity 8-12.xls**. Click **OK**.

4. Use Print Preview to confirm that the header and footer appear on the worksheet. Print the worksheet. Save again using the same name. Close the worksheet.

TEXT ALIGNMENT

The **alignment**, or justification, commands move text to the right or left margin, or place text in the horizontal center of the page. Centering is often used for headings and at other times when you wish to set text apart from the rest of the document. You can change the alignment in the Paragraph dialog box with simple voice commands.

You can also change the alignment of lines of text by speaking the names of options on the Formatting toolbar. To align text in Voice Command mode, use these commands:

- Center
- Align Left
- Align Right

You speak commands to align text using the default tab stops or tab stops you have set in a document. Simply say `Tab` to move text to the next tab stop. In *Office XP*, you can use the Tab command in Dictation mode. For Tablet PCs, switch to Voice Command before using the Tab command.

LINE SPACING

Word offers several options for spacing in a document such as single, 1.5 lines, and double. Single spacing is used for letters and memos. Double spacing is often used for reports. Line spacing can be set in the Paragraph dialog box as marked in Figure 7-11. You can access this dialog box with simple voice commands.

To open the Paragraph dialog box with voice commands, say: `Format, Paragraph, Indents and Spacing` (to bring the tab to the front if necessary). After a dialog box has been opened, simply say the commands you see in the dialog box, such as `Alignment/Center` or `Line Spacing/Double`.

Figure 7-13
The Paragraph Dialog Box

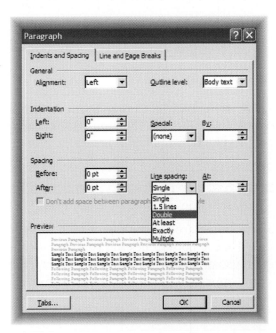

Scaling

Scaling options provided in the Print Setup dialog box allow you to decrease or increase the size of your printed document. You might use 90% as the scaling setting, for example, to make your worksheet fit within 1 inch margins. You can also use the Fit to option to specify the number of pages tall and wide you want your document to be. Use the scaling feature with caution. Do not make a document so small that it is hard to read.

ACTIVITY 8-10 SET ORIENTATION AND SCALING

1. Start *Excel*. Open *Activity 8-8* that you created earlier.

2. Choose **Page Setup** from the File menu. On the Margins tab, enter 1 for the top, bottom, left, and right margins. Check the boxes to center the data vertically and horizontally.

3. On the Page tab, choose **Landscape** orientation. Under scaling, select the **Adjust to** option and set **120%** for the scaling. For print quality, choose the lowest print quality available. Click **OK**.

4. Preview the document to confirm that it is centered and fits on one page. Print the worksheet. Save the worksheet as *Activity 8-10* in the *DigiTools your name\Chapter 8* folder. Close the worksheet.

ACTIVITY 8-11 SET A PRINT AREA

1. Start *Excel*. Open *Activity 8-10* that you created earlier.

2. Save the worksheet as *Activity 8-11* in the *DigiTools your name\Chapter 8* folder.

3. Select Cells A4:C19. Choose **Print Area** from the File menu. Choose **Set Print Area**.

4. Use Print Preview to view the data. Only the data from the Date, Receipt No., and Balance columns should appear. The data should be centered vertically and horizontally on the page. Print the worksheet.

5. Save again using the same name.

6. Practice clearing the print area. Choose **Print Area** from the File menu. Choose **Clear Print Area**. Print Preview the file. The entire document should appear. Close the file without saving.

HEADERS AND FOOTERS

HELP KEYWORDS
Header
Add headers and footers for printing

A header is information placed at the top of each page in a document. A footer is information placed at the bottom of each page in a document. Several standard headers and footers are provided, which you can select to appear on your printed worksheet. You can also create your own "custom" header or footer.

ACTIVITY 7-13 — FORMAT TEXT

1. Open *Word* and display the Language bar. Tablet PC users, display the Input Panel's Speech bar. Dictate the following paragraph.

   ```
   Benjamin Franklin was an inventor and a statesman. Mark
   Twain was the author of several popular books, such as
   The Adventures of Tom Sawyer. President Bush was in
   office after President Clinton. What do all these men
   have in common?  They were famous in their day.
   ```

2. Select the paragraph by saying `Select Paragraph`. Align the paragraph to the center of the page by saying `Center`.

3. Select the paragraph again and move it to the right side of the page. Say `Align Right`.

4. Select the paragraph again and move it to the left side of the page. Say `Align Left`.

5. Select the paragraph again. Change the line spacing to double spacing. Say `Format, Paragraph, Indents and Spacing, Line Spacing, Double, OK`.

6. Select the name of the book and make it italic. Select the name of each person and make the names bold.

7. Go to the top of the document. Dictate a title for the document: `Famous Men`. Select the title and change it to bold format, all uppercase letters, and center alignment.

8. Go to the beginning of the paragraph. Say `Tab` to indent the paragraph to the first default tab stop. (You may need to switch to Voice Command mode before saying the command.) Close the document without saving.

APPLICATION 7-2 — REPORT

> **DigiTip**
> See page App-7 in Appendix A to review unbound report format.

1. Use your dictation skills to create the unbound report shown on pages 215–216. Dictate all the text. Correct errors in recognition. Correct other errors such as for capitalization. Then apply formatting.

2. Format the report title for all caps, bold, and center alignment. Leave three blank lines after the title. Change the line spacing for the paragraphs to double. Indent the paragraphs to the first default tap stop, 0.5 inches. Underline the side headings. Apply italic to the book titles each time they appear.

3. Place the References on a separate page. Format the References title for all caps, bold, and center alignment. Leave three blank lines after the title. Use single line spacing for the reference paragraphs. Use 0.5 inch hanging indent for the paragraphs. Number pages 2 and 3.

4. Save the document as *Application 7-2* in your *DigiTools your name\Chapter 7* folder. Print the document.

SETTING A PRINT AREA

DigiTip

If you want to print a portion of a worksheet only once, you do not have to set a print area. Simply select the range of cells you want. Choose **Print** from the File menu. Choose **Selection** and click **OK**.

By default, *Excel* prints an entire worksheet when you give the Print command. You can tell *Excel* to print only a part of the worksheet. A part of a worksheet specified for printing is called a **print area**. To set a print area, select the cells you want to include. Choose **Print Area** from the File menu. Then choose **Set Print Area**. Later, you may want to print the entire worksheet or set a different print area. To clear the print area setting, choose **Print Area** from the File menu. Then choose **Clear Print Area**.

The print area can also be set on the Sheet tab of the Page Setup dialog box. This tab allows you to change other settings such as:

- Print titles (Rows and/or columns to appear on each page)
- Print options (Gridlines, Draft quality, Row and column headings)
- Page order

PAGE OPTIONS

As explained earlier, the Page Setup dialog box can be accessed from the File menu. The Page tab of the Page Setup dialog box provides options for

- Setting the page orientation
- Scaling the size of the printed document
- Selecting a paper size and a print quality
- Setting the number used on the first page

Page orientation and scaling are discussed further below. The options available for paper size and print quality will vary depending upon the printer used. You might want to choose a lower print quality when printing drafts for proofreading or documents for your own use. Use a higher print quality when the document must make a good impression on other readers.

HELP KEYWORDS

Orientation
Print landscape or portrait

Page Orientation

Page orientation is the direction in which text is placed on the page relative to the shape of the page. A standard size piece of paper used for documents is 8.5" by 11". The short side of the paper is placed at the top and text runs down the page. This layout is called portrait orientation. Worksheets are often too wide to fit attractively on an 8.5" by 11" page. You can change the layout so the long side of the paper is placed at the top. This is called landscape orientation.

To set the page orientation, choose **Page Setup** from the File menu or **Setup** on the Print Preview toolbar. On the Page tab, choose the desired orientation and click **OK**.

Figure 8-16
Page Options

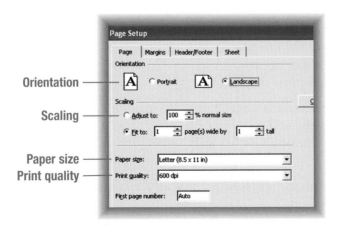

Samuel Clemens

Samuel Clemens was one of America's most renowned authors. The colorful life he led was the basis for his writing. Although his formal education ended when he was 12 years old with the death of his father, his varied career interests provided an informal education that was not unlike many others of his generation. Clemens brought these rich experiences to his life in writing.

Sam Clemens was recognized for his fiction as well as for his humor. It has been said that, "... next to sunshine and fresh air Mark Twain's humor has done more for the welfare of mankind than any other agency." (Railton, "Your Mark Twain," 1999) By cleverly weaving fiction and humor, he developed many literary masterpieces. Some say his greatest masterpiece was "Mark Twain," a pen name (pseudonym) Clemens first used in the Nevada territory in 1863. This fictitious name became a kind of mythic hero to the American public. (Railton, "Sam Clemens as Mark Twain," 1999) Some of his masterpieces that are among his most widely read books are The Adventures of Tom Sawyer and Adventures of Huckleberry Finn.

The Adventures of Tom Sawyer

The Adventures of Tom Sawyer was first published in 1876. Such characters as Tom Sawyer, Aunt Polly, Becky Thatcher, and Huck Finn have captured the attention of readers for generations. Boys and girls, young and old, enjoyed Tom Sawyer's mischievousness. Who can forget how Tom shared the privilege of whitewashing Aunt Polly's fence? What child isn't fascinated by the episode of Tom and Becky lost in the cave?

Adventures of Huckleberry Finn

Adventures of Huckleberry Finn was first published in 1885. Many of the characters included in The Adventures of Tom Sawyer surface again in Huckleberry Finn. Children are able to enjoy many fun times with Tom and Huck.

(continued on next page)

SETTING MARGINS AND CENTERING

HELP KEYWORDS

Margins
Set margins for printing

By default, *Excel* uses .75 inches for the side margins for a worksheet. The default top margin is 1 inch. The default bottom margin is 1 inch. You can change the margins by accessing the Page Setup dialog box. To change margins, choose **Page Setup** from the File menu or **Setup** on the Print Preview toolbar. On the Margins tab, enter the desired margins and click **OK**.

You can also select options on the Margins tab to center the worksheet horizontally and/or vertically.

Figure 8-15
Page Setup Dialog Box

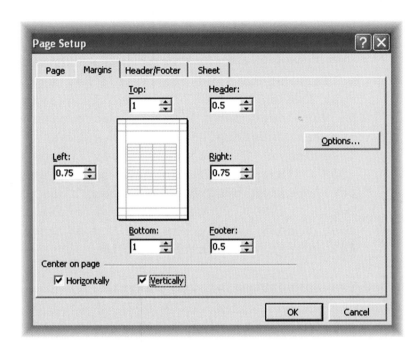

ACTIVITY 8-9

SET MARGINS AND CENTERING

1. Start *Excel*. Open the file *Activity 8-8* that you created earlier. Delete the worksheet for next year.

2. Choose **Page Setup** from the File menu. Click the **Margins** tab. Enter 1 for the left margin and 1 for the right margin. Check the boxes by Horizontally and Vertically to center the data. Click **OK**.

3. Save the file as *Activity 8-9* in the *DigiTools your name\Chapter 8* folder. Preview the worksheet. It should be centered vertically and horizontally on the page. Print the worksheet. Close the file.

References

Railton, Stephen. "Your Mark Twain."
    http://etext.lib.virginia.edu/railton/sc_as_mt/yournt13.html
    (24 September 1999).

Railton, Stephen. "Sam Clemens as Mark Twain."
    http://etext.lib.virginia.edu/railton/sc_as_mt/cathompg.html
    (24 September 1999).

Peer Check Compare your report with a classmate's report. Discuss any differences and make corrections if needed.

DIGITOOLS
DIGITAL WORKBOOK — **CHAPTER 7**

Open the data file *CH07 Workbook*. Complete these exercises in your *DigiTools Digital Workbook* to reinforce and extend your learning for Chapter 7:

- Review Questions
- Vocabulary Reinforcement
- Math Practice: Calculating Deposit Totals
- From the Editor's Desk: Sentence Fragments
- Keyboarding Practice: Opposite Hand Reaches and Speed Building

9. In Cell H2, enter the head Total. Bold and center the head. In H3, enter a formula to add the sales units for January through June. Use Fill to copy the formula to H4:H9. Adjust the worksheet title so it is centered over the data.

10. Format the numbers in B3:H8 for Comma style with no (zero) decimal places. **Hint:** Click the **Decrease Decimal** button one or more times until no decimals appear in the numbers.

11. Choose **Worksheet** from the Insert menu to insert a new worksheet. Rename the new worksheet using next year as the name, for example, 2005.

12. Return to the worksheet for the current year. Select A1:H8. Click **Copy** on the Standard toolbar. Move to the worksheet for next year. Click in A1. Click **Paste** on the Standard toolbar.

13. In next year's worksheet, select B3:G8. Choose **Clear** from the Edit menu. Choose **Contents**. This will remove the data from the cells but leave the Comma, 0 Decimals number format in the cells. A hyphen will appear in cells in the Total column because data needed for the formulas has been deleted. However, the formulas are still in place in the cells.

14. Return to Cell A1 of this year's worksheet. Save the workbook as *Activity 8-8* in the *DigiTools your name\Chapter 8* folder. Print both worksheets in the workbook. (**Hint:** Go to this year's worksheet. Hold down **Shift** and click next year's worksheet tab to select both worksheets for printing.) Close the workbook and *Excel*.

LESSON 8-4 PRINTING

OBJECTIVES

In this lesson you will:

1. Center a worksheet horizontally and vertically.
2. Set margins and print a worksheet.
3. Set a print range.
4. Change page orientation.
5. Print worksheet headers and footers.

Excel provides several printing options you can set to make your printed worksheet more attractive or easy to read. You can set the side, top, and bottom margins for printed documents. You can choose a page orientation (portrait or landscape) for worksheets and charts. You can add headers and footers to print on worksheets or charts.

Tooling Up!

In these unit exercises, you will go online and use a variety of DigiTools to learn, think, and write about career alternatives, business issues, and trends. The skills and knowledge you have learned in the previous chapters will help you complete these exercises.

CAREER CLUSTERS

Scientific Research and Engineering

Scientific research and engineering careers are among the most exciting and demanding of any career cluster. They require high levels of education, training, and experience. Scientists and engineers generally specialize in one specific area of engineering or research, such as marine engineering, hydrology, or astronomy.

Some of these career categories have more available job openings than others. For example, there are many thousands of civil and electrical engineers, but only a few thousand microbiologists or astronomers. Examine other career specialties in science and engineering in the table below.

Careers in Scientific Research and Engineering

Position and Yearly Salary		Position and Yearly Salary	
Aerospace Engineer	$71,380*	Astronomer	$76,390*
Agricultural and Food Scientists	$52,290*	Electrical Engineer	$68,630*
Agricultural Engineer	$54,300*	Medical Scientist	$62,650*
Biochemist or Biophysicist	$61,680*	Marine Engineer	$66,960*
Biomedical Engineer	$63,330*	Hydrologist	$54,570*
Computer Scientist	$76,970*	Nuclear Engineer	$80,200*
Chemical Engineer	$72,780*	Microbiologist	$54,500*
Conservation Scientist	$49,460*	Mining and Geological Engineer	$65,370*
Civil Engineer	$61,000*	Zoologist and Wildlife Biologist	$47,410*

*Salaries based on 2001 industry averages, Bureau of Labor Statistics, www.bls.gov, downloaded January 2003.

Scientific research and engineering careers require an in-depth understanding of math and complex scientific subjects. Nearly all require college degrees in scientific fields of study. There are various levels of college and university degrees. Coursework usually starts with general education courses. Studies become more and more specialized after two, four, six, eight, and even more years of learning.

To insert a worksheet, click a worksheet tab. Choose **Worksheet** from the Insert menu. The new worksheet will be inserted to the left of the existing sheet. Rename the new worksheet if desired.

To delete a worksheet, right-click the worksheet tab. Choose **Delete** from the menu.

8-14
Delete Worksheet

ACTIVITY 8-8   PRACTICE FILL AND MODIFY A WORKSHEET

In this activity, you will practice using Fill and Fill Series. You will also practice inserting and deleting worksheets.

1. Start *Excel*. Open a new blank worksheet.

2. The workbook contains several worksheets. To delete Sheet 2, right-click the sheet name. Choose **Delete** from the menu. Leave Sheet 1, but delete all other sheets in the workbook.

3. Rename Sheet 1 using the current year as the name, for example 2004.

4. In Cell A3, enter `Product 1`. In Cell A4, enter `Product 2`. Select A1:A2. Click and drag the Fill handle down to Cell A8. You should now have *Product 1* through *Product 6* in the cells.

5. In Cell B2, enter `January`. Select B2. Click and drag the Fill handle to the right to Cell G2. You should now have *January* through *June* in the cells.

6. In Cell A1, enter UNIT SALES. Bold the title. Use Merge and Center to center the title over the data.

7. Change the height of Row 2 to 20. Center and bold the column heads in Row 2.

8. Enter data for the unit sales shown below.

	A	B	C	D	E	F	G
1				UNIT SALES			
2		January	February	March	April	May	June
3	Product 1	782	750	675	650	680	700
4	Product 2	953	874	899	925	753	650
5	Product 3	450	439	487	477	399	400
6	Product 4	321	356	389	345	332	381
7	Product 5	98	102	135	126	119	106
8	Product 6	498	378	245	478	358	495

Review these four major levels of college and university degrees for scientists and engineers:

- **Associate** or technical degrees. These two-year degrees prepare people for careers assisting in areas of scientific research and engineering.
- Bachelor's degrees, including **Bachelor of Science** (**B.S.**) or **Bachelor of Arts** (**B.A.**). These four-year degrees go beyond associate or technical degrees. Bachelor candidates study general education topics and focus on a specialty, called a major. Majors related to science or engineering may include biology, astronomy, engineering, mathematics, computer science, agriculture, or conservation science.
- Master's degrees, including **Master of Science** (**M.S.**) or **Master of Arts** (**M.A.**), usually require two years of additional study. Studies are more focused on some area of science or engineering.
- Doctoral degrees, including **Doctor of Philosophy** (**Ph.D.**), usually require four years of additional education. Top scientists and engineers often pursue this level of education. Studies focus on a very specialized area of science or engineering.

EXERCISE 2-1

LEARN ABOUT DEGREE OPPORTUNITIES IN SCIENCE AND ENGINEERING

TEAMWORK

INTERNET

1. Work in a team with three or four classmates to complete this activity. Each team member, choose a different career from the Careers in Scientific Research and Engineering table. For example, say you pick astronomer, another member chooses civil engineer, another chooses marine engineer, and your final team member chooses biochemist.

2. Use a search engine such as *Google* (www.google.com) or *Yahoo!* (www.yahoo.com) to learn a little more about these careers and what these career professionals do on the job. Each team member, write a brief description (a half page in length) for one of the four scientific and engineering-related careers.

3. Research educational institutions, colleges, and universities that offer degrees for the careers you chose. For each career, identify a college or university that provides bachelor's, master's, or more advanced degrees for the careers you chose.

Hint: As you did in Exercise 1-2, begin close to home. Start by visiting the Web sites of universities and colleges in your community or state. Discover if they offer degrees in the engineering and science subjects selected by your team. Then expand your search to other states.

4. List each college or university and the degrees offered to students for each of the careers being researched.

5. Each team member, write a summary of what you learned and share your findings with the team in a printed document. Save the document as *Exercise 2-1* in the *DigiTools your name\Unit 2* folder. Print four copies each. Read what each member of your team has written. Discuss whether or not any members of your team are interested in pursuing any of these college or university degree options.

Health Science

Healthcare-related careers are among the fastest growing of any career cluster. And why not? Health science professionals are involved in a great work—helping people feel their best and live healthy.

FILL AND FILL SERIES

You can use *Excel's* Fill command to copy the contents of a cell to other cells in the same row or column. For example, if you have a formula in Cell B2, you can use Fill to copy it to cells B3:B6 or to Cells C2:F2. You can fill cells by using the Fill command found on the Edit menu or by using the Fill handle. To fill cells using the Fill handle, select the cell that contains the data you want to copy. Drag the Fill handle down or to the right to copy the data. In Figure 8-12, the formula in Cell C4 is being copied to C5:C7.

Figure 8-12
Use Fill to copy cell contents to the same row or column.

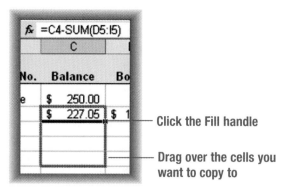

- Click the Fill handle
- Drag over the cells you want to copy to

HELP KEYWORDS

Fill
About filling in data based on adjacent cells

You can also use Fill to automatically continue a series of numbers, number/text combinations, dates, or time periods based on data you enter. For example, if you enter January, you can use Fill to automatically enter February, March, and April. If you enter 1 in A1 and 2 in A2, Fill can automatically enter 3 in A3, 4 in A4, and so on.

To continue a series using Fill, enter a number, date, or time in a cell. For numbers, enter at least the first two numbers in the series. Select the cell (two cells for numbers). Drag the Fill handle down or to the right to extend the series of data. In Figure 8-13, a series of numbers is being extended. Notice that the last number in the series is shown in a box below the selection. This lets you see how far you have extended the series as you drag over cells.

Figure 8-13
Fill Series can be used to extend a series of numbers or dates.

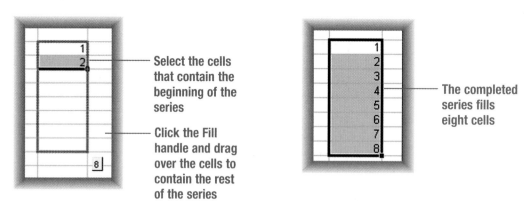

- Select the cells that contain the beginning of the series
- Click the Fill handle and drag over the cells to contain the rest of the series
- The completed series fills eight cells

INSERTING A WORKSHEET

You can move between worksheets in a workbook and change the name of worksheets to make them easier to use. You have practiced both these procedures in previous activities. You can also delete and insert worksheets to modify a workbook.

When most people think of healthcare, they think of their local doctors or nurses caring for patients in a hospital. However, this career cluster is much broader than that. This career includes athletic trainers for sports teams, salespeople for drug companies, and cancer researchers. Microbiologists for the Center for Disease Control, emergency medical technicians (EMTs), and assistants who maintain medial records are also included. Look at the variety of career choices in the table below.

Careers in Health Science

Position and Yearly Salary		Position and Yearly Salary	
Occupational Health and Safety Specialist and Technician	$46,190*	Pharmacist	$72,830*
		Audiologist	$49,700*
Obstetrician	$133,430*	Optometrist	$88,100*
Athletic Trainer	$35,380*	Registered Nurse	$48,240*
Internist	$126,940*	Nutritionist	$41,070*
Medical Records and Information Technician	$25,370*	Surgeon	$137,050*
		Dentist	$110,820*
Family or General Practitioner	$110,020*	Pediatrician	$116,550*
Emergency Medical Technician or Paramedic	$25,450*	Chiropractor	$76,870*
		Respiratory Therapist	$39,870*
Anesthesiologist	$131,680*	Podiatrist	$94,500*
Occupational Therapist	$52,210*		

*Salaries based on 2001 industry averages, Bureau of Labor Statistics, www.bls.gov, downloaded January 2003.

Health and occupational safety careers, like science and engineering careers, require considerable education and advanced training. For example, medical doctors must pursue as many as eight years of post-high-school education. They must also complete residency work under careful supervision at hospitals and clinics before they are allowed to practice medicine. Doctors must be licensed by the state in which they work.

EXERCISE 2-2 UNTANGLING CAREERS IN HEALTHCARE

1. The terms used to describe medical specialties can be confusing! Give yourself a quiz. Scan this list of occupations and write your best guess as to what you think each of these people do.

- Anesthesiologist
- Audiologist
- Cardiologist
- Chiropractor
- Dermatologist
- Emergency Medical Technician
- Endocrinologist
- Gerontologist
- Immunologist
- Internist
- Neurologist
- Nutritionist
- Occupational Therapist
- Ophthalmologist
- Optometrist
- Pediatrician
- Pharmacist
- Podiatrist
- Respiratory Therapist
- Speech Pathologist

4. Change the head for Column I from *Miscellaneous* to `Other`. Center the column heads. Select A4:I5 and click the **Center** button on the Formatting toolbar.

5. Click and drag the column borders to adjust the column widths. Make the columns wide enough to fit the longest entry and leave a small amount of space between data in the columns.

6. Left-align the dates. Select A6:A19. Click the **Align Left** button on the Formatting toolbar. Center-align the data in Cells B6:B19.

7. Format the dollar amounts in the worksheet. Select C6:I21. Click the **Currency** button on the Formatting toolbar.

8. Make the worksheet title bold. Click in Cell A1. Click the **Bold** button on the Formatting toolbar. Click in Cell A2 and make the date bold also. Select the column heads in Cells A4:I5. Make the column heads bold.

DigiTip

For Currency number format, the position of the dollar sign varies depending on whether the format is set using the Currency Style button or the Format Cells dialog box.

9. Add top and bottom borders to the column head cells. Select A4:I5. Click the down arrow on the **Borders** button. Choose the **Top and Bottom Border** option.

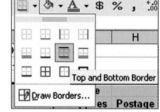

10. Place a single border above and a double border below the item totals. Select A21:I21. Click the down arrow on the Borders button. Choose the **Top and Double Bottom Border** option.

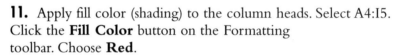

11. Apply fill color (shading) to the column heads. Select A4:I5. Click the **Fill Color** button on the Formatting toolbar. Choose **Red**.

12. Red shading may be too dark when the worksheet is printed. Click **Undo** on the Standard toolbar to remove the shading.

13. Apply light green fill color to the column heads. Select A21:I21. Apply the same light green fill color to the item totals.

14. You have decided that Cells A21:C21 would look better without fill color and borders. Select A21:C21. Choose **Clear** from the Edit menu. Choose **Format**.

15. Save the worksheet as *Activity 8-7* in the *DigiTools your name\Chapter 8* folder. Preview and print the worksheet. The worksheet will print on two pages. In the next lesson, you will learn to set print options and change margins for more attractive printed worksheets. Close the worksheet.

2. Check your guesses. Access the Internet. Use a search engine to search for a medical dictionary, such as MedicineNet.com, or a general dictionary. Look up each career option listed in Step 1. Determine what each of these careers is all about. Information about several of the careers can also be found in the *Occupational Outlook Handbook* at http://www.bls.gov/oco/.

3. Write a short description in your own words of what this person does in his or her career. Save the document as *Exercise 2-2* in the *DigiTools your name\Unit 2* folder. Print the document.

Peer Check Compare the information you found with a classmate's career descriptions. What did you learn from this discussion?

Manufacturing

Manufacturers make many of the goods and products we use, drive, and wear. They may also make packaging for the food we eat. When some think of manufacturing, they think of people working on assembly lines making products like cars, computers, or processed foods. Such manual labor is still necessary to make some products. Today, however, automation is used in manufacturing many products. Automation makes use of machinery, computers, and robotics to make products. It replaces work formerly done by hand.

Automation has increased output while decreasing manufacturing costs. Automation also requires higher educational levels and technical training by those who work in today's high-tech manufacturing facilities. Decreasing manufacturing costs makes the goods and products we use more affordable. This applies to everything we buy from cell phones, to PDAs, and even the food we eat. Look at a few manufacturing career options and sample salaries in this table.

Careers in Manufacturing

Position and Yearly Salary		Position and Yearly Salary	
Production Manager	$44,740*	Butcher and Meat Cutter	$26,510*
Purchasing Manager	$61,260*	Industrial Engineer	$61,940*
Process Control Programmer	$38,830*	Baker	$21,830*
Cost Estimator	$50,450*	Mechanical Drafter	$41,750*
Forging Machine Operator	$28,230*	Engine Assembler	$29,610*
Computer Scientist	$76,970*	Mechanical Engineering Technician	$43,980*
Machinist	$32,880*	Electronic Equipment Assembler	$23,900*
Fish Cutter	$18,310*	Graphic Designer	$39,670*
Electrical Engineer	$68,630*	Production Finisher	$23,740*

*Salaries based on 2001 industry averages, Bureau of Labor Statistics, www.bls.gov, downloaded January 2003.

Modern automated manufacturing takes organizational skill. Manufacturers must analyze every step of the work process carefully. They must plan each step, organize the delivery of supplies and parts, examine costs, design and customize machines, program robots, and learn exactly how long it will take to complete each step of the process. This allows them to deliver their products on time, on budget, and in the most efficient manner.

To set a specific row height:

- Choose **Row** on the Format menu.
- Click **Height**.
- Enter a number for the height.
- Click **OK**.

CLEARING CELLS

HELP KEYWORDS

Clear
Clear cell formats or contents

The **Delete** key can be used to delete data from a cell. However, pressing the **Delete** key will not remove formatting or comments for a cell. The Clear command allows you to remove contents (data), formatting, or comments from a cell or to remove all these items at once. To clear cell contents, choose **Clear** from the Edit menu. Choose the desired option.

Figure 8-11
The Clear command is found on the Edit menu.

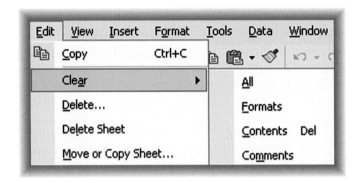

DigiTip The *Excel* Comment feature allows you to attach notes to a cell. A comment can help you remember what a formula does, for example. When you move the cell pointer over a cell with a comment, the comment pops up in a small box. To learn more, search for *Comments* in *Excel* Help.

ACTIVITY 8-7

FORMAT A WORKSHEET

In this activity, you will format the petty cash record you updated in the last activity.

1. Start *Excel*. Open the file *Activity 8-6* that you saved earlier.

2. Select A1:I1. Click the **Merge and Center** button to merge the cells and center the worksheet title. Merge and center cells to center the date over the worksheet. Change the font size of the worksheet title. Click Cell A1. Choose **12** from the Font Size drop-down list.

3. Change the height of Row 6 to leave some extra space between the column heads and the first row of data. Click in Cell A6. Choose **Row** from the Format menu. Choose **Height**. Enter 20 and click **OK**.

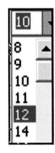

EXERCISE 2-3

TEAM WORK

OBSERVE AND FLOWCHART A MANUFACTURING PROCESS

To explain the manufacturing planning process, consider a simple example that everyone can relate to. Consider the steps you must take to write a research report. The process starts by creating a list of the required steps. (See the table below.) For the initial list, don't worry about placing the steps in any particular order. Simply brainstorm and list each possible step no matter how small. Then, record the approximate time required to complete each step as seen here.

Step	Estimated Time Required
1. Search for sources of information on the report topic	2 days
2. Proofread and print a final copy	½ day
3. Choose a topic	1 day
4. Write the first draft	3 days
5. Revise based on peer reviews	1 day
6. Get peer reviews of report	1 day
7. Read information from sources	5 days
8. Format the report in the required style	½ day
9. Write a report outline	1 day

1. A type of flowchart called a Gantt chart can be used to place the steps in a logical order. Open the data file *Gantt chart*. This chart shows how long it should take to perform the entire task—in this case, 15 days. After listing the steps, the manufacturing planner indicates with arrows and bars what steps come first, second, third, and so on. Notice that in our example, the process starts with Step 3, goes to Step 1, and so on, until the final step in the process is complete. The width of each bar on the chart represents the amount of time needed to complete each step. After this analysis has been completed, manufacturers can find ways to streamline and improve the process.

2. Apply the flowcharting process to a real manufacturing situation. Work in a team with three or four other students to complete this exercise. As a team, obtain permission from a local company to observe the steps of their manufacturing process. This can be as simple as observing the process of making hamburgers or other fast food, baking at a donut shop, bottling at a soda pop factory, assembly at an auto plant, or manufacturing at a computer company. If the product is very large or complicated, you might work with just a part of the product, such as installing the exhaust system in a car.

3. Observe each step of the process. Make a complete list of each step. Do not leave out anything, no matter how small or insignificant the step may seem at the time.

4. Time each step. (**Note:** For the manufacturing of a hamburger, you will record minutes and seconds, whereas certain steps in an automobile manufacturing process may require hours to complete.)

5. As a team, plan a Gantt chart to organize all of the steps of the process and their times of completion. Draw a rough sketch of the chart on paper.

6. Each team member, create the chart in a table similar to the example data file, *Gantt chart*. See the notes below the sample chart for tips on formatting the table and creating the arrows on the chart. Save the document as *Exercise 2-3* in the *DigiTools your name\Unit 2* folder. Print the chart.

FORMATTING DATA

The data in cells can be formatted to make the worksheet attractive and easy to read. You can use buttons on the Formatting toolbar or options on the Format menu.

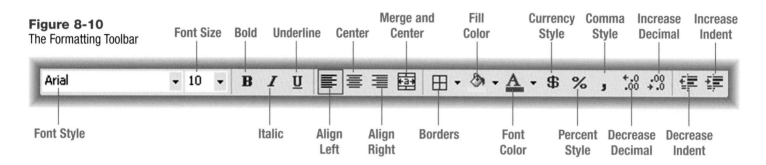

Figure 8-10
The Formatting Toolbar

HELP KEYWORDS

Format
About formatting worksheets and data

Some ways you can format cells using the Formatting toolbar are described below:

- **Font style and size.** To change the font style, select the cell(s) you want to format. Choose a font style and size from the drop-down lists on the Formatting toolbar.

- **Character format.** To apply bold, italic, or underline format, select the cell(s). Click the **Bold**, **Italic**, or **Underline** button on the Formatting toolbar.

- **Alignment.** To change the alignment of data, select the cell(s). Click an alignment button (**Align Left, Center,** or **Align Right**) on the Formatting toolbar.

- **Borders.** To apply borders, select the cell(s). Click the down arrow on the **Borders** button and choose an option.

- **Fill color.** To apply shading (fill color), select the cell(s). Click the down arrow on the **Fill Color** button and choose an option.

- **Font color.** To change the color of text, select the cell(s). Click the down arrow on the **Font Color** button and choose an option.

- **Number format.** To change the number format or style, select the cell(s). Choose a number style (**Currency**, **Percent**, or **Comma**) from the Formatting toolbar. To change the number of decimals displayed in the number, click the **Decrease Indent** or **Increase Indent** buttons. Additional number formats can be accessed by choosing **Cells** from the Format menu.

CHANGING ROW HEIGHT

When you change the font size of data to a larger size, the row height automatically adjusts to fit the data. However, you may want to change the row size to make the worksheet more attractive. You can change the row height by clicking and dragging the top or bottom border of the row heading. You can also set the row height using the Row Height dialog box.

Transportation, Distribution, and Logistics

Logistics has often been called the science of transport and supply. Just like manufacturers, professionals in transportation and distribution must be extremely organized. They must plan each step of the transportation and distribution process.

For example, logistics is required to bring together all of the supplies and materials necessary to keep manufacturing plants working efficiently. When Ford Motor Company makes a car, all the parts that go into a vehicle must be made—often in some other state or country—and sent to the manufacturing plant at just the right time, not too early and not too late. It would be very costly to hold up the manufacture of thousands of cars because a shipment of headlights was late or went to the wrong place! Look at a few career options and sample salaries in this table.

Careers in Transportation, Distribution, and Logistics

Position and Yearly Salary		Position and Yearly Salary	
Aircraft Cargo Handling Supervisor	$40,720*	Bus Driver	$29,420*
Ship Captain	$49,660*	Shuttle Car Operator	$76,390*
Manager of Transportation Vehicle Operator	$44,210*	Tractor-trailer Truck Driver	$33,690*
		Rail Yard Engineer	$40,130*
Transportation Inspector	$46,530*	Local Delivery Truck Driver	$25,630*
Airline Pilot	$99,400*	Crane Operator (Loading Ships)	$36,770
Ship Loader	$33,880*	Train Locomotive Engineer	$47,300*
Air Traffic Controller	$83,350*	Sailor	$30,270*
Freight, Stock, and Material Mover	$21,170*		

*Salaries based on 2001 industry averages, Bureau of Labor Statistics, www.bls.gov, downloaded January 2003.

After a product has been made, it must then be transported to stores, dealerships, or directly to customers. For instance, a car manufacturer can ship cars to dealers on trucks, trains, and even by boat. Some products, like fresh fish from Maine, may be packed in ice and travel quickly by air to restaurants in Iowa or Utah.

The distribution and transportation of products is what keeps the economy moving forward. Without logistics, the economy would be paralyzed. Supermarkets would soon run out of food. Medicines and surgical supplies at hospitals and pharmacies would soon run out. Most of the products we normally buy would disappear from store shelves.

EXERCISE 2-4   LOGISTICAL PLANNING

After manufacturers have created products, they must transport and distribute their products to customers. In this exercise you will find the fastest and then the least costly methods of shipping a product from your home to Mason, Ohio.

1. Open the data file *Logistics*. This document contains information about two shipments you must make.

2. Read the description for each shipment. Choose the shipping service (U.S. Postal Service or a private mail delivery company) that will deliver the package in the time required for the lowest cost. Indicate the delivery method/company, the delivery time you choose, and the cost of shipping the package for each shipment.

3. Save the information as *Exercise 2-4* in the *DigiTools your name\Unit 2* folder. Print the information.

Figure 8-9
Unformatted Worksheet

	A	B	C	D	E	F	G	H	I	J
1	PETTY CASH RECORD									
2	November 30, 20--									
3										
4							Office			
5	Date	Receipt No.	Balance	Books	Taxi	Parking	Supplies	Postage	Miscellaneous	
6	1-Nov	Balance	250							
7	6-Nov	39	227.05	12.95		10				
8	8-Nov	40	218.55		8.5					
9	8-Nov	41	198.85			12	7.7			
10	11-Nov	42	188.1					10.75		
11	17-Nov	43	181.65						6.45	
12	19-Nov	44	162.9	18.75						
13	21-Nov	45	139.9		23					
14	22-Nov	46	128.4		11.5					
15	24-Nov	47	88.25	30.15		10				
16	25-Nov	48	76.8					11.45		
17	26-Nov	49	58.3				18.5			
18	29-Nov	50	45.3			13				
19	30-Nov	Balance	45.3							
20										
21			Item Totals	61.85	43	45	26.2	22.2	6.45	

13. Save the worksheet as *Activity 8-6* in the *DigiTools your name\Chapter 8* folder. Do not print the worksheet. You will format the worksheet in the next activity and then print it. Close the worksheet.

Self Check Compare your worksheet with Figure 8-9. Do you have the correct amounts in the correct cells? If not, update your worksheet.

Stewardship of Company Funds

A **steward** is a person morally responsible for the careful use of money, time, talents, or other resources. Companies expect employees to use company funds only for approved purposes. Misusing company funds is unethical. Consider the following situation.

You are the person responsible for the petty cash fund in your department. The fund is maintained at $1,000 because of the many small payments that must be made during the month. One Wednesday, shortly after the fund had been replenished, one of your friends in the department says to you: "I certainly didn't budget my money very well this week; I'm down to my last $5. You know that I'm dependable. Would you let me borrow $25 from the petty cash fund? You have almost $1,000 just lying there! If I had only $25, I'd be able to take care of my expenses until Friday, which is payday."

In this situation, you are a steward of company funds. You must decide whether loaning money from the petty cash fund is good stewardship.

Follow-up

1. Would you lend your friend money from the petty cash fund? Why or why not?

2. What might be the consequences of making an unauthorized payment from the fund?

3. What other action could you take in this situation to help your friend?

BUSINESS TRENDS AND ISSUES

Global Marketplace

The area in which a company does business is called its marketplace. In the past, many U.S. companies sold their goods or services only in the domestic marketplace, meaning within the United States. Many of these companies now produce or sell their products in countries around the world, in the global marketplace. Some companies have moved into the global marketplace using only traditional sales methods, such as retail stores. Other companies have used ecommerce to enter the global marketplace. **Ecommerce** means business conducted electronically, as in making purchases or selling products using the World Wide Web.

Selling in the global marketplace affects companies in many ways. Sometimes, employees must travel to other countries. Communications, too, must be international. Information for Web sites, advertising materials, and product instructions must be available in many languages. All personnel must be sensitive to customs and methods from other cultures. Employees must be aware of time zone differences. They must know how to access information about travel and places for travelers to stay and work in other countries.

EXERCISE 2-5

INTER N E T

INTERNATIONAL TRAVEL ARRANGEMENTS

You are an employee of a company that sells in the global marketplace. You need to make travel arrangements for a trip next month to the company's new branch office.

1. Choose a large foreign city to which you will travel. (Assume the new branch office is in this city). You will travel from your home to the company's new branch office. Your travel date to that city is one month from today. Your return travel date is one week later.

2. Using a printed airline guide or an airline or travel Web site, research airline flights to that city. Choose the flights you think are most appropriate, considering the costs and the schedule. Pretend that you have reserved these flights. Make a note of the flight information or print the information from the Web site if possible.

3. Search the Web to find a hotel in the branch office city. Choose the room and rates you think are most appropriate. Pretend that you have reserved a room. Make a note of the information or print the information from the Web site if possible. Pretend you have been given the confirmation number Vl379XA.

4. A company representative, Ms. Kitty How, will meet you at the airport and provide transportation during your stay. She will take you to the airport for your return flight.

5. Open the data file *Sample Itinerary*. Study the format and content of this document as a guide for creating your own itinerary.

6. Create an itinerary for your trip to include travel details and the scheduled activities shown on page 224. Attach any reservation information you have printed from Web sites for the airline and hotel. If you did not print information, key notes about each reservation and attach the notes to the itinerary.

HELP KEYWORDS

Insert column
  Insert blank cells, rows, or columns

1. Start *Excel*. Open the data file *Petty Cash*.

2. You no longer want the company name to appear on the petty cash record. Click the row heading for Row 2 to select the row. Choose **Delete** from the Edit menu to delete the row.

3. A payment made on November 19 was not entered in the worksheet. You need to insert a row for the data. Click a cell in Row 12. Choose **Rows** from the Insert menu to insert a new blank row. Enter this data in Row 12:

Date: `November 19`
Receipt No.: `44`
Books: `18.75`

4. The formulas below Cell C12 will no longer be correct. Select C13:C19. Press the **Delete** key. Copy the formula from C11 to C12:C19.

5. No entries are in the Meals column because the company has changed its policy regarding meal expenses. All expenses for meals must now be submitted on each employee's expense report. You need to delete the Meals column. Click the column heading to select Column H. Choose **Delete** from the Edit menu to delete the column.

6. Previously, expenses for parking have been included in the Miscellaneous column. You need to create a column to show parking expenses separately. Click in a cell in Column F. Choose **Columns** from the Insert menu to insert a new blank column. In Cell F5, key the column heading `Parking`.

7. On the entry for November 6, the number 10 that is in Miscellaneous should be in Parking. Select Cell I7. Click the **Cut** button on the Standard toolbar. Select Cell F7. Click the **Paste** button on the Standard toolbar.

8. The Miscellaneous amount for November 8 should also be in Parking. Click Cell I9. Move the pointer to the left cell border until it becomes a four-sided arrow. Click and drag the data to Cell F9.

9. The Miscellaneous amounts for November 24 and 29 should also be in Parking. Drag cell contents or use Cut and Paste to update the worksheet.

10. When you use Cut and Paste or drag numbers to different cells, the formulas may change. For example, click in Cell C7. Before you cut and pasted the data in this row, the formula in Cell C7 was =C6-SUM(D7:I7). Notice that now the formula sums cells D7:H7. That formula is fine for this worksheet. However, you want to be able to use this worksheet again and update it next month. In December you might have a Miscellaneous amount for this row. The formula would not sum an amount in Row I. Change the formula in C7 to =C6-SUM(D7:I7). Copy this formula to Cells C8:C19.

11. Sum the numbers in Column D to find the total amount spent for books. The results should appear in Cell D21. (Select Cells D6:D21. Click the **AutoSum** button.)

12. Sum the numbers in the Taxi, Parking, Office Supplies, Postage, and Miscellaneous columns to find the totals spent for each of these items. Place the results in Row 21.

Day 1	Travel to destination city
Day 2	9:30 a.m. — 11:30 a.m. Tour of new office
	12 noon — 2 p.m. Lunch with branch manager, Mr. Lou
	2:30 p.m. — 4:30 p.m. Prepare meeting room and materials
Days 3, 4, and 5	9:30 a.m. — 4:30 p.m. Provide training to employees at the branch office (1 hour lunch break starting around noon)
Day 6	9:30 a.m. — 11:30 a.m. Meeting with department managers to discuss additional training needs
Day 7	Travel to home city

7. Save the document as *Exercise 2-5* in the *DigiTools your name\Unit 2* folder. Print the document.

CRITICAL THINKING

Telecommuting

The practice of working and communicating with others from a home office or other remote location is called **telecommuting**. A worker who telecommutes shares information with clients or coworkers using a telephone, e-mail, and Internet or intranet connections. Some people who work at a remote location are able to participate in teleconferences with persons at the company offices or other locations. Occasionally, the worker may meet in person with other company personnel or clients.

EXERCISE 2-6

EVALUATING JOB OPPORTUNITIES

Assume that you have completed your studies and are seeking your first full-time job. You have been interviewed by personnel at two companies. Each company has offered you a position. You like both companies as far as the nature of the work, the salary, and the employee benefits. However, there is a difference in where and how you will work.

In Company A, you would be expected to come to the company offices each day. The company has nice offices with current computers and other equipment. The supervisor seems very helpful and friendly.

In Company B, you would be telecommuting. Company B would provide you with all the equipment and furniture for your workstation at home. You would have access to the supervisor via e-mail or telephone. From time to time—possibly no more than once in three weeks—you would be expected to attend a training session or a team meeting at the company offices. Which position would you accept?

1. In a *Word* document, make a list of the factors you would consider in making a decision. Think about both the advantages and the disadvantages of working in each location. Choose Company A or Company B. Write a brief paragraph in which you discuss your decision. Give reasons for your choice.

2. Save the document as *Exercise 2-6* in the *DigiTools your name\Unit 2* folder. Print the document.

LESSON 8-3

EDITING AND FORMATTING WORKSHEETS

OBJECTIVES *In this lesson you will:*

1. Delete and insert columns and rows.
2. Change cell formats and clear cells.
3. Modify row size.
4. Apply borders and shading.

DELETING AND INSERTING COLUMNS AND ROWS

After creating a worksheet, you may find that you need to add or delete rows or columns because you have changed the data the sheet contains. To delete a row or column, select the row or column by clicking on the row or column heading. Choose **Delete** from the Edit menu to delete the row or column.

Figure 8-8
Worksheet with Row and Column Selected

Click a column heading to select the column

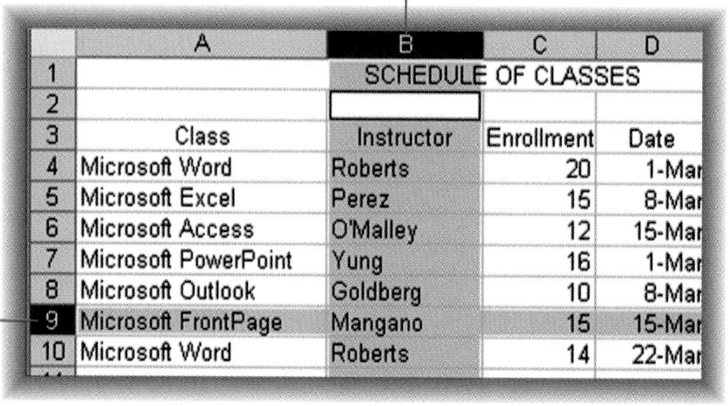

Click a row heading to select the row

	A	B	C	D
1		SCHEDULE OF CLASSES		
2				
3		Class	Instructor	Enrollment	Date
4	Microsoft Word	Roberts	20	1-Mar
5	Microsoft Excel	Perez	15	8-Mar
6	Microsoft Access	O'Malley	12	15-Mar
7	Microsoft PowerPoint	Yung	16	1-Mar
8	Microsoft Outlook	Goldberg	10	8-Mar
9	Microsoft FrontPage	Mangano	15	15-Mar
10	Microsoft Word	Roberts	14	22-Mar

To insert a column, click in the column to the right of where you want the new column to appear. Choose **Columns** from the Insert menu. To insert a row, click in the row below where you want the new row to appear. Choose **Rows** from the Insert menu. To insert several rows at once, highlight the number of rows you want to insert. Choose **Rows** from the Insert menu.

ACTIVITY 8-6 — MODIFY A WORKSHEET

In some companies, cash is needed occasionally to pay for small expenses such as delivery charges, taxi fares, or postage. To pay for such expenses, departments may be given a small sum of money called a petty cash fund. Records must be kept to show how money from the petty cash fund is spent. You will update and format a petty cash record in this activity.

WRITING

Possessives, Numbers in Text, and Sentence Structure

In Exercise 2-7 you will apply what you have learned about using possessives, numbers in text, and sentence structure while editing a FAX transmittal.

EXERCISE 2-7 — FIXING A FAX

1. Access the Corporate View intranet. Go to the Style Guide. (Choose **Corporate Communications, Style Guide**.) Choose the **FAX** link. Read the information at the top of this Web page that explains what a FAX is. Look over the sample FAX. Read the information that follows the sample under the **FAX Transmittal Information** link.

2. Open the data file *Shipping Confirmation*, which contains information for a FAX. Proofread the document, looking especially for errors in possessive use, number use, and sentence structure. Correct all the mistakes or rewrite sentences to make them correct.

3. Save the document as *Exercise 2-7* in the *DigiTools your name\Unit 2* folder. Print the document.

Peer Check Compare your corrected FAX document with a classmate's document. What did you learn from this comparison? Make corrections, if needed, after your discussion.

3. Create an *Excel* workbook to record the data about gas mileage. Enter an appropriate title for your worksheet.

4. In Column A, enter the car names. In Column B, enter the average gas mileage (in miles per gallon) for each car. Enter appropriate column heads.

5. The sales representative who will use the car will drive about 30,000 miles per year. In Column C for each car, enter a formula to find the number of gallons of gas needed to drive 30,000 miles. (Divide 30,000 by the average miles per gallon for the car.) Enter an appropriate head for Column C.

6. The average gas price (for comparison purposes) will be $2 per gallon. In Column D, enter a formula to calculate the annual fuel cost for each car. (Multiply the gallons per year by $2.) Enter an appropriate head for Column D.

7. Center the worksheet title across the data. Change the column widths as needed to create an attractive worksheet.

> **DigiTip**
> To review memo format, see page App-5 in Appendix A.

8. Start *Word* and open a new blank document. Write a memo to your manager, James Barnes. Use the current date and an appropriate subject heading. Tell Mr. Barnes which car has the lowest fuel costs per year. Mention the name of the Web site(s) you used to find this information. Suggest that Mr. Barnes might want to visit this site(s) if he wants to view more details about the cars.

9. In *Excel*, select the range of cells that contain your data. Click the **Copy** button. Move to your *Word* memo. Position your insertion point below the last paragraph. Click the **Paste** button to paste the data into your memo. Center-align the table and make other adjustments as needed for an attractive format.

10. Save the workbook as *Application 8-1 Worksheet* in the *DigiTools your name\Chapter 8* folder. Close the workbook and *Excel*.

11. Save the memo as *Application 8-1 Memo* in the *DigiTools your name\Chapter 8* folder. Print and close the document. Close *Word*.

> **Peer Check** Compare the data you found about gas mileage for each car with the data another team found. If the data are different, determine why. For example, did your teams use different Web sites as sources of information?

UNIT 3

Increasing Productivity with DigiTools

In Unit 3, you will learn to use strategies and DigiTools to be a more productive student and future employee. In Chapter 8, you will calculate, analyze, and chart data using *Microsoft Excel*. In Chapter 9, you will store, report, and manage data using *Microsoft Access*. You will learn to develop and deliver presentations and create multimedia slide shows using *Microsoft PowerPoint* in Chapter 10. Chapter 11 provides instruction in basic HTML and Web site design. In Chapter 12, you will learn to be more productive by managing time and records effectively.

WHAT YOU SHOULD KNOW ABOUT
Company Goals

Many companies create annual reports that provide information about company activities during the past year. These annual reports may also include goals for the future. A **goal** is a result that one strives to attain. In some instances, goals are simply expressed as a long-term vision statement. For example, the head of one computer company declared that the company's vision was to have a computer on every workplace desk and in every home. Other goals make predictions about level of earnings, new markets, new products, or improved customer service. The goals set by the company affect the work of all employees.

Many companies develop strategic plans. A **strategic plan** outlines the overall direction in which a company wants to move to achieve its goals. Strategic plans are long-term in nature (from three to five years). They usually are developed by top management. Examples of strategic planning decisions include introducing or discontinuing products or expanding into new markets.

The success a business has in achieving its goals rests largely on how well the resources of the business are used. **Resources** are personnel or assets such as property, goods, money, or equipment. A **budget** is a plan detailing how a business intends to manage its resources. A budget is typically prepared for a 12-month period.

Businesses compare their financial progress to their goals. They use financial statements such as the income statement and the balance sheet. An **income statement** is a report that lists revenues, expenses, and the income or loss of a business for the reporting period. A **balance sheet** is a report that lists the assets, liabilities, and owner's equity for a company on a specific date. Investors and lenders use these reports to make judgments about the financial health of the business and about whether goals are being achieved.

DigiTip
Liabilities are debts of a company. Owner's equity is the owner's share of the worth of a company, also called capital.

APPLICATION 8-1   COMPARING FUEL COSTS

You work for a small insurance agency. The company is planning to purchase a new car for use by a sales representative. You have been asked to research and compare fuel costs for several different cars that are being considered. Work with a classmate to complete this activity.

INTERNET
TEAMWORK

1. Open your browser and access the Internet. Search the Internet to find sites that give information about car gas mileage. Use a search term such as *new cars* or search for the name of the car maker such as *Ford*.

2. Find the average gas mileage (miles per gallon) for the current model year of the cars listed below. Note the name and URL of each Web site you use to find information.
   - Ford Crown Victoria
   - Honda Accord
   - Chevrolet Cavalier
   - Ford Focus
   - Toyota Corolla
   - Dodge Neon

CHAPTER 8

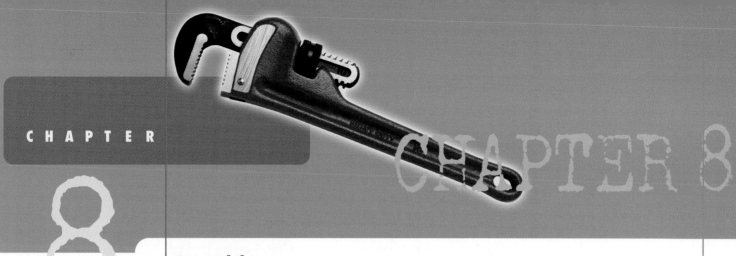

Spreadsheets

OBJECTIVES *In this chapter you will:*

1. Create worksheets and enter data.
2. Create and revise formulas.
3. Edit and format worksheets.
4. Save and print worksheets.
5. Create, save, and print charts.

Excel is application software that helps users manage, present, and analyze numbers and related data. *Excel* and similar programs are often called spreadsheet software. In this chapter, you will create worksheets by entering text, values, formulas, and functions. You will format worksheets to make them attractive and easy to read. You will also learn to create charts to help you analyze and explain your data more clearly.

LESSON 8-1: WORKSHEET BASICS

OBJECTIVES *In this lesson you will:*

1. Open a blank worksheet.
2. Learn the parts of a worksheet.
3. Enter data in a worksheet.
4. Navigate within a worksheet.
5. Change column widths.
6. Save and print a worksheet.

REVIEW

Excel shares many of the features that are common to all of the applications of *Office XP*. You are already familiar with the title bar, the menu bar, the Standard toolbar, the Formatting toolbar, the Task Pane, and the Drawing toolbar. Your previous experience will make it easy for you to use these features in *Excel*.

Cell D4 that instructs *Excel* to add to the two cells to the left, Cells B4 and C4. (=B4+C4). Then suppose you copy that formula down one cell to Cell D5. *Excel* will automatically adjust the formula to add the two cells to the left, Cells B5 and C5. (=B5+C5) Much of the time, you want your cell references to be relative because this makes copied formulas work well.

At other times, you want a formula always to refer to a particular cell, even when the formula is copied to another location. An **absolute cell reference** always refers to one particular cell. Absolute cell reference have a $ symbol before the column letter and the row number. $B$1 is an absolute cell reference in the formula =$B$1+D4.

In a **mixed cell reference**, only the column or the row reference is absolute. In the reference $A1, the column reference is absolute but the row number is not. The row number will change if the formula is copied.

ACTIVITY 8-5 USE ABSOLUTE CELL REFERENCES

An income statement is a financial document that shows a company's profit or loss for a specific period of time. It answers the question, "How successful was the company during the period of time?" A company may also use an income statement to compare certain expenses to sales. This allows the company to see how closely the expense budgets were followed.

1. Start *Excel*. Open the data file *Income Statement*.

2. To calculate the percent advertising expense is of the sales amount:
   - Click in Cell H12 where the results should display.
   - Enter a formula to divide the advertising expense amount by the sales amount. The cell reference for the sales amount should be absolute. In Cell H12, enter =F12/$G$6. The amount $0.00 will display.
   - Change the format of the cell to Percentage and 2 decimal places. Now 0.33% will display in the cell.

3. Next, find the percentage each of the other expenses amounts is of the sales amount. Copy the formula in Cell H12 to Cells H13:H18. Click in one or more of the cells in this range. Note that the reference to Cell G6 remains constant.

4. Save the workbook as *Activity 8-5* in the *DigiTools your name\Chapter 8* folder. Preview the worksheet and correct any errors. Print the worksheet and close *Excel*.

ƒx	=F13/$G$6	
F	G	H
S		
500		0.33%
1,000		0.67%
900		0.52%

The absolute cell reference remains the same when the formula is copied

Peer Check Compare your percentage amounts in Column H with those on a classmate's worksheet. Are the amounts the same? If not, examine your formulas to determine why they are different.

The features and commands you learned in Chapter 1 are listed below. Many of the commands are the same for *Excel* and *Word*. Review these commands now, if needed, so you will be ready to learn new commands in this lesson.

Command	Page Reference
Start *Excel*	Activity 1-4, page 16
Close *Excel*	Activity 1-4, page 16
Save a file	Activity 1-6, page 19
Open a file	Activity 1-7, page 20
Close a file	Activity 1-7, page 20
Print a file	Activity 1-8, page 22
Access Help	Activity 1-9, page 23

PARTS OF A WORKSHEET

Each *Excel* file is called a **workbook**. A workbook contains one or more worksheets. A **worksheet** is the primary document that you use in *Excel* to store and work with data. A worksheet is also called a spreadsheet. A worksheet consists of cells that are organized in columns and rows. Sheet tabs, located at the bottom of the worksheet, identify all the worksheets in a particular workbook. When you start *Excel*, a worksheet will be displayed on the screen. Refer to Figure 8-1 as you learn about a worksheet.

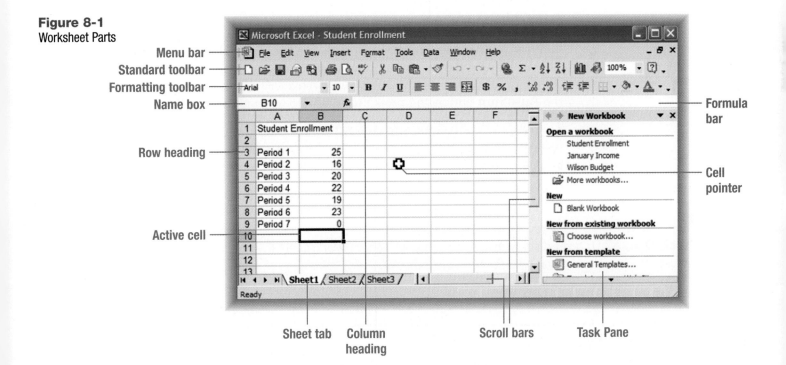

Figure 8-1
Worksheet Parts

HELP KEYWORDS

Function
About functions

The table on page 238 shows only a few of the functions available in *Excel*. To access more functions, click the **Insert Function** button on the Formula bar. The Insert Function dialog box will display and provide access to all functions. You can enter formulas using the Insert Function dialog box or simply type formulas with functions in the active cell.

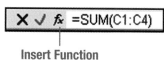

Insert Function

ACTIVITY 8-4 USE FUNCTIONS IN A WORKSHEET

Once again you need to update the software classes worksheet. For each month, you will find the total number of students attending classes and the average number of students in each class. You will also find the total of the fees for the classes for each month.

1. Start *Excel*. Open the file *Activity 8-3* that you created earlier.

2. On the March worksheet, make Cell C11 the active cell. Enter the formula =SUM(C4:C10) and click **Enter**. The total students in all classes for March now displays in Cell C11. In Cell A11, enter the label Total.

3. On the March worksheet, make Cell C12 the active cell. Enter the formula =AVERAGE(C4:C10) and click **Enter**. The average number of students in all classes for March now displays in Cell C12.

4. Change the number format for Cell C12 to Number, 0 decimal places. The number 15 will display in Cell C12.

5. In Cell A12, enter the label Average.

6. Now you need to find the total of all fees for March classes. You can do so quickly using the AutoSum button. **AutoSum** adds numbers in a group of adjoining cells above or to the left of the active cell. Click in Cell F11. Click the **AutoSum** button and press **Enter**. $2,015.00 displays in Cell F11.

7. Move to the April worksheet. Enter formulas and labels and change the number format for the April worksheet as you did for the March worksheet.

8. Save the workbook as *Activity 8-4* in the *DigiTools your name\Chapter 8* folder. Preview the worksheets and correct any errors. Print the worksheets and close *Excel*.

Self Check Does $2,015.00 display in Cell F11 in your March worksheet? If you got a different answer, recheck your formula.

ABSOLUTE AND RELATIVE CELL REFERENCES

Formulas may have three types of cell references: relative, mixed, and absolute cell references. By default, cell references are relative. The cell references you used in formulas in Activity 8-4 were relative. When you copy a formula with a relative cell reference, the cell references are automatically adjusted. For example, suppose you have a formula in

The *Excel* menu bar provides access to all commands. The Standard toolbar and the Formatting toolbar contain command buttons. Worksheets have these parts:

- **Column:** Vertical list of information. Columns are labeled alphabetically from left to right (A, B, C, etc.).

- **Row:** Information arranged horizontally. Rows are numbered from top to bottom (1, 2, 3, etc.).

- **Cell:** An intersection of a column and row. Each cell has its own address consisting of the column letter and the row number such as A1 or C36. A1 refers to Column A, Row 1; C36 refers to Column C, Row 36. The Name box displays the address of the active cell. The **active cell** is the one that is selected and is ready for data to be entered. The active cell is surrounded by a dark border. The contents of the active cell show in the Formula bar.

- **Cell pointer:** Cross-shaped pointer that appears when the mouse is over a cell. Similar to the arrow in *Word*, the cell pointer does not indicate the active cell. Remember that the cell that is highlighted with the dark border is the active cell. To select a specific cell, click or tap inside it.

- **Cell range:** A group of cells. The cells can be next to one another or in different parts of the worksheet. A range of cells is indicated by a colon between the cell references such as A1:D4. Cell ranges can also have names, such as Sales. The names can be referenced in formulas. In Figure 8-2, the cell range A3:C6 is highlighted. To select a range of cells, click and drag over the cells.

Figure 8-2
A Range of Cells

	A	B	C	D
1	SCHEDULE OF CLASSES			
2				
3	Class	Instructor	Enrollment	Date
4	Microsoft Word	Roberts	20	1-Mar
5	Microsoft Excel	Perez	15	8-Mar
6	Microsoft Access	O'Malley	12	15-Mar
7	Microsoft PowerPoint	Yung	16	1-Mar

DATA TYPES

Information is entered in a worksheet as either text (words), values (numbers and dates), or formulas.

Text is any combination of numbers, spaces, and nonnumeric characters. For example, Sales Staff, 1991 Woodlake Drive, 206-555-0122, and 239TXY are all examples of text. Text automatically aligns at the left. You will learn how to change the default alignment in a later lesson.

A **number** is a constant value that can include numeric characters (1, 2, 46, 2.653) as well as some symbols (+, -, ( ), /, %). If the cell is not large enough to display a number, #### will display. The data will display on the screen when you increase the column width. Numbers automatically align at the right.

ACTIVITY 8-3 — CREATE FORMULAS IN A WORKSHEET

A fee must be charged to cover the cost of materials for the software classes. In this activity, you will update the workbook to add data for fees.

1. Start *Excel*. Open the file *Activity 8-2* that you created earlier.

2. On the March worksheet, make Cell E3 the active cell. Enter a column heading `Fee`. Change the alignment for the column head to Center.

3. The fee for the Outlook class is $10. The fee for the FrontPage class is $25. The fee for all other classes is $20. Enter the fee for each class in the appropriate cell in Column E on the March worksheet.

4. Repeat Steps 2 and 3 for the April worksheet.

5. On the March worksheet, enter the column heading `Total Fees` in Cell F3. Change the alignment for the column head to Center.

6. On the March worksheet in Cell F4, enter a formula to multiply the fee by the number of students in the Microsoft Word class. Your formula should be =C4*E4. The result that displays in F4 should be 400.

7. To copy the formula in Cell F4 to the other classes, click in Cell F4. Click the **Copy** button on the Standard toolbar. Select Cells F5:F10 and click the **Paste** button on the Standard toolbar. Press **Escape**.

8. To format the numbers for fees as currency, select Cells E4:F10 on the March worksheet. Choose **Cells** from the Format menu. On the Number tab, choose **Currency** under Category. Leave the default of **2** decimal places. Click **OK**.

9. The worksheet title is no longer centered because you have added data to the worksheet. To move the title back to A1, click in A1. Click the **Merge and Center** button. To center the title, select Cells A1:F1. Click the **Merge and Center** button.

10. Move to the April worksheet. Add a column head and formulas for Column F as you did on the March worksheet. Change the fee amounts to Currency format. Center the title over the data.

11. Save the workbook as *Activity 8-3* in the *DigiTools your name\Chapter 8* folder. Preview the worksheets and correct any errors. Print the worksheets and close *Excel*.

USING FUNCTIONS

Excel provides several predefined formulas called **functions**. For example, functions can be used to add a column of numbers, find the average of numbers, or find the maximum number in a range of cells. Some commonly used functions are shown in the table below.

Function	Formula	Meaning
SUM	=SUM(A5:A12)	Adds all cells from A5 through A12
AVERAGE	=AVERAGE(C5:F5)	Finds the average of all cells from C5 through F5
MIN	=MIN(A2:A20)	Finds the smallest number in all cells A2 through A20
MAX	=MAX(A2:A20)	Finds the largest number in all cells A2 through A20
DATE	=now()	Inserts the current date and time

A **date** or **time** in *Excel is* treated as a value. Because dates and times are values, they can be used in formulas. Also because they are values, dates and times do not always display as you key them. For example, *3-Mar* will display if you key *March 3*. The data can be displayed in traditional date format (March 3) by changing the format options.

A **formula** is a set of instructions for calculating values in cells. For example, the formula =C2+C4 would add the numbers in Cells C2 and C4. You will learn about entering formulas in the next lesson.

ENTERING DATA

When data is keyed in a worksheet, it appears in both the active cell and the Formula bar; however, it is not yet entered into the cell. Data is entered into the active cell by keying the data and then striking the **Enter** key or clicking the **Enter** button on the formula bar. Press the **Delete** key or click the **Cancel** button to delete data that appears on the Formula bar but has not yet been entered into the cell.

Figure 8-3
The Cancel and Enter buttons are found on the Formula bar.

Key a cell reference in the Name box and press Enter to go to that cell.

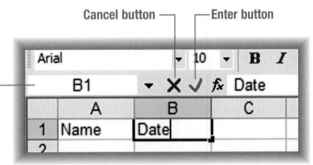

Sometimes the text entered in a cell is too long to fit within the cell width. In this case, the excess characters display over the next cell if it is blank. If the next cell is not blank, the screen only displays a portion of the text. By default, the column width is 8.43 characters. You will learn to adjust the width of columns later in this lesson.

You can change the format of cell contents, change the width of columns and rows, and add borders and shading to make an attractive worksheet. You will learn how to do this in a later lesson.

NAVIGATING WITHIN A WORKSHEET

HELP KEYWORDS
Scroll
Scroll through a worksheet

If a worksheet is large, you will not be able to view all of it on the screen at one time. To move within a worksheet, use the scroll bars or the keystrokes shown below.

To Move	Strike
Up, down, left, or right one cell	↑, ↓, ←, →
Right one column	**Tab**
Left one column	**Shift+Tab**
Up or down one window	**PgUp or PgDn**
To beginning of a row	**Home**
To beginning of the sheet (Cell A1)	**Ctrl+Home**
To last cell containing data	**Ctrl+End**

To enter a formula in a worksheet:

- Click in the cell where you want the formula results to appear.
- Key an equal sign (=).
- Enter numbers or cell references and calculation operators. (You can click on a cell to enter its cell reference in the formula.)
- Press or click **Enter**.

In Figure 8-6, a formula is being entered in Cell F4. First look at the illustration on the left. The user has keyed an equal sign, clicked in Cell C4, keyed a multiplication sign (*), and clicked in Cell E4. Notice that at this point, the formula displays in the active cell and in the Formula bar. Once the user clicks Enter, the formula will be entered into the cell as shown in the illustration on the right. The results ($400) will display in Cell F4. The formula will display in the Formula bar.

Figure 8-6
An *Excel* Formula

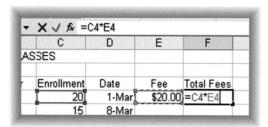

HELP KEYWORDS

Formula
Create a formula

When creating formulas, enclose a part of the formula to be calculated first in parentheses. The formula = (6+4)/2 would result in the answer 5 because the addition inside the parentheses is performed first.

To edit a formula, click the cell that displays the formula results. The formula will display on the Formula bar. Click in the Formula bar. Change numbers, cell references, or calculation operators as needed. Click **Enter**.

NUMBER FORMATS

By default, numbers in *Excel* are displayed with the General format. You can change the format of numbers to formats such as Currency, Percent, or Number. You may want to change formats to display a dollar sign for dollar amounts or to display numbers with a certain number of decimal points. To change a number format, select the cell(s) you want to format. Choose **Cells** from the Format menu. Choose the desired format and other options on the Format Cells dialog box.

Figure 8-7
Format Cells Dialog Box

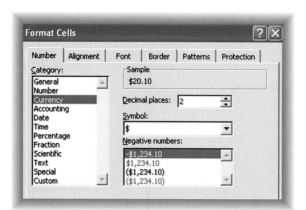

CHAPTER 8   Lesson 8-2   237

The active cell does not move to the area displayed when the scroll bars are used. Use your mouse or digital pen to click or tap in a cell to make it the active cell. You can also key a cell reference in the Name box and press **Enter** to go to a particular cell. (See Figure 8-2.)

CHANGING COLUMN WIDTH

HELP KEYWORDS
Column
Change column width and row height

Sometimes the data you key in a cell is too long to display in the cell. The data will display when you increase the width of the column. You can change the width of a column by clicking and dragging the column border. You can set a specific column width by using the Column, Width commands from the Format menu. You will practice these methods for changing column widths in this lesson.

ACTIVITY 8-1   CREATE A SIMPLE WORKSHEET

The business teachers at your local high school are planning to teach software classes on selected Saturdays as a service to the community. You have been asked to create a worksheet containing information about the classes.

1. To start *Excel*, click **Start, (All) Programs, Microsoft Excel**. A blank worksheet will appear and Cell A1 will be the active cell.

2. In Cell A1, key SCHEDULE OF CLASSES and strike **Enter**.

3. Click in Cell A3 to make it the active cell. Beginning in Cell A3, key the information from the table below, moving across the rows. Strike **Tab** after each entry until you reach the end of a row of data. At the end of each row of data, strike **Enter** to move to the next row. Some of the entries are too wide to fully display in the cells. You will change the column widths in a later step. Notice that the format of the dates appears different from what you keyed. Continue keying until you have entered all the data.

Class	Instructor	Enrollment	Date
Microsoft Word	Roberts	20	3-1
Microsoft Excel	Perez	15	3-8
Microsoft Access	O'Malley	12	3-15
Microsoft PowerPoint	Chen	16	3-1
Microsoft Outlook	Goldberg	10	3-8
Microsoft FrontPage	Mangano	15	3-15

Self Check Check your work by comparing it to the table above. It is very important that your data are correct. This worksheet will be used in the next few activities.

13. Save the file using the same name.

14. Both the March and April worksheets should still be selected. (If they are not, click on the March sheet tab. Hold down the **Ctrl** key and click the April sheet tab.)

15. Click the **Print Preview** button on the Standard toolbar to preview the selected worksheets. Click the **Next** and **Previous** buttons on the Print Preview toolbar to view both worksheets. Proofread your worksheets carefully. If you find errors, click the **Close** button to return to the worksheet. Correct the errors and save again.

16. Select the March and April worksheets if they are not already selected. To print the selected worksheets, click the **Print** button on the Print Preview toolbar (or on the Standard toolbar if you have closed Print Preview). Close the workbook and *Excel*.

LESSON 8-2: CREATING AND EDITING FORMULAS

OBJECTIVES *In this lesson you will:*

1. Create and revise formulas in a worksheet.
2. Change number formats.
3. Use functions in a worksheet.
4. Use relative, absolute, and mixed cell references.

CREATING FORMULAS

A formula is a set of instructions to perform calculations in a cell. All formulas begin with an equal sign (=). All formulas have a number or cell reference (e.g., 12 or A5) and a calculation operator that indicates what to do. In the formula =5+12, the numbers 5 and 12 are added. In the formula =A5+C2, the values in Cells A5 and C2 are added. The cell reference may be a cell address such as D4 or a range of cells (D4:D8). Study the calculation operators in the table below. The value in Cell B4 is 10. The value in Cell D4 is 2.

Operation	Example	Meaning	Result
Addition (+)	=B4+D4	Adds the values in B4 and D4	10+2=12
Subtraction (-)	=B4-D4	Subtracts the value in D4 from the value in B4	10-2=8
Division (/)	=B4/D4	Divides the value in B4 by the value in D4	10/2=5
Multiplication (*)	=B4*D4	Multiplies the values in B4 and D4	10*2=20
Percent (%)	=B4*8%	Calculates 8% of the value in B4	10*8%=.8

4. Save the file. Choose **Save As** from the File menu. Navigate to the *DigiTools your name\Chapter 8* folder. Key *Activity 8-1* in the File name box. **Microsoft Excel Workbook (*.xls)** should appear in the Save As type box. Click **Save**.

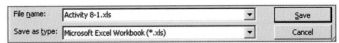

5. You need to change the width for Column A to display all the contents. Click on the right border of the column heading and drag to the right to change the column width. Make the column wide enough to display the longest entry in the column.

Click and drag a column heading border to resize a column
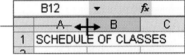

6. Practice setting an exact column width. Go to Cell B1. Choose **Column** from the Format menu. Click **Width**. Key 12 in the Column width box. Click **OK**.

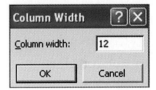

7. Use this same procedure to change the width of Column A to 20.

8. Center-align the column heads for an attractive table. Click in a cell that contains a column head such as Cell A3, Class. Click the **Center** button on the Formatting toolbar. Click in Cell B3 and drag to the right to select the range of Cells B3:D3. Click the **Center** button on the Formatting toolbar. All the column heads should now be centered.

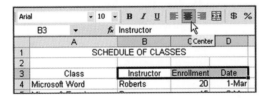

HELP KEYWORDS

Merge
Merge or split cells or data

9. Center the worksheet title, SCHEDULE OF CLASSES, over the data in the worksheet. Click in Cell A1 and drag to the right to select Cells A1:D1. Click the **Merge and Center** button on the Formatting toolbar. Your worksheet should look like Figure 8-4.

Figure 8-4
Worksheet with Centered Column Heads

	A	B	C	D
1	SCHEDULE OF CLASSES			
2				
3	Class	Instructor	Enrollment	Date
4	Microsoft Word	Roberts	20	1-Mar
5	Microsoft Excel	Perez	15	8-Mar
6	Microsoft Access	O'Malley	12	15-Mar
7	Microsoft PowerPoint	Chen	16	1-Mar
8	Microsoft Outlook	Goldberg	10	8-Mar
9	Microsoft FrontPage	Mangano	15	15-Mar

3. Make Cell A10 the active cell. Click the **Paste** button on the Standard toolbar. The class name and instructor are copied to Row 10. Press the **Escape** key to remove the blinking border from the copied selection.

4. In Cell C10, enter `14` for the enrollment. In Cell D10, enter `March 22` for the date.

5. Save the worksheet as *Activity 8-2* in the *DigiTools your name\Chapter 8* folder.

6. You need to make a copy of the sheet for March classes and edit it to include data for April. To copy Sheet 1, choose **Move or Copy Sheet** from the Edit menu. On the Move or Copy dialog box, check the **Create a copy** option. Click **OK**.

7. A copy of the worksheet named Sheet 1 (2) now appears. Click the **Sheet 1** tab and look at the data. Click the **Sheet 1 (2)** tab and look at the data. Note that the same data appears on both sheets.

> **DigiTip**
>
> When you copy data from one worksheet to another using the Copy command, any changes you have made to the column widths do not copy to the new worksheet. Use Move or Copy Sheet when you want to copy all data and the worksheet format.

8. Rename the sheets to make them easier to use. Right-click the **Sheet 1 (2)** tab. Choose **Rename** from the menu. Enter `March` for the sheet name and press **Enter**.

9. Right-click the **Sheet 1** tab. Choose **Rename** from the menu. Enter `April` for the sheet name and press **Enter**. Now you can easily identify which sheet has data for March or April.

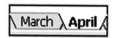

10. On the April worksheet, edit the data to show the April classes. Select the data in Cells A10-D10. Press the **Delete** key to delete the data. Change the enrollment and dates in Rows 4–9 to those shown below. To change the data, click in a cell and begin keying the new data. The old data will automatically be deleted and the new data inserted when you press **Enter** or **Tab**.

	A	B	C	D
3	Class	Instructor	Enrollment	Date
4	Microsoft Word	Roberts	21	April 5
5	Microsoft Excel	Perez	16	April 12
6	Microsoft Access	O'Malley	14	April 19
7	Microsoft PowerPoint	Chen	15	April 12
8	Microsoft Outlook	Goldberg	8	April 19
9	Microsoft FrontPage	Mangano	20	April 5

11. Click the **March** sheet tab to make it the active worksheet. In Cell B7, change the name of the instructor from Cheng to `Yung`.

12. Spell-check the worksheets. By default, *Excel* checks the active sheet. To check all sheets that have data, you must first select the sheets. Click in Cell A1 on the March sheet. Hold down the **Ctrl** key and click the **April** sheet tab. Both March and April sheets are now selected. Click the **Spelling** button on the Standard toolbar. Choose an appropriate option for each possible error. Click **OK** to close the dialog box when the check is complete.

10. Click the **Save** button on the Standard toolbar to save the file using the same name.

11. Click the **Print Preview** button on the Standard Toolbar to preview the worksheet. Click the **Zoom** button on the Print Preview toolbar to view the text in a larger size. Proofread your worksheet table carefully. If you find errors, click the **Close** button to return to the worksheet. Correct the errors and save again.

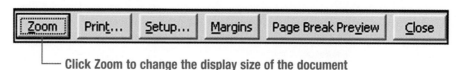
Click Zoom to change the display size of the document

> **DigiTip**
> In a later activity, you will learn to center the worksheet table on the page.

12. To print the worksheet table, click the **Print** button on the Print Preview toolbar (or on the Standard toolbar if you have closed Print Preview).

13. Close the workbook by clicking the **Close** button on the workbook window or by choosing **Close** from the File menu. Close *Excel* by choosing **Exit** from the File menu.

CUT, COPY, AND PASTE COMMANDS

The Cut, Copy, and Paste commands are useful for editing data in a worksheet. Data in a worksheet can be cut or copied and pasted to a new location. When you cut data, the data is removed from the original location and pasted to the new location. When you copy data, the data remains in the original location and is pasted to the new location. Data can be pasted to another location in the worksheet, to another worksheet in the workbook, or to another program.

> **DigiTip**
> Use the Cut and Paste commands when you want to move data to another location. Use the **Delete** key to remove data from the worksheet.

- To cut data, select the data. (Click and drag over the cells.) Click the **Cut** button on the Standard toolbar or choose **Cut** from the Edit menu.

- To copy data, select the data. Click the **Copy** button on the Standard toolbar or choose **Copy** from the Edit menu.

- To paste data, move to the location where you want the data to appear. Click in a cell to make it active. Click the **Paste** button on the Standard toolbar or choose **Paste** from the Edit menu.

Cut Copy Paste

WORKSHEET TABS

> **HELP KEYWORDS**
> Worksheet
> Rename a sheet

By default, worksheets are named Sheet 1, Sheet 2, and so on. The name of the active worksheet appears in darker type than the other worksheets. You can change a worksheet name to a name that is more meaningful and describes the data the worksheet contains. To change the name of the active worksheet, right-click on the sheet name. Choose **Rename**. Key a new name for the sheet and press **Enter**.

Right-click on a sheet name and choose Rename

The name of the active sheet appears in dark type

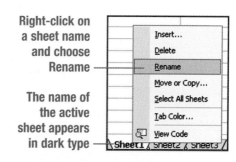

SPELL CHECK

HELP KEYWORDS
Check
  Check spelling

You can use *Excel's* Spelling feature to help you locate and correct spelling errors. To check spelling in a worksheet, choose **Spelling** from the Tools menu or click the **Spelling** button on the Standard toolbar. The Spelling dialog box will display.

Figure 8-5
Spelling Dialog Box

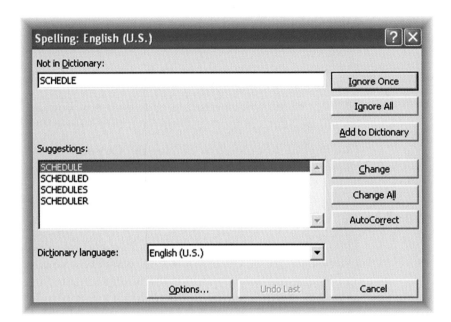

Excel will highlight possible spelling errors. For each instance, click an appropriate option:

- If the word is correct, click **Ignore Once** or **Ignore All**.
- If the word is not correct, key the correct word or choose from the list of suggested words. Then click **Change**.
- If you want to add this word to *Excel's* dictionary, click **Add to Dictionary**.
- Click **Change All** to make the change every time the word appears in the document.

Click **OK** to close the box that tells you the spelling check is complete.

ACTIVITY 8-2

EDIT A WORKBOOK

The business teachers at your local high school have decided to offer software classes in April as well as in March. You will edit the workbook to include new information for the April classes and updates for the March classes.

1. Start *Excel*. Open the file *Activity 8-1* that you created earlier.

2. Add data for an additional *Microsoft Word* class to be offered in March. Click and drag to select Cells A4:B4. Click the **Copy** button on the Standard toolbar.

Reference Guide

PROOFREADERS' MARKS
Proofreaders' marks are used to mark corrections in keyed or printed text that contains problems and/or errors. As a keyboard user, you should be able to read these marks accurately when revising or editing a rough draft. You also should be able to write these symbols to correct the rough drafts that you and others key. The most-used proofreaders' marks are shown below.

Mark	Meaning
∥	Align copy; also, make these items parallel
¶	Begin a new paragraph
Cap ≡	Capitalize
⌒	Close up
ℓ	Delete
⌒#	Delete space
No ¶	Do not begin a new paragraph
∧	Insert
⌃	Insert comma
⊙	Insert period
⌄⌄	Insert quotation marks
#>	Insert space
⌄	Insert apostrophe
stet	Let it stand; ignore correction
lc	Lowercase
⊔	Move down; lower
⊏	Move left
⊐	Move right
⊓	Move up; raise
sp	Spell out
∽ tr	Transpose
—	Underline or italic

E-MAIL FORMAT AND SOFTWARE FEATURES
E-mail format varies slightly, depending on the software used to create and send it.

E-mail Heading
Most e-mail software includes these features:
Attachment: line for attaching files to an e-mail message
Bcc: line for sending copy of a message to someone without the receiver knowing
Cc: line for sending copy of a message to additional receivers
Date: month, day, and year message is sent; often includes precise time of transmittal; usually is inserted automatically
From: name and/or e-mail address of sender; usually is inserted automatically
Subject: line for very brief description of message content
To: line for name and/or e-mail address of receiver

E-mail Body
The message box on the e-mail screen may contain these elements or only the message paragraphs (SS with DS between paragraphs).
- Informal salutation and/or receiver's name (a DS above the message)
- Informal closing (e.g., "Regards," "Thanks") and/or the sender's name (a DS below the message). Additional identification (e.g., telephone number) may be included.

Special E-mail Features
Several e-mail features make communicating through e-mail fast and efficient.
Address list/book: collection of names and e-mail addresses of correspondents from which an address can be entered on the To: line by selecting it, instead of keying it.
Distribution list: series of names and/or e-mail addresses, separated by commas, on the To: line.
Forward: feature that allows an e-mail user to send a copy of a received e-mail message to others.
Recipient list (Group): feature that allows an e-mail user to send mail to a group of recipients by selecting the name of the group (e.g., All Teachers).
Reply: feature used to respond to an incoming message.
Reply all: feature used to respond to all copy recipients as well as the sender of an incoming message.
Signature: feature for storing and inserting the closing lines of messages (e.g., informal closing, sender's name, telephone number, address, fax number).

Figure E-38
Use *OneNote's* Find feature to locate text in notes.

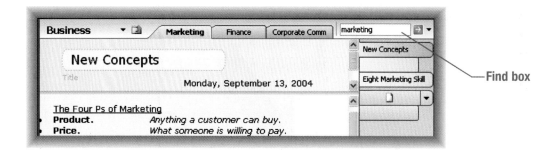

3. *OneNote* finds and highlights the first instance of the word **Marketing**. Click or tap the **Forward** and **Back** arrows to move to other instances of the word. See Figure E-39.

Figure E-39
Find Results and View List

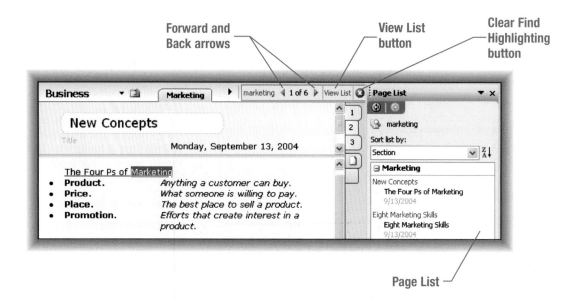

4. Click the **View List** button. This will display a list of the pages that contain the word *Marketing*. Click each page on the list to view the words. Click the **Clear Find Highlighting** button to remove the highlighting. Close *OneNote*.

ENVELOPE GUIDES

Return Address
Use block style, SS, and Initial Caps or ALL CAPS. If not using the Envelopes feature, begin as near to the top and left edge of the envelope as possible—TM and LM about 0.25".

Receiver's Delivery Address
Use USPS (postal service) style: block format (SS), ALL CAPS, no punctuation. Place city name, two-letter state abbreviation, and ZIP Code +4 on last address line. One space precedes the ZIP Code.

If not using the Envelopes feature, tab over 2.5" for the small envelope and 4" for the large envelope. Insert hard returns to place the first line about 2" from the top.

Mailing Notations
Key mailing and addressee notations in ALL CAPS.

Key mailing notations, such as SPECIAL DELIVERY and REGISTERED, below the stamp and at least three lines above the envelope address.

Key addressee notations, such as HOLD FOR ARRIVAL or PERSONAL, a DS below the return address and about three spaces from the left edge of the envelope.

If an attention line is used, key it as the first line of the envelope address.

Standard Abbreviations
Use USPS standard abbreviations for states (see list below) and street suffix names, such as AVE and BLVD. Never abbreviate the name of a city or country.

International Addresses
Omit postal (ZIP) codes from the last line of addresses outside the U.S. Show only the name of the country on the last line. Examples:

```
MR HIRAM SANDERS
2121 CLEARWATER ST
OTTAWA ONKIA OB1
CANADA

MS INGE D FISCHER
HARTMANNSTRASSE 7
4209 BONN 5
FEDERAL REPUBLIC OF GERMANY
```

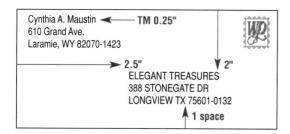

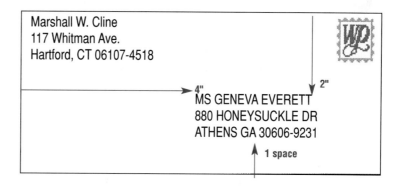

Folding Procedures

Small Envelopes (Nos. 6 3/4, 6 1/4)
1. With page face up, fold bottom up to 0.5" from top.
2. Fold right third to left.
3. Fold left third to 0.5" from last crease.
4. Insert last creased edge first.

Large Envelopes (Nos. 10, 9, 7 3/4)
1. With page face up, fold slightly less than one-third of sheet up toward top.
2. Fold down top of sheet to within 0.5" of bottom fold.
3. Insert last creased edge first.

Window Envelopes (Letter)
1. With page face down, top toward you, fold upper third down.
2. Fold lower third up so address is showing.
3. Insert sheet into envelope with last crease at bottom.
4. Check that address shows through window.

6. Select the title **Style Guide** and the first paragraph of text under the title. Choose **Edit**, **Copy** from the menu bar. Click in the *OneNote* page. Choose **Edit**, **Paste** from the menu bar. The text should appear in *OneNote* as shown in Figure E-37.

7. Close *Internet Explorer*. Maximize your *OneNote* window. Clean up your notes. For example, move the containers underneath each other as shown in Figure E-37. Resize and reposition the graphics and text so they fit comfortably on your *OneNote* page.

> **HELP KEYWORDS**
> Insert document
> Insert a document or file

8. Practice inserting a document as a picture in *OneNote*. Click **Insert** on the menu bar. Choose **Document as Picture**. Navigate to your *Corporate View* data files. Select the *Plante Bio* file and click **Insert**. The document will be inserted as a graphic on a new page.

9. Practice inserting a picture into *OneNote* from a file. On the Plante Bio page, click **Insert** on the menu bar. Choose **Picture**, **From File**. Navigate to your Corporate View data files. Select the file *CV_Banner* and click **Insert**.

10. Arrange the elements on the Plante Bio page. Place the Corporate View graphic at the top of the page. Place the document graphic under it. Place the note with the source information about the document under the document graphic.

> **DigiTip**
> If you use the Copy and Paste commands to copy online material, the corresponding Web address is automatically inserted below the text you copy. This feature is very helpful if you need to revisit a site or must prepare references for a report.

SEARCHING NOTES

> **HELP KEYWORDS**
> Find
> Tips for searching notes

A convenient way to search notes is by using *OneNote's* **Find** feature. This feature will search both typed and handwritten notes, provided you have written them clearly. Learn how to search for text in your notes as you complete the next activity.

ACTIVITY E-19   SEARCH NOTES

1. Open *OneNote* if it is not already open. Open your **Business** folder.

2. Choose **Edit**, **Find** from the menu bar. A Find text box will open as shown in Figure E-38. Key `Marketing` in the text box and click the arrow beside the box.

State and Territory Abbreviations

Alabama	AL
Alaska	AK
Arizona	AZ
Arkansas	AR
California	CA
Colorado	CO
Connecticut	CT
Delaware	DE
District of Columbia	DC
Florida	FL
Georgia	GA
Guam	GU
Hawaii	HI
Idaho	ID
Illinois	IL
Indiana	IN
Iowa	IA
Kansas	KS
Kentucky	KY
Louisiana	LA
Maine	ME
Maryland	MD
Massachusetts	MA
Michigan	MI
Minnesota	MN
Mississippi	MS
Missouri	MO
Montana	MT
Nebraska	NE
Nevada	NV
New Hampshire	NH
New Jersey	NJ
New Mexico	NM
New York	NY
North Carolina	NC
North Dakota	ND
Ohio	OH
Oklahoma	OK
Oregon	OR
Pennsylvania	PA
Puerto Rico	PR
Rhode Island	RI
South Carolina	SC
South Dakota	SD
Tennessee	TN
Texas	TX
Utah	UT
Vermont	VT
Virgin Islands	VI
Virginia	VA
Washington	WA
West Virginia	WV
Wisconsin	WI
Wyoming	WY

DOWNLOADING TO ONENOTE

OneNote allows you to download information from the Internet easily. You can download pictures, graphics, charts, and data of all kinds from the Web and from other applications by using the Copy and Paste commands. You can also drag selected items from one program window to another.

OneNote allows you to insert a picture from a file into a note and insert documents from other applications as graphics. In this exercise, you will download graphics and text from the Corporate View intranet site using both the drag and drop and the Copy and Paste methods. You will also learn to insert pictures and documents as graphics.

DigiTip
Before continuing, complete Activities 2-1 and 2-2 in your *DigiTools* textbook, pp. 36 and 38.

ACTIVITY E-18 — DOWNLOAD TEXT AND GRAPHICS

1. Open *OneNote* and position the window so that it is on the right-hand side of your screen. Size the window so that it only covers from one half to three quarters of the screen.

2. Open your **Business**, **Corporate Communications**, **New Concepts** page.

3. Open *Internet Explorer*. Position *Internet Explorer* on the left-hand side of your screen. Size the window so that it only occupies about one-half to three-quarters of the screen. *OneNote* and *Internet Explorer* can overlap a bit.

4. Open the Corporate View intranet. (See Activities 2-1 and 2-2 in your *DigiTools* textbook.) Choose **Corporate Communications** followed by **Style Guide**.

5. Select the Corporate Communications graphic at the top of the intranet page and drag it from the *Internet Explorer* window to the *OneNote* page. See Figure E-37.

Figure E-37
Copy and paste graphics and text from one application to another.

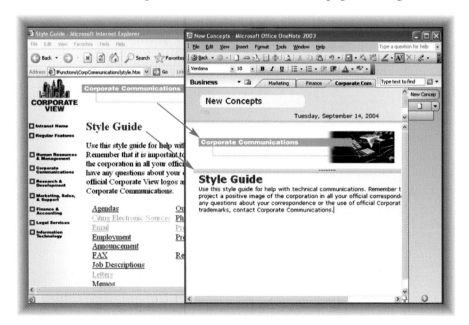

REPORT DOCUMENTATION

Good report writing includes proof that the reported statements are sound. The process is called **documenting.**

Most school reports are documented in the body and in a list. A reference in the body shows the source of a quotation or paraphrase. A list shows all references alphabetically.

In the report body, references may be noted (1) in parentheses in the copy (textual citations or parenthetical documentation); (2) by a superscript in the copy, listed on a separate page (endnotes); or (3) by a superscript in the copy, listed at the bottom of the text page (footnotes). A list may contain only the sources noted in the body (REFERENCES or Works Cited) or include related materials (BIBLIOGRAPHY).

Two popular documenting styles are shown: *Century 21* and MLA (Modern Language Association).

Century 21
Examples are listed in this order: (1) textual citation, (2) endnote/footnote, and (3) References/Bibliography page.

Book, One Author
(Schaeffer, 1997, 1)
   [1]Robert K. Schaeffer, Understanding Globalization, (Lanham, MD: Rowman & Littlefield Publishers, Inc., 1997), p. 1.
Schaeffer, Robert K. Understanding Globalization (Lanham, MD: Rowman & Littlefield Publishers, Inc., 1997).

Book, Two or Three Authors
(Prince and Jackson, 1997, 35)
   [2]Nancy Prince and Jeanie Jackson, Exploring Theater (Minneapolis/St. Paul: West Publishing Company, 1997), p. 35.
Prince, Nancy, and Jeanie Jackson. Exploring Theater. Minneapolis/St. Paul: West Publishing Company, 1997.

Book, Four or More Authors
(Gerver, et al., 1998, 9)
   [3]Robert Gerver, et al., South-Western Geometry: An Integrated Approach (Cincinnati: South-Western Educational Publishing, 1998), p. 9.
Gerver, Robert, et al. South-Western Geometry: An Integrated Approach. Cincinnati: South-Western Educational Publishing, 1998.

Encyclopedia or Reference Book
(Encyclopedia Americana, 1998, Vol. 25, p. 637)
   [4]Encyclopedia Americana, Vol. 25 (Danbury, CT: Grolier Incorporated, 1998), p. 637.
Encyclopedia Americana, Vol. 25. "Statue of Liberty." Danbury, CT: Grolier Incorporated, 1998.

Journal or Magazine Article
(Harris, 1993, 755)
   [5]Richard G. Harris, "Globalization, Trade, and Income," Canadian Journal of Economics, November 1993, p. 755.
Harris, Richard G. "Globalization, Trade, and Income." Canadian Journal of Economics, November 1993, 755–776.

Web Site
(Railton, 1999)
   [6]Stephen Railton, "Your Mark Twain," www.etext.lib.virginia.edu/railton/sc_as_mt/yourmt13.html (September 24, 1999).
Railton, Stephen. "Your Mark Twain." www.etext.lib.virginia.edu/railton/sc_as_mt/yourmt13.html (24 September 1999).

Modern Language Association
Examples include reference (1) in parenthetical documentation and (2) on Works Cited page.

Book, One Author
(Schaeffer 1)
Schaeffer, Robert K. Understanding Globalization. Lanham, MD: Rowman & Littlefield, 1997.

Book, Two or Three Authors
(Prince and Jackson 35)
Prince, Nancy, and Jeanie Jackson. Exploring Theater. Minneapolis/St. Paul: West Publishing, 1997.

Book, Four or More Authors or Editors
(Gerver et al. 9)
Gerver, Robert, et al. South-Western Geometry: An Integrated Approach. Cincinnati: South-Western, 1998.

Encyclopedia or Reference Book
(Encyclopedia Americana 637)
Encyclopedia Americana. "Statue of Liberty." Danbury, CT: Grolier, 1998.

Journal or Magazine Article
(Harris 755)
Harris, Richard G. "Globalization, Trade, and Income," Canadian Journal of Economics. Nov. 1993: 755–776.

Web Site
(Railton)
Railton, Stephen. Your Mark Twain Page. (24 Sept. 1999) www.etext.lib.virginia.edu/railton/sc_as_mt/yourmt13.html.

2. Using the ink tools found on the toolbar at the bottom of the side note window, write the following text. Click or tap the **Scroll Down by Half Page** button if you run out of room to write.

```
Market.  Any group, large or small, that may purchase a
product.

Marketplace.  Any location or space where business
transactions take place.
```

3. Click and drag the right border of the *OneNote* window to make it larger. The full *OneNote* window will appear. You can also click the **Maximize** button to open the full window. Now you can see the various folders and sections.

4. Choose the **My Notebook** down arrow to see where side notes are stored in relation to the other folders and sections. See Figure E-36. Click off the list to close it.

Figure E-36
Side Notes in the My Notebook List

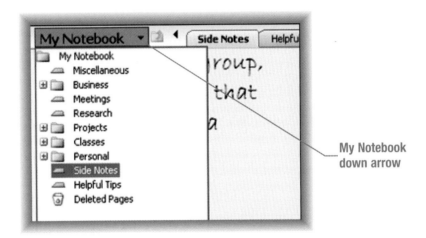

5. Select all the text you have written in your side note. Choose **Edit, Copy** from the menu bar.

6. Go to your **Business** folder. Open your **Marketing** section and choose your **New Concepts** page. Click below the last note on the page. Choose **Edit, Paste** from the menu bar. You have now placed information from your side note into another page.

> **DigiTip**
> If you have a Tablet PC, you can select any text you have handwritten and convert it into typed text. To do so, select the words and then choose **Tools, Convert Handwriting to Text**.

7. Go back to your side note window. Select all the text you have written in your side note. Choose **Edit, Copy** from the menu bar.

8. Open a new blank document in *Word*. Choose **Edit, Paste** from the menu bar. You have now placed information from your side note into a *Word* file. Close the *Word* file without saving. Close *OneNote*.

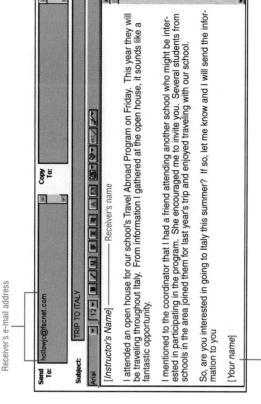

E-mail Message

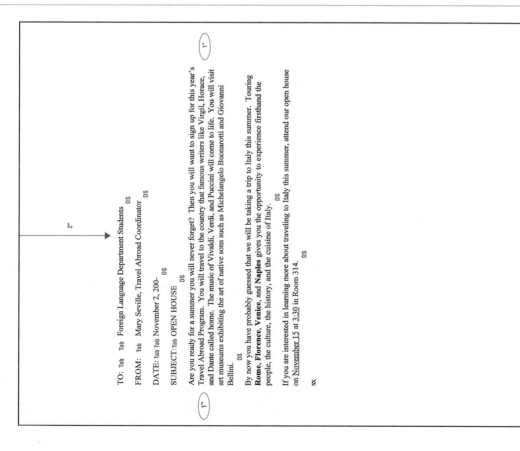

Interoffice Memo

Figure E-34
Subpage with Drawing of the Marketing Circle of Success

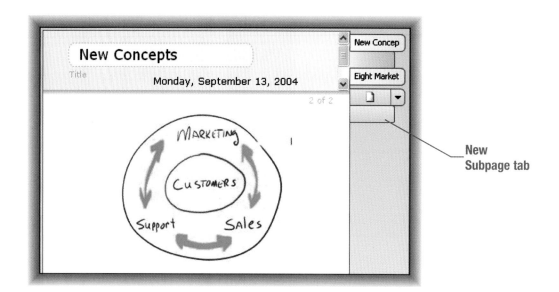

DigiTip
When drawing, you may want to remove the rule lines from the page.

USING SIDE NOTES

HELP KEYWORDS
Side note
  About side notes

You will often find that you need to write a note quickly while you are working in another program. The best way to do this is in a side note. A **side note** is simply *OneNote* running in a small window. You can open a side note by clicking the Open New Side Note icon on the taskbar.

After you take notes in a side note, you may want to copy the information into other folders or sections. You can do this easily using the Copy and Paste commands. In the next activity, you will create a side note and then copy notes from a side note to another page.

ACTIVITY E-17   CREATE A SIDE NOTE

1. Open a side note by clicking the **Open New Side Note** icon on your taskbar as shown in Figure E-35. (If you can't find the icon on the taskbar, open *OneNote*. Choose **Window** from the menu bar. Select **New Side Note Window**.)

Figure E-35
Side Note Icon and Window

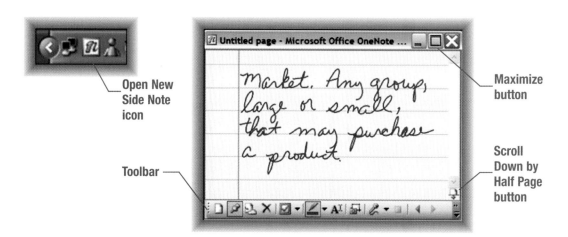

APPENDIX E: MICROSOFT *ONENOTE*

Personal-Business Letter

Return address: 230 Glendale Ct.
Brooklyn, NY 11234-3721
DS
Date: February 15, 200-
QS

Letter address: Ms. Julie Hutchinson
1825 Melbourne Ave.
Flushing, NY 11367-2351
DS

Salutation: Dear Julie
DS

Body: It seems like years since we were in Ms. Gerbig's keyboarding class. Now I wish I would have paid more attention. As I indicated on the phone, I am applying for a position as box office coordinator for one of the theaters on Broadway. Of course, I know the importance of having my letter of application and resume formatted correctly, but I'm not sure that I remember how to do it.

Since you just completed your business education degree, I knew where to get the help I needed. Thanks for agreeing to look over my application documents; they are enclosed. Also, if you have any suggestions for changes to the content, please share those with me too. This job is so important to me; it's the one I really want.

Thanks again for agreeing to help. If I get the job, I'll take you out to one of New York's finest restaurants.
DS

Complimentary close: Sincerely
QS

Writer: Rebecca Dunworthy
DS

Enclosure notation: Enclosures

(1" LM, 1" RM, 2" top margin)

Business Letter in Block Format

Wyoming Women's Historical Museum Foundation
P.O. Box 900 ★ Cody, Wyoming 82414-6392 ★ (307) 555-6115 ★ museum@wyoming.com

February 20, 200-

MR AND MRS ERIC RUSSELL
PO BOX 215
MOORCROFT WY 82721-2342

Dear Mr. and Mrs. Russell

Wyoming women were the first women in the United States to have the right to vote (1869). Ester Morris of South Pass City became the first woman judge in 1870. Wyoming was the first state to elect a woman to state office when Estelle Reel was elected State Superintendent of Public Instruction in 1894. Nellie Tayloe Ross became the first female governor in the United States when she was elected governor of Wyoming in 1925.

It is time to honor women such as these for the role they played in shaping Wyoming and U.S. history. A Wyoming Women's Historical Museum is being planned. With your help, the museum can become a reality.

Our community would benefit from the increased tourist activity. Thousands of tourists visit the nation's first national monument, Devil's Tower, each year. Since Moorcroft is only 30 miles from Devil's Tower, a museum would draw many of them to our city as they travel to and from the Tower.

National and state funds for the project are being solicited; however, additional funding from the private sector will be required. Please look over the enclosed brochure and join the Wyoming Women's Historical Museum Foundation by making a contribution.

Sincerely

William P. Shea

xx

Enclosure

Figure E-33
Inking Tools

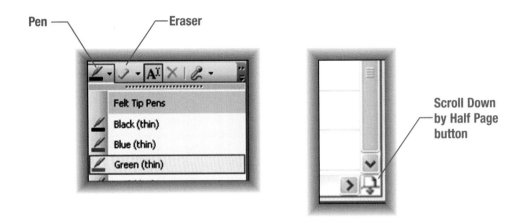

3. Below the existing note, use cursive or printed writing to write the text below. If you make a mistake, select the **Eraser** button. Then move the eraser over the area of the note you want to remove. Select the **Eraser** tool again to turn it off or press **Esc**.

```
The Marketing Circle of Success
1.   Bring customer-oriented products to market.
2.   Sell products to customers effectively.
3.   Support products after the sale.
```

4. If needed, add extra paper at the bottom of your page by clicking the **Scroll Down by Half Page** button at the bottom of the vertical scroll bar. See Figure E-33.

5. Close *OneNote* or continue to the next activity.

CREATING SUBPAGES FOR DRAWINGS

A **subpage** is a page added underneath a primary page. Subpages are used to avoid creating endlessly long pages for any single subject. For example, the **New Concepts** page in the **Marketing** section is getting a bit long. Perhaps it would be better to add a subpage when you want to add more information. In the next activity, you will create a subpage to hold a drawing you will create.

ACTIVITY E-16   CREATE A DRAWING ON A SUBPAGE

1. Open *OneNote*. Choose the **Business** folder, the **Marketing** section, and the **New Concepts** page.

2. Click or tap the **New Subpage** button as shown in Figure E-34.

3. Using various pen widths, colors, and styles, sketch the diagram shown in Figure E-34 on your subpage. Use the **Eraser** tool to correct mistakes. Close *OneNote* when you have finished the drawing.

2" TM

Title

SAMUEL CLEMENS

DS

"Mark Twain"

QS

Report body

Samuel Clemens was one of America's most renowned authors. The colorful life he led was the basis for his writing. Although his formal education ended when he was 12 years old with the death of his father, his varied career interests provided an informal education that was not unlike many others of his generation. Clemens brought these rich experiences to life in his writing.

1" LM 1" RM

DS

Sam Clemens was recognized for his fiction as well as for his humor. It has been said that, " . . . next to sunshine and fresh air Mark Twain's humor has done more for the welfare of mankind than any other agency." (Railton, "Your Mark Twain," 1999) By cleverly weaving fiction and humor, he developed many literary masterpieces. Some say his greatest masterpiece was "Mark Twain," a pen name (pseudonym) Clemens first used in the Nevada Territory in 1863. This fictitious name became a kind of mythic hero to the American public. (Railton, "Sam Clemens as Mark Twain," 1999) Some of his masterpieces that are among his most widely read books are *The Adventures of Tom Sawyer* and *Adventures of Huckleberry Finn.*

Textual citation

Textual citation

DS

Side heading

The Adventures of Tom Sawyer

DS

The Adventures of Tom Sawyer was first published in 1876. Such characters as Tom Sawyer, Aunt Polly, Becky Thatcher, and Huck Finn have captured the attention of readers for generations. Boys and girls, young and old, enjoy Tom Sawyer's mischievousness. Who can

Unbound Report (p. 1) with Textual Citations

Page number

2

forget how Tom shared the privilege of whitewashing Aunt Polly's fence? What child isn't fascinated by the episode of Tom and Becky lost in the cave?

DS

Side heading

Adventures of Huckleberry Finn

DS

Adventures of Huckleberry Finn was first published in 1885. Many of the characters included in *The Adventures of Tom Sawyer* surface again in *Huckleberry Finn.* Children are able to live vicariously through Huck. What child hasn't dreamed of sneaking out of the house at night and running away to live a lifestyle of their own making?

1" LM 1" RM

QS

REFERENCES

QS

List of references

Railton, Stephen. "Your Mark Twain." http://etext.lib.virginia.edu/railton/sc_as_mt/yourmt13.html (24 September 1999).

DS

Railton, Stephen. "Sam Clemens as Mark Twain." http://etext.lib.virginia.edu/railton/sc_as_mt/cathompg.html (24 September 1999).

Unbound Report (p. 2) with References

Note: References are often placed on a separate page.

LESSON 5: USING INKING, DOWNLOADING, AND SEARCHING TOOLS

OBJECTIVES *In this lesson you will:*

1. Take notes using digital pen and ink.
2. Create subpages.
3. Create drawings and insert a picture from a file.
4. Create side notes and copy text to other pages.
5. Copy from *OneNote* to a *Word* document.
6. Download graphics and text from Web pages.
7. Insert a document from another application.
8. Use *OneNote's* Find feature to search for text.

INKING NOTES

OneNote allows you to use your handwriting and **digital ink** to take notes and draw sketches. In this lesson, you will update your marketing notes using various colors of ink. If you're using a traditional desktop or laptop PC, you can use a digital handwriting tablet for this exercise. If you do not have a digital tablet, you can use your mouse or key the information. Using a mouse for inking can be somewhat awkward and isn't recommended for a long period of time. If you're using a Tablet PC, you are in your element. Tablet PCs are made for inking.

ACTIVITY E-15 — INK NOTES USING THE PEN TOOLS

HELP KEYWORDS

Ink
Show or hide drawing and writing tools

1. Open *OneNote*. Choose the **Business** folder, the **Marketing** section, and the **New Concepts** page.

DigiTip
Prepare your page for ink note-taking by showing the rule lines.

2. Select the **Pen** button's down arrow as shown in Figure E-33. Choose one of the thin felt tip pen colors such as **Green**.

INFORMATION SUPERHIGHWAY

Technology has a significant impact on our lives and will have an even greater impact in the future. During the early 1990s the term "Information Highway" was starting to be used to describe the next wave of technological advancements. As always there were those who were skeptical about what impact, if any, the "Information Highway" would have on our lives.

One writer, as late as December 1993, indicated that he was not holding his breath. The information superhighway may or may not become a truly transforming technology. He felt that if it did, it may take years and compared the acceptance of the information superhighway to the acceptance of the car.

It takes time for breakthrough technologies to make their mark. Consider the car. In 1908 Henry Ford began selling the Model T. One early effort of low-cost cars was to rid cities of horses. A picture of a New York street in 1900 shows 36 horse carriages and one car; a picture of the same street in 1924 shows 40 cars and one carriage. This was a big deal. In 1900, horses dumped 2.5 million pounds of manure onto New York streets every day. Still, the car culture's triumph was slow.[1]

Other writers during this same time period were much more optimistic about the value of the superhighway and began predicting what it would mean to all of us in the near future.

The Information Superhighway is going to affect your life, whether you want it to or not. In the very near future you will talk to your friends and family, send letters, go shopping, get news, find answer to your questions. . . .[2]

ENDNOTES

[1] Robert J. Samuelson, "Lost on the Information Highway," Newsweek, (December 20, 1993), p. 111.

[2] Laurence A. Canter and Martha S. Siegel, How to Make a FORTUNE on the Information Superhighway (New York: Harper-Collins Publishers, Inc., 1994) p. 1.

[3] "AOL Poised for Record Subscriber Growth," http://dailynews.yahoo.com/h/nm/19990921/wr/aol_3.html. (21 September 1999).

ACTIVITY E-14 — APPLY BOLD, ITALIC, AND UNDERLINE

1. Open *OneNote* if it is not already open. Open your **Business** folder. Select the **Marketing** section tab and the **New Concepts** page.

2. Select the word **Product**. Click the **Bold** button to apply bold to this word. Apply bold to the words **Price**, **Place**, and **Promotion**.

3. Select the text **Anything a customer can buy**. Click the **Italic** button to apply italic to this text. Apply italic to the definitions for the other three words.

4. Select the text **The Four Ps of Marketing**. Click the **Underline** button to underline the words.

5. Close *OneNote* or continue to the next activity.

Peer Check Move to a classmate's computer and check his or her work on the screen. Ask your classmate to check your work. Discuss any differences and make corrections, if needed.

APPLICATION E-4 — APPLY APPEARANCE CHANGES TO THE MARKETING SECTION

1. Open *OneNote* if it is not already open. Open your **Business** folder. Select the **Marketing** section tab and the **Eight Marketing Skills** page.

2. Apply bold and italic to the list of eight marketing skills in the note as you did in Activity E-14. (Use bold for the skills and italic for the definitions.)

3. Select all the text in the note. Change the font to Arial, 14-point size, and blue color.

4. Change the Marketing section color to a color of your choice.

5. Select the words **Marketing Managers** in the first sentence. Highlight the words in yellow.

6. If you have created additional pages, sections, or folders for practice that you do not need, delete them now. Close *OneNote*.

Print Resume

Douglas H. Ruckert

8503 Kirby Dr.
Houston, TX 77054-8220
(713) 555-0121
dougr@suresend.com

OBJECTIVE: To use my computer, Internet, communication, and interpersonal skills in a challenging customer service position.

EDUCATION: Will be graduated from Eisenhower Technical High School in June 2004, with a high school diploma and business technology emphasis. Grade point average is 3.75.

Relevant Skills and Courses:

- Proficient with most recent versions of Windows and Office, including Word, Excel, Access, PowerPoint, and FrontPage.
- Excelled in the following courses: Keyboarding, Computer Applications, Business Communication, and Office Technology.

Major Accomplishments:

- Future Business Leaders of America: Member for four years; vice president for one year. Won second place in Public Speaking at District Competition; competed (same event) at state level.
- Varsity soccer: Lettered three years and served as captain during senior year.
- Recognition: Named one of Eisenhower's Top Ten Community Service Providers at end of junior year.

WORK EXPERIENCE: Hinton's Family Restaurant, Server (2003-present): Served customers in culturally diverse area, oriented new part-time employees, and resolved routine customer service issues.

Tuma's Landscape and Garden Center, Sales (2002-2003): Assisted customers with plant selection and responsible for stocking and arranging display areas.

REFERENCES: Will be furnished upon request.

Electronic Resume

Douglas H. Ruckert
8503 Kirby Dr.
Houston, TX 77054-8220
(713) 555-0121
dougr@suresend.com

SUMMARY

Strong communication and telephone skills; excellent keyboarding, computer, and Internet skills; and good organizational and interpersonal skills.

EDUCATION

Will be graduated from Eisenhower Technical High School in June 2004, with a high school diploma and business technology emphasis. Grade point average is 3.75.

Relevant Skills and Courses:

Proficient with most recent versions of Windows and Office, including Word, Excel, Access, PowerPoint, and FrontPage.

Excelled in the following courses: Keyboarding, Computer Applications, Business Communication, and Office Technology.

Major Accomplishments:

Future Business Leaders of America: Member for four years; vice president for one year. Won second place in Public Speaking at District Competition; competed (same event) at state level.

Varsity soccer: Lettered three years and served as captain during senior year.

Recognition: Named one of Eisenhower's Top Ten Community Service Providers at end of junior year.

WORK EXPERIENCE

Hinton's Family Restaurant, Server (2003-present): Served customers in culturally diverse area; oriented new part-time employees and resolved routine customer service problems.

Tuma's Landscape and Garden Center, Sales (2002-2003): Assisted customers with plant selection and responsible for stocking and arranging display areas.

REFERENCES

Will be furnished upon request.

Figure E-31
Change Font Settings

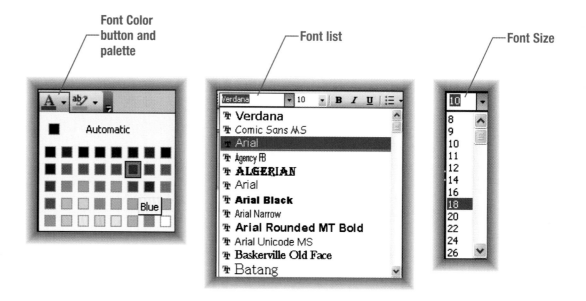

5. Change the font color for the six lines under **Financial Benefits** to **red**. Change the font color for the five lines under **Investments** to **green**.

> **DigiTip**
> After you make a color selection, that color will remain available on the Font Color button until you change it.

6. Select all of the lines in the note. Click the down arrow for the **Font** list as shown in Figure E-31. Scroll through the various font choices and select a font from the list. That font will then be applied to your notes.

7. Select all of the lines in the note. Click the down arrow for the **Font Size** as shown in Figure E-31. Scroll through the various font sizes. Choose a font size from the list, such as **14** or **18**. That size will then be applied to the text.

8. Close *OneNote* or continue to the next activity.

Applying Font Attributes

Sometimes, changing colors and fonts can actually make notes harder to read! Applying more traditional font attributes, such as **bold**, *italic*, and underline, can often be a better choice for emphasizing text. Buttons for these font attributes can be found on the Formatting toolbar as shown in Figure E-32.

Figure E-32
Bold, Italic, and Underline Buttons

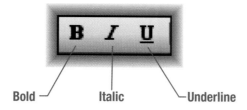

Employment Application Form

Application for Employment
Regency Insurance Company

An Equal Opportunity Employer

PERSONAL INFORMATION

NAME (LAST FIRST)	SOCIAL SECURITY NO.	CURRENT DATE	TELEPHONE NUMBER
Ruckert, Douglas H.	368-56-2890	5/22/04	(713) 555-0121

ADDRESS (NUMBER, STREET, CITY, STATE, ZIP CODE)	U.S. CITIZEN	DATE YOU CAN START
8503 Kirby Dr., Houston, TX 77054-8220	☒ Yes ☐ No	6/08/04

ARE YOU EMPLOYED NOW?	IF YES, MAY WE INQUIRE OF YOUR PRESENT EMPLOYER?	IF YES, GIVE NAME AND NUMBER OF PERSON TO CALL
☒ Yes ☐ No	☒ Yes ☐ No	James Veloski, Manager (713) 555-0182

POSITION DESIRED	SALARY DESIRED	STATE HOW YOU LEARNED OF POSITION
Customer Service	Open	From Ms. Anne D. Salgado, Eisenhower Business Technology Teacher

HAVE YOU EVER BEEN CONVICTED OF A FELONY?
☐ Yes ☒ No   IF YES, EXPLAIN.

EDUCATION

	NAME AND LOCATION OF SCHOOL	YEARS ATTENDED	DID YOU GRADUATE?	SUBJECTS STUDIED
COLLEGE				
HIGH SCHOOL	Eisenhower Technical High School, Houston, TX	2000 to 2004	Will be graduated 6/04	Business Technology
GRADE SCHOOL				
OTHER				

SUBJECTS OF SPECIAL STUDY/RESEARCH WORK OR SPECIAL TRAINING/SKILLS DIRECTLY RELATED TO POSITION DESIRED

Windows and Office Suite, including Word, Excel, Access, PowerPoint, and FrontPage

Office technology course with telephone training and interpersonal skills role-playing

FORMER EMPLOYERS (LIST LAST POSITION FIRST)

FROM – TO (MTH & YEAR)	NAME AND ADDRESS	SALARY	POSITION	REASON FOR LEAVING
9/03 to present	Hinton's Family Restaurant, 2204 S. Wayside Ave., Houston, TX 77023-8841	$6.85/hr.	Server	Want full-time position in my field
6/02 to 9/03	Tuma's Landscape and Garden Center, 10155 East Hwy., Houston, TX 77029-4419	$5.75/hr.	Sales	Employed at Hinton's

REFERENCES (LIST THREE PERSONS NOT RELATED TO YOU, WHOM YOU HAVE KNOWN AT LEAST ONE YEAR)

NAME	BUSINESS ADDRESS	TELEPHONE NUMBER	TITLE	YEARS KNOWN
Ms. Anne D. Salgado	Eisenhower Technical High School, 100 W. Cavalcade, Houston, TX 77009-2451	(713) 555-0134	Business Technology Instructor	Four
Mr. James R. Veloski	Hinton's Family Restaurant, 2204 S. Wayside Ave., Houston, TX 77023-8841	(713) 555-0182	Manager	One
Mrs. Helen T. Landis	Tuma's Landscape and Garden Center, 10155 East Hwy., Houston, TX 77029-4419	(713) 555-0149	Owner	Three

I UNDERSTAND THAT I SHALL NOT BECOME AN EMPLOYEE UNTIL I HAVE SIGNED AN EMPLOYMENT AGREEMENT WITH THE FINAL APPROVAL OF THE EMPLOYER AND THAT SUCH EMPLOYMENT WILL BE SUBJECT TO VERIFICATION OF PREVIOUS EMPLOYMENT DATA PROVIDED IN THIS APPLICATION, ANY RELATED DOCUMENTS, OR DATA SHEET. I KNOW THAT A REPORT MAY BE MADE THAT WILL INCLUDE INFORMATION CONCERNING ANY FACTOR THE EMPLOYER MIGHT FIND RELEVANT TO THE POSITION FOR WHICH I AM APPLYING, AND THAT I CAN MAKE A WRITTEN REQUEST FOR ADDITIONAL INFORMATION AS TO THE NATURE AND SCOPE OF THE REPORT IF ONE IS MADE.

Douglas H. Ruckert
SIGNATURE OF APPLICANT

Employment Application Letter

8503 Kirby Dr.
Houston, TX 77054-8220
May 10, 2004

Ms. Jenna St. John
Personnel Director
Regency Insurance Company
219 West Greene Rd.
Houston, TX 77067-4219

Dear Ms. St. John:

Ms. Anne D. Salgado, my business technology instructor, informed me of the customer service position with your company that will be available June 15. She speaks very highly of your organization. After learning more about the position, I am confident that I am qualified and would like to be considered for the position.

Currently I am completing my senior year at Eisenhower Technical High School. All of my elective courses have been in computer and business-related courses. I have completed the advanced computer application class where we integrated word processing, spreadsheet, database, presentation, and Web page documents by using the latest suite software. I have also taken an office technology course that included practice in using the telephone and applying interpersonal skills.

My work experience and school activities have given me the opportunity to work with people to achieve group goals. Participating in FBLA has given me an appreciation of the business world.

The opportunity to interview with you for this position will be greatly appreciated. You can call me at (713) 555-0121 or e-mail me at dougr@suresend.com to arrange an interview.

Sincerely,

Douglas H. Ruckert

Enclosure

Highlighting Notes and Changing Fonts

At times it may prove helpful to **highlight**—surround in color—key portions of your notes. You can use *OneNote's* Highlight feature to add color to parts or all of a note.

You may also want to change the style or size of the text you use in notes. To change the style of the letters in a note, change the **Font** setting. To change the size of letters in a note, change the **Font Size** setting. Use the **Font Color** setting to change the color of words and letters. Buttons for these settings are found on the Formatting toolbar.

ACTIVITY E-13 — WORK WITH FONTS AND COLOR

1. Open *OneNote* if it is not already open. Open your **Business** folder. Open the **Finance** section and the **New Concepts** page.

2. Select **Departments**, the first major heading in the outline. Click the down arrow on the **Highlight** button. A color palette will appear as shown in Figure E-30. Choose a highlight color, such as yellow, from the palette. That color will be applied to the word *Departments*.

Figure E-30
Highlight Text in a Note

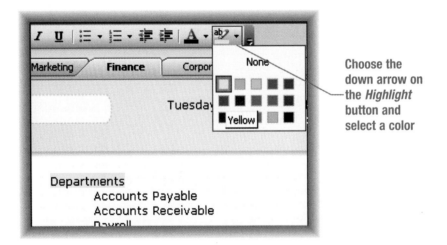

Choose the down arrow on the *Highlight* button and select a color

3. Apply a **turquoise** highlight color to the words **Financial Benefits**. Apply a **bright green** highlight color to the word **Investments**.

4. Select words beginning with **Accounts Payable** down through **Purchasing**. Click the down arrow on the **Font Color** button. A color palette will appear as shown in Figure E-31. Choose a **blue** font color from the palette.

Using Your Equipment and Media Properly and Safely

Follow these guidelines when using your computer hardware and software to avoid a variety of problems and prevent harm to you, others, and the hardware, software, disks, and data.

Operating Electrical Equipment

Follow these rules to operate all electrical equipment safely.

1. Do not unplug equipment by pulling on the cord. Instead, grasp the plug near the outlet.

2. Do not stretch an electrical cord across an aisle where someone might trip over it.

3. Do not touch frayed electrical cords. Instead, immediately report them to your teacher.

4. Do not drop books or other objects on or near the equipment.

5. Do not take food and liquids (including aerosol sprays or cleaners) near the equipment. If your computer does get wet, turn it off immediately and report the situation to your teacher.

6. Do not move equipment without permission and proper assistance.

Using Your Computer Safely

Follow these guidelines for using a computer, monitor, and keyboard and other accessories safely.

1. Keep the area around the equipment's air vents open and unblocked to prevent overheating.

2. Place your keyboard, mouse, writing tablet, or digital pen where it will not be bumped off the desk.

3. Adjust the angle, brightness, and focus of the monitor for comfortable viewing and to reduce glare.

4. Adjust the angle of a writing tablet for comfortable writing. Grip a mouse or digital pen firmly but not too tightly.

5. Do not remove or insert computer cables or attach any device to your computer without your teacher's permission and without first turning off the equipment, if required.

6. Keep the monitor and writing pad free of dust and fingerprints by dusting it often with a soft, lint-free cloth.

7. Do not put fingers, pencils, pens, paper clips, etc. into disk drives.

8. Do not attempt to clear paper jams in a printer unless you have been trained to do so and have your teacher's permission to clear jams.

9. Use correct keying techniques and good posture to prevent repetitive stress injury.

10. Static electricity can cause damage to your computer equipment or data. Static electricity may build up in your body from walking across some types of carpeting or in very dry room conditions. To prevent a static discharge that can damage equipment, ground yourself by touching a metal object such as a desk or table before touching the computer.

Protecting Software and Data

Follow these precautions to avoid damage to programs, drives, data files, and storage media.

1. Do not press down on a CD-ROM or DVD drive when opening or closing it. Keep the tray closed when you are not using it.

2. Handle floppy disks and compact discs (CDs) by their labels or edges. Do not touch the shutter of a 3.5" disk.

ACTIVITY E-12: CHANGE THE APPEARANCE OF A PAGE

1. Open *OneNote* if it is not already open. Open your **Business** folder and choose the **Finance** section tab.

2. Right-click the **Finance** section tab to open the pop-up menu. Choose **Section Color** from the pop-up menu. Then choose the color **Green**. See Figure E-28.

Figure E-28
Change Section Color

3. Choose the **Show/Hide Rule Lines** button to reveal lines in the note-taking area as shown in Figure E-29.

Figure E-29
Use the Show/Hide Rule Lines to display lines on the page.

Show/Hide Rule Lines

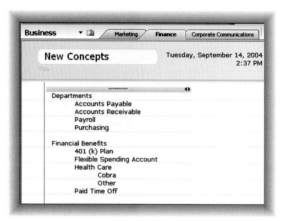

4. Choose the **Show/Hide Rule Lines** button again to remove the lines. Close *OneNote* or continue to the next activity.

Self Check Decide whether or not you like the rule lines. Use them if you think they are helpful.

3. Do not insert or remove a disk while the drive's in-use light is on.

4. Keep disks away from extreme heat or cold, direct sunlight, static electricity, and magnetic fields.

5. To prevent data loss, exit any application program and save/close files you have open before you "shut down" your computer.

6. Do not change a filename extension (the last three characters after the period). Doing so may make the file unusable.

7. Do not delete or move files that are part of an installed program. Doing so may make the program unusable.

8. Do not disable virus-scanning software unless you have your teacher's permission to do so. (Disabling virus-scanning software may be necessary when installing some commercial software programs.)

9. Computers can be damaged by magnets or magnetic fields. Do not place magnets on or near computer equipment. Do not place your monitor or computer near equipment that has a magnetic field. Some telephones and radios and speakers not designed for use with computers produce a magnetic field.

Working with Printers

Your documents will look better if you take good care of your printer and learn to operate it properly. One of the most common tasks related to using your printer is loading paper. Follow these guidelines to get good print quality and avoid paper jams.

1. Use the type of paper recommended for your printer (inkjet, laser, or dot matrix) or an all-purpose paper. For some printers, you can choose a particular paper type in the printer control program. For example, you might choose the "glossy/photo paper" option when you want to print a photograph.

2. Store paper in a cool, dry place to keep it in good condition before use.

3. Fan the paper to separate any pages that may be stuck together before placing the paper in the printer.

4. Place only the recommended amount of paper in the paper feed tray. Overloading the feed tray can cause paper jams.

5. Do not use paper smaller or larger than the size recommended for your printer by the manufacturer. You may need to feed envelopes and paper cut in small sizes (such as note cards) manually into the printer.

6. Follow the manufacturer's or your teacher's instructions for clearing paper jams. Many printers have a Help program to provide operating and troubleshooting information. An example of how to clear a paper jam in an inkjet printer is shown in the following box.

Typical Procedure for Clearing a Paper Jam

- Open the operator panel and remove any paper from the document path.
- If paper is still jammed in the printer, open the cartridge access door and remove any paper from the printer paper path.
- Close the cartridge access door. Then close the operator panel firmly until it snaps into place.
- Press Stop/Clear.

Cleaning a Printer. Perhaps the most important thing you can do to maintain your printer is to keep it clean—both inside and out. Before cleaning your printer, turn it off and unplug the power cable. Wipe the outside of the printer with a slightly damp cloth. If your printer came with a cleaning brush, use it according to the manufacturer's directions. You can clean some parts of the inside of the printer with a dry, lint-free cloth. For hard-to-reach areas, use a can of compressed air to remove paper dust or toner. (Do not touch the drum inside a laser printer. The oils from your skin can cause problems with print quality. Do not touch the contact on the ink cartridge for an inkjet printer.) Be sure to plug in the printer when you are finished cleaning it.

4. Right-click the **Delete 2** page tab. Choose **Delete** from the pop-up menu as shown in Figure E-27 to delete the page.

DigiTip
Tablet PC users, rest the tip of your digital pen directly on the tab name. After the mouse appears, lift the pen up and the pop-up menu will be displayed.

Figure E-27
Delete a Page

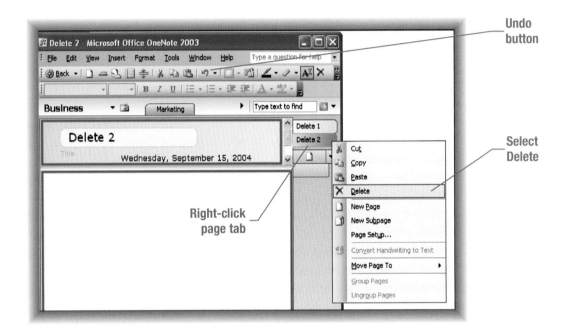

5. Click the **Undo** button (marked in Figure E-27) to restore the page.

6. Delete the **Delete 1** page. Delete the **Delete 2** page.

7. Now you will delete the **Practice** section. Right-click the **Practice** folder tab. Choose **Delete** from the pop-up menu. Choose **Yes** in the dialog box that asks if you want to delete the file.

8. Practice deleting a section. Carefully delete the **Human Resources** section in your **Business** folder. Close *OneNote* or continue to the next activity.

CHANGING THE APPEARANCE OF PAGES

OneNote allows you to change the appearance of your pages. For example, you might want to use different color schemes in different folders. The color can be a visual cue to alert you if you happen to choose the wrong folder. Adding rule lines is one of the most helpful changes you can make to the page appearance. In later activities, you will use your own handwriting to enter notes. The rule lines can help you keep your handwriting neat and organized as you ink your notes across the page.

Changing Cartridges or Toner. Periodically, you may need to change the ink cartridges or add toner to your printer. Many printers will give you a low toner or low ink warning on the computer monitor or the printer display before all the ink or toner is used. Follow the manufacturer's instructions to remove and replace ink cartridges or place toner in the printer. You may need to print a test page after changing cartridges in an inkjet printer. The test page allows you to see the print quality and make alignment adjustments if needed.

Cost-effective Printing. To use your printer in a cost-effective way, view documents on screen before printing. Use color only when needed because printing in black is less expensive than printing in color. Some printers offer you a choice of print qualities. You may be able to print documents for proofreading purposes in a "draft" or "quick" mode, which uses less ink than the normal print mode. Some printers allow you to print on both sides of a page. This option is helpful for saving paper or for creating booklets. Become familiar with the options available for your printer.

Sharing a Printer. Be considerate of others with whom you share a printer. When you pick up your documents from the printer, make sure you have only your documents. Place other documents in a tray or basket provided for that use. Respect the privacy of the people with whom you share a printer. Do not read documents printed by others. Print very long documents when others are least likely to need the printer, if possible. You may be responsible for helping to maintain a shared printer. Check the paper level when you retrieve your printed documents. Add paper to the printer if needed. Check for printer warning lights and report problems to the appropriate person.

Caring for Peripherals

You may have several other computer peripherals in addition to your printer, such as a writing pad and digital pen, a scanner, an external storage drive, or a digital camera.

When using these peripherals, follow the general guidelines given above for using electrical equipment safely. Follow your teacher's or the manufacturer's instructions for using and caring for these devices properly.

Understanding What You Have Learned

Write a short paragraph describing each of the following:

1. The action you will take if your computer becomes wet.

2. The action you will take if your printer becomes jammed.

3. How you should remove dust and fingerprints from your monitor.

4. How you will arrange the monitor, keyboard, and speakers (if any) within your work area.

5. How you will hand a CD-ROM to a classmate.

6. What may happen if you change the filename extension of a file.

7. Steps you can take to help avoid paper jams in a printer.

8. Computer peripherals (other than printers) that you have and how you can care for them properly.

Computer and Media Use Checklist

Your teacher may use a checklist similar to the one shown on the following page to rate your ability to use and care for your computer equipment and media properly.

5. Inside the **Departments** section, move the lines to place them in alphabetical order. (**Accounts Payable** should be first in the list and **Purchasing** should be last.) Close *OneNote*.

LESSON 4: DELETING ELEMENTS AND CHANGING APPEARANCE

OBJECTIVES

In this lesson you will:

1. Create new sections and pages.
2. Delete sections and pages.
3. Change the color for a section.
4. Show and hide rule lines on a page.
5. Highlight notes with color.
6. Change the font, font color, and font size used in notes.
7. Format notes with bold, italic, and underline.

In this lesson, you will continue to learn to be more productive using *OneNote*. You learned earlier how to add pages. In this lesson you will learn how to delete unneeded pages and sections. You will learn to highlight key points in notes by using color. You will also practice formatting notes by changing fonts and using bold, italic, and underline.

DELETING SECTIONS AND PAGES

No matter what method you use to record information, being able to find the information again when you need it is important. This is true for notes you create with *OneNote*. Deleting sections and pages that you no longer need will make finding information in remaining sections easier. You will learn to delete sections and pages in the following activity.

ACTIVITY E-11   ADD AND DELETE SECTIONS AND PAGES

HELP KEYWORDS

Delete page
Add or delete a page or subpage

1. Open *OneNote*. Open your **Business** folder and choose the **Marketing** section tab.

2. Create a new section in the **Business** folder. (Choose **Insert, New Section**.) Name the new section **Practice**. In the header title area for the untitled page, enter **Delete 1**.

3. Create a new page in the **Practice** section. (Click the **New Page** button.) In the header title area for the untitled page, enter **Delete 2**. You should now have a new section and two new pages. You will use the new section and pages to practice deleting.

Computer and Media Use Check Sheet

Rating and Grades

Excellent	4 points	(3.5–4.0)	A	93–100
Good	3 points	(2.6–3.5)	B	85–92
Average	2 points	(1.6–2.5)	C	78–84
Poor	1 point	(0.6–1.5)	D	70–77

Rating Periods

	Rating	1	2	3	4	5	6	7	8	9	10	11	12
Operating electrical equipment													
1. Unplugs equipment properly													
2. Does not place electrical cords across aisles													
3. Reports frayed or damaged cords as needed													
4. Keeps food and liquids away from equipment													
5. Does not drop books or other objects on or near equipment													
6. Does not move equipment without permission													
Using computers safely													
1. Keeps equipment air vents open and unblocked													
2. Places keyboard and monitor properly													
3. Does not remove or add cables and devices without permission													
4. Keeps screen free of dust and fingerprints													
5. Does not place fingers, pencils, etc. in disk drives													
6. Uses correct keying techniques and posture													
7. Discharges static electricity before touching computer													
Protecting software and data													
1. Does not press down on a CD-ROM or DVD drive													
2. Handles floppy disks and compact discs (CDs) by their labels or edges													
3. Does not insert or remove a disk while the drive's in-use light is on													
4. Keeps disks away from extreme heat or cold, direct sunlight, static electricity, and magnetic fields													
5. Exits programs and saves/closes files before shutting down computer													
6. Does not change filename extensions													
7. Does not disable virus-scanning software													
8. Does not place monitor or computer near magnetic fields													
Working with peripherals													
1. Uses the proper type and sizes of paper for printer; loads the paper correctly													
2. Reports paper jams or clears jams with teacher's permission													
3. Cleans equipment correctly (if instructed to do so)													
4. Installs ink cartridges, toner, or ribbons correctly (if instructed to do so)													
5. Uses printer and other devices in a cost-effective way													
6. Is considerate of others with whom a printer or other device is shared													
7. Helps maintain a shared printer or other device													
	Total												

- Distributing <Tab> Bringing products to places where customers can conveniently buy them. <Enter>

- Promoting <Tab> Activities such as branding, advertising, and other efforts to build product awareness. Includes the positioning of products relative to the competition. Requires the understanding of the media. <Enter>

- Selling <Tab> The process of making the sale or closing the deal with the customer. <Enter>

- Supporting <Tab> Maintaining customer satisfaction after the sale of a product. <Enter>

5. Close *OneNote* or continue to the next activity.

APPLICATION E-3 EDIT YOUR NOTES

1. Open *OneNote* if it is not already open. Open your **Business** folder and choose the **Marketing** section tab. Select the **New Concepts** page.

2. Click in the note directly before the word **Anything**. Press the **Tab** key. Press **Tab** before the second word for the remaining lines in the list to create a columnar outline. Your note should look similar to Figure E-26.

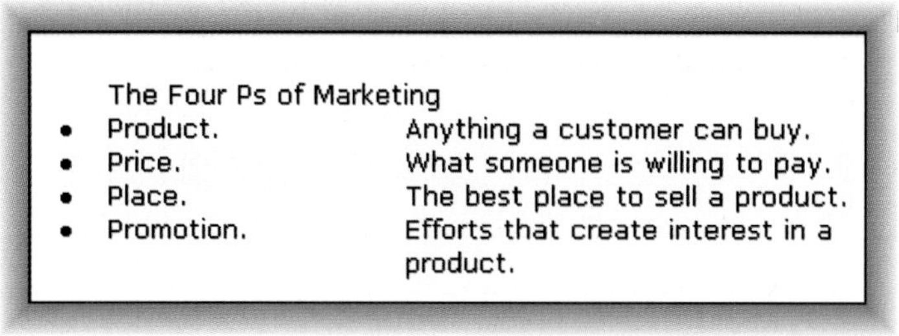

Figure E-26
Improve the Four Ps Concept List

> The Four Ps of Marketing
> - Product.   Anything a customer can buy.
> - Price.   What someone is willing to pay.
> - Place.   The best place to sell a product.
> - Promotion.   Efforts that create interest in a product.

3. Go to the **Financial** section in your **Business** folder. Choose the **New Concepts** page.

4. Select all the lines in the **Financial Benefits** section of the outline. Move the lines up to come before the **Investments** section of the outline. This places the main sections of the outline in alphabetical order.

Skillbuilding

LESSON 1

OBJECTIVES *In this lesson you will:*

1. Improve technique on individual letters.
2. Improve keying speed on 1' and 2' writings.

1A • 5'
Conditioning Practice

Key each line twice SS; then key a 1' writing on line 3; determine *gwam*.

alphabet 1 Jack Faber was amazingly quiet during the extensive program.
spacing 2 it has | it will be | to your | by then | in our | it may be | to do the
easy 3 Jan may make a big profit if she owns the title to the land.

gwam 1' | 1 | 2 | 3 | 4 | 5 | 6 | 7 | 8 | 9 | 10 | 11 | 12 |

1B • 18'
Technique: Individual Letters

Key each line twice SS (slowly, then faster); DS between 2-line groups.

Goal:
To keep keystroking action limited to the fingers.

Emphasize continuity and rhythm with curved, upright fingers.

A 1 Anna Haas ate the meal, assuming that her taxi had departed.
B 2 Bobby Barber bribed Bart to buy the baseball, bat, and base.
C 3 Chuck Cusack confiscated a raccoon and a cat from my clinic.
D 4 Donald doubted that Todd could decide on the daily dividend.
E 5 Ellen and Steven designed evening dresses for several years.

F 6 Felicia and her friend split their fifer's fees fifty-fifty.
G 7 Garn Taggart haggled with Dr. Gregg over the geography exam.
H 8 The highest honors for Heath were highlighted on each sheet.
I 9 Heidi Kim is an identical twin who idolizes her twin sister.
J 10 Janet and Jody joined Jay in Jericho in West Jordan in July.

K 11 Karl kept Kay's knickknack in a knapsack in the khaki kayak.
L 12 Molly filled the small holes in the little yellow lunch box.
M 13 Mr. Mark murmured about the minimal number of grammar gains.

gwam 1' | 1 | 2 | 3 | 4 | 5 | 6 | 7 | 8 | 9 | 10 | 11 | 12 |

APPENDIX C: SKILLBUILDING

Peer Check Move to a classmate's computer and check his or her work on the screen. Ask your classmate to check your work. Discuss any differences and make corrections, if needed.

ACTIVITY E-10   CREATE A COLUMNAR OUTLINE FROM A BULLETED LIST

1. Open *OneNote* if it is not already open. Open your **Business** folder and choose the **Marketing** section tab. Select the **Eight Marketing Skills** page.

2. Click or tap in the note directly after the word **Researching**. Press the **Tab** key.

3. Key the text shown below. Notice how the words wrap around in a column as you key. See Figure E-25.

```
The study of marketing forces, economic trends, and the
analysis of the competition to identify markets for new
products. <Enter>
```

Figure E-25
Note with Columnar Outline

4. Follow the same procedure to enter the text shown below after the remaining bullet items.

```
    • Planning <Tab>    The organization of the marketing
                        process, including marketing plans,
                        budgets, and schedules. <Enter>

    • Developing <Tab>  Developing products, services, or
                        solutions for the market. <Enter>

    • Pricing <Tab>     Determining competitive pricing
                        on both the wholesale and retail
                        levels. <Enter>
```

1C • 15'
Speed Check: Paragraphs

1. Key a 1' writing on each paragraph (¶); determine *gwam* on each writing.

2. Using your better *gwam* as a base rate, select a goal rate and key two 1' guided writings on each ¶ as directed below.

Note:
Copy used to build or measure skill is triple-controlled for difficulty.

EASY all letters used | *gwam* 2'

```
          •    2    •    4    •    6    •    8    •
    Are you one of the people who often look from       5
     •   10    •   12    •   14    •   16    •   18    •
    the copy to the screen and down at your hands? If   10
     •   20    •   22    •   24    •   26    •   28    •
    you are, you can be sure that you will not build a  15
     •   30    •   32    •   34    •   36    •   38    •
    speed to prize. Make eyes on copy your next goal.   20
          •    2    •    4    •    6    •    8    •
    When you move the eyes from the copy to check       24
     •   10    •   12    •   14    •   16    •   18    •
    the screen, you may lose your place and waste time  30
     •   20    •   22    •   24    •   26    •   28    •
    trying to find it. Lost time can lower your speed   35
     •   30    •   32    •   34    •   36    •   38    •
    quickly and in a major way, so do not look away.    39
```

gwam 2' | 1 | 2 | 3 | 4 | 5 |

Quarter-Minute Checkpoints

gwam	1/4'	1/2'	3/4'	Time
16	4	8	12	16
20	5	10	15	20
24	6	12	18	24
28	7	14	21	28
32	8	16	24	32
36	9	18	27	36
40	10	20	30	40

Guided (Paced) Writing Procedure
Select a practice goal

1. Key a 1' writing on ¶ 1 of a set of ¶s that contain word-count dots and figures above the lines, as in 1C above.

2. Using the *gwam* as a base, add 4 *gwam* to determine your goal rate.

3. Choose from Column 1 of the table at the left the speed nearest your goal rate. In the quarter-minute columns beside that speed, note the points in the copy you must reach to maintain your goal rate.

4. Determine the checkpoint for each quarter minute from the dots and figures in ¶ 1. (Example: Checkpoints for 24 *gwam* are 6, 12, 18, and 24.)

Practice procedure

1. Key two 1' writings on ¶ 1 at your goal rate, guided by the quarter-minute signals (1/4, 1/2, 3/4, time). Try to reach each of your checkpoints just as the guide is called.

2. Key two 1' writings on ¶ 2 of a set of ¶s in the same way.

3. If time permits, key a 2' writing on the set of ¶s combined, without the guides.

1D • 7'
Keying Technique

Key each line twice.

Double letters

1 bill foot berry letter deep pool groom egg balloon
2 Matt will look at a free scanner tomorrow at noon.

Balanced hands

3 risk usual to maid the corn box did pan rifle dish
4 He may go with me to the city by the lake to work.

Shift keys

5 Los Angeles Dodgers|PowerPoint|The Book of Virtues
6 The New York Yankees will play the Boston Red Sox.

Space Bar

7 to do a be box and the or was it see me by ten ask
8 As near as I can tell, it is five or six days old.

ACTIVITY E-9

INSERT SPACE AND ADD TO A NOTE

HELP KEYWORDS

Insert space
Add more space to a page

Figure E-24
Use the Insert Extra Writing Space button to add space to a note.

1. Open *OneNote* if it is not already open. Open your **Business** folder and select the **Finance** section tab. Select the **New Concepts** page if it does not already appear.

2. Choose the **Insert Extra Writing Space** button as shown in Figure E-24.

Insert Extra Writing Space

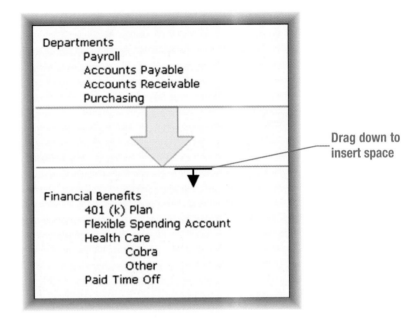

Drag down to insert space

3. Click on the blank line between **Purchasing** and **Financial Benefits** in the note. Drag down about 1.5" to make room for additional notes. An arrow will appear as you do this as shown in Figure E-24.

4. Enter the following new lines in the added space between the two sections of the outline.

```
Investments <Enter>
<Tab>Dividends <Enter>
     Liquid Assets <Enter>
     <Tab> Cash Flow <Enter>
          Interest Rates <Enter>
<Backspace>
     Retirement Plans <Enter>
<Enter> <Backspace>
```

DigiTip
Use your **Backspace** and **Enter** keys to add or remove line spaces as needed.

5. Close *OneNote* or continue to the next activity.

Lesson 2

OBJECTIVES *In this lesson you will:*

1. Improve technique on individual letters.
2. Improve keying speed on 1' and 2' writings.

2A • 5'
Conditioning Practice

Key each line twice SS; then key a 1' writing on line 3; determine *gwam*.

alphabet 1 Jim quickly realized that the beautiful gowns are expensive.
spacing 2 did go|to the|you can go|has been able|if you can|to see the
easy 3 Dick and the girls may go downtown to pay for the six signs.

gwam 1' | 1 | 2 | 3 | 4 | 5 | 6 | 7 | 8 | 9 | 10 | 11 | 12 |

2B • 18'
Technique Mastery: Individual Letters

Key each line twice SS (slowly, then faster); DS between 2-line groups.

Goal:
To keep keystroking action in the fingers.

Emphasize continuity and rhythm with curved, upright fingers.

N 1 Neither John nor Ned wanted a no-nonsense lesson on manners.
O 2 One out of four people openly oppose our opening more docks.
P 3 Phillip chomped on apples as the puppy slept by the poppies.
Q 4 Quin quickly questioned the queen about the quarterly quota.
R 5 Ray arrived at four for a carriage ride over the rural road.

S 6 Steve sold six pairs of scissors in East Sussex on Saturday.
T 7 The tot toddled into the store to pet a cat and two kittens.
U 8 Usually you use undue pressure to persuade us to use quotas.
V 9 Vivian survived the vivacious vandal who wore a velvet veil.
W 10 When will the worker be allowed to wash the new west window?

X 11 The tax expert explained the extensive excise tax exemption.
Y 12 You usually yearn to play with Mary day after day after day.
Z 13 Zoro's zippy zigzags dazzled us but puzzled a zealous judge.

gwam 1' | 1 | 2 | 3 | 4 | 5 | 6 | 7 | 8 | 9 | 10 | 11 | 12 |

2C • 5'
Skill Building

Key each line twice SS; DS between 2-line groups.

Space Bar

1 is it to go me see was you she pool turn they were next best
2 I will be able to try to fix the computer next week for you.

Word response

3 they did may auto form make both them soap held the ham busy
4 I may make a big sign to hang by the door of the civic hall.

Double letters

5 school butter took sell hood green foot current room stubborn
6 Will was a little foolish at the football assembly this week.

gwam 1' | 1 | 2 | 3 | 4 | 5 | 6 | 7 | 8 | 9 | 10 | 11 | 12 |

APPENDIX C: SKILLBUILDING

The columnar outlining feature allows subtopics to be placed in a column to the side of the main topic. You can see how this feature can make your notes more accessible and easier to study by looking at Figure E-25 on page App-136.

ACTIVITY E-8 — CREATE A VERTICAL OUTLINE

HELP KEYWORDS

Outline
Structure notes as outlines

1. Open *OneNote* if it is not already open. Open your **Business** folder and choose the **Finance** section tab.

2. On the **New Concepts** page, click in the upper-right corner of the note-taking area to create a new note container.

3. Key or write the information shown below. When you see <**Enter**>, strike or choose **Enter**. When you see <**Tab**>, strike or choose **Tab**. When you see <**Backspace**>, strike or choose **Backspace**. Your notes should take the form of a simple outline as shown in Figure E-23.

```
Departments<Enter>
<Tab>Payroll <Enter>
     Accounts Payable <Enter>
     Accounts Receivable <Enter>
     Purchasing <Enter>
<Enter> <Backspace>
Financial Benefits<Enter>
<Tab>401(k) Plan <Enter>
     Flexible Spending Account <Enter>
     Health Care <Enter>
     <Tab>COBRA <Enter>
          Other
<Enter> <Backspace>
     Paid Time Off <Enter>
<Enter> <Backspace>
```

4. Close *OneNote* or continue to the next activity.

ADDING SPACE AND DATA

Have you ever discovered that you left something out when taking notes on a pad of paper? Have you ever wished you could just make more paper in the middle of your notes? *OneNote* allows you to add space in the middle of a note. This makes inserting missing information in just the right place quick and easy. To insert space in a note, use the Insert Extra Writing Space button on the toolbar.

2D • 10'
Handwritten Copy (Script)

Each sentence at the right is from a U.S. president's inaugural address. Key each sentence; then key it again at a faster pace.

"How far have we come in man's long pilgrimage from darkness toward light?"

* * * * *

"We must hope to give our children a sense of what it means to be a loyal friend, a loving parent, a citizen who leaves his home, his neighborhood and town better than he found it."

* * * * *

"If we fail, the cause of free self-government throughout the world will rock to its foundations."

* * * * *

"Ask not what your country can do for you—ask what you can do for your country."

* * * * *

"So, first of all, let me assert my firm belief that the only thing we have to fear is fear itself."

2E • 12'
Speed Building

1. Key one 1' unguided and two 1' guided writings on ¶ 1.
2. Key ¶ 2 in the same way.
3. Key two 2' unguided writings on ¶s 1–2 combined; determine *gwam*.

Quarter-Minute Checkpoints

gwam	1/4'	1/2'	3/4'	1'
20	5	10	15	20
24	6	12	18	24
28	7	14	21	28
32	8	16	24	32
36	9	18	27	36
40	10	20	30	40
44	11	22	33	44
48	12	24	36	48
52	13	26	39	52
56	14	28	42	56

AVG all letters used — gwam 2'

It is okay to try and try again if your first efforts do not bring the correct results. If you try but fail again and again, however, it is foolish to plug along in the very same manner. Rather, experiment with another way to accomplish the task that may bring the skill or knowledge you seek.

If your first attempts do not yield success, do not quit and merely let it go at that. Instead, begin again in a better way to finish the work or develop more insight into your difficulty. If you recognize why you must do more than just try, try again, you will work with purpose to achieve success.

6. On the Marketing, New Concepts page, select the lines that describe the four Ps (all lines except the first line in the notes). Click the **Bullets** button to change this text to a bulleted list. Click the **Increase Indent** button, if desired, to move the bullets to align to the right of the title.

Peer Check Move to a classmate's computer and check his or her work on the screen. Ask your classmate to check your work. Discuss any differences and make corrections, if needed.

LESSON 3: OUTLINING WITH ONENOTE

OBJECTIVES *In this lesson you will:*

1. Create a vertical outline.
2. Use the columnar feature to enhance outlines.
3. Add note-taking space to an outline.

ONENOTE'S OUTLINING FEATURES

An **outline** is a document that organizes facts and details by topics and subtopics or headings and subheadings. Outlining is an important note-taking activity. *OneNote* allows you to create outlines and to reorganize outlines after they have been entered.

With *OneNote*, you can create both *vertical* and *columnar* outlines quickly and easily. A vertical outline lists each concept or subheading on a separate line underneath its main idea or heading. A vertical outline is shown in Figure E-23.

Figure E-23
Note with Vertical Outline

```
Departments
    Payroll
    Accounts Payable
    Accounts Receivable
    Purchasing

Financial Benefits
    401 (k) Plan
    Flexible Spending Account
    Health Care
        Cobra
        Other
    Paid Time Off
```

LESSON 3

OBJECTIVES — *In this lesson you will:*

1. Build straight-copy speed and control.
2. Improve keying technique.

3A • 5'
Conditioning Practice

Key each line twice SS; then key a 1' writing on line 3; determine *gwam*.

alphabet	1	Kevin can fix the unique jade owl as my big prize.
caps lock	2	JAY used the CAPS LOCK key to key CAPITAL letters.
easy	3	The small ornament on their door is an ivory duck.
gwam 1'		1 | 2 | 3 | 4 | 5 | 6 | 7 | 8 | 9 | 10 |

3B • 15'
Technique: Response Patterns

1. Key each line twice SS (slowly, then faster); DS between 2-line groups.
2. Key a 1' writing on lines 3, 6, 9, and 12.

TECHNIQUE HINT

Combination response: Most copy requires word response for some words and letter response for others. In such copy (lines 7–9), use top speed for easy words, lower speed for words that are harder to key.

	1	In we up be my are pin tar lip car him sad joy set
letter response	2	were you|at my|red kiln|as you see|you are|fat cat
	3	My cat darted up a tree as we sat in Jim's garage.
	4	it do am me so men did and lap fit ham pan got hen
word response	5	to us|by the|it is|to go|she may|for me|to fix the
	6	She may fix the dock if I do the work for the man.
	7	he as is my to in is no am we by on it up do at or
combination response	8	to be|is up|to my|or up|is at|go in|do we|if we go
	9	Steve and Dave may be by my dock; we may see them.
letter	10	Jon was up at noon; Rebecca gave him my red cards.
combination	11	Jay was the man you saw up at the lake in the bus.
word	12	I may go to the lake with the men to fix the door.

3C • 8'
Speed Check: Sentences

1. Key a 30" writing on each line. Your rate in *gwam* is shown word-for-word above and below the lines.
2. Key another 30" writing on each line. Try to increase your keying speed.

```
         2    4    6    8   10   12   14   16   18   20   22
```
1. He may go with us to the city.
2. Pamela may do half the work for us.
3. Ruth may go with us to the city to work.
4. Sign the forms for the firm to pay the girls.
5. Jan may make all the goal if she works with vigor.
6. He may sign the form if they make an audit of the firm.

```
gwam 30"  2    4    6    8   10   12   14   16   18   20   22
```

Note: If you finish a line before time is called and start over, your *gwam* is the figure at the end of the line PLUS the figure above or below the point at which you stopped.

APPENDIX C: SKILLBUILDING

6. Choose the remaining bulleted list items and touch and drag them up or down to place them in the order shown in Figure E-21.

7. Move the note that contains the list into the bottom of the other note. Your merged note should look like Figure E-22. Close *OneNote* or continue to the next activity.

Figure E-22
Merged Note with Bulleted List

> Marketing Managers must identify customer needs and tailor their product planning, sales, and support efforts to meet those needs. Marketing Managers direct the day-to-day details of the marketing, sales, and customer support activities. They must master eight important skills.
>
> - Researching
> - Planning
> - Developing
> - Pricing
> - Distributing
> - Promoting
> - Selling
> - Supporting

Self Check Does your note look like the one in Figure E-22? If not, try merging the notes again.

APPLICATION E-2   CREATE NEW SECTIONS AND PAGES

HELP KEYWORDS

Create section
Create a section or folder

In this application, you will practice what you have learned by creating new sections, naming pages, and formatting text as a bulleted list.

1. Open *OneNote* if it is not already open. Open your **Business** folder.

2. Create a new section in the Business folder. (Choose **Insert**, **New Section**.) Name the new section **Finance**. In the header title area of the untitled page, key or write New Concepts.

3. Create another new section in the Business folder. Name the section **Corporate Communications**. In the header title area of the untitled page, key or write New Concepts.

4. Create another new section in the Business folder. Name the section **Human Resources**. In the header title area of the untitled page, key or write New Concepts.

5. Choose the **Marketing** tab. Choose the **Definitions** page. Change the name of the page in the header title area to **New Concepts**.

DigiTip

If you make a mistake, you can rename a section. Right-click on the section tab and choose **Rename** from the pop-up menu.

3D • 10'
Handwritten Copy (Script)

Key each quotation twice (slowly, then faster); DS between 2-line groups.

1. "No man is rich enough to buy back his past."

     * * * * *

2. "Nothing great was ever achieved without enthusiasm."

     * * * * *

3. "Keep your face to the sunshine and you cannot see the shadow."

     * * * * *

4. "It is the greatest of all advantages to enjoy no advantage at all."

     * * * * *

5. "If you want something said, ask a man; if you want something done, ask a woman."

     * * * * *

6. "Man does not live by words alone, despite the fact that sometimes he has to eat them."

3E • 12'
Speed Building

1. Key one 1' unguided and two 1' guided writings on each ¶.
2. Key two 2' unguided writings on ¶s 1–2 combined; determine *gwam*.

AVG all letters used gwam 2'

As you build your keying power, the number of errors you make is not very important because most of the errors are accidental and incidental. Realize, however, that documents are expected to be without flaw. A letter, report, or table that contains flaws is not usable until it is corrected. So find and correct all errors.

The best time to detect and correct errors is immediately after you finish keying the copy. Therefore, just before you print or close a document, proofread and correct any errors you have made. Learn to proofread carefully and to correct all errors quickly. To do the latter, know ways to move the pointer and to select copy.

Quarter-Minute Checkpoints

gwam	1/4'	1/2'	3/4'	1'
20	5	10	15	20
24	6	12	18	24
28	7	14	21	28
32	8	16	24	32
36	9	18	27	36
40	10	20	30	40
44	11	22	33	44
48	12	24	36	48
52	13	26	39	52
56	14	28	42	56

After you have created a list, you may want to change the order of the items in the list. You can do so easily by clicking and dragging the paragraph move handle for the item. You will practice creating and editing a bulleted list in the next activity. A numbered list is created in a similar manner.

ACTIVITY E-7 — CREATE A BULLETED LIST

HELP KEYWORDS

Bullet list
Create a bulleted list

1. Open *OneNote* if it is not already open. Click or tap the **Business** folder tab and then the **Marketing** section tab. Open the **Eight Marketing Skills** page by clicking on its tab.

2. Choose the **Type/Selection Tool** if it is not already selected. Click or tap outside and well below the note you entered earlier.

3. Select the **Bullets** button located on the Formatting toolbar. See Figure E-20.

4. Key or write the following bulleted list items. Choose or strike **Enter** at the end of each item.

- Supporting
- Promoting
- Researching
- Planning
- Selling
- Pricing
- Distributing
- Developing

5. Now you will change the order of items in the list. Click or tap the first word in the bulleted list in the note. A paragraph move handle will appear to the left of the line as shown in Figure E-21. Click and drag the paragraph move handle down to make the word **Supporting** the last item in the list.

Figure E-21
Reorder Items in a List

Paragraph move handle

- Supporting
- Promoting
- Researching
- Planning
- Selling
- Pricing
- Distributing
- Developing

New Order

- Researching
- Planning
- Developing
- Pricing
- Distributing
- Promoting
- Selling
- Supporting

APPENDIX E: MICROSOFT *ONENOTE*

LESSON 4

OBJECTIVES *In this lesson you will:*

1. Build straight-copy speed and control.
2. Improve keying technique.

4A • 5'
Conditioning Practice

Key each line twice SS; then key a 1' writing on line 3; determine *gwam*.

alphabet 1 J. Fox made five quick plays to win the big prize.
spacing 2 It will be fun for us to try to sing the old song.
easy 3 The sorority may do the work for the city auditor.
gwam 1' | 1 | 2 | 3 | 4 | 5 | 6 | 7 | 8 | 9 | 10 |

4B • 12'
Difficult-Reach Mastery

1. Key each line twice SS; DS between 2-line groups.
2. Note the lines that caused you difficulty; practice them again to increase rate.

Adjacent (side-by-side) keys (lines 1–4) can be the source of many errors unless the fingers are kept in an upright position and precise motions are used.

Long direct reaches (lines 5–8) reduce speed unless they are made without moving the hands forward and downward.

Reaches with the outside fingers (lines 9–12) are troublesome unless made without twisting the hands in and out at the wrist.

Adjacent keys

1 Jerry and Jason were not ready to buy a newspaper.
2 Polly dropped my green vase on her last trip here.
3 Marty opened the carton to retrieve the power saw.
4 Bert and I were there the week before deer season.

Long direct reaches

5 My niece may bring the bronze trophy back to them.
6 Manny broke his thumb when he spun on the bicycle.
7 Betty is under the gun to excel in the ninth race.
8 They must now face many of the facts I discovered.

Reaches with 3d and 4th fingers

9 A poet told us to zip across the road to get away.
10 Zack saw the sapodilla was almost totally sapless.
11 A poet at our palace said to ask for an allowance.
12 Was it washed when you wore it to our school play?

4C • 13'
Script Copy

1. Key the ¶s twice DS (slowly, then faster).
2. Key a 1' writing on ¶ 1.
3. Key a second 1' writing on ¶ 1, trying to key two additional words.
4. Key a 1' writing on ¶ 2.
5. Key a second 1' writing on ¶ 2, trying to key two additional words.
6. To determine *gwam*, count the words in partial lines.

gwam 1'

Thomas Jefferson was a very persuasive writer. Perhaps his most 13
persuasive piece of writing was the Declaration of Independence, which he was 29
asked to prepare with John Adams and Benjamin Franklin to explain the need 44
for independence. 47

We all should recognize parts of that document. For example, "We 13
hold these truths to be self-evident, that all men are created equal, that they are 30
endowed by their Creator with certain unalienable Rights, that among these are 46
Life, Liberty and the pursuit of Happiness." 54

APPENDIX C: SKILLBUILDING

CHANGING PAGE NAMES

OneNote makes changing page names easy. You simply select the old page name and key or write a new name. The page name will change on the page tab when it is changed in the page title header. You will practice changing a page name in the next activity.

ACTIVITY E-6

CHANGE A PAGE NAME

> **HELP KEYWORDS**
> **Page name**
>   Rename a page or subpage

1. Open *OneNote* if it is not already open. Click or tap the **Business** folder tab and then the **Marketing** section tab. Open the **Marketing Goals** page by clicking its tab.

2. Select the text **Marketing Goals** in the header title area. Key `Eight Marketing Skills` and strike **Enter**. The new page title should appear as shown in Figure E-19.

Figure E-19
Eight Marketing Skills Page Header

> **DigiTip**
> The name on a page tab changes when the header title changes.

3. Close *OneNote* or continue to the next activity.

CREATING AND EDITING LISTS

Lists are used to emphasize information or make data easier to read and understand. When items in a list require a certain order, a numbered list is used. When the order of the items is not important, a bulleted list is used.

To create a list in a note, select the Numbering or Bullets button on the toolbar. Key or write the items in the list, striking or selecting **Enter** after each item. Select the Numbering or Bullets button again to turn off numbering or bullets. You can also add bullets or numbering after you have keyed the items. Simply select the items and click the Numbering or Bullets button.

Figure E-20
Bullets and Numbering Buttons

Bullets   Numbering

APPENDIX E: MICROSOFT *ONENOTE*

4D • 12'
Technique: Response Patterns

1. Key each line twice SS; DS between 2-line groups.
2. Key a 1' writing on lines 10–12 to increase speed; find *gwam* on each line.

TECHNIQUE HINT

Letter response (lines 1–3): Key the letters of these words steadily and evenly.

Word response (lines 4–6): Key these easy words as words—instead of letter by letter.

Combination response (lines 7–9): Key easy words at top speed; key harder words at a lower speed.

letter response
1 milk faced pill cease jump bread join faster jolly
2 were you|up on|are in fact|my taxes are|star gazed
3 My cat was in fact up a tree at my estate in Ohio.

word response
4 oak box land sign make busy kept foal handle gowns
5 go to the|it may work|did he make|she is|he may go
6 Did he make a big profit for the six formal gowns?

combination response
7 is pin when only their dress forms puppy kept care
8 when fate|east of|right on|nylon wig|antique cards
9 Pam was born in a small hill town at the big lake.

letter 10 Edward gave him a minimum rate on state oil taxes.
combination 11 Their eager neighbor may sign up for a tax rebate.
word 12 He may work with the big firms to fix the problem.

gwam 1' | 1 | 2 | 3 | 4 | 5 | 6 | 7 | 8 | 9 | 10 |

4E • 8'
Skill Building

1. Key a 1' writing on each ¶; determine *gwam*.
2. Key two 2' writings on ¶s 1–2 combined; determine *gwam*.

EASY all letters used gwam 2'

Do you think someone is going to wait around 5
just for a chance to key your term paper? Do you 10
believe when you get out into the world of work that 15
there will be someone to key your work for you? 20
Think again. It does not work that way. 24

Even the head of a business now uses a keyboard 29
to send and retrieve data as well as other informa- 34
tion. Be quick to realize that you will not go far 39
in the world of work if you do not learn how to key. 44
Excel at it and move to the top. 47

gwam 2' | 1 | 2 | 3 | 4 | 5 |

Quarter-Minute Checkpoints

gwam	1/4'	1/2'	3/4'	Time
16	4	8	12	16
20	5	10	15	20
24	6	12	18	24
28	7	14	21	28
32	8	16	24	32
36	9	18	27	36
40	10	20	30	40

ACTIVITY E-5

CREATE A NEW ONENOTE PAGE

HELP KEYWORDS

New Page
Add or delete a page or subpage

1. Open *OneNote* if it is not already open. Click or tap the **Business** folder tab and then the **Marketing** section tab. The Definitions page should be on the screen.

2. Click or tap the **New Page** button. The button is shown in Figure E-17.

Figure E-17
Select the **New Page** button to add a new page to a section.

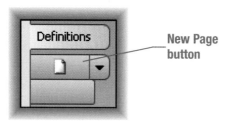

DigiTip
OneNote will allow you to add as many pages as you need to any section.

3. Key or write the words `Marketing Goals` in the header title area on the new page as shown in Figure E-18.

Figure E-18
Marketing Goals Page

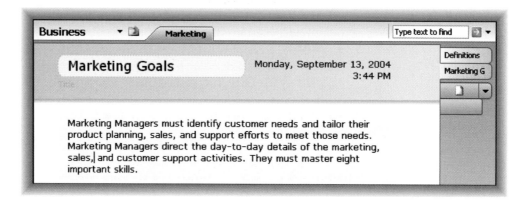

4. Tap in the top left-hand corner of the note-taking area. Key or write the following paragraph. Notice that when you run out of space on one line, more space will be added below.

```
Marketing Managers must identify customer needs and
tailor their product planning, sales, and support
efforts to meet those needs.  Marketing Managers direct
the day-to-day details of the marketing, sales, and
customer support activities.  They must master eight
important skills.
```

5. Close *OneNote* or continue to the next activity.

LESSON 5

OBJECTIVES *In this lesson you will:*

1. Improve technique on individual letters.
2. Improve keying speed on 1' and 2' writings.

5A • 5'
Conditioning Practice
Key each line twice SS; then key a 1' writing on line 3; determine *gwam*.

alphabet 1 Jack liked reviewing the problems on the tax quiz on Friday.
figures 2 Check #365 for $98.47, dated May 31, 2001, was not endorsed.
easy 3 The auditor may work with vigor to form the bus audit panel.

gwam 1' | 1 | 2 | 3 | 4 | 5 | 6 | 7 | 8 | 9 | 10 | 11 | 12 |

5B • 18'
Technique Mastery: Individual Letters
Key each line twice SS (slowly, then faster); DS between 2-line groups. Take 30" writings on selected lines.

Technique Goals:
- curved, upright fingers
- quick-snap keystrokes
- quiet hands and arms

Emphasize continuity and rhythm with curved, upright fingers.

A 1 Aaron always ate a pancake at Anna's annual breakfast feast.
B 2 Bobby probably fibbed about being a busboy for the ballroom.
C 3 Cody can check with the conceited concierge about the clock.
D 4 The divided squad disturbed Dan Delgado, who departed today.
E 5 Pete was better after he developed three new feet exercises.

F 6 Jeff Keefer officially failed four of five finals on Friday.
G 7 Her granddaughter, Gwen, gave me eight gold eggs for a gift.
H 8 Hans helped her wash half the cheap dishes when he got home.
I 9 I investigate the significance of insignias to institutions.
J 10 Judge James told Jon to adjourn the jury until June or July.

K 11 Knock, khaki, knickknack, kicks, and kayak have multiple Ks.
L 12 Lillian left her landlord in the village to collect dollars.
M 13 The minimum amount may make the mission impossible for many.

gwam 1' | 1 | 2 | 3 | 4 | 5 | 6 | 7 | 8 | 9 | 10 | 11 | 12 |

Figure E-14
Definitions Notes

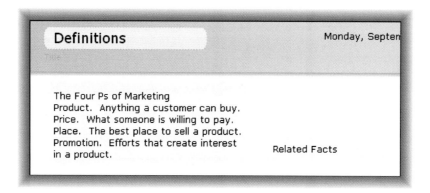

8. Now you will delete the Related Facts note. Move the cursor over the Related Facts note to display the handle. Click the handle to select the note. A selected note has a dotted line around it as shown in Figure E-15. Press or choose **Delete** to delete the note.

Figure E-15
Selected Note

9. Now you will resize the remaining note. Move the cursor over the note to display the handle. Move the cursor over the arrows at the right of the handle until the cursor becomes a two-sided arrow. See Figure E-16.

Figure E-16
Resize a Note Container

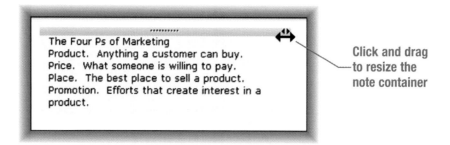

Click and drag to resize the note container

10. When the cursor becomes a two-sided arrow, click and drag to the left to make the container narrower. Click and drag to the right to make the container wider. Practice resizing the container a few times. Close *OneNote* or continue to the next activity.

DigiTip
Notice that the length of the container changes as you change the width.

5C · 12'
Handwritten Copy (Script)

Key each quotation twice (slowly, then faster); DS between 2-line groups.

1. "Every man I meet is in some way my superior."

2. "Find out what you like doing best and get someone to pay you for doing it."

3. "I can resist everything except temptation."

4. "I have always thought the actions of men the best interpreters of their thoughts."

5. "A teacher affects eternity; he can never tell, where his influence stops."

6. "I'm a great believer in luck, and I find the harder I work the more I have of it."

5D · 15'
Speed Building: Guided Writing

1. Key one 1' unguided and two 1' guided writings on each ¶; determine *gwam*.
2. Key two 2' unguided writings on ¶s 1–2 combined; determine *gwam*.

Quarter-Minute Checkpoints

gwam	1/4'	1/2'	3/4'	1'
20	5	10	15	20
24	6	12	18	24
28	7	14	21	28
32	8	16	24	32
36	9	18	27	36
40	10	20	30	40
44	11	22	33	44
48	12	24	36	48
52	13	26	39	52
56	14	28	42	56

AVG — all letters used — gwam 2'

To move to the next level of word processing power, you must now demonstrate certain abilities. First, you must show that you can key with good technique, a modest level of speed, and a limit on errors. Next, you must properly apply the basic rules of language use. Finally, you must arrange basic documents properly.

If you believe you have already learned enough, think of the future. Many jobs today require a higher level of keying skill than you have acquired so far. Also realize that several styles of letters, reports, and tables are in very common use today. As a result, would you not benefit from another semester of training?

APPENDIX C: SKILLBUILDING

Self Check Do you have five notes entered in five containers? Your screen should appear similar to Figure E-12.

Reorganizing Notes

OneNote permits you to reorganize and change your notes after you have created them. You can move, merge, resize, and split *OneNote* containers. You can also delete containers and the notes they hold. In the next activity, you will practice these procedures.

ACTIVITY E-4

MERGE, SPLIT, DELETE, AND RESIZE NOTES

HELP KEYWORDS

Move notes
Rearrange notes on a page

1. Open *OneNote* if it is not already open. Click or tap the **Business** folder tab and then the **Marketing** section tab. The Definitions page should be on the screen.

2. Click or tap the handle of the container that holds the note: **Product. Anything a customer can buy**. A four-sided arrow will appear. As it does, drag this note directly into the bottom of the **The Four Ps of Marketing** note. The two notes will be combined as shown in Figure E-13.

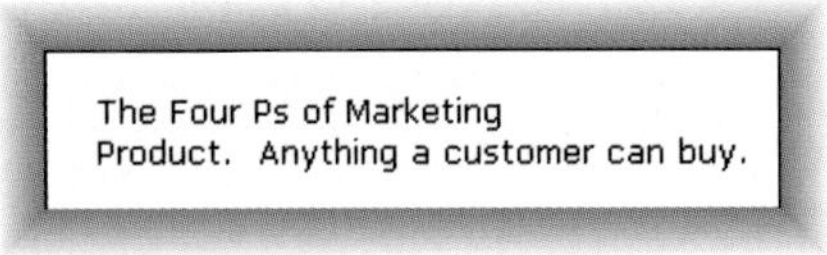

Figure E-13
Combined Notes

> The Four Ps of Marketing
> Product. Anything a customer can buy.

3. Click the handle of the **Price. What someone is willing to pay** note. Drag this note into the bottom of the **The Four Ps of Marketing** note.

4. Click the handle of the **Place. The best place to sell a product** note. Drag this note into the bottom of the **The Four Ps of Marketing** note.

5. Click the handle of the **Promotion. Efforts that create interest in a product** note. Drag this note into the bottom of the **The Four Ps of Marketing** note. The five notes you created earlier should now be combined as one note as shown on the left of Figure E-14. (You have not yet created the Related Facts note shown in the figure.)

6. Click at the end of the combined note. Strike **Enter**, if needed, to create a new line. Key Related Facts.

7. Click and drag to select the text **Related Facts**. Drag the text out of the note container and release. The text should now be split into a separate note. See Figure E-14.

APPENDIX E: MICROSOFT *ONENOTE*

Lesson 6

OBJECTIVES — In this lesson you will:
1. Improve technique on individual letters.
2. Improve keying speed on 1' and 2' writings.

6A • 5'
Conditioning Practice

Key each line twice SS; then key a 1' writing on line 3; determine *gwam*.

alphabet 1 Wayne gave Zelda exact requirements for taking the pulp job.
fig/sym 2 Add tax of 5.5% to Sales Slip #86-03 for a total of $142.79.
easy 3 The six girls at the dock may blame the man with their keys.

gwam 1' | 1 | 2 | 3 | 4 | 5 | 6 | 7 | 8 | 9 | 10 | 11 | 12 |

6B • 18'
Technique Mastery: Individual Letters

Key each line twice SS (slowly, then faster); DS between 2-line groups. Take 30" writings on selected lines.

Technique Goals:
- curved, upright fingers
- quick-snap keystrokes
- quiet hands and arms

Emphasize continuity and rhythm with curved, upright fingers.

N 1 Ann wants Nathan to know when negotiations begin and finish.
O 2 Robert bought an overcoat to go to the open house on Monday.
P 3 Philippi purchased a pepper plant from that pompous peddler.
Q 4 Quincy quickly questioned the adequacy of the quirky quotes.
R 5 Our receiver tried to recover after arm surgery on Thursday.

S 6 Russ said it seems senseless to suggest this to his sisters.
T 7 Tabetha trusted Tim not to tinker with the next time report.
U 8 She was unusually subdued upon returning to our summerhouse.
V 9 Vivian vacated the vast village with five vivacious vandals.
W 10 Warren will work two weeks on woodwork with the wise owners.

X 11 Six tax experts expect to expand the six extra export taxes.
Y 12 Yes, by year's end Jayme may be ready to pay you your money.
Z 13 Zelda quizzed Zack on the zoology quiz in the sizzling heat.

gwam 1' | 1 | 2 | 3 | 4 | 5 | 6 | 7 | 8 | 9 | 10 | 11 | 12 |

6C • 5'
Skill Building

Key each line twice SS; DS between 2-line groups.

Space Bar
1 day son new map cop let kite just the quit year bay vote not
2 She may see me next week to talk about a party for the team.

Word response
3 me dye may bit pen pan cow sir doe form lamb lake busy their
4 The doorman kept the big bushel of corn for the eight girls.

Double letters
5 Neillsville berry dollar trees wheels sheep tomorrow village
6 All three of the village cottonwood trees had green ribbons.

OneNote Containers

OneNote organizes notes in areas called **containers**. Think of these containers as blocks of text. Each container can be moved around as needed. In Figure E-12, notes have been entered in five containers. When you click or tap in a container or move the pointer over a container, a handle (colored bar) appears at the top. See Figure E-12. This handle allows you to move or resize the container. In Activity E-3, you will practice writing notes in containers.

Figure E-12
Note containers can be placed randomly on the page.

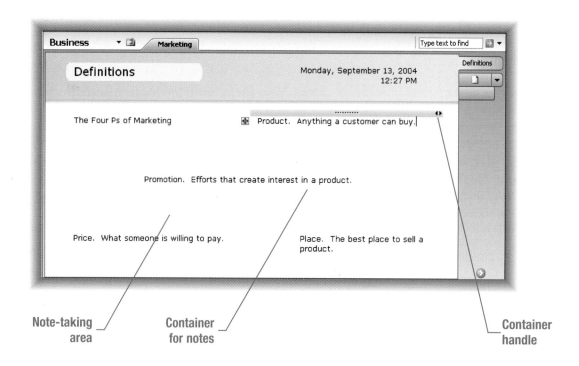

ACTIVITY E-3 CREATE NOTES

DigiTip
Choose the **Text/Selection Tool** to complete the next few activities. See Figure E-1.

HELP KEYWORDS
Write notes
Write or draw notes on a page

1. Open *OneNote*. Click or tap the **Business** folder tab and then the **Marketing** section tab. Click or tap in the top-left corner of the note-taking area. Key or write `The Four Ps of Marketing`.

2. Click or tap in the top-right corner of the note-taking area. Key or write `Product.  Anything a customer can buy.`

3. Click or tap in the bottom-left corner of the note-taking area. Key or write `Price.  What someone is willing to pay.`

4. Click or tap in the bottom-right corner of the note-taking area. Key or write `Place. The best place to sell a product.`

5. Click or tap in the center of the note-taking area. Key or write `Promotion.  Efforts that create interest in a product.`

6. Close *OneNote* or continue to the next activity.

6D • 10'
Skill Transfer

1. Key a 1' writing on each ¶; determine *gwam* on each.
2. Compare rates. On which ¶ did you have highest *gwam*?
3. Key two 1' writings on each of the slower ¶s, trying to equal your highest *gwam* in Step 1.

Note:

Relative speeds on different kinds of copy:
- highest—straight copy
- next highest—script copy
- lowest—statistical copy

To determine *gwam*, use the 1' *gwam* scale for partial lines in ¶s 1 and 2, but count the words in ¶ 3.

AVG all letters/figures used

You should try now to transfer to other types of copy as much of your straight-copy speed as you can. Handwritten copy and copy in which figures appear tend to slow you down. You can increase speed on these, however, with extra effort.

An immediate goal for handwritten copy is at least 90 percent of the straight-copy rate; for copy with figures, at least 75 percent. Try to speed up balanced-hand figures such as 26, 84, and 163. Key harder ones such as 452 and 980 more slowly.

Copy that is written by hand is often not legible, and the spelling of words may be puzzling. So give major attention to unclear words. Question and correct the spacing used with a comma or period. You can do this even as you key.

6E • 14'
Speed Building

1. Key a 1' writing on each ¶; determine *gwam* on each writing.
2. Add 2–4 *gwam* to better rate in Step 1 for a new goal.
3. Key three 1' writings on each ¶ trying to achieve new goal.

AVG all letters used

When you need to adjust to a new situation in which new people are involved, be quick to recognize that at first it is you who must adapt. This is especially true in an office where the roles of workers have already been established. It is your job to fit into the team structure with harmony.

Learn the rules of the game and who the key players are; then play according to those rules at first. Do not expect to have the rules modified to fit your concept of what the team structure and your role in it should be. Only after you become a valuable member should you suggest major changes.

APPENDIX C: SKILLBUILDING

Lesson 2: TAKING AND ORGANIZING NOTES

OBJECTIVES *In this lesson you will:*

1. Place notes randomly on a page.
2. Reorganize, merge, and split notes.
3. Create new pages for notes.
4. Change the name of a page.
5. Create bulleted lists.
6. Reorder list items.

You have made some important changes to *OneNote*. By clicking the *My Notebook* drop-down arrow, you can display a list of all of the folders and sections in *OneNote*. Locate the list shown in Figure E-11. You can open any one of the folders or sections simply by choosing its name from this list. Look for the Miscellaneous and Meetings sections in Figure E-11. Notice that your Business folder can be distinguished from a section by the folder icon.

Figure E-11
My Notebook List

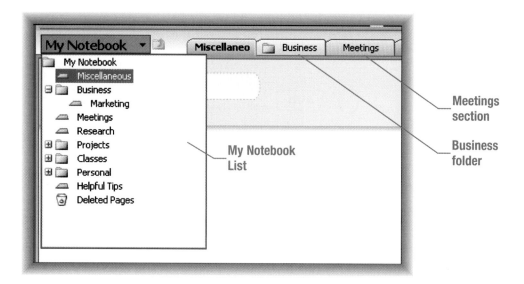

TAKING NOTES IN *ONENOTE*

OneNote will allow you to enter your notes in any order, even in a random fashion all around the *OneNote* screen. This is a great technique to use when you're trying to write facts and ideas quickly before you forget what you are trying to record!

Enhancing Personal Productivity with PDAs

OBJECTIVES *In this appendix you will:*

1. Study PDA operating systems.
2. Input data with *Graffiti*, *Graffiti 2*, natural handwriting, or onscreen keyboards.
3. Review alternative input methods such as speech recognition or special keyboards.
4. Learn to take notes or memos.
5. Create contact or address lists.
6. Update calendars or date books.
7. Create task or to do lists.
8. Make mathematical and financial calculations with either the calculator, an expense program, or a spreadsheet.
9. Learn about multimedia on PDAs.
10. Learn to share information by beaming, synchronizing, downloading, or uploading.

Kaoru Takase works at Corporate View, a simulated (not real) Colorado-based company. Kaoru was recently promoted to the position of Product Marketing Manager in the company's TeleView division. She wants to make the best impression possible. Before starting her new job, she visited an electronics store and purchased a **PDA (Personal Digital Assistant)**. A PDA is a handheld mobile computing device. Kaoru believes a PDA will be useful in keeping track of meetings, contacts, and activities with which she will become involved. Little does she know that her PDA will become an essential business productivity tool. Lately she's been asking the question, "How did I ever get along without a PDA?"

Like Kaoru, it's time for you to embrace these exciting digital communication DigiTools.

APPLICATION E-1 | MINIMIZE, MAXIMIZE, RESTORE, AND CLOSE *ONENOTE*

In this application, you will use the Minimize, Maximize, Restore, and Close commands in *OneNote*. These are the same commands you learned in Chapter 1, Activity 1-3. Review the commands quickly by looking at Figure E-10. Then follow the steps below.

Figure E-10
Minimize, Maximize, Restore, and Resize *OneNote*

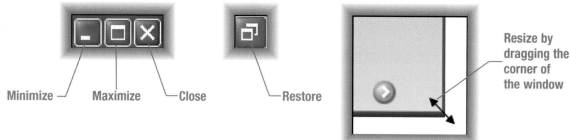

DigiTip
OneNote takes good care of your notes for you! Every 30 seconds and each time you exit, all changes are saved.

1. Click or tap the **Minimize** button to reduce the *OneNote* window to a button on the taskbar.

2. Click the **Microsoft Office OneNote** button on the taskbar to restore the program to its original size.

3. Click or tap the **Maximize** button to enlarge the window to fill the entire screen.

4. Click or tap the **Restore** button to change the *OneNote* window to a smaller size.

5. Click and drag the bottom corner of the *OneNote* window to make the *OneNote* window fill about 75 percent of the screen.

6. Click the **Close** button to close *OneNote*. Don't worry about saving. *OneNote* saves automatically as you exit.

WHAT YOU SHOULD KNOW ABOUT Smart Phones for the Pros on the Go!

Workplace productivity depends on the careful management of time. Schedules, phone conversations, e-mail, contacts, to do lists, expenses, notes, and other business data are critical. PDAs are perfectly suited to manage this kind of data. For this reason, many professionals carry their PDAs wherever they go. However, busy professionals also need to carry their portable phones. This means carrying two handheld devices. To help lighten the load, PDAs and phones were merged into a single DigiTool—the smart phone.

Combining PDA and phone features is a natural merger. The advantages are obvious. For example, if you have a contact's phone number in your PDA, wouldn't it be convenient to have it on your cell phone too? Then you can place a call without having to look up a number on your PDA first.

Today, phones have gone way beyond two-way voice conversations. They can send movies, pictures, live images, and text messages. They also can be used as walkie-talkies. These "smart" phones keep getting smarter. According to Michael J. Miller of *PC Magazine* (July 2003), "Intel predicts that by the end of the decade there will be . . . 2.5 billion phones with more processing power than today's PCs." With that much power in hand, in the future, when people talk about their PDAs they'll probably be talking about their phones! These phones will also be speech-recognition capable. If you need to take notes or send an e-mail message, you can input your information by talking instead of by entering characters letter by letter, as is done today. Some of the things that you can expect from the newest generation of smart phones include:

- Contact lists with phone numbers, addresses, and full contact information
- Synchronization with a desktop PC
- Calendar features
- Web access
- Text messaging
- Digital pictures
- Music files played as ring tones
- Games
- Downloadable business applications such as expense software
- Built-in cameras
- Multimedia messaging using animations, photos, or music
- Walkie-talkie features
- E-mail

Michael J. Miller. *PC Magazine* Web site. "Rejecting the Tech Doomsayers." 20 May 2004. http://www.virtualdevices.net/.

4. The name of the new section will be highlighted. To rename the new section, key `Marketing` and strike **Enter**. (If you make a mistake or want to rename a section later, right-click the tab name. Choose **Rename** from the pop-up menu. Then key a new name.) The Marketing section tab is shown in Figure E-7.

Figure E-7
New Marketing Section

5. Click in the page header title box. Key `Definitions` as shown in Figure E-8. Notice that the date and time when the page is created is automatically inserted into the header. Also notice that the name you enter in the header becomes the name on the page name tab.

Figure E-8
Definitions Page Header

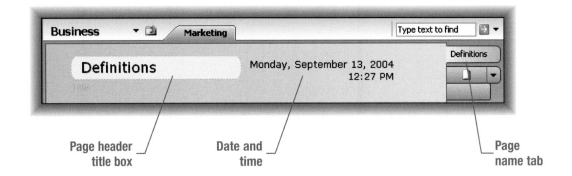

6. Page tabs can be displayed by number or by header titles. You can change views by clicking the Hide/Show Page Titles button at the bottom right of the screen. The button is shown in Figure E-9. Click or tap the **Hide/Show Page Titles** button twice to see both views. Click the **Hide/Show Page Titles** button again, if needed, to display the full page titles.

Figure E-9
Hide/Show Page Titles Buttons

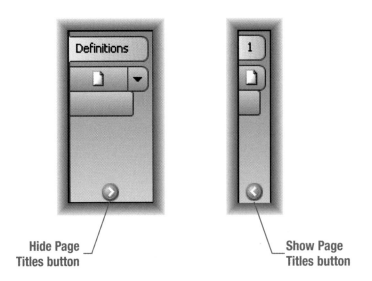

What You Should Know About

Smart Phones for the Pros on the Go!

continued

Figure D-1
Many professionals carry their smart phones wherever they go.

LESSON 1 — UP AND GOING

OBJECTIVES *In this lesson you will:*

1. Explore various models of PDAs.
2. Learn about the two most popular PDA operating systems.
3. Discover how to set up a PDA for first-time use.
4. Access, investigate, open, and close programs with hard buttons, soft buttons, and icons.
5. Learn about the four major applications accessed by hard buttons.
6. Discover and catalog additional applications found on a PDA.
7. Use the scroll buttons.
8. Compare and contrast PDAs with PCs.

Kaoru Takase wants to improve her job performance. As a newly-hired Product Marketing Manager for Corporate View, she knows she must stay on top of her game. One of the tools that can help her stay organized is her PDA.

PDAs are powerful DigiTools, or digital communications devices. PDAs are called by a variety of names: Palms, Palm Pilots, handheld computers, iPAQs, or Pocket PCs. The different names reflect the many different PDA makers and the models they offer.

To learn more about these small DigiTools, Kaoru visits a nearby electronics store and asks many questions about which model of PDA will best meet her needs. She quickly learns that there are two main operating systems for PDAs: *Palm OS* and *Windows Mobile for Pocket PC*.

A section is like any other computer file, with any number of related pages in it. In the next activity, you'll create a new folder and add several new sections.

ACTIVITY E-2

CREATE A FOLDER AND A SECTION

HELP KEYWORDS

Folders
About sections and folders

1. To create a new folder, choose **Insert** from the menu bar. Choose **New Folder**. A new folder will appear with the name *New Folder*. The name will be highlighted as shown in Figure E-4.

Figure E-4
New Folder

2. To rename the folder, key `Business` and strike **Enter**. See Figure E-5.

Figure E-5
New Folder Renamed as Business

Click or tap Business to open the folder

DigiTip

If you strike **Enter** before keying the new name, you can still rename the folder. To do so, right-click the folder name to open the pop-up menu. Choose **Rename** from the pop-up menu. Key the new name and strike **Enter**.

3. Click or tap the **Business** folder tab. You will be greeted by the message shown in Figure E-6. Click or tap in the middle of the message on the screen. A new section will be created.

Figure E-6
Click or tap in the middle of the screen's message to begin a new section.

Figure D-2
PDAs are powerful DigiTools.

OPERATING SYSTEMS FOR PDAS

An **operating system (OS)** is the essential software that allows a PDA to work. *Palm OS*, created by PalmSource, first caught people's imagination in 1996. Visit www.palmsource.com to learn more about this OS. *Windows Mobile for Pocket PC* is a product of Microsoft Corp. This OS also became popular in the late 1990s. Visit www.microsoft.com/windowsmobile/ to learn more about this OS. While there are other operating systems available, *Palm OS* and *Windows Mobile for Pocket PC* are the most widely accepted.

Figure D-3
A stylus is an input device for a PDA.

With both operating systems, a pen-like **stylus** is used to input information into the miniature computer inside. PDAs are small, mobile, and literally fit in the palm of your hand. PDA capabilities have even been added to smart cell phones and wristwatches. Some smart phones and PDAs also have built-in digital cameras and video recorders.

PDAs are much less powerful than desktop, laptop, or Tablet PCs. They have slower processors and less available memory than PCs. Still, these small, mobile devices can do some amazing things. PDAs continue to become more powerful. Who knows? A few years from now, a tiny PDA may be as powerful as today's Tablet PCs, laptops, or desktop computers!

DigiTip
A stylus is a digital pen-like input device.

THE STANDARD AND FORMATTING TOOLBARS

OneNote gives one-tap or one-click access to key features from the Standard and Formatting toolbars. When you first open a new copy of *OneNote*, the Standard and Formatting toolbars appear on one line, and many buttons are hidden. Moving the toolbars to separate lines will reveal more commands.

ACTIVITY E-1 — DISPLAY THE STANDARD AND FORMATTING TOOLBARS

1. Start *OneNote* by choosing **Start**, **All Programs**, **Microsoft Office**, **Microsoft Office OneNote**.

2. Tap or click the down arrow at the end of either toolbar and choose **Show Buttons on Two Rows** as shown in Figure E-2.

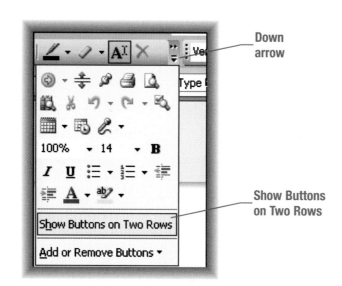

Figure E-2
Show Buttons on Two Rows

DigiTip
If you don't see your Standard and Formatting toolbars, click or tap **View**, **Toolbars**. Place a checkmark next to both **Standard** and **Formatting**.

ONENOTE FOLDERS, SECTIONS, AND HEADERS

Information in *OneNote* is organized in **folders**. Folders you create in *OneNote* are subfolders of the *My Notebook* folder. You can identify folders by a folder icon as shown in Figure E-3. Inside folders you'll find **sections**. A section can have as many pages as you care to create.

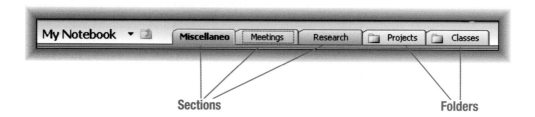

Figure E-3
Folders and Sections in the *My Notebook* Folder

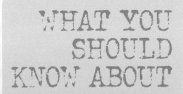

WHAT YOU SHOULD KNOW ABOUT
SPOT Technology

You may soon decide to carry your PDA around your wrist or on your keychain. Thanks to Microsoft's Smart Personal Objects Technology (SPOT), people can use a SPOT enabled wristwatch, radio, alarm clock, keychain, or other small DigiTool to receive their:

- News
- Sports schedules and scores
- Traffic reports
- Stock quotes
- Weather
- Personal schedule and calendar
- E-mail
- Personal reminders
- Games
- Television guides

Figure D-4
Microsoft offers a subscription service for SPOT devices.

With SPOT DigiTools you can receive the most accurate information wherever you are at the exact time you need it. For instance, SPOT devices can automatically go online to reset their time to the exact millisecond. They can also notify their users of important meetings with unique chimes and alarms.

A subscription service for SPOT devices, like MSN Direct, will constantly monitor, upload, and update all of your personal SPOT information using wireless technologies.

GETTING TO KNOW THE BUTTONS

As you examine a PDA, you will notice all sorts of buttons and icons. PDAs generally have hard buttons and soft buttons or icons for programs. **Hard buttons** are physical buttons on the PDA that you can press to start the PDA or launch programs. **Soft buttons** or icons appear on the PDA screen. They are also used to launch programs. Several types of buttons are marked in Figure D-5 for Pocket PC and Figure D-6 for *Palm OS* devices.

LESSON 1

TAKING NOTE OF ONENOTE

OBJECTIVES — *In this lesson you will:*

1. Explore the key features of *OneNote*.
2. Display the Standard and Formatting toolbars on two rows.
3. Create a new *OneNote* folder.
4. Create a new section.
5. Name a page.
6. Show and hide page titles.
7. Minimize, maximize, restore, and close *OneNote*.

Begin your study of *OneNote* by exploring the program's interface as shown in Figure E-1. *OneNote* looks and works much like other *Microsoft Office* products such as *Word*, *PowerPoint*, and *Excel*. For example, notice the familiar menu bar and the Standard and Formatting toolbars shown in Figure E-1. As you work through the next few activities, you will customize the program interface to serve you better.

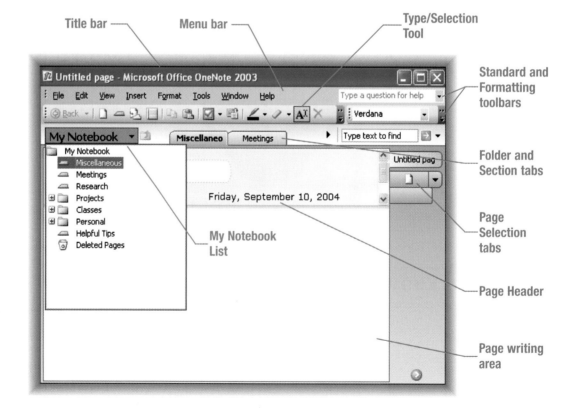

Figure E-1
Microsoft OneNote Application Window

DigiTip

Choose the **Type/Selection Tool** to complete the next few activities. The Selection Tool will allow you to type, handwrite, or dictate text into your *OneNote* notes.

APPENDIX E: MICROSOFT *ONENOTE*

Figure D-5
Pocket PC PDA

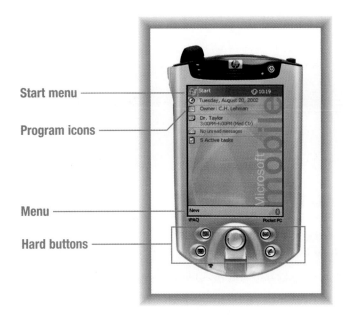

Figure D-6
Palm PDA

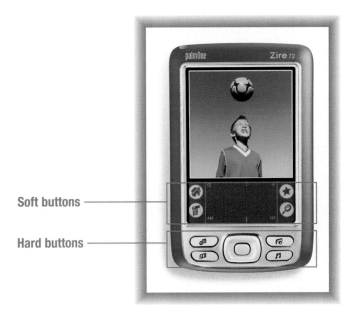

ACTIVITY D-1   COMPLETE SETUP

When you first turn on a brand-new PDA, you will need to complete some basic setup steps. You will calibrate your stylus, set the date and time, and adjust your time zone options. If you are the very first person to use the PDA, turn it on and follow the initial setup instructions exactly.

The first step is to tap the stylus on several targets that appear on the screen. These targets will adjust your digitizer settings. A **digitizer** is a device that converts data into a digital format that can be input, stored, and displayed by a computer. The digitizer rec-

Microsoft® OneNote®

OBJECTIVES *In this appendix you will:*

1. Explore tasks you can do with *OneNote*.
2. Create *OneNote* folders, sections, and pages.
3. Use *OneNote's* vertical and columnar outlining features.
4. Collect notes and information from the Web and other sources.
5. Manipulate and format *OneNote* notes.
6. Use ink tools to take notes.
7. Use side notes.
8. Use *OneNote* with other applications.
9. Search *OneNote* notes.

OneNote is an exciting note-taking program for personal computers. Just as its name suggests, *OneNote* provides one place to record your notes, thoughts, and ideas. No more lugging around paper-filled binders in a backpack. With *OneNote*, you will never run out of paper!

You can use *OneNote* effectively on mobile Tablet PCs, traditional laptops, or even stationary desktop PCs. You can create notes with:

- A keyboard
- A digital pen and handwriting recognition
- A headset and speech recognition

OneNote is a powerful research tool. *OneNote* allows you to collect information from the Internet, e-books, *Word*, *Excel*, *PowerPoint*, or just about anywhere else you may find something useful. You can create notes for years and search them later with *OneNote's* powerful keyword searching tools.

Microsoft and *OneNote* are registered trademarks of Microsoft Corporation in the United States and/or other countries.

ognizes where you touch the screen with your stylus. Read and follow the steps for your PDA's operating system.

Pocket PC

Note: If initial settings have already been set on your PDA, read this activity for background information. Then do Activity D-2 to begin working with your Pocket PC.

1. Turn on your Pocket PC. Wait a few seconds for it to startup. Tap your stylus in the middle of the screen to let your PDA know you are ready to begin.

Figure D-7
Align the stylus on a Pocket PC by tapping the screen.

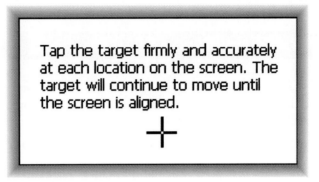

2. If your PDA has not been used before or has been reset for your use, you will need to align the screen. Tap a series of targets accurately with your stylus, as shown in Figure D-7.

3. Continue following the instructions on the screen and make changes as necessary. (**Note:** Pay attention to special features as they are explained, such as cutting and pasting, new software updates, and possible uses of your PDA.)

Figure D-8
Changing Settings on a Pocket PC

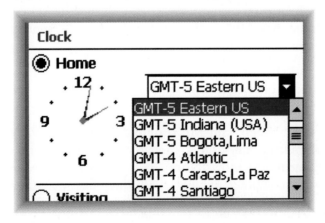

DigiTip
To make adjustments, like changing the time zone option, tap the down arrow to display a list.

2. How many units must Corporate View sell to make $1.5 million in profit if the phones are sold for the following amounts?

$75.00 each
$87.50 each
$100.00 each

3. If the phones are sold for $100 each, what will be the profit margin percentage?

4. Do you think Corporate View should try to sell more smart phones at a lower price or fewer phones at a higher price? What factors might affect your opinion?

APPLICATION D-13 — USE POCKET POWERPOINT PRESENTATIONS

Many PDA users download presentation software, which allows them to make quality presentations to groups using a PDA instead of a personal computer. If you have this software available on your PDA, prepare a short presentation for Kaoru.

1. Create a presentation called **Smart Phones and PDAs**.

2. Create slides that point out the differences between smart phones and PDAs (which you wrote about in Application D-11).

3. Include the financial information you calculated while preparing Application D-12. Include your own thoughts based upon what you considered in Step 4 in Application D-12.

4. Deliver the presentation to a group using your PDA.

APPLICATION D-14 — EXPERIMENT WITH OTHER PDA APPLICATIONS

What other powerful applications do you have on your PDA? Search for and experiment with the following applications:

- **E-Book Reader:** Specialized e-book software, such as *Microsoft Reader*, allows you to download books, magazines, and newspapers to your PDA. Use this software to download a book and read it while using the scroll features to move from page to page.

- **Camera, Video, and Multimedia Tools:** Many PDAs and smart phones now come with built-in cameras. Use these tools to create photos. Then add the photos to a multimedia slideshow. (**Note:** To do this, you may need to upload the images to your personal computer and work with them first in a program like *Photoshop*.)

- **Instant Messaging:** PDAs can send instant messages just like personal computers. Download a popular instant messaging program and use it to communicate with a list of buddies or contacts.

- **Phone Features:** As you have learned, phone features are being added to PDAs and PDA features are being added to smart phones. Use the phone features of your PDA if they are available. For example, you can use text messaging or add phone numbers to your *Contacts* or *Address Book* applications.

Palm OS

Note: If initial settings have already been set on your PDA, read this activity for background information. Go to the next activity to begin working with your PDA.

1. Turn on your PDA.

2. If your PDA has not been used before or has been reset for your use, you will need to align the stylus to the screen. Tap the target accurately with your stylus, as shown in Figure D-9. You can also access the target screen at any time to align your stylus. To do so, tap the **Preferences** icon. Then choose **Digitizer**.

Figure D-9
Align the stylus on a Palm PDA by tapping the screen.

3. Continue to follow the instructions on the screen. To make changes, tap on the boxes next to the headings to reveal the change options. For example, tap the text box next to **Location** or **Time Zone**. (The options will vary depending on the version of *Palm OS* that your PDA uses.) See Figure D-10 for two examples.

4. Tap on a correct option for your time zone and then tap **OK**.

Figure D-10
Change Time Zone Settings

APPENDIX D: ENHANCING PERSONAL PRODUCTIVITY WITH PDAS

APPLICATION D-11 — WRITE AND FORMAT A REPORT

Your PDA may already have a scaled-down word processing program, such as *Pocket Word* or *Word To Go*. These applications provide a limited number of formatting tools, which are unavailable in notes or memos programs. For example, you can bold and center text. You can even add bullets. Under the Tools menu, you can run a Word Count or a Spell Check. Your files can be saved and uploaded to a personal computer where they can be opened by *Microsoft Word* and edited further.

Kaoru is preparing a position paper of 100 words or more comparing PDAs to smart phones. Help her prepare by sharing with her your thoughts on the subject.

1. Open your word processing application. Key the report heading shown below. Use the Insert Date feature to include today's date.

2. Key the paragraphs shown and continue with your thoughts.

```
                PDAS VERSUS SMART PHONES

Current Date

PDAs come in all shapes and sizes. Recently, however,
many cell phones have added PDA features and PDAs have
added phone features. Are they becoming the same
DigiTool?  In this position paper, the following
DigiTools will be compared:

•  PDAs
•  Smart phones

In my opinion...
```
(Continue with your own thoughts comparing the strengths and weaknesses of each DigiTool. Do research on the Internet or using other resources to learn more about smart phones.)

APPLICATION D-12 — USE POCKET SPREADSHEET PROGRAMS

PDA spreadsheet programs, like *Pocket Excel*, provide basic spreadsheet capabilities. They are great for simple calculations. Read the following scenario and prepare a spreadsheet accordingly. If you do not have a spreadsheet, you may use your calculator.

Kaoru needs some preliminary sales figures for a meeting she has with Casey. This information will be used to help determine the price of the Corporate View smart phone. Make the following calculations.

During the production meeting, the staff determined that the smart phone could be manufactured and shipped to the United States from Hong Kong for $50.00 each. What must Corporate View charge for this phone to make a profit? Calculate the price points of the smart phone.

1. The target profit margin for this smart phone is 25 percent. To make a 25 percent profit margin, what selling price must be used for the phone? How many units of the phone will Corporate View need to sell to make $1.5 million in profit?

ACTIVITY D-2

CHANGE DATE AND TIME SETTINGS

After setup is complete, your PDA will take you to its beginning screen. Review the various options that are available.

Pocket PC

1. After setup, your Pocket PC should take you to the *Today* screen. If the *Today* screen does not appear, tap **Start**. Choose **Today** from the most frequently used programs list, as seen in Figure D-11.

Figure D-11
Access the Today screen from the Start menu.

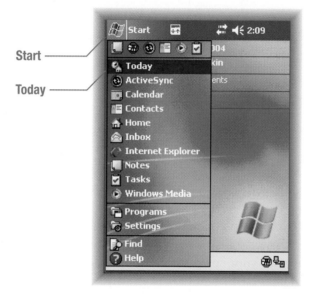

DigiTip
The *Today* screen is the starting point for Pocket PC users. It alerts you to any new messages, appointments, or tasks, along with today's current time and date.

2. You can view the current time at the top of the *Today* screen. The date and time may be correct. Nevertheless, practice changing the settings. Tap the **Clock** icon shown in Figure D-11. This will open the time and date settings dialog window.

3. Change the date and time settings to show the current date and time in your area.
- Setting the time is easy. Tap the hour or the minute boxes and use the up or down arrows to increase or decrease the time. Also, be sure you tap **AM** or **PM** as needed.
- Set the month by tapping the down arrow next to the month to reveal a calendar. Tap the forward or backward arrows to move ahead or back month by month. Tap on the correct day. See Figure D-12.
- Reset the time zone setting if necessary. To reset the time zone, tap the down arrow to reveal a list of zones. Tap the zone for your location.

Figure D-101
Enter your e-mail message.

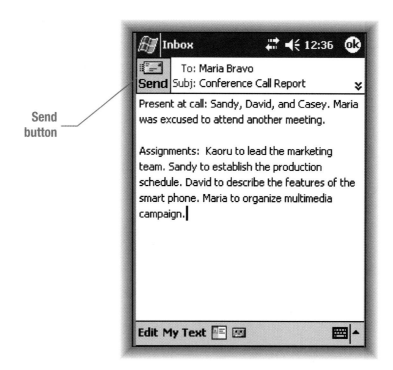

Send button

8. Tap the **Send** button to send your e-mail. Close the *Inbox* program.

Note: If you have a wireless connection, your e-mail is sent when you tap Send. However, if you connect your PDA with a cable to a PC, you will need to synchronize your PDA with your computer. Your new message will be uploaded to your PC's e-mail program. The message will be sent from your PC whenever you choose to send and receive messages.

OTHER POWERFUL PDA APPLICATIONS

Many more applications are available for PDAs. Many already come with your PDA, and others can be downloaded from the Internet. Now that you know how to use your PDA effectively, you should be able to figure out how to use most of the following applications. These other programs include:

- Word processing
- Spreadsheet
- Electronic presentation software
- E-book reader software
- Camera, video, and multimedia software
- Games
- Instant messaging
- Phone features (found on smart phones)

Try some of these exciting applications on your PDA. Learn to use a few of them by completing the following application exercises. If you do not have these applications, then skip or adapt the application exercises to the software you have available on your PDA.

4. Tap **OK.** Tap **Yes** to save the settings.

Figure D-12
Change the Time and Date Settings

Self Check Does your *Today* screen show the current date and time? If not, follow the directions in Step 3 to make changes.

Palm OS

1. After setup, tap the soft **Home** or **Applications** button. (The options will vary depending on the version of *Palm OS* that your PDA uses.) This will display the most commonly used programs on your PDA, as shown in Figure D-13.

Figure D-13
Tap **Home** or **Applications** to see program icons.

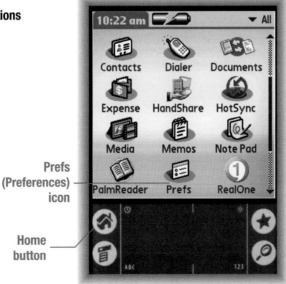
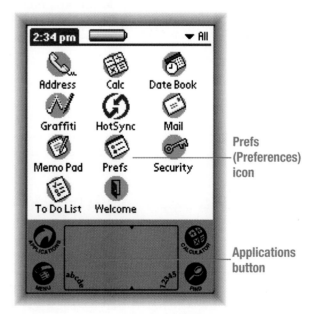

Figure D-100
A typical e-mail message on a PDA.

Delete Message icon

Reply icon

Back and Forward Arrows

5. Now, it is your turn to send an e-mail message. Close any open messages. Create a message to Maria Bravo. Maria was unable to participate in the conference call. She has asked you for a brief summary of the responsibilities that were assigned at the meeting.

6. Choose **New** at the bottom of the screen to open a new e-mail message. The e-mail address you entered for Maria in Lesson 4 is not a real address. Use your own e-mail address or an address provided by your instructor in the **To** field.

7. Enter the subject and body of the e-mail message shown below. Your completed message should look similar to Figure D-101 (with a different To address).

Subject: Conference Call Report

Present at call: Sandy, David, and Casey. Maria was excused to attend another meeting.

Assignments: Kaoru to lead the marketing team. Sandy to establish the production schedule. David to describe the features of new smart phone. Maria to organize multimedia marketing campaign.

2. At the top of the screen you will see the current time. The date and time may be correct. Nevertheless, practice changing the settings. Tap the **Prefs** (Preferences) icon. See Figure D-13. Tap **Date & Time**. This will open the Date & Time window, as shown in Figure D-14.

Figure D-14
The Date & Time Window

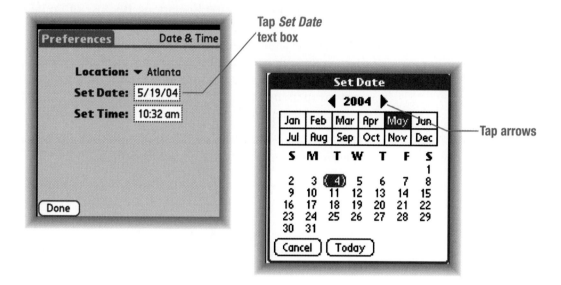

3. Change the time and date settings as necessary.
- Setting the time is easy. Tap the **Set Time** text box. This will open the Set Time window. Tap the up or down arrows to increase or decrease the time. Tap **AM** or **PM** as needed. Tap **OK** to close the box.
- Set the date by tapping the **Set Date** box to reveal a calendar. See Figure D-14. Tap the arrows to move ahead or back by year. Tap a month. Tap on the correct day. Tap **Today** to close the window.
- Reset the time zone setting if necessary. Tap the **Location** arrow or **Set Time Zone** box. Choose a correct option for your time zone.

4. Tap **Home** or **Applications** to return to the list of program icons.

Self Check Access your calendar page that shows information for one day. Does the screen show the current date and time? If not, follow the directions in Step 3 to make changes.

INVESTIGATING PROGRAMS AND APPLICATIONS

Applications are software programs that allow the user to accomplish certain tasks. On a personal computer, you are probably familiar with word processors, spreadsheets, Web browsers, and games. These programs are examples of applications. Many of the same

ACTIVITY D-29 USE E-MAIL WITH A PDA

Kaoru discovers that she can download her e-mail to her PDA and read it on the go. She often spends part of her lunch hour reviewing her most important e-mail. She soon realizes that there are some limitations. Because of limited memory found on PDAs, e-mail programs often only download a portion of each e-mail message. They also don't immediately show attachments. Nevertheless, Kaoru finds that she can quickly scan her e-mails and answer the most important messages before she returns to her PC.

Note: The *Inbox* application for Pocket PC provides the example for these step-by-step instructions. Your e-mail software should have very similar steps and procedures. Adapt these steps to the specific e-mail software available for your use.

1. If you have synchronized your computer to your Inbox on your PC, you should be able to read e-mail. Open your *Inbox* application on your PDA.

2. Choose **Inbox** from the Show menu. See Figure D-99.

Figure D-99
Inbox Folders

3. Double-tap on a message to open it. A typical message can be seen in Figure D-100 on page 111.

> **DigiTip**
> If you can't read your messages, check with your system administrator and see what settings need to be adjusted to make reading e-mail possible.

4. Notice the icons at the bottom of the screen (marked in Figure D-100). The Back and Forward arrows can be used to move to the previous or next e-mail message. Choosing the Reply icon opens a menu where you can choose to reply or forward a message. Choosing the Delete icon deletes the message. Practice using the Back and Forward arrows to access other messages.

applications can be found on PDAs. Here's a short list of commonly used PDA applications:

- Notes or Memos
- Contacts or Address Lists
- Calendars or Date Books
- Tasks or To Do Lists
- Camera and Video Tools
- E-mail
- Instant Messaging
- Phone Dialer

- Word Processor
- Spreadsheet
- Calculator
- Expense Software
- E-book Reader
- Backup or Synchronization Tools
- Games
- Personal Alarm Clock

ACTIVITY D-3 VIEW PROGRAMS

Kaoru soon realizes that using a PDA is much like using a personal computer. Like a PC, a PDA has many programs to choose from. Each has been created to solve a problem, complete a task, or increase productivity. For example, when using a *Windows* PC, choosing the Start menu will reveal the most frequently used programs. Pocket PCs and *Palm OS* PDAs work in much the same way. Let's investigate further. Follow the instructions for your PDA OS.

Pocket PC

1. Tap the **Start** button and choose **Notes**. See Figure D-15. Close Notes by tapping the **Close** button (or **OK, Close**).

Figure D-15
Start and Programs Options

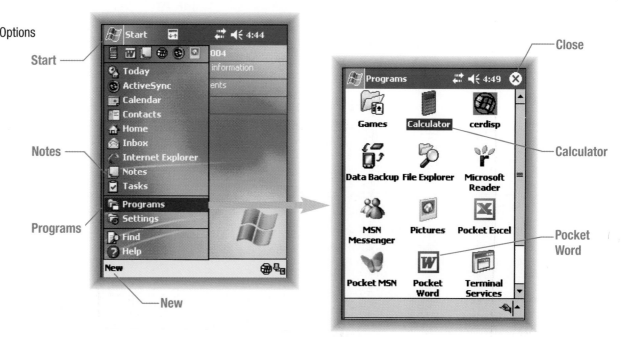

APPLICATION D-10 ENTER BANKING RECORDS ON YOUR PDA

In Activity D-28, you learned to use your financial software. Apply what you learned by entering the remaining banking transactions for January, 20--.

1. Open your financial application (such as *Microsoft Money*). Open the Checking account you created in Activity D-28.

2. Record the additional transactions, both withdrawals and deposits, shown in the table below. On a sheet of paper, write the date for each transaction and the new Balance amount after the transaction.

Peer Check Compare your Balance figure for the last transaction on January 31 to that of a classmate. Do your figures agree? If not, recheck your transactions and make corrections.

CHECKING ACCOUNTS TRANSACTIONS
January 1 - 31, 20--

Date of Transaction	Payee or Description	Withdrawals	Deposits	Balance
1/2/20--	Beginning Balance			$500.00
1/2/20--	Little Pizza Place	$12.50		$487.50
1/2/20--	Little Pizza Place	$14.50		$473.00
1/5/20--	Corporate View		$1,256.45	$1,729.45
1/5/20--	Electrical Power Company	$44.56		
1/5/20--	Gas Company	$139.89		
1/5/20--	Food Market	$46.79		
1/10/20--	Phone Company	$75.80		
1/10/20--	Deposit (from Savings)		$1,000.00	
1/10/20--	Pikes Peak Community College	$860.00		
1/18/20--	Home Mortgage Company	$944.00		
1/18/20--	Food Market	$44.93		
1/20/20--	Corporate View		$1,256.45	
1/21/20--	Springville City	$29.90		
1/21/20--	Car Loan Bank	$254.60		
1/21/20--	Mike's Gas Station	$22.98		
1/21/20--	Al's Auto Lube	$26.50		
1/25/20--	Food Barn	$8.50		
1/25/20--	Internet Service Company	$10.50		
1/25/20--	Credit Card Payment	$356.00		
1/31/20--	Bank Service Charge	$8.50		
1/31/20--	Checking Interest		$2.08	

2. Open various other programs. Tap **Start**, **Programs**. Then tap the **Pocket Word** icon. Tap the **New** button to open a new document if a blank document does not appear. Close *Word* by tapping the **OK** button and then the **Close** button.

3. Tap the **Calculator** icon to open the Calculator. Tap the number **7**. Tap the **+** symbol. Tap the numbers for **45**. Tap the **+** symbol. The sum of these two numbers, 52, should appear on the screen. Tap the **C** (Clear) button to clear the numbers.

4. Practice adding, subtracting, multiplying, and dividing numbers as you would using a typical calculator. Close the Calculator by tapping the **Close** button.

5. Close the *Programs* screen by tapping the **Close** button as marked in Figure D-15.

Palm OS

1. Access the list of program icons. (Tap the **Home** or **Applications** button.)

Figure D-16
Palm OS Applications Categories

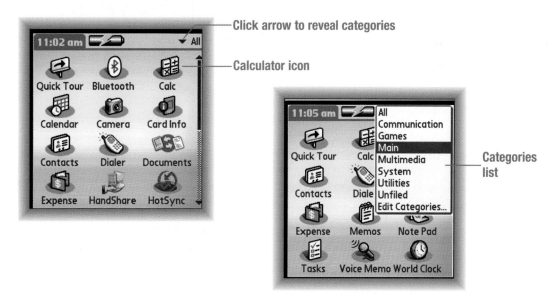

2. Tap the down arrow at the top of the screen. This will reveal a list of categories of applications on the PDA. Choose **Main**.

3. Tap the down arrow again and choose **System** to reveal the system applications. These applications help fine tune your operating system.

4. Try this neat trick. Tap the **Home** or **Applications** button several times to reveal various lists. Continue tapping until you come to the **All** list again.

5. Open various applications. Tap your stylus on the **Memos** (also **Memo/Memo Pad**) icon.

6. Don't worry about closing **Memos**. Simply tap the **Home** or **Applications** button again. Then choose the **Calculator** application. Tap the number **7**. Tap the **+** symbol. Tap the numbers for **45**. Tap the **+** symbol. The sum of these two numbers, 52, should appear on the screen. Tap the **C** (Clear) button to clear the numbers.

Figure D-98
Enter in a Deposit

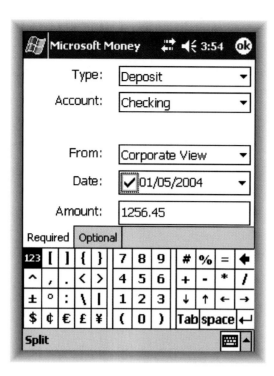

Self Check Check your account balance. At this stage your balance should be **$1,719.45**. If it is not, go back and reconcile your account by checking the accuracy of your transactions.

11. Your bank statement showing transactions for January has arrived. When reviewing it, you noticed a **discrepancy** (lack of agreement) between your bank statement and the balance shown in your *Microsoft Money* program.

12. The transaction you recorded for Arnie's purchase of pizza should have been $14.50 to the Little Pizza Place. Change the amount for this transaction from *$24.50* to **$14.50**. Change the *Pizza Pan Pizza Shop* to **Little Pizza Place**. Choose **OK** to record the entry.

DigiTip
Tap the transaction to open it for editing.

13. Close the financial application.

7. Practice adding, subtracting, multiplying, and dividing numbers as you would using a typical calculator. Tap the **Home** or **Applications** button to close the Calculator.

8. Finally, open (tap) the **Quick Tour** (or **Welcome**) application. View and read the screens. Learn as much as you can from the tour.

APPLICATION D-1 — CALCULATE EXPENSES

Use your PDA's calculator to sum Kaoru's expenses for a recent business trip to Chicago.

Hotel	$348.28
Ground Transportation	80.00
Airline Ticket	258.00
Food	98.57
Total	$

Peer Check Compare answers with a classmate. If your answers do not agree, add the expenses again to determine which answer is correct.

USING THE HARD BUTTONS

DigiTip
Buttons may be arranged in a circle or other pattern around the scroll wheel. The buttons can be reprogrammed to open most applications.

Hard buttons make it easy to access commonly used programs. Typically, four hard buttons and a scroll button are found on a PDA. Hard buttons are programmable. You can change the applications that appear when you press the buttons. By default, Pocket PC buttons normally access the *Calendar, Contacts, Inbox,* and *Home* programs. See Figure D-17.

Figure D-17
Pocket PC Buttons

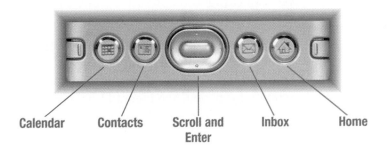

Calendar — Contacts — Scroll and Enter — Inbox — Home

5. Imagine you are having a pizza party for members of your team. You make your first withdrawal from your checking account to pay for pizza. Imagine you have paid $12.50 to the Little Pizza Place using your debit card. Enter this transaction. Choose **Withdrawal** as the Type, and **Checking** as the Account.

6. Enter the Payee `Little Pizza Place` and the amount `12.50`. See Figure D-97.

Figure D-97
Record a Withdrawal

7. In the date field, use **January 2** of the current year. Choose **OK** to record the transaction.

8. Imagine that a lot more people show up to the party than you expected. You lend your debit card to your best friend, Arnie. Arnie makes a quick pizza run for you. He reports back that he paid $24.50 to the Pizza Pan Pizza Shop on January 2, 20--.

9. Record the transaction for Arnie's pizza purchase. Choose **Withdrawal** as the Type and **Checking** as the Account. Enter the remaining transaction information. Choose **OK** to record the withdrawal.

10. Make your first deposit for a paycheck of **$1,256.45** on **January 5, 20--**. The check is drawn on **Corporate View** for a part-time job you have with the company. Choose **Deposit** as the Type and **Checking** as the Account. Enter the remaining transaction information. See Figure D-98. Choose **OK** to record the transaction.

The hard buttons on *Palm OS* PDAs vary widely. The hard buttons for a typical PDA are labeled in Figure D-18.

Figure D-18
Palm OS Buttons

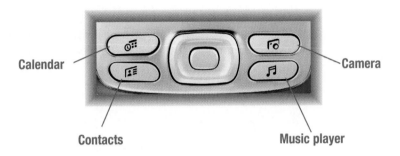

ACTIVITY D-4

ACCESS THE HARD BUTTONS

Kaoru learned that she can open her most important applications quickly by using the hard buttons. She practiced pressing each button repeatedly until she memorized the program that each button opens. In this activity, you will practice using the buttons on your PDA.

Both Pocket PC and Palm OS

1. Start your PDA.

2. Press the first hard button to the left on your PDA. What program does this button open? Note the purpose of the button for future reference.

3. Press each remaining hard button on your PDA, working from left to right (and top to bottom). Note the purpose of each button.

4. Open the *Notes* or *Memos (Memo Pad)* application using a hard button on your PDA. (Open the program using the programs icon if a hard button is not available for this program.) This program allows you to create typed or handwritten notes, drawings, and voice recordings.

5. Open your *Contacts* or *Addresses* application using a hard button on your PDA. (Open the program using the program icon if a hard button is not available for this program.) This program allows you to organize the contact information for your friends, associates, and family members.

6. Open the *Calendar* or *Date Book* program using a hard button on your PDA. (Open the program using the program icon if a hard button is not available for this program.) This application allows you to keep track of your appointments, meetings, and personal schedule.

7. Open your *Tasks* or *To Do List* program using a hard button on your PDA. (Open the program using the program icon if a hard button is not available for this program.) This program allows you to track everything you must do throughout the day, week, month, or year.

DigiTip
You may also open programs by tapping the **Home** or **Application** button on a *Palm OS* PDA or by tapping **Start**, **Programs** on a Pocket PC.

APPENDIX D: ENHANCING PERSONAL PRODUCTIVITY WITH PDAS

This helps her monitor how much she is spending and how much she is being paid. In this activity, you will practice using financial software as Kaoru does.

Note: *Microsoft Money* provides the example for these step-by-step instructions. Your financial software should have very similar steps and procedures. Adapt these steps to the specific financial software available for your use.

1. Open your financial application. (For example, open *Microsoft Money* by tapping **Start**, **Programs**, **Microsoft Money**.)

2. You will create a new checking account. Tap **New** to begin.

3. Enter the following information. See Figure D-96. Choose **OK** to create the account.

Account Name:	Checking
Account Type:	Bank
Opening Balance:	500.00

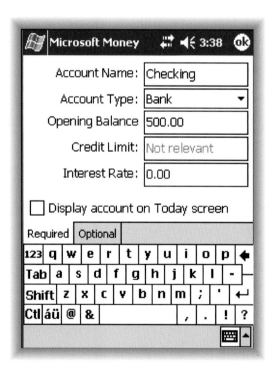

Figure D-96
Create a checking account with a $500 opening balance.

4. Open your checking account by tapping directly on the **Checking** account name. Then choose **New** at the bottom of the screen.

8. Press the *Date Book* or *Calendar* button several times to scroll through the different options and displays. For example, you can display daily, weekly, monthly, or yearly views in your *Calendar* or *Date Book* program. See Figure D-19.

Figure D-19
Calendar View Options for Pocket PC

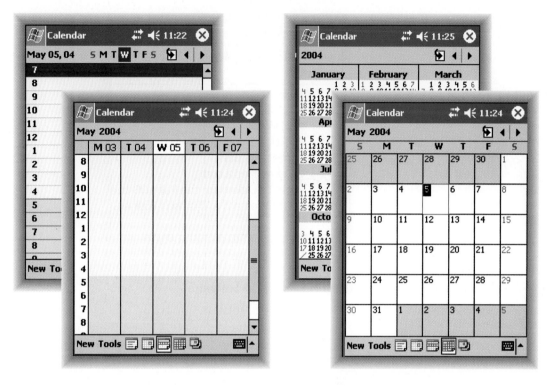

DigiTip
Pressing the button in the center of your scroll button on many PDAs is like pressing **Enter** or **Return** on a personal computer.

9. Practice using your scroll wheel or buttons. Open the *Calendar* or *Date Book* program. Access the screen that shows a monthly calendar. See Figure D-20. Use the scroll buttons to move forward and backward through the dates. Try each scroll button to see its effect. After you have practiced using the scroll buttons, close the program.

Figure D-20
Monthly Calendar Screen

Palm OS

Pocket PC

7. Return to your Favorites list by clicking the Favorites icon at the bottom of the screen. Then access the most interesting Web pages you have saved by selecting the pages from the list.

Figure D-95
Select sites from a Favorites list.

Using Financial PDA Software

Powerful financial software programs, like *Quicken* and *Money*, have moved from desktop PCs down to handheld devices. These programs can help you manage your budget, keep track of business expenses, and even pay bills online.

Financial programs allow you to create *accounts* and manage *transactions*. You can create many accounts: checking, credit card, savings, and investment. You can then manage your money in each of these accounts. A transaction includes any entry into the software, such as a withdrawal or a deposit. A **withdrawal** takes money out of the account. Whenever you pay for something you buy, you are making a withdrawal. A **deposit** puts money back into your account. For example, depositing a paycheck will add to your account.

You can synchronize your finances from your PDA to your financial software on your PC. You can also synchronize your financial data with the computers at a bank or credit union. When you synchronize your financial account with a PC and bank, it is called *reconciling*. **Reconciling** is a process whereby the financial data at the bank is compared with your personal financial records. Reconciling helps avoid errors. If you are overcharged or if somebody makes an unauthorized withdrawal against your account, you can catch the mistake during the reconciling process.

ACTIVITY D-28 — USE FINANCIAL APPLICATIONS ON YOUR PDA

For Kaoru, keeping track of her business and personal expenses is important. She downloaded *Microsoft Money for Pocket PC* from the Microsoft Web site. She then purchased a copy of *Microsoft Money* for her PC. She synchronizes data between the two programs.

10. Depending on the model of your PDA, programs may close or remain open when you switch to another program. Close any open programs.

ACTIVITY D-5 — DISCOVER AND CATALOG YOUR APPLICATIONS

The hard buttons on a PDA will access only a few applications. There are many more applications on most PDAs. Kaoru quickly took inventory of the programs available on her PDA. You will do the same in this activity. To answer the questions that follow, write on a piece of paper, key answers in *Microsoft Word*, or record your answers in *Microsoft OneNote* if it is available on your PC.

1. Start your PDA and the list of programs.
   - *Pocket PC* users, tap **Start** and a list of the most frequently used programs will appear. When you tap the **Programs** icon, additional available programs will be displayed.
   - *Palm OS* users, tap the **Home** or **Applications** icon. Select the **All** category as practiced in Activity D-3.

2. Catalog all the applications on your PDA. Describe briefly what each program does to help a PDA user. If you cannot determine the purpose of a program, make a note of the program name. Consult the documentation that came with your PDA or ask your instructor about this program.

Example:

	Program	**Description**
1.	Calculator	Helps the user solve mathematical problems
2.		

DigiTip
Pocket PC users can learn about some programs by using the Help feature. Tap **Start**, **Help**. Choose the name of the application if it appears in the list.

Peer Check Compare your answers with those of a classmate. Do you have the same programs listed?

APPLICATION D-2 — COMPARE PDAS AND PCS

You probably have some experience with a personal computer. How would you compare a desktop, laptop, or Tablet PC to a handheld *Palm OS* or a Pocket PC PDA? Think about three strengths of each and record your answers.

1. A personal computer is better than a PDA in the following ways:

2. A PDA is better than a personal computer in the following ways:

1. Learn from your system administrator or instructor how your PDA connects to the Internet.

2. Open your PDA's browser software (such as *Internet Explorer* or *Netscape*). Enter a Web address that will take you to a search engine, such as www.google.com, www.yahoo.com, or www.ask.com. See Figure D-93.

Figure D-93
Access a search engine to look for financial software for PDAs.

3. Enter a search term to locate information about PDA financial software. Search for:
   - *Microsoft Money for Pocket PC*
   - *Pocket Quicken*
   - PDA Expense applications

4. Tap the hyperlinks that seem most likely to provide the information you want. Review the pages that you access. Learn as much as you can about the programs.

5. Return to the most helpful Web pages you have found. Bookmark or add those pages to your Favorites list. Start by choosing the **Favorites** icon (the star) at the bottom of the screen.

6. Choose **Add/Delete** at the bottom of the screen. Choose **Add**, then **Add** again. Choose **OK** to close the Favorites window and return to your browser screen.

Figure D-94
Add an item to *Internet Explorer* Favorites.

LESSON 2: INPUT TOOLS

OBJECTIVES *In this lesson you will:*

1. Use a variety of input tools.
2. Use the onscreen alphanumeric, special characters, and international characters keyboards.
3. Practice *Graffiti* or *Graffiti 2* and natural handwriting input.
4. Write notes or memos and tasks or to do lists.
5. Learn about voice input for PDAs.

Data can be input into a PDA in several ways. Some of the methods or tools you can use to enter data are listed below.

- Onscreen alphanumeric keyboard
- *Graffiti* in the Writing Area (*Palm OS*), Block Recognizer, or Letter Recognizer (Pocket PC)
- Natural Handwriting Recognition or *Transcriber*
- Voice Recognition
- Docked to a keyboard
- Beaming Tools
- The Internet (to download applications and data)

Throughout this appendix, you will learn about input methods. In this lesson, you will focus on the first three input methods listed above. These methods can be used with almost every PDA.

ONSCREEN KEYBOARDS

One of the most convenient ways to input information into your PDA is with an **onscreen keyboard**. An onscreen keyboard looks like the traditional keyboard used with a personal computer, but it appears on the PDA screen. To use an onscreen keyboard, simply tap the letters, numbers, punctuation marks, or special characters that you wish to enter. Tap the **Enter** key to move down one line and the **Backspace** key to delete letters as you would on a PC keyboard. Onscreen keyboard entry is a bit slow; but at times, it proves to be the most accurate method. The onscreen keyboard will not activate unless a program that accepts text input has been opened.

Both the *Palm OS* and Pocket PC devices have very similar keyboard layouts. You have several keyboards to choose from:

- The **Standard** keyboard has lowercase letters and numbers.
- The **Shifted** keyboard has capital letters and symbols.

If your personal computer is already connected to the Internet, you can connect to the Internet through your PC. Just change a few settings in your synchronization software. See Figure D-92. You can then surf the Internet as it passes through the PC to your PDA through your synchronization cable or infrared connection.

> **DigiTip**
> To create a pass though connection to the Internet for a Pocket PC, choose the **Options** button on the *Microsoft ActiveSync* status window on your PC. Choose the **Rules** tab and select **The Internet** from the **Pass Through** connection options. See Figure D-92.

Figure D-92
Set the Pass Through connection options.

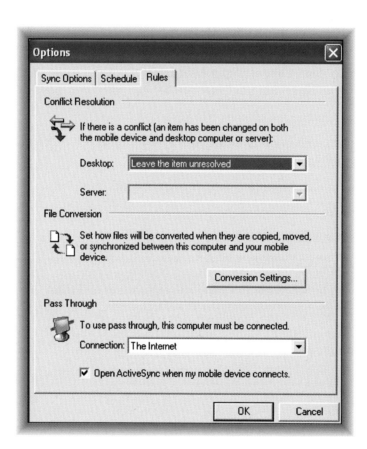

ACTIVITY D-27   SEARCH THE INTERNET WITH YOUR PDA

Kaoru hears that some of her friends are using PDA financial software to keep track of personal finances. Some of her friends were using *Pocket Quicken*. Others were using *Microsoft Money for Pocket PCs*. Still others were using other expense programs they found on the Web. She decides to go online and learn more about possible financial software options. Kaoru is hoping that she can download a financial software package to her computer and then install it on her PDA. She'll search the Web using her PDA. You give it a try, too!

- The **Numeric** or Numbers keyboard has numbers, special characters, and symbols.
- The **International** characters keyboard has letters often associated with languages such as Spanish, French, or German.

Examples of three keyboards are shown in Figure D-21.

Figure D-21
Onscreen Keyboards

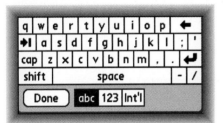
Palm Standard Keyboard

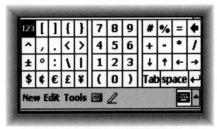
Pocket PC Numeric Keyboard

Pocket PC International Keyboard

WHAT YOU SHOULD KNOW ABOUT
Very Cool Keyboards for PDAs

Let's face it—PDAs are small! That's part of their charm. But their diminutive size does make it awkward to use traditional keyboard input. To solve this problem, some very clever people came up with a variety of solutions. The first was the foldout keyboard, which simply unfolds and attaches or "docks" to a PDA. Another company created a soft keyboard that can literally roll up into a ball. This keyboard can be pulled out of a travel bag just in time to type something important into a PDA. Other alternatives include wireless keyboards, fabric keyboards, keyboards you use with your thumbs, and onscreen keyboards.

The innovation doesn't stop there, however. **Projection keyboards**, sometimes called phantom keyboards or virtual keyboards, use a laser to project a full-size keyboard onto a desk or other surface. Virtual keyboards may also be projected into thin air. This keyboard accepts finger movements and turns them into keystrokes with a high degree of accuracy and no physical contact with a desktop surface. Optical infrared sensing is used to detect hand movements. Sensors monitor finger movements and translate those actions into keystrokes. Advocates believe projection keyboards will replace several million foldout keyboards used on PDAs. They may also be handy for Tablet PC or smart phone users.

Projection systems can also be used to project virtual knobs and switches and can even create a virtual mouse out of thin air. This may be very handy in hospitals where sterile data entry is essential.

Projection keyboards may give new life to keyboarding. They could alter the ergonomics and injury risk factors associated with current typing if people begin to use projected keyboards.

Virtual keyboarding sounds exciting, but it may add to the cost of PDAs, smart phones, or Tablet PCs. There is also the projection device to consider. If the projection system is not built directly into the PDA, then a separate projection device is required.

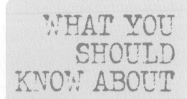

WHAT YOU SHOULD KNOW ABOUT Upgrading to a New PDA

You may decide to upgrade from your current PDA to a new PDA or smart phone. When you do, your synchronization software can help you avoid reentering all of your personal data by hand.

First, simply synchronize your old PDA with your PC. Then, synchronize your new PDA or smart phone with *Microsoft Outlook* or another PIM application. Your new PDA should now have all your current data. You have transferred your data from your old PDA to your new one in a few easy steps!

SEARCHING FOR FINANCIAL APPLICATIONS ONLINE

A great number of programs and applications are made for PDAs. Some of the most important include financial software. This software will help you keep track of your personal finances. Some programs are free, and some cost a little money. Some of the best software packages will synchronize with your financial software on your personal computer. They can even go one step further and synchronize your financial data with your bank or credit union.

The Internet is a great place to learn more about financial packages and other applications that may interest you. Go online with your PDA to find out more.

How the Internet Works with a PDA

PDAs are becoming more and more like personal computers. Many PDAs can be connected to the Internet, just like a personal computer. Regardless of how you connect, you'll need the permission of your system administrators to use the Internet. They can show you the settings you need to change to establish an Internet connection. There are several ways to link up:

- Use a Wi-Fi 802.11 connection.
- Use a wireless phone service on a smart phone/PDA.
- Use Bluetooth.
- Use a pass through connection on a PC already connected to the Internet.

Wi-Fi or **802.11** is a popular wireless protocol or communications system. It connects personal computers to the Internet without wires or cables. Wi-Fi uses specially designed receiving antennas to send and transmit data to and from the Internet. If you are within approximately 100 feet of a Wi-Fi hub, you can connect to the Net.

If you have a smart phone, you can sign up with any number of companies that provide wireless services and Internet access. Some examples of companies that provide these services are AT&T Wireless, Cingular, T-Mobile, and Sprint.

Bluetooth is a short-range wireless system. If you have a Bluetooth-enabled Internet access point connected to the Internet, you can use this technology to go online.

What You Should Know About

Very Cool Keyboards for PDAs

continued

Another alternative includes two hand-mounted devices that sense hand movements. A sensitive board and wristband uses hand movements to measure fingertip movements. This could be another solution that people will find attractive.

ACTIVITY D-6

USE ONSCREEN KEYBOARDS WITH NOTES OR MEMO PAD

DigiTip
R&D stands for *research and development*.

Kaoru Takase will have her first chance to use her new PDA on the job. As she arrives at work, Kaoru receives a call from her boss, Casey Jones, on her mobile phone. Casey is at a trade show in Orlando, Florida. She wants Kaoru to set up a conference call with David Wu concerning a new smart phone/PDA that David's R&D team has designed. It's important to get the new product into manufacturing and begin a marketing campaign. The product is expected to have $1.5 million in sales in its first year and even greater sales in the years to follow. Kaoru must also contact Maria Bravo, a specialist in Corporate Communications, and Sandy Frank, a Product Manager in Manufacturing, who will direct the manufacture of the new PDA-enabled smart phone.

Pocket PC

1. Open the *Notes* program. (Tap **Start**, then **Notes**.) Choose **New** at the bottom of the screen to open a new note.

2. Open the onscreen keyboard by tapping the up arrow in the bottom right-hand corner of the screen. This will reveal the various input options. Choose **Keyboard**.

4. Choose the **Calendar** view. Enter the following appointments for the upcoming week.

```
Date:        Next Thursday
Subject:     Administrative Assistant Interviews
Location:    Robin Mills' Office
Starts:      2:30 p.m. (Thursday Afternoon)
Ends:        4:00 p.m.
Occurs:      Once
Reminder:    Remind me (25 minutes)
Attendees:   Robin Mills


Date:        Next Friday
Subject:     Progress Report
Location:    Casey Jones's Office
Starts:      9:30 a.m. (Friday morning)
Ends:        10:30 a.m.
Occurs:      Once
Reminder:    Remind me (15 minutes)
Attendees:   Casey Jones
```

5. Choose the **Tasks** view. Enter the following tasks.

```
Set up working lunch with Robin
Set up meeting with Marketing department
Order airline tickets for business trip (set for High
priority)
```

6. Synchronize your PDA with your PC.

Self Check Check your *Contacts* or *Address*, *Calendar* or *Date Book*, and *Task* or *To Do List*. Have all of the new entries transferred successfully to your PDA?

Figure D-22
Choose Keyboard from the Options list.

3. Tap the **Shift** key several times to see how the letters change from uppercase to lowercase letters. (Tapping the **Shift** key reveals special symbols such as the **@**. To locate a **/** (slash) or a **-** (hyphen), you must tap the **Shift** key again.) Also, locate the **Enter** key, as you will need to tap it several times in the next step.

Figure D-23
Pocket PC Standard Keyboard

4. Enter the following information using the Standard and Shifted keyboards:

```
Conference Call              <Tap the Enter key.>
                             <Tap the Enter key.>
Smart Phone/PDA              <Tap the Enter key.>
E-mail dwu@corpview.com      <Tap the Enter key.>
```

5. Switch to the Numeric keypad by tapping the **123** key as marked in Figure D-23. Enter the following information:

```
$1,500,000.00      <Tap the Enter key.>
```

> **DigiTip**
>
> Click the **Business** button under Addresses. A **Check Address** box will open. This box can make it easier to enter the mailing address. Choose **OK** when the address has been entered.

2. Click the **New** button on the toolbar or choose **File**, **New**. Enter the following contact information for Robin Mills. See Figure D-91. To enter the department name, click the **Details** tab. Enter the other information on the General tab.

Full name:	Robin Mills
(Job) Title:	Director of Human Resources
Department:	Human Resources
Company:	Corporate View
Work:	303-555-0116
Fax:	303-555-0115
E-mail:	rmills@corpview.com
Address:	One Corporate View Drive
City:	Boulder
State:	CO
ZIP Code:	80303
Country:	USA

3. Choose the **Save and Close** button to save the contact. See Figure D-91.

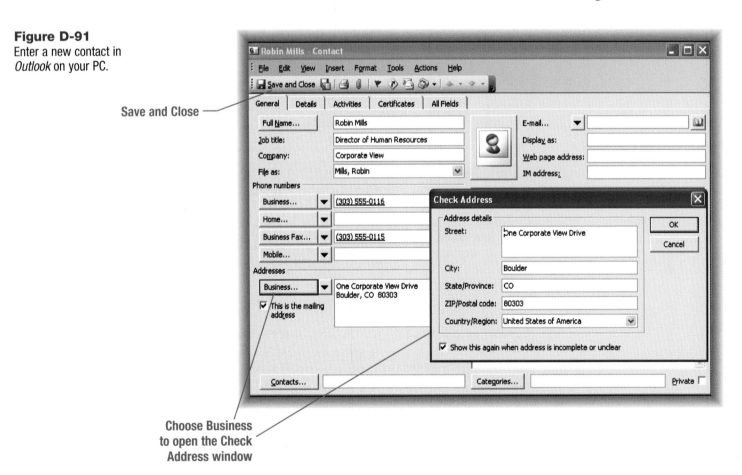

Figure D-91
Enter a new contact in *Outlook* on your PC.

6. Enter the name María. You will need to use the International keyboard to key the Spanish accented letter **í**. Return to the onscreen keyboard by tapping the **123** key. Tap the **áü** key marked in Figure D-23. This will reveal the International keyboard, allowing you to tap the accented **í**.

```
María     <Tap the Enter key.>
```

7. Enter the names of the additional team members that must participate in the conference call.

```
Sandy     <Tap the Enter key.>
David     <Tap the Enter key.>
```

8. Choose **OK** to close the Note. *Conference Call* should now be in your list on the *Notes* menu. As you create future notes, they will be listed in alphabetical order. The first line or phrase from each note is used as a title. Tap the **Close** button to exit *Notes*.

Self Check Does *Conference Call* appear in your list on the *Notes* menu?

Palm OS

1. Open the *Memos* (or *Memo Pad*) application. Choose **New** to open a new memo.

2. Tap the **ABC** area of the input panel in the bottom left corner of the screen, as shown in Figure D-24.

Figure D-24
Tap ABC on the input panel to open the keyboard.

3. Tap the **Shift** key several times to see how the letters change from uppercase to lowercase letters. (To locate a / (slash) or a - (hyphen), you must tap the **Shift** key again.) Also, locate the **Enter** key, as you will need to tap it several times in the next step.

4. Enter the following information using the Standard and Shifted keyboards:

```
Conference Call    <Tap the Enter key.>
                   <Tap the Enter key.>
Smart Phone/PDA    <Tap the Enter key.>
```

Figure D-90
Contacts Displayed in *Outlook*

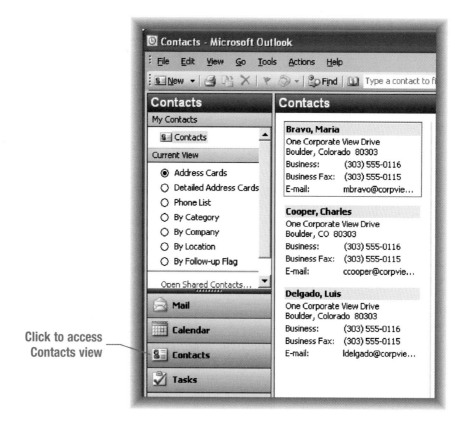

Click to access Contacts view

3. Choose the **Calendar** (Date Book) view in the Navigation pane. View the copy of the appointments and calendar items that have been transferred to your PDA.

4. Choose the **Tasks** (To Do list) view in the Navigation pane. View how your items have transferred to your computer.

Entering Data into a PC for Your PDA

In many ways, contacts, addresses, and appointments are much easier to enter on a PC than into a PDA. You can use a full-size keyboard or speech recognition to input an address or a task. This can be much faster than using *Graffiti 2*, an onscreen touch keyboard, or transcriber on a PDA. Using the synchronization features, the new information can then be downloaded from *Outlook* to a PDA.

ACTIVITY D-26 ENTER INFORMATION INTO OUTLOOK

Kaoru is constantly looking for ways to save time and be more productive. She quickly learns that it's often faster to enter her contacts, appointments, and tasks into *Microsoft Outlook* on her computer. Then she can synchronize the data with her PDA or smart phone. You try it!

1. Open *Microsoft Outlook*. Choose the **Contacts** view.

Figure D-25
The Palm Standard Keyboard

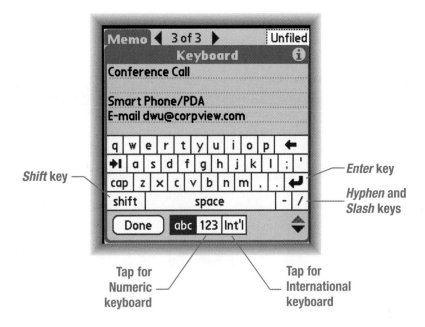

5. To enter characters such as **@** and **–** (hyphen), switch back and forth from the Standard keyboard to the Numeric keyboard. Tap the letters **abc** and numbers **123** as marked in Figure D-25. Enter the following information:

```
E-mail dwu@corpview.com    <Tap the Enter key.>
$1,500,000.00              <Tap the Enter key.>
```

6. Enter the name María. You will need to use the International keyboard (**Int'l**) to key the Spanish accented letter **í**. Tap the **Int'l** button. This will reveal the International keyboard, allowing you to tap the accented **í**.

```
María     <Tap the Enter key.>
```

7. Enter the names of the additional team members that must participate in the conference call.

```
Sandy     <Tap the Enter key.>
David     <Tap the Enter key.>
```

8. Tap **Done** to close the onscreen keyboard. Tap **Done** again to save the memo. *Conference Call* will now appear in your list on the *Memos* menu. As you create future notes, they will be listed in numerical order. The first line or phrase from each note is used as a title. Tap the **Home** or **Applications** button to exit *Memos*.

Self Check Does *Conference Call* appear in your list on the *Memos* menu?

Set Up Synchronization (*Palm OS*)

When you first connect your PDA to your computer, you may need to work through a series of choices.

1. Normally, you should choose **Local** as your connection type.

2. Choose **Cradle/Cable** as the means of communicating to your computer. See Figure D-89.

Figure D-89
HotSync Setup Options

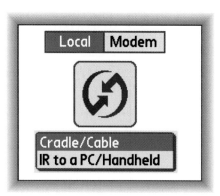

3. After your choices have been made, tap the **HotSync** button marked in Figure D-87 on page App-101. Synchronization can begin and the *HotSync Progress* screen will appear as shown in Figure D-88 on page App-101.

Displaying PDA Data on a Computer

A variety of personal information management (PIM) applications can display your PDA data on your personal computer. For example, *Palm OS* users can download *Palm Desktop* for both *Macintosh* and *Windows* computers. This popular PIM can be downloaded from the PalmOne web site at www.palmone.com.

Another very popular PIM is called *Microsoft Outlook*. *Outlook* can be used with both Pocket PC and *Palm OS* PDAs. For this reason, *Outlook* will be used in examples in this lesson. You can use any PIM, however, to complete the next activity. (**Note:** *Palm OS* support for *Microsoft Outlook* users can also be found at www.palmone.com.)

ACTIVITY D-25 VIEW YOUR PDA DATA IN YOUR PIM

Kaoru has just synchronized her PDA with her Tablet PC. She is a bit anxious, not knowing for sure if the information has transferred correctly. Because she is also concerned about how the information will appear, she opens her PIM to view the information. You will do the same in this exercise.

1. Open *Microsoft Outlook* or your PIM software.

2. Choose the **Contacts** (Address) view by clicking **Contacts** in the Navigation pane at the left of the screen. View a copy of your personal and business contacts that have been transferred to your PC. See Figure D-90.

GRAFFITI AND GRAFFITI 2

Graffiti is a program that allows users to input information into a PDA by handwriting on the screen. *Graffiti 2* is a later version of the *Graffiti* program. Created by Palm, *Graffiti* allows you to use a shorthand way of writing letters of the alphabet. It can be used on Pocket PCs as well as Palm PDAs. Many people find *Graffiti* to be more convenient than the onscreen keyboards when inputting text into a PDA. However, you must practice to write the letters accurately.

ACTIVITY D-7 — IDENTIFY YOUR HANDWRITING PROGRAMS

Kaoru found the onscreen keyboards somewhat cumbersome. During her lunch break, she explored her PDA to learn what handwriting system she could use. Explore as Kaoru did to find the handwriting programs on your PDA.

Pocket PC

1. Tap the **Start** menu and choose **Settings**. Choose **Input**.

2. Tap the down arrow for Input Method and choose **Block Recognizer**. Note that the Block Recognizer option allows you to input letters using *Graffiti* writing strokes. See Figure D-26.

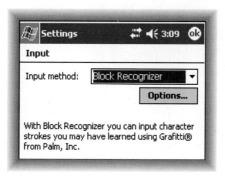

Figure D-26
Pocket PC users can choose an input method such as Block Recognizer.

3. Tap the down arrow for Input Method and note the other choices available. *Letter Recognizer* allows you to write using strokes similar to *Graffiti 2*. *Transcriber* allows you to write in cursive, print, or mixed handwriting styles.

4. You will practice writing using one or more of the handwriting options in a later activity. Tap **OK**, then **Close**.

Palm OS

1. Access the list of programs available on your PDA. Open the *Quick Tour* or *Welcome* program.

2. Choose the **Entering Data** option or tap the **Next** button to move through the screens. The Entering Data portion of the tour will tell you about the *Graffiti* or *Graffiti 2* program on your PDA. See Figure D-27.

Palm OS

1. Confirm that *Palm Desktop* software has been loaded onto your computer. (Click **Start**, **All Programs** and look for the program name.) If the program is not present, consult your teacher for instructions.

2. Connect the cable between your PDA and your personal computer.

> **DigiTip**
> By default, programs such as *Calendar*, *Contacts*, *Tasks*, and *Memos* are synchronized. You can change these options if you wish using the *Palm Desktop* program on your PC.

3. Turn on your PDA. Let your computer and your PDA locate each other and begin the synchronization process. If synchronization does not start immediately, start it manually by tapping the **HotSync** icon or the **Sync** soft button. See Figure D-87. (If synchronization still does not begin, follow the steps below under *Set Up Synchronization*.)

Figure D-87
Palm Synchronization

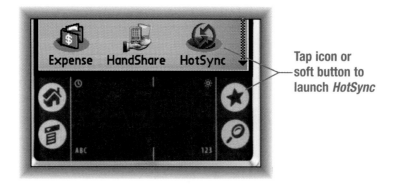

Tap icon or soft button to launch *HotSync*

4. Once synchronization begins, the *HotSync Progress* screen will appear. This screen will track and report on the synchronization process. See Figure D-88.

Figure D-88
HotSync Progress Screen

Figure D-27
Graffiti 2 allows you to handwrite on the PDA screen.

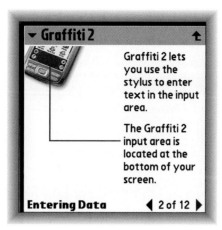

3. Just read the screens for now. You will practice writing using this program in a later activity. Tap the **Done** button to close the tour.

WHAT YOU SHOULD KNOW ABOUT Voice Recognition on a PDA

Have you ever thought it would be wonderful to input information into your PDA just by talking? Well, in a limited way, you can do that already. If you buy and install a program for your PDA such as *Voice Command* by Microsoft, you can give verbal commands to a PDA. Other companies such as IBM and ScanSoft also offer voice command programs. These commands will take the place of tapping commands with your stylus. You will be able to call up contacts instantly by name, dial a number on a smart phone, or check off a task item just by talking.

Writing Graffiti or Graffiti 2

The writing area on a PDA, also called the **input panel**, has different areas for writing letters and numbers. Letters are entered using specific gestures on the left side while numbers are written on the right. Input panels for Pocket PC and Palm PDAs are similar, as shown in Figure D-28.

Set Up Synchronization (Pocket PC)

When you first connect your PDA to your computer, you will need to work through a series of setup screens. The first will have you establish a partnership between your PDA and your computer. A partnership establishes synchronization rights. These rights detail what can and can't be uploaded and downloaded during synchronization. Normally, a Standard Partnership should be your choice.

1. Confirm that *Microsoft ActiveSync* has been loaded onto your computer. (Click **Start**, **All Programs** and look for the program name.) If the program is not present, consult your teacher for instructions.

2. Connect the cable between your PDA and your personal computer if it is not already connected.

3. On your PC, choose **Start**, **ActiveSync**. Select **Standard Partnership**, as shown in Figure D-86. Choose **Next** to continue.

Figure D-86
New Partnership Screen

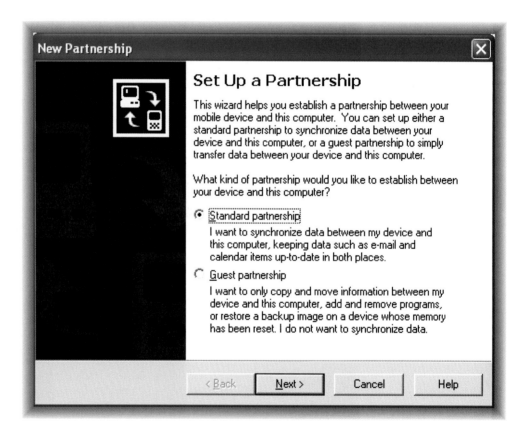

4. Work through the other setup screens as directed. For example, you should choose the option that allows you to synchronize with your computer. You may also be asked to accept or rename your PDA for synchronization purposes.

5. After you have established a partnership, synchronization can begin and the *Microsoft ActiveSync* status screen will appear. This screen will track and report on the synchronization process. See Figure D-85 on page App-99.

Figure D-28
PDA Input Panels

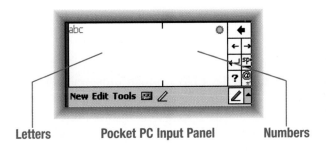
Letters — Pocket PC Input Panel — Numbers

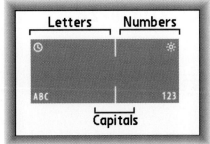
Palm Input Panel

On most Palm devices, the input panel is always visible. Pocket PC users may need to open the input panel. Pocket PC users can open the Block Recognizer by choosing the up arrow in the bottom right corner of the Pocket PC screen. This arrow appears when a program that can use Block Recognizer is open, such as *Notes*. See Figure D-29.

Figure D-29
Block Recognizer is an option on the Pocket PC.

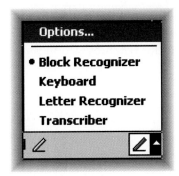

Graffiti Gestures

Special gestures or strokes can be used when writing *Graffiti* or *Graffiti 2*. These gestures allow you to add space between words, backspace to erase letters, and move to a new line. Some commonly used gestures are shown in the following table.

> **DigiTip**
> A **gesture** is a quick motion with the stylus on a PDA screen producing a letter, symbol, or action.

Gesture	Action
■—	Backspace or erase character to the left *Touch your stylus and make a quick stroke to the left.*
—■	Space between words or characters *Touch your stylus then slide horizontally to the right.*
╲■	Move to a new line (Enter or Return) *Touch your stylus and drag diagonally from right to left.*

When using *Graffiti 2*, you write on a certain area of the input panel to create a capital letter. When using the *Graffiti* program, you use a Shift stroke before a letter to indicate a capital. The example on the left in Figure D-30 will produce a capital letter *A*. If you need to make a series of capital letters, use the Caps Lock stroke (two upward gestures) to lock the capital letters. This is like pressing the **Caps Lock** key on a keyboard. The right example in Figure D-30 will produce the acronym *PDA*.

ACTIVITY D-24: SYNC BETWEEN A PDA AND A COMPUTER

Kaoru knows that however valuable her PDA hardware may be, the business and personal data stored on her PDA is much more valuable. This important information may be lost if her PDA crashes, loses battery power, is somehow damaged, or becomes lost or stolen. To protect her data, Kaoru learns how to synchronize or "sync" her PDA with her personal computer. She uses her synchronization software and a USB cable to connect her PDA to her computer. Then she makes a backup copy of her data on her Tablet PC. It's time for you to do the same.

Pocket PC

1. Connect the cable between your PDA and your personal computer.

2. Turn on your PDA. Let your computer and your PDA locate each other and begin the synchronization process. If synchronization does not start immediately, follow the steps below under *Set Up Synchronization* to start it manually.

3. Once synchronization begins, the *Microsoft ActiveSync* status screen will appear. This screen will track and report on the synchronization process. See Figure D-85.

> **DigiTip**
> By default, *Calendar*, *Contact*, *Tasks*, *Favorites*, and *Inbox* (e-mail) data are synchronized. You can add other options later if you wish.

Figure D-85
Microsoft ActiveSync Window

APPENDIX D: ENHANCING PERSONAL PRODUCTIVITY WITH PDAS

Figure D-30
Gestures for Creating Capital Letters

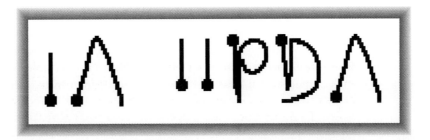

ACTIVITY D-8　　PRACTICE GRAFFITI OR GRAFFITI 2

Kaoru decided to learn more about using *Graffiti*. A program on her PDA shows how to write letters, numbers, punctuation marks, and helpful gestures. After this initial training, Kaoru was ready to practice writing in the *Memos* or *Notes* application. In this exercise, you will learn more about *Graffiti* or *Graffiti 2* and practice writing letters and numbers.

Pocket PC

1. Learn more about and practice *Graffiti* or *Graffiti 2*. Open the *Notes* program. Open a new note.

2. Tap the up arrow at the lower right of the screen to choose an input option. Choose **Block Recognizer** to use *Graffiti* strokes or choose **Letter Recognizer** to use *Graffiti 2* strokes.

3. Tap the **Help** button as marked in Figure D-31. Choose **About Letter Recognizer** or **About Block Recognizer** depending on which program you want to learn about.

Figure D-31
Pocket PC Help Screen

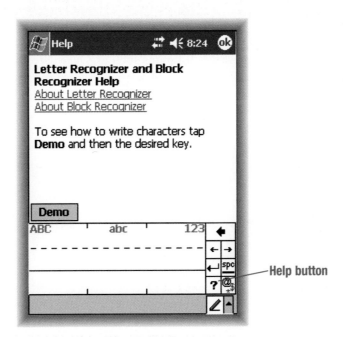

Help button

data. Microsoft's *ActiveSync* for Pocket PCs and *HotSync* for *Palm OS* are examples of programs you can use to upload your vital data to your personal computer. Uploaded data can be displayed on your PC in a personal information management (PIM) application like *Microsoft Outlook* or *Palm Desktop*. For example, in Figure D-84, all of the business contacts from Lesson 4 have been uploaded from a PDA to *Outlook*.

Figure D-84
Microsoft Outlook can display data uploaded from your PDA.

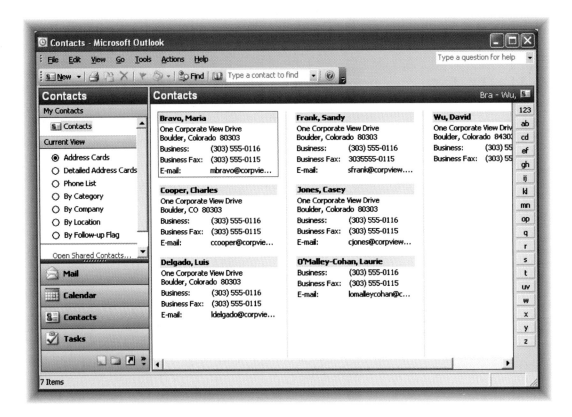

You can also use PIM programs such as *Outlook* to download contacts, tasks, calendar items, and e-mail to your PDA as you synchronize. In this way, your PC and PDA can work together to increase your effectiveness and productivity.

ActiveSync (for Pocket PC), *HotSync* (for *Palm OS*), or other synchronization software must be installed on your PDA and your PC before you can synchronize the two devices. The software usually comes already installed on your PDA. A CD-ROM is usually included that can be used to install the software on your PC.

DigiTip
Synchronization software can also be downloaded from Microsoft at www.microsoft.com, from Palmsource at www.palmone.com, or from the manufacturer of your PDA.

4. Choose and read each menu option for Letter Recognizer or Block Recognizer. Tap the **Back** and **Forward** buttons to navigate. Tap the **Demo** button to see how letters and other gestures are written. After you finish reading all the screens, tap **OK** to close Help and return to the *Notes* screen.

5. Now practice writing using Block Recognizer or Letter Recognizer. On the first line of the note, write `Alphabet Practice`. Use the Enter gesture twice to move down two lines (or tap the **Enter** key on the right of the input panel).

6. Write each letter of the alphabet in lowercase (small) letters. Use the Space gesture to make a space between letters. Use the Backspace gesture to backspace and erase any errors. Then try writing the letter again. Access the Help screens again if you have trouble with a particular letter. Use the Enter gesture once to move down one line.

7. Write the entire alphabet in capital letters. If you are using Letter Recognizer, write in the area of the input panel under the capital letters *ABC*. If you are using Block Recognizer, use the Shift stroke before writing each letter for the first few letters. Then use the Caps Lock stroke and write the remaining letters. See Figure D-30 to review these gestures.

8. Practice writing each number (0 – 9) three times. Use the Space gesture to space after each number. Use the Backspace gesture to erase mistakes. Close the note.

Palm OS

1. Learn more about and practice *Graffiti* or *Graffiti 2*. Access the list of programs available on your PDA. Open the *Quick Tour* or *Welcome* program. (You may also have a *Graffiti 2* program icon that will provide the same information.)

2. Choose the **Entering Data** option or tap the **Next** button to move through the screens. The Entering Data portion of the tour will tell you about the *Graffiti* or *Graffiti 2* program. On the *Letters* or *Practice Letters* screen, practice writing letters. (You may have to tap a **Try It!** button to reach this screen.) See Figure D-32. Practice writing lowercase (small) letters at the left of the screen.

3. Continue through the practice screens. Practice writing capital letters and numbers.

> **DigiTip**
> To turn off Caps Lock mode, use the Shift stroke to set your input area back to lowercase.

Figure D-32
Letters Practice Screen

Date:	Next Thursday
Subject:	Administrative Assistant Hiring
Location:	Robin Mills' Office
Starts:	3:30 p.m.
Ends:	4:00 p.m.
Occurs:	Once
Reminder:	Remind me (25 minutes)
Attendees:	Robin Mills

LESSON 5: SYNCHRONIZING YOUR PDA; USING ONLINE, FINANCIAL, E-MAIL, AND OTHER APPLICATIONS

OBJECTIVES

In this lesson you will:

1. Synchronize your PDA data with a personal computer.
2. Display PDA data in personal information management software on a PC.
3. Network a PDA with Wi-Fi, Bluetooth, or pass through connections.
4. Send and receive e-mail from your PDA.
5. Download applications for your PDA.
6. Enter a Web address and navigate the Internet through hyperlinks.
7. Use a Web search tool.
8. Set up a personal checking account using PDA financial software.
9. Add, edit, and reconcile banking transactions using PDA financial software.
10. Try word processing, spreadsheet, presentation, reader, instant messaging, and multimedia applications.

In the first four lessons of Appendix D, you learned to use some basic PDA applications. You learned to input appointments, contacts, addresses, phone numbers, and to do or task lists. These tasks, however, just scratch the surface of what you can do with your PDA or smart phone. Now that you know the basics, you can take full advantage of your PDA's power. To do so, however, you must link your PDA to your personal computer, tie it into the global Internet, and synchronize all of your valuable data.

SYNCHRONIZING WITH A PC

Wouldn't it be a shame if your PDA's battery died and you accidentally lost all of your important contacts, appointments, and tasks? To prevent this, you can synchronize your PDA data with your desktop or portable computer. **Synchronize** means to exchange data between certain programs on your PDA and your PC so that they contain the same

4. After you finish all of the practice screens, tap **Done** to close the program.

5. Open the *Memos* (or *Memo Pad*) application. Choose **New** to open a new memo. For the first line of the note, write `Alphabet Practice`. Use the Enter gesture twice to move down two lines.

6. Write each letter of the alphabet in lowercase (small) letters. Use the Space gesture to make a space between letters. Use the Backspace gesture to backspace and erase any errors. Then try writing the letter again. Use the Enter gesture once to move down one line.

7. Write the entire alphabet in capital letters. If you are using *Graffiti 2*, write in the area of the input panel for capital letters (the center). If you are using *Graffiti*, use the Shift stroke before writing each letter for the first few letters. Then use the Caps Lock stroke and write the remaining letters. See Figure D-30 to review these gestures.

8. Practice writing each number (0 – 9) three times. Use the Space gesture to space after each number. Use the Backspace gesture to erase mistakes.

9. Practice opening Help screens with information about *Graffiti* or *Graffiti 2*. Tap on the input panel at the center bottom of the screen and drag the stylus all the way to the top of the screen in one stroke. Use the up and down arrows to move through the screens. Tap **Done** to close Help. Close the memo.

Punctuation and Special Characters

For the most part, you can use standard handwriting to create punctuation marks as you need them. However, sometimes it may be difficult to enter punctuation marks by hand. Both the Pocket PC and *Palm OS* provide options to help make entering punctuation simpler.

In Letter Recognizer for Pocket PC, tap the **Symbols** button to access the Symbols keyboard, as shown in Figure D-33. Use this keyboard to enter punctuation marks. To close the keyboard, tap the **Symbols** button again.

> **DigiTip**
> To turn off Caps Lock mode, use the Capital gesture to set your input area back to lowercase.

Figure D-33
Pocket PC Symbols Keyboard

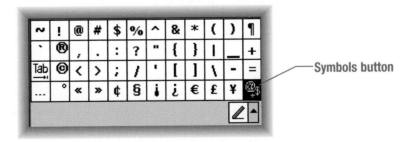

For Palm devices, enter punctuation marks using the online keyboard. See Figure D-34. Tap **123** in the bottom right corner of the input area to access the Numeric keyboard. Use the onscreen keyboard for any punctuation marks that give you trouble. Tap **Done** to close the keyboard.

Pocket PC

Open your *Calendar* application. Open the specific meeting you wish to beam by tapping on it. Choose **Tools, Beam Appointment**.

Palm OS

Open your *Calendar* or *Date Book* application. Choose the meeting you wish to beam by tapping on it. Tap the date (for example: **Jun 07, 04**) at the top of the screen and choose **Beam Event** from the menu that appears. (You can also tap the **Menu** soft button to access the menu.)

Peer Check Check with the classmate to whom you tried to beam your appointments. Was the data received? If not, try the beaming activity again.

APPOINTMENTS

```
Date:       Next Monday
Subject:    Advertising Campaign
Location:   Maria's Office
Starts:     1:00 p.m. (Monday afternoon)
Ends:       5:00 p.m.
Occurs:     Weekly, every Monday
Reminder:   Remind me (15 minutes)
Attendees:  Maria Bravo


Date:       Next Tuesday
Subject:    Web Site Meeting
Location:   Luis Delgado's Office
Starts:     8:00 a.m. (Tuesday morning)
Ends:       9:30 a.m.
Occurs:     Once
Reminder:   Remind me (15 minutes)
Attendees:  Maria Bravo, Luis Delgado


Date:       Next Tuesday
Subject:    Hong Kong Production Schedule
Location:   Conference Room 101
Starts:     1:00 p.m. (Tuesday afternoon)
Ends:       3:00 p.m.
Occurs:     Weekly, every Tuesday
Reminder:   Remind me (15 minutes)
Attendees:  Sandy Frank, David Wu
```

Figure D-34
Palm Numeric Keyboard

Tap to view symbols and special characters

APPLICATION D-3 — UPDATE A NOTE OR MEMO

Kaoru received another voice message from Casey about the time for the conference call concerning the new smart phone/PDA. She will be available at 1:30 p.m. Eastern Standard time. Kaoru knows that every conference call should be well planned. A few minutes before the call, she jots down a few notes about things that must be discussed. With her PDA handy, she will certainly remember every item and question she has concerning the new smart phone/PDA. She quickly learns that notes should be written cryptically; that is, leaving out any unnecessary words. Update the note or memo with the new information.

1. Open the *Notes* or *Memos* program. Open the **Conference Call** note or memo that you created earlier.

2. Use the writing system you practiced in Activity 1-8 to update the note or memo. Use the Symbols or Numeric keyboards for any symbols you have trouble writing. Tap on the line below *Conference Call* and write `1:30 p.m. EST`.

3. Add the last names for the participants followed by their phone number extensions at their Corporate View offices.

```
María Bravo 0175
Sandy Frank 0183
David Wu 0223
```

Figure D-83
Set the Alarm to 15 minutes.

7. Choose the **Note** icon (**Note** button) at the bottom of the screen. Enter the following into the note that is attached to this calendar item.

```
Attending:
Charles Cooper and David Wu

Review legal documents, warranty
information, and patents and
copyright issues.
```

8. Choose **Done** to record the event.

Self Check Does your event have a Note icon and an Alarm icon beside it?

APPLICATION D-9 — ENTER AND BEAM APPOINTMENTS

TEAMWORK

Kaoru visits with Maria to plan and schedule additional meetings. They quickly realize they can save time by entering half of the meetings each and then beaming the information to each others' PDAs or smart phones.

1. In this activity, work with a teammate just as Maria and Kaoru did. Enter data in your *Calendar* or *Date Book* application for two of the meetings shown on pages App-96–App-97. Have your partner enter the other two meetings. Refer to Activity D-23 for information on how to enter new *Calendar* items.

2. Beam the meetings you entered to your partner. Have your partner beam meetings to you. Beaming *Calendar* events is similar to beaming *Contacts* entries. Review the procedures below for your PDA.

DigiTip
Did you know that you can enter birthdays into your PDA? You can be reminded several days in advance so you will have time to buy or make a card or gift.

APPENDIX D: ENHANCING PERSONAL PRODUCTIVITY WITH PDAS

DigiTip
Pocket PC users, write your punctuation marks in the number section.

4. Enter the following notes below the names.

```
Discussion Items:

Brainstorm exciting name!
Discuss schedule, timeline, and costs.
Who is on product team?
```

5. Close the note or memo.

Self Check Compare the phone extension numbers in your note or memo with those listed in Step 3. Are the numbers correct? If not, make changes as needed.

NATURAL HANDWRITING RECOGNITION

The next innovation in PDA input utilizes natural handwriting recognition technology. This technology is similar to the handwriting technologies found on Tablet PCs. **Natural handwriting technology** allows you to write words and phrases normally on a PDA screen using either cursive or printed characters. The Pocket PC version of this DigiTool is called *Transcriber*.

Transcriber allows you to write complete words and phrases instead of individual letters. Many people find using *Transcriber* a big improvement over using *Graffiti* and onscreen keyboards. For handwriting recognition to be accurate, you must practice writing clearly. Here are some things to remember to improve your accuracy:

- Write as big as you can.
- Write as clearly as you can.
- Don't crunch words together. Leave a space between each word, or write only one word per line. See Figure D-36 on page App-59.
- Use either printed or cursive letters, whichever you find to be most accurate.

DigiTip
Generally, cursive writing will give more accurate input than printing. However, both work as long as they are written very legibly.

3. Tap **10:00** (followed by **New** if necessary). The *Set Time* screen will appear. In the **Start Time** box, make sure **10:00 am** is displayed. Change the end time. Tap **11:00 am** to select the time. Tap **30** in the right column. This will change the end time to **11:30 am**. See Figure D-81. Tap **OK**.

Figure D-81
Set Time Screen

4. Notice the little bracket that appears to the left of the 10:00 to 11:30 time span. This indicates that an event has been set for this time period.

5. On the line beside 10:00, enter the event description `Legal Meeting, Conf. Room 101`. See Figure D-82.

Figure D-82
Enter an Event Description

6. Choose **Details** at the bottom of the screen. Tap the box next to **Alarm** to place a check mark in the box. Increase the time on the alarm. Enter **15** by **Minutes**, as shown in Figure D-83.

ACTIVITY D-9 — USING NATURAL HANDWRITING FOR TASKS

Kaoru decided it was time to use her to do or task list. This list will help remind her of all the important things that she must accomplish. The idea behind to do or task lists is to record each item and then check it off after it has been completed. In this exercise, practice with the Transcriber tool while creating Kaoru's task list.

Pocket PC

1. Open the *Tasks* program (tap **Start**, **Tasks**). Tap the up arrow at the bottom of the screen and choose **Transcriber**.

2. Tap the phrase **Tap here to add a new task**, as shown in Figure D-35.

Figure D-35
Create a Task

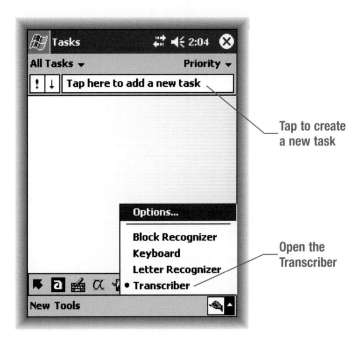

3. Open the Microsoft Transcriber Keyboard by tapping on its icon. See Figure D-36. This tool brings up a keyboard that will allow you to make deletions (with the Backspace key) and to enter difficult punctuation marks.

4. Write a name for the task, `Call David`. Use either cursive or printed writing style. Tap anywhere below the entry to add this new item to the task list.

6. Tap on the **Attendees** line. A list of your contacts will be available. Place a check mark next to **Charles Cooper** and **David Wu**. Tap **OK**.

7. In the Reminder field, choose **Remind me**. Set the timer for **15 minutes**.

8. Choose the **Notes** tab at the bottom of the screen. Enter the following text for the note: `Review legal documents, warranty information, and patents and copyright issues.`

9. Choose **OK** to close the note. Choose **OK** to record the entry. You may get a message that says, "Inform attendees about meeting changes?" Choose **No** for this simulation.

> **DigiTip**
> In a real business environment, in Step 9, you may want to choose **Yes**. Selecting this option will automatically send an e-mail message to each person you marked. The message will contain a reminder about the meeting that you have just scheduled.

Self Check Tap on your 10:00 a.m. appointment to open it. Does your appointment display an Alarm icon near the top right of the screen? Does your note appear correctly?

Palm OS

1. Open the *Calendar* (*Date Book*) application. Choose the **Weekly** view. Move ahead to next week by clicking the **Forward Arrow** as marked in Figure D-80. (**Note:** These arrows advance one full week at a time.)

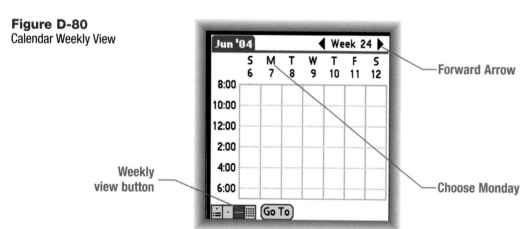

Figure D-80
Calendar Weekly View

2. Tap **M** (for Monday). By tapping on that day, you will automatically be taken to the Daily view.

Figure D-36
Enter a Task Name

5. Tap the phrase **Tap here to add a new task** again. Using either cursive or printed words, enter the next task item `Look up time zones`. Tap anywhere below the entry to add this new item to the task list.

DigiTip
Words can be written anywhere on the screen, even over the transcriber's keyboard.

6. Enter three additional task items:

   ```
   Check e-mail
   Pick up dry cleaning
   Update Web page
   ```

7. When a task item has been completed, tap the box next to it. The item will be marked with a check mark. Practice by placing a check mark next to **Pick up dry cleaning**. See Figure D-37.

DigiTip
If you make a mistake while entering a task, touch and drag over the item. Pause slightly while still touching the screen and the item will be highlighted. You can then rewrite the correct words.

8. Close the *Tasks* application.

Figure D-78
Calendar Daily View

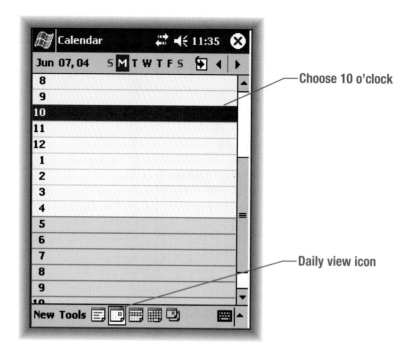

3. Tap **10** in the list of times at the left of the screen to choose a beginning time for the event. (Choose the **10** near the top of the screen for 10 a.m.)

4. Choose **New** at the bottom of the screen. On the Subject line, enter `Legal Meeting`. On the Location line, enter `Conf. Room 101`. See Figure D-79.

Figure D-79
Calendar Appointment Information

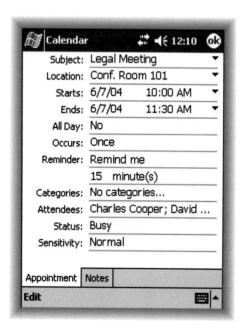

5. In the Starts box, make sure **10:00 AM** is displayed. In the Ends box, change **11:00 AM** to **11:30 AM**.

Figure D-37
Place a check mark next to completed items.

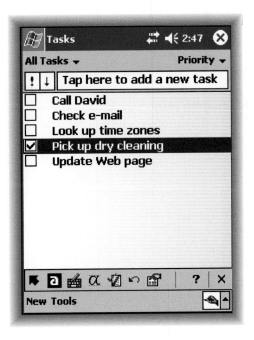

Palm OS

Natural handwriting recognition is available on some PDAs that use the *Palm OS*. If your Palm device does not have a natural handwriting recognition program, complete this activity using *Graffiti 2* or the onscreen keyboards.

1. Open the *Tasks* or *To Do List* application. Open your natural handwriting recognition application or use another input tool for this exercise.

2. Choose **New**. Enter a task name, `Call David`. See Figure D-38. Tap anywhere below the entry to add this new item to the task list.

Figure D-38
Create a Task

ACTIVITY D-23 — ENTER AN APPOINTMENT

Product marketing managers are often assigned a single product. They shepherd the product from research and development through production and into the marketplace. While all of this is going on, they must also plan marketing campaigns to help sell the new product. This very demanding job requires efficient personal and team organization. This is where Kaoru's *Calendar* program on her new PDA comes in handy.

Kaoru's assignments continue to grow. During the conference call, Casey outlined several critical meetings that need to be set up within the next week. Casey wants all of these meetings scheduled before her return. It's time to juggle a few schedules and make these meetings happen.

Pocket PC

1. Open the *Calendar* application. Choose the **Weekly** view. Move ahead to next week by clicking the **Forward Arrow** as marked in Figure D-77. (**Note:** These arrows advance one full week at a time.)

Figure D-77
Calendar Weekly View

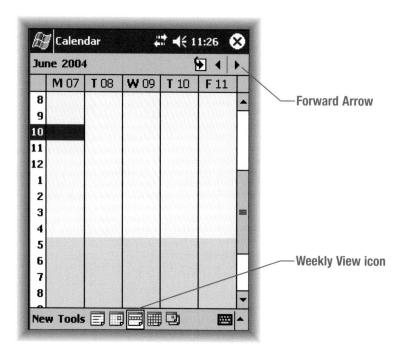

2. Tap **M** (for Monday). By tapping on that day, you will automatically be taken to the Daily view. Notice that the Daily view icon is now selected. See Figure D-78.

3. Tap **New** again. Enter each task listed below. (Remember to tap anywhere below the entry to add a new item to the list.)

```
Look up time zones
Check e-mail
Pick up dry cleaning
Update Web page
```

4. When a task item has been completed, tap the box next to it. The item will be marked with a check mark. Practice by placing a check mark next to **Pick up dry cleaning**. In some versions of *Palm OS*, the checked item will move to the top of the list. See Figure D-39.

Figure D-39
Palm Task List

APPLICATION D-4 — FIND TIME ZONE DIFFERENCES WITH YOUR PDA

Casey Jones is in Orlando, Florida, in the Eastern time zone. However, she wants to have a conference call with colleagues in Boulder, Colorado, in the Mountain time zone.

1. If it is 11 a.m. in Orlando, Florida, what time is it in Boulder, Colorado? Use your PDA to find the time difference between the two time zones.

Pocket PC

Go to the *Today* screen. Tap the **Clock** icon. Set the time zone to a city in the Eastern time zone. Note the time. Set the time zone to a city in the Mountain time zone. Note the time. How many hours separate the two time zones?

Palm OS

Open the *Prefs* (Preferences) application and select **Date and Time**. (You can also use the *World Clock* program if it is available on your PDA. See Figure D-40.) Set the time zone (Location) to a city in the Eastern time zone. Note the time. Set the time zone to a city in the Mountain time zone. Note the time. How many hours separate the two time zones?

Palm OS

1. Tap the **Find** button marked in Figure D-75. Enter **l** on the *Find* line and tap **OK**.

Figure D-75
Find Soft Button

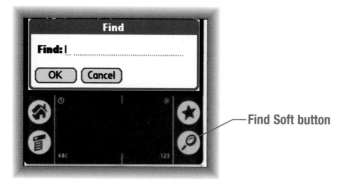

2. Notice that only two or three matches are shown. The search results are limited to only those contacts with their first or last names beginning with the letter L. See Figure D-76.

Figure D-76
Contacts Find Box

3. Tap **Cancel**. Repeat Steps 1-2, but this time enter the letters **lu**. If necessary, spell out the entire first name **Luis** until only the entry for Luis Delgado is visible. Tap **Cancel**.

CALENDAR OR DATE BOOK PROGRAM

A calendar or date book can help you organize all of your activities. Activities can be scheduled daily, weekly, and monthly. Kaoru finds it necessary to add appointments and keep track of meetings, deadlines, and projects using her PDA. She learns how to create appointments and edit entries in her *Date Book* or *Calendar* application. She also sets alarms to remind her of important meetings and events. In the next activity, you will learn to enter a meeting or appointment as Kaoru does.

Figure D-40
World Clock Program

2. Find the time differences for other cities in other time zones. If it is 11 a.m. in Orlando, Florida, what time is it in:
   a. Honolulu, Hawaii
   b. London, United Kingdom
   c. Berlin, Germany
   d. Hong Kong, People's Republic of China

Peer Check Compare your answers with those of a classmate. If your answers do not agree, determine which ones are correct.

LESSON 3: BASIC PERSONAL PRODUCTIVITY TOOLS, PART 1

OBJECTIVES *In this lesson you will:*

1. Change the name of a memo or note.
2. Delete a memo or note.
3. Prioritize, check off, and delete tasks or to do items.
4. Add attachments to tasks or to do items.
5. Set priorities and categories for tasks or to do items.
6. Categorize memos or notes.
7. Write notes using the ink feature if available.

DigiTip
Activities D-8 and D-9 in Appendix D Lesson 2 are prerequisites for this lesson. Complete each of them before continuing if you have not already done so.

Self Check How many contacts are stored in your Personal category? You should have at least three contacts (friends or family) in this category.

ACTIVITY D-22   SEARCH WITH THE FIND FEATURE

After you have stored data for many contacts, locating the exact contact you need can take some time. Fortunately, the *Find* feature makes locating contacts as easy as tapping a few letters with your onscreen keyboard. Kaoru must quickly locate the contact information for Luis Delgado. Follow the steps for your OS to see how she accomplished that task.

Pocket PC

1. Open your *Contacts* application.

2. Show all of your contacts by choosing the **All Contacts** option. Tap in the **Find a name** box. See Figure D-73.

Figure D-73
Find Contacts

3. Enter the letter **l**. Notice that only two or three contacts remain visible, including those with first names beginning with the letter L. See Figure D-74.

Figure D-74
Contacts Find Box

4. Tap the **Clear** arrow to clear your list to begin again. See Figure D-74.

5. Display only the contact for Luis Delgado. Enter the letters **lu**. That should be enough to display just the contact for Luis. Close the application.

> **DigiTip**
> You may also spell out the entire first name of **Luis** until only the entry for Luis Delgado is visible.

The applications that first made PDAs valuable were:

- Notes or Memos
- Tasks or To Do Lists
- Contacts or Address Lists
- Calendars or Date Books
- Calculators
- Beaming

In the next few lessons, you will master these core PDA programs. As you learn to use them, you will also learn ways to increase your personal productivity. After learning to use these applications, learning any new programs you may add to your PDA will be easier.

MANAGING MEMOS OR NOTES

Memos or notes programs are great for jotting down facts, thoughts, quotes, or ideas. Keeping your notes organized is important. Notes and memos may be listed in alphabetical order on your system. If so, this is helpful for finding a note quickly. The names you choose for your memos and notes will help you rapidly find the information you are looking for. Use descriptive titles. A **descriptive title** is a phrase that briefly and accurately describes a note's contents. These titles will remind you what the notes or memos you have written are about. Deleting old, unneeded notes is also important because they can use valuable memory space and make it harder to locate newer notes.

ACTIVITY D-10 — CHANGE THE NAMES OF MEMOS AND NOTES

Note: Practice writing or inserting punctuation before continuing if you have not already done so. See the Punctuation and Special Characters section of Appendix D Lesson 2.

Marketing managers, like Kaoru, often use the term "talking points" to describe a list of issues they wish to discuss. In this exercise, you will change the name of the **Conference Call** note or memo to **Talking Points**.

Pocket PC

1. Open the *Notes* application (tap **Start**, **Notes**).

2. Rename the **Conference Call** note (created in Activity D-6). Move your stylus directly on the title, **Conference Call**. A series of circles will appear and then a pop-up menu will be displayed, as shown in Figure D-41. Choose **Rename/Move** from the menu. The Rename/Move dialog box will appear.

Self Check How many contacts are stored in your Business category? You should have at least seven contacts (Corporate View employees) in this category.

Palm OS

1. Open your *Contacts* (*Address*) application.

2. Open the first contact for which you wish to change the category. Choose **Edit** from the bottom of the screen. Choose **Details**. Tap the down arrow by **Categories** and choose **Business**. Tap **OK**. Tap **Done** to return to the *Contacts* list. Repeat the process for each business contact.

3. To view all of the Business contacts, tap the down arrow for the **Categories** menu at the top right of the screen. Choose **Business** from the menu. See Figure D-72.

Figure D-72
View Only Business Contacts

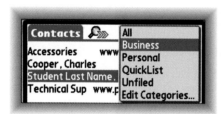

Self Check How many contacts are stored in your Business category? You should have at least seven contacts (Corporate View employees) in this category.

APPLICATION D-8 ENTER AND CATEGORIZE THREE PERSONAL CONTACTS

Now that you know how to enter contacts into a PDA, enter data for your friends and family.

1. Open your *Contacts* (*Address*) application.

2. Choose three friends or family members. Create a new contact for each person. Enter each person's contact information.

3. Set the category for each new contact to **Personal**.

4. To view how your contacts have been organized, switch between the show **All**, show **Business**, and show **Personal** contacts options. If you have forgotten how to do this operation, review Activity D-21 Categorize Business Contacts.

Figure D-41
Rename a Note

3. In the **Name** box, delete the previous title, **Conference Call**, and replace it with the new title, **Talking Points**, as shown in Figure D-41. Tap **OK** to return to your list of notes.

Self Check The name should now appear as **Talking Points** in the *Notes* list.

Palm OS

1. Open the *Memos* (or *Memo Pad*) application. Open the **Conference Call** memo (created in Activity D-6).

2. Highlight the old title **Conference Call** by touching and dragging over the name, as shown in Figure D-42.

Figure D-42
Rename a Memo

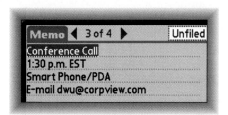 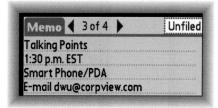

3. With the old name selected, write or key the new title, **Talking Points**, as shown in Figure D-42. Tap **Done** to return to your memos list.

Self Check The name should now appear as **Talking Points** in the *Memos* list.

ACTIVITY D-20 EDIT A CONTACT

Kaoru noticed something strange about Charles Cooper's e-mail address. Everyone else at Corporate View used the initial of his or her first name followed by the last name (mbravo@corpview.com). Charles did the reverse (charlesc@corpview.com). Kaoru wondered if she had made a mistake. When she checked the Corporate View intranet, she found that the correct e-mail address is *ccooper@corpview.com*. Learn to edit a contact item as you update the e-mail address.

Both Pocket PC and Palm OS

1. Open your *Contacts (Address)* program. Open the contact you wish to edit, **Cooper, Charles**.

2. Choose **Edit** from the bottom of the screen. Change the e-mail address from `charlesc@corpview.com` to `ccooper@corpview.com`.

3. Tap **OK** or **Done** to return to the *Contacts* list.

ACTIVITY D-21 CATEGORIZE BUSINESS CONTACTS

Kaoru finds having her business contacts on her PDA quite handy. She decides she would like to have contact data for all of her friends and family recorded in her PDA as well. Before entering data for personal contacts, however, she should move all of her current contacts into the Business category. This will make it easier for her to separate business from personal contacts in the future. Follow Kaoru's example and place all of the current Corporate View contacts in the Business category.

Pocket PC

1. Open your *Contacts* application.

2. Open the first contact for which you wish to change the category. Choose **Edit** from the bottom of the screen. Scroll down to the **Categories** item. Tap **No categories…**. Place a check mark next to **Business**. Tap **OK** to return to the *Contacts* list. Repeat the process for each business contact.

3. To view all of the Business contacts, choose **Business** from the **Show** menu located in the top left corner of the screen, as shown in Figure D-71.

Figure D-71
View Only Business Contacts

ACTIVITY D-11

DELETE MEMOS AND NOTES

Kaoru knows that she must delete old notes or memos in order to keep her lists from becoming cluttered. Follow her lead by deleting the old memo or note **Alphabet Practice** (that you created in Activity D-8).

Pocket PC

1. Open the *Notes* application (tap **Start**, **Notes**).

2. To delete the note, rest your stylus directly on the title **Alphabet Practice**. A series of circles will appear and then a pop-up menu will be displayed, as shown in Figure D-43.

Figure D-43
Notes Pop-Up Menu

3. Choose **Delete**. You will be asked if you want to permanently delete the note. Choose **Yes**.

Self Check The note title will no longer appear in the *Notes* list.

Palm OS

1. Open the *Memos* (or *Memo Pad*) application. Open the **Alphabet Practice** memo.

2. Tap the **Details** button at the bottom of the memo screen. Choose **Delete**.

3. You will be asked if you want to permanently delete the memo. Remove the check mark from the **Save archive copy on PC** box if one appears. See Figure D-44. Tap **OK**.

> **DigiTip**
> Routinely, you may want to save a copy of each memo on your PC before deleting it. However, since this memo is unimportant, you can choose not to save a copy on your PC.

Figure D-69
Choose Select Business Card

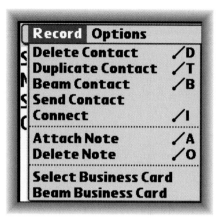

4. In the Select Business Card dialog box choose **Yes**, as shown in Figure D-70.

Figure D-70
Select Business
Card Dialog Box

5. Align the infrared ports between your PDA and another person's PDA. Send your business card by holding down the **Contacts** (**Address**) button on your PDA. Remember to keep the button depressed until your PDA indicates that the information has been sent successfully.

> **DigiTip**
> *Palm OS* users can also send their business card by opening the *Contacts* menu and choosing **Beam Business Card**.

Peer Check Check with a classmate to whom you tried to beam your business card. Was the data received? If not, try the beaming activity again.

APPENDIX D: ENHANCING PERSONAL PRODUCTIVITY WITH PDAS

Figure D-44
Delete Memo Dialog Box

Self Check The note title will no longer appear in the *Memos* list.

ACTIVITY D-12 — MARK COMPLETED TASKS OR TO DO ITEMS

Kaoru just finished talking with David. With that task complete, she can check off her *Call David* item. The check mark means the task or to do item has been completed.

Both Pocket PC and Palm OS

1. Open your *Tasks* or *To Do List* application.

2. Tap in the box next to the *Call David* item until a check mark appears. See Figure D-45.

Figure D-45
A check mark indicates a completed task.

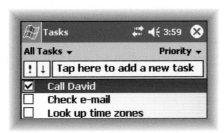

Pocket PC

Palm OS

ATTACHMENTS

Attachments are additional information in the form of a document or other file. Attachments became popular with e-mail as people began to attach photos, files, and other information to e-mail messages. Attachments on PDAs work in a similar way. For example, if you have a task that needs further explanation, you can attach a note and link it to your task or to do item so they will appear together.

PDA PERSONAL BUSINESS CARDS

Beaming is an important business skill. It is very common today for businesspeople to hand business cards to each other as they meet. Now these business cards can be electronically sent from one PDA to another. For instance, if you have your personal electronic business card saved on your PDA, you can beam that business card to a prospective client or to a colleague. The contact information will be stored on the client's or colleague's PDA or smart phone.

ACTIVITY D-19 — CREATE AND BEAM A PERSONAL BUSINESS CARD

> **DigiTip**
> Do not enter any personal information you do not want widely shared with others. You may want to include an office or a school phone number rather than your personal number. You may also want to use an office or school address rather than your personal home address.

Kaoru meets new business contacts every day. Many of these people have PDAs or smart phones. She realizes that it would be convenient for them if she could beam her contact information to their PDAs. That way, they wouldn't need to enter her contact information by hand. Follow Kaoru's example and create your own electronic business card.

Pocket PC

1. Open the *Contacts* application. Tap **New**.

2. Enter your information for this contact. Tap **OK** after you have entered the data. See the DigiTip at the left regarding personal data.

3. Beam your contact information to one or more classmates.

Peer Check Check with a classmate to whom you tried to beam your business card. Was the data received? If not, try the beaming activity again.

Palm OS

Many Palm PDAs have a feature that will allow you to send your personal business card anytime you want by simply pressing and holding down the *Contacts* (*Address*) button. To use this feature, however, you must change a few settings.

1. Open the *Contacts* application. Tap **New**.

2. Enter your information for this contact. Tap **OK** after you have entered the data. See the DigiTip above regarding personal data.

3. Open the contact for your name. Tap **Contact** (**Address**) at the top of the screen to access a menu. Choose the **Select Business Card** option. See Figure D-69.

DIGITAL INK

Pocket PC users can use digital ink input. If you have a recent version of *Palm OS*, you may also have the ability to use digital ink. **Digital ink** allows you to insert pictures or handwritten text into a document. There are two ways to use ink. You can draw pictures or you can write notes in ink, which preserves your original handwriting. This input method is handy when you must capture information quickly and don't have the time to worry about clear penmanship or correcting recognition errors. To use this option, tap on the Handwriting or Pen Tool icon.

ACTIVITY D-13: ATTACH NOTES TO TASKS OR TO DO ITEMS

Kaoru has another important task to perform before the much-anticipated conference call can take place. She must determine the exact time of a conference call for each participant. One of the team members is in Orlando, Florida, in the Eastern time zone. The other members are in the Mountain time zone. She will attach this information as a note to her tasks or to do item.

Pocket PC

1. Open your *Tasks* application. Tap in the middle of the **Look up time zones** task to open it.

2. Tap **Edit** located at the bottom of the screen. Tap **Notes**, also located at the bottom of the screen. Tap the **Drawing** or **Ink** icon. See Figure D-46.

Figure D-46
Notes Written with Digital Ink

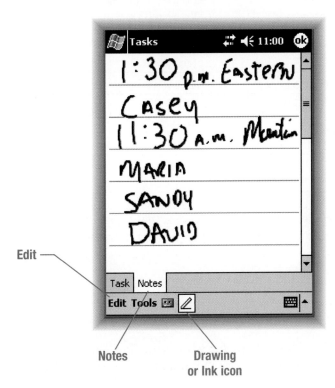

2. Enter half of the contacts below. Your partner should enter the other half of the contacts. Change the field names selected on your *Contacts* screen, if needed, to match the data you must enter. (*Palm OS* users, use the **Other** field for the department name.)

3. Beam your contacts to your partner. Ask your partner to beam contacts to you.

Contacts Information

The following information is the same for each contact. Enter this information (along with the other information shown in the table below) for each contact.

Company:	Corporate View
Work Telephone:	303-555-0116
Fax:	303-555-0115
Address:	One Corporate View Drive
City:	Boulder
State:	CO
ZIP Code:	80303
Country:	USA

NAME	TITLE (JOB)	DEPARTMENT	E-MAIL
Bravo, Maria	Communications Specialist	Corporate Communications	mbravo@corpview.com
Delgado, Luis	Intranet Webmaster	Information Technology	ldelgado@corpview.com
Frank, Sandy	Production Manager	Manufacturing	sfrank@corpview.com
Jones, Casey	Director of Marketing and Sales	Marketing and Sales	cjones@corpview.com
O'Malley-Cohen, Laurie	Customer Support Coordinator	Customer Support	lomalleycohen@corpview.com
Wu, David	Assistant Director	Research and Development	dwu@corpview.com

Self Check How many contacts are stored on your PDA? You should have seven contacts who are Corporate View employees.

3. Write the information shown below in the note:

   ```
   1:30 p.m. Eastern
   Casey
   11:30 a.m. Mountain
   Maria
   Sandy
   David
   ```

4. Tap **OK** to close your task with its attached handwritten note.

Self Check Reopen the **Look up time zones** item note. Your handwritten text should appear in the note.

Palm OS

1. Open your *Tasks* or *To Do List* application.

2. Tap the **Look up time zones** item to select it. Tap **Details** located at the bottom of the screen. Tap the **Note** icon or button, also located at the bottom of the screen. See Figure D-47.

Figure D-47
Attach a Note

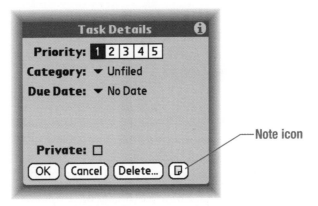

3. If it is available, choose the handwriting (ink) option. See Figure D-48. If you do not see this icon and handwriting is not available, enter your data using either *Graffiti 2* or your onscreen keyboards.

Figure D-67
Beam a Contact to Another PDA

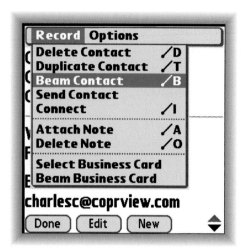

4. A screen will appear that will indicate the progress of your beaming. You will see messages that specify *searching*, *sending*, then *disconnecting* from the target PDA.

5. After the data has been sent, the receiving device will display a message asking if the information is to be accepted. When the user taps **Yes**, the information will be saved on the target PDA. See Figure D-68.

Figure D-68
Accept a Contact from Another PDA

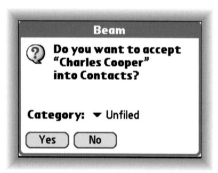

APPLICATION D-7   ENTER AND BEAM BUSINESS CONTACTS DATA

TEAM WORK

Kaoru realizes that she has several contacts to enter and wonders if there is a way to download the information quickly. Fortunately, Maria Bravo has a smart phone and knows several tricks. They sit down together and Maria beams to Kaoru her **electronic business card**. This business card is simply a *Contacts* entry that contains Maria's contact information. They soon realize that they can easily share contacts data quickly. This will avoid the necessity of hand-entering each entry individually. Try this quick trick yourself!

Both Pocket PC and Palm OS

1. Work with a classmate to complete this activity. Together, review the directions for how to beam from one PDA to another in Activity D-18.

Figure D-48
The Ink/Handwriting Option in Notes

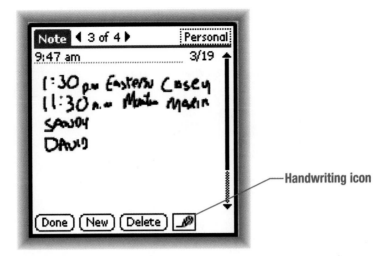

4. Write or key the information shown below in the note. If you use handwriting, your note will look similar to Figure D-48. If you use another input tool, your note will look similar to Figure D-49.

```
1:30 p.m. Eastern
Casey
11:30 a.m. Mountain
Maria
Sandy
David
```

Figure D-49
Note Attached to Tasks Item

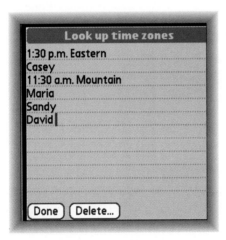

5. Tap **Done** to close your note.

Self Check To see how your note appears, tap the **Note** icon beside the task name on the *Tasks* list. (On some versions of *Palm OS*, select the **Look up time zones** item, choose **Details**, then **Note**.)

4. A screen will appear that lists devices nearby that you can beam to. If multiple devices are listed, select the device you want. For example, in Figure D-65, Kaoru is beaming from her PDA to Maria's SmartPhone. Choose **Tap to send** to start the beaming. This screen will also indicate the progress of your beaming effort.

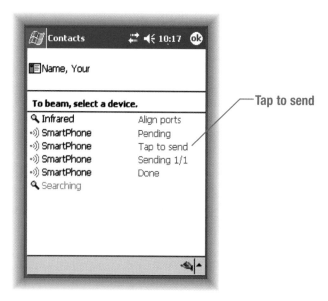

Figure D-65
Beam a Contact to Another PDA

5. After the data has been sent, the receiving device will display a message asking if the information is to be accepted. When the user taps **Yes**, the information will be saved on the target PDA. See Figure D-66.

Figure D-66
Accept a Contact from Another PDA

Palm OS

1. Team up with another PDA user. Locate the infrared ports on your PDAs. Align the two infrared ports so there are no obstructions between the two devices. **Align** in this sense means to position two unobstructed infrared devices facing each other.

2. Open your *Contacts* (*Address*) application by tapping its icon.

3. Open the contact you wish to beam. Tap **Contact** (**Address**) at the top of the screen to access a menu. Choose **Beam Contact** (**Beam Address**) from the menu. See Figure D-67.

TASKS OR TO DO ITEM PRIORITIES

Some tasks or to do items are more urgent or important than others. To help organize a task or to do list by **priority** (order of importance), PDAs come with a prioritization feature. Learn how priorities work in the next activity.

ACTIVITY D-14 — FLAG AND PRIORITIZE TASKS OR TO DO ITEMS

During the conference call, Kaoru was assigned the task of setting up a series of meetings for next week. This is obviously a high priority item! Everyone must be informed, scheduled, and confirmed for the meetings. Because of its importance, Kaoru gave this new item the highest possible priority. Indicating a priority can be done by changing a few simple settings. The program will then "flag" the item as high priority, causing it to attract attention and stand out from the regular list of items.

Another way to be reminded of the importance of an item is to give the item a due date. The due date will remind a PDA user that something must be done before a deadline is reached. You will use both the priority and due date methods to flag an item as high priority.

Pocket PC

1. Open your *Tasks* list. Create a new item called **Schedule Meetings**.

2. Tap the **Schedule Meetings** item to open it. Select **Edit** at the bottom of the screen. Tap on the **Priority** section. Choose **High** from the list.

3. Choose the **Due Date** option and select a date one week from today on the calendar that appears. See Figure D-50. Choose **OK** to close the window.

Figure D-50
Set the Due Date for a Task

BEAMING DATA

DigiTip
After you learn to beam contact information, you can beam other data as well (such as tasks, notes, programs, and games) from one PDA to the other.

Beaming is a process that uses infrared sensors to send information to and from PDAs. This is very similar to how you send information to your television with a remote control. Beaming can be very useful. For example, in the next few activities you will enter data for a list of contacts. Wouldn't it be nice if you could team up with another person to share the task of entering the data? This is just what you will do. After you and your teammate enter the data, you can beam the data back and forth to each other. Using this procedure, you will have your contacts list completed in record time.

ACTIVITY D-18

 WORK

BEAM DATA TO ANOTHER PDA

Follow the steps in this activity to learn to beam contacts data from one PDA to another.

Pocket PC

1. Team up with another PDA user. Locate the infrared ports on your PDAs. Align the two infrared ports so there are no obstructions between the two devices. **Align** in this sense means to position two unobstructed infrared devices facing each other.

2. Open your *Contacts* application by tapping the **Contacts** button or by choosing **Start**, **Contacts**.

3. Open the contact you wish to beam. After the contact is open, choose **Tools**, **Beam Contact**. See Figure D-64.

DigiTip
An alternative way to beam contacts is to touch a contact name with your stylus. Hold on the name until a pop-up menu appears. Then choose **Beam Contact**.

Figure D-64
Beam a Contact to Another PDA

4. On the *Tasks* list window, tap the **Sort By** down arrow near the upper right corner of the screen. Choose **Priority**. See Figure D-51.

Figure D-51
Change the way tasks are sorted using the Sort By menu.

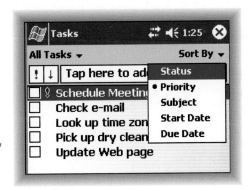

Self Check The high priority items should be placed at the top of the list.

Palm OS

1. Open your *Tasks* or *To Do List* application. Create a new item called **Schedule Meetings**.

2. Tap the new **Schedule Meetings** task to select it. Choose **Details** at the bottom of the screen. This will open the Task Details (To Do Item Details) dialog box.

3. Tap 1 in the **Priority** section, as shown in Figure D-52.

Figure D-52
Change the Priority of a Task

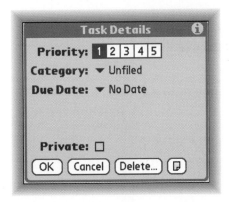

4. Choose the **Due Date** option and select the **In 1 week** option. Choose **OK** to close the window.

DigiTip
When you set a due date or a deadline for an item, your PDA will automatically remind you to complete the task before it is too late!

2. Enter the last name `Cooper` and the first name `Charles`. For the company, enter `Corporate View`. In the **Work** field, enter the work phone number as shown in Figure D-62.

Figure D-62
Enter the contact information for Charles Cooper.

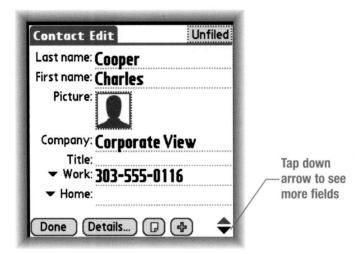

Tap down arrow to see more fields

3. Tap the down arrow by **Home**. Choose **Fax** to change the field name. Enter the fax number `303-555-0115` in this field.

4. Enter `Corporate Counsel` in the **Title** field.

5. Tap the down arrow indicated in Figure D-62 to reveal more entry fields. Enter the e-mail address `charlesc@corpview.com`.

6. Enter the work address information as shown in Figure D-63. Tap **Done** when you have completed the entry.

Figure D-63
Enter the address information for Charles Cooper.

Self Check *Cooper, Charles* should now appear on your *Contacts* list.

5. If you have the *Tasks* program, the window shows completed and uncompleted items. The priority is shown for each task. Dates completed or due dates are also shown. On some versions of *Palm OS,* you must set the information you want to show on the screen. See Figure D-53. If you have the *To Do List* program, choose **Show**. Then choose **Show Due Dates** and **Show Priorities**. Choose **OK**.

Figure D-53
To Do Preferences Screen

6. Choose each of the other items in your list and change the priority to a 2.

Self Check The high priority items should be placed at the top of the list.

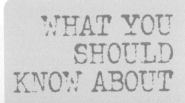

WHAT YOU SHOULD KNOW ABOUT
Setting Priorities

Most people need to do more things during a day than they have time to do them. Professional businesspeople as well as students must set priorities. Priorities help assign a value for each task or activity.

In your academic life, you'll need to make choices. Should you study for a history test or finish your math assignment? You'll need to make choices in your personal life also. Should you wash your car or go to that movie you've wanted to see for over a week? These are important choices because time is limited.

Using the priorities feature on your PDA can help you set a value for each activity and an order in which you will complete activities. Setting priorities is a powerful way to increase your personal efficiency and productivity.

2. Tap the down arrow next to the **Name** field to open a pop-up box. This box will allow you to enter Charles Cooper's name. Enter the first name and last name for `Charles Cooper` as shown in Figure D-60. Tap outside the box to close it.

Figure D-60
Enter a name for Charles Cooper.

Tap down arrow to reveal name entry box

DigiTip
When you view this contact later, you may find that the program has changed *USA* to *United States*.

3. Enter Charles' job title in the **Job title** field.

4. Tap the down arrow next to the **Work addr** (Work address) field to open a pop-up box. Enter the business address information for Charles Cooper. See Figure D-61.

Figure D-61
Enter the work address for Charles Cooper.

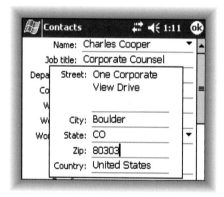

5. Enter the other information for Charles Cooper, such as his department, company, e-mail, work fax, and work phone number. Tap **OK** to return to your *Contacts* list.

Self Check *Cooper, Charles* should now appear on your *Contacts* list.

Palm OS

1. Open the *Contacts* or *Address* application. Tap **New**.

CATEGORIES FOR TASKS

Imagine what would happen if you have 20 personal items, 20 business items, and a dozen miscellaneous items to worry about. Your tasks or to do list would certainly become difficult to manage, even if all of the items were prioritized properly. A **category** is a class or group of items. On a PDA, categories allow you to sort tasks or to do items into logical groupings. The default groupings usually include Business and Personal. Other categories, such as Holiday or Unfiled, may also be included. You can create other categories if you wish.

ACTIVITY D-15 — CATEGORIZE TASKS OR TO DO ITEMS

Kaoru has noticed that the longer her tasks list becomes, the more jumbled items appear. She decides to use categories to group tasks by type. This will help her organize her activities.

Pocket PC

1. Open your *Tasks* list. Tap the **Schedule Meetings** item to open it. Select **Edit** at the bottom of the screen.

2. Choose the **Categories** option. Place a check mark in the box next to **Business**. See Figure D-54. Tap **OK** to close the Categories window. Tap **OK** again to return to the *Tasks* list.

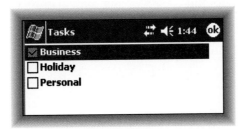

Figure D-54
Tasks can be assigned to a category such as Business.

3. Place the following two items in the Business category.

Look up time zones
Update Web page

4. Place the following two items in the Personal category.

Check e-mail
Pick up dry-cleaning

5. Change the setting to display only Business tasks. Select the **Show** down arrow. Choose **Business** from the list. See Figure D-55. Notice that only **Business** items are displayed.

ADDRESS OR CONTACT LISTS

One of the most helpful programs on a PDA is the *Address* or *Contacts* application. This application provides a powerful way to organize your personal and business contacts. The *Address* or *Contacts* program allows you to collect the following types of information:

- Names
- Addresses
- Phone and fax numbers
- E-mail addresses
- Mailing and shipping addresses
- Other personal contact information

In the next activity, you will use your PDA to organize contact information for several Corporate View employees.

ACTIVITY D-17 — ENTER A BUSINESS CONTACT

Kaoru learned quickly how to enter new contacts into her PDA. She finds her PDA extremely useful when keeping track of all the new contacts and people she meets. Below you can see the first contact she entered. Charles is a lawyer in the Legal Services Department. He advises Kaoru on the many legal aspects of the smart phone/PDA release. She was able to pull the following information about Charles from the Corporate View intranet database.

Contact Information for Charles Cooper

Last Name:	Cooper
First Name:	Charles
(Job) Title:	Corporate Counsel
Department:	Legal Services
Company:	Corporate View
Work Telephone:	303-555-0116
Work Fax:	303-555-0115
E-mail:	charlesc@corpview.com
Work Address:	One Corporate View Drive
City:	Boulder
State:	CO
ZIP Code:	80303
Country:	USA

Enter the contact information for Charles Cooper by following the steps for your PDA.

Pocket PC

1. Open the *Contacts* application and tap **New**.

Figure D-55
Access the Show menu to display different categories of tasks.

6. Change the setting to show only the **Personal** items.

7. Access the Show menu again and choose **All Tasks**. This setting will display both Personal and Business categories.

> **DigiTip**
> Notice how you can also display all completed tasks or just the active tasks. Active tasks are those that have yet to be completed.

Palm OS

1. Open the *Tasks* or *To Do List* application. Tap the **Schedule Meetings** item to select it. Choose **Details** at the bottom of the screen. This will open the Task Details (To Do Item Details) dialog box.

2. Choose the **Category** option. Select **Business** from the list. See Figure D-56. Tap **OK** to close the window.

Figure D-56
Tasks can be assigned to a category such as Business.

3. Place the following two items in the Business category.

Look up time zones
Update Web page

APPLICATION D-6 BEGIN A PERSONAL JOURNAL

Busy businesspeople take the time to record personal notes and thoughts. Many keep personal journals on their PDA. These journals may contain notes or memos that can serve as reminders as to what they were thinking and doing on a particular day.

1. Use your *Notes* or *Memos* application to begin a personal journal. Record notes in the journal for five days or longer. For example, you might list the main tasks you completed each day, ideas you have for activities being planned (birthday party), or items you need to purchase on your next shopping trip.

2. Begin each day's note or memo with "Journal" and the current date. Using this system, your notes or memos will be organized by date. For example: Journal May 1, 20--, Journal May 2, 20--, and so on.

3. Place each of your journal notes in the **Personal** folder or category.

LESSON 4: BASIC PERSONAL PRODUCTIVITY TOOLS, PART 2

OBJECTIVES

In this lesson you will:

1. Enter and edit business and personal contacts.
2. Beam business contacts.
3. Create and beam a personal business card.
4. Categorize business and personal contacts.
5. Sort contacts by personal and business categories.
6. Search contacts with the *Find* feature.
7. Create, edit, customize, and schedule appointments, meetings, or events.
8. Beam appointments.
9. Set alarms for the date in *Calendar* or *Date Book*.

In this lesson, you will learn to use your PDA's *Contacts* or *Address* program and *Calendar* or *Date Book* program. These tools will help you organize personal and business information, manage your schedule, plan events, and arrange activities. You will also learn how to share information with others by beaming data to another PDA.

4. Place the following two items in the Personal category.

Check e-mail
Pick up dry-cleaning

5. Change the setting to display only Business tasks. Tap **Category** at the top of the screen. Tap the down arrow under Category. Choose **Business** from the list. See Figure D-57. Notice that only **Business** items are displayed.

Figure D-57
Access the Category menu to display different categories of tasks.

6. Change the setting to show only the **Personal** items.

7. Tap **All** at the top of the screen. This setting will display items in all categories.

ACTIVITY D-16 — CATEGORIZE BUSINESS MEMOS OR NOTES

You can group memos or notes by placing them in different folders or categories. In this activity, you will place the **Talking Points** note or memo in the Business folder or category.

Pocket PC

1. Open your *Notes* application.

2. Open the **Talking Points** note that you created earlier. Tap **Tools** at the bottom of the screen. Choose **Rename/Move**.

3. Tap the **Folder** option and choose **Business**. See Figure D-58. Tap **OK** to close the window. Close the note.

Figure D-58
Select a Folder for Notes

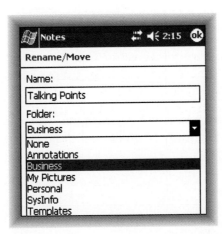

4. On the *Notes* screen, tap the **Show** down arrow and choose **Business**. Your **Talking Points** note should be displayed. Change the **Show** option back to **All Folders**.

Palm OS

1. Open your *Memos (Memo Pad)* application.

2. Open the **Talking Points** memo that you created earlier. Tap the category name at the top of the screen. The category name should be **Unfiled** (unless you have experimented with this option.)

3. Choose **Business** from the list. See Figure D-59. Tap **Done** to close the memo.

Figure D-59
Select a Category for Memos

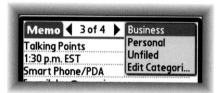

4. On the *Memos* screen, tap the **Category** down arrow at the top of the screen and choose **Business**. Your **Talking Points** note should be displayed. Change the **Category** option back to **All**.

APPLICATION D-5 — CREATE AND ORGANIZE YOUR PERSONAL TASKS

You probably have a very busy life. It's time to organize it!

1. Open your *Tasks* or *To Do List* application.

2. Add five new task items for things that you need to accomplish within the next few days.

3. Set the category for each new item to **Personal**.

4. Prioritize the items by type (High, Normal, or Low) or number (1, 2, 3) depending on the settings available on your PDA.

5. Check each item on your list as you complete it.

Peer Check Discuss the priority you have given each task with a classmate. Ask whether your classmate agrees with your priority assignments. Make changes to the priority of an item, if desired, after your discussion.

INDEX

0–9 keys, 110–114

A

A key, 71–74
absolute cell referencing, 239–240
acceptable use policy (AUP) for network, 9–10, 11
acceptance letters, career building and, 450
Access (See Microsoft Access)
access points, 6, 7
active cells, spreadsheets and, 229
addressing an envelope, 141–142
agriculture and natural resources careers, 62
Airborne Express, 152–154
alcohol use, 448
aligning text, 132, 136–137; speech recognition and, 213
alphanumeric keys, 70
Alt key, 70
AltaVista, 57
Ampersand (&) key, 117–118
anchor tags, HTML and Web site design in, 329–331
animation, presentations and, 304–307

Answer Wizard help, 23
Apostrophe (') key, 104–105
appearance, professional, 309
Apple Macintosh, 12
AppleWorks, 15
application letter, job, 420–423, 438
applications, job, 3, 407–409
architecture and construction careers, 63
arranging files, file management, 29
arrow keys, 70
arts, audio/video technology, communications careers, 378–380
Asterisk (*) key, 118–119
At (@) key, 117–118
attachments to email, 55
attitude, in telephone communications, 360
attributes and values, in HTML and Web site design, 333–334
audience response, presentations and, 310–312
audio for presentations, 293–294, 304–307
audio/video technology, communications careers, 378–380

AutoFit feature, tables and, 147
Autoforms, databases and, 278–280
Automatic Recognition feature, handwriting recognition and, 177–178
AutoShapes, 168

B

B key, 89–90
Backslash (\) key, 118–119
Backspace key, 18, 70, 101–103; speech recognition and, 202
balance sheets, 241
bibliography (See also reference page), 154
biography, 466
blank document, 20
block format letter, 138–139
block paragraphs, 98
block print, handwriting recognition and, 180–181
body language and gestures, 311–312
boldface, 130–131; speech recognition and, 211–212
bookmarks, 47–48
Bracket keys, 118–119
browser software, 35–36

budgets, 241
bulleted lists, 157
business and administration careers, 456–457
business letter, 140
business trends and issues, 65–66, 223, 384–385, 463

C

C key, 85–86
Cancel button, 14
capitalization and changing case, speech recognition and, 208–209
Caps Lock key, 70, 99–100
career and academic skills, viii
career clusters and pathways to success, 31–32, 60–65, 217–225, 378–384, 456–466
career strategies, 5, 389–466; acceptance letters in, 450; career clusters and pathways to success in, 31–32, 60–65, 217–225, 378–384, 456–466; changing jobs and, 412; choosing a career and, 389–392; chronological resume design in, 418–420; corporate Web sites for

INDEX

employment opportunities in, 400–401; e-mail resume services, 416–417; employee handbook and, 454; employee rights and obligations in, 452–455; employee right-to-know laws and, 453–454; evidence to support your resume in, 430–431; follow-up letters in, 449–450; hiring practices and discrimination in, 452–453; investigating career paths in, 391–392; job analysis database for, 404–405; job descriptions and, 392–393, 396–397, 402, 418–419; job interviews in, 440–449; letter of application in, 420–423, 438; letters of recommendation in, 433–435; not getting the job, 451; *Occupational Outlook Handbook* searches for, 397–398; online applications and, 424–427; online job search for, 397–401; organizing and managing job search data in, 404–405; personal employment journal for, 413–415; personal Web page for, 423–424, 439; placement centers and services, 403; portfolios creation in, 428–433, 439; presentations and multimedia shows in, 435–436; qualifications vs. employer needs in, 410–411, 426–427; resumes, resume building in,
407–409, 412–415, 418–420, 437–438; salaries and wages in, 394–396; scanning of electronic resumes, 425; starting your new career, 428–455; temporary employment agencies and, 400, 403; traditional sources of job leads for, 402–403; Web hosting services and, 417; Web sites for employment opportunities in, 399–400; carpal tunnel syndrome (CTS), 108–109, 191 case, changing, 141 cell pointer, spreadsheets and, 229 cell range, 229 cells; spreadsheets and, 229, 231, 243–251; tables and, 144, 146–148, 149–150 cellular phones, 359 centering text, 132, 136–137, 213, 252 central processing unit (CPU), 6 Change Case command, 141 changing case, speech recognition and, 208–209 changing jobs, 412 character formats (bold, italic, underline), 130–131, 211–212 charts and graphs, 257–261; column type, 257–259; creating, 258–259; modifying, 260, pie type, 260–261 chronological reminder files, 353 chronological resume design, 418–420 citations (See textual citations)
citing online sources, 67–68 clearing cells, spreadsheets and, 247 Click and Type feature, 170 clip art, 165, 167–168 client software, email, 52 closing documents, *Microsoft Office*, 20–21 closing programs, *Windows*, 16–17 closing remarks, for presentations, 314 code of ethics (*See also* ethics), 65–66 coding records, records management and, 374 collect calls, 367 Colon (:) key, 97–98 color, 296–297; hexadecimal values for colors in, 334–335; HTML and Web site design in, 332, 346 column charts, 257–259 columns, rows, and cells in tables, 144, 146–148 columns, rows, and cells in spreadsheets, 229, 231, 243–251 Comma (,) key, 95–96 Command key, 70 commands, speech recognition and, dictating, 194–196 commitments, 103 communication, 1; video conferences and, 49–50; flow of information in, 3–4 communications careers, 378–380 compact disks (CDs), 24, 372 company goals, 241 computer crime, 9 computer interface, 12–15 computers, 2, 6
D
D key, 71–74 data files, ix data types; databases and, 263–264; spreadsheets and, 229–230, 237, 246,
confidential information, 159 confidential memos, 128 conflict resolution at work, 66–67 construction careers, 63 Contents help, 23 continuous improvement, 385 continuous speech recognition speeds (CSR), 198 Control (CTRL) key, 70 Copy command, 134–135, 233 copying and pasting Web information, 45 copying files and folders, 28–29 copying paper records, 375–377 copyright, 46–47, 338 Coral Word Perfect Office, 15 Corporate View intranet, ix Correction button, handwriting recognition and, 180 Correction command, speech recognition and, 202–204 counteroffers, in salary negotiation, 394 criminal activity, 448 currency data/signs, spreadsheets and, 248 cursive writing style, handwriting recognition and, 182 customer service, 384–385 Cut command, 134–135, 233

databases, 4, 262–282; Autoforms from, 278–280; changing table structure and fields in, 267–272; creating, 263, 264–267; data types in, 263–264; entering records/data in, 269–271; fields in, 263, 268; filtering records in, 273–274; forms from, 277–280; HTML and Web site design in, document conversion for, 342–345; job analysis database example, 404–405; navigating in, 264; projects by leader query example in, 276; querying in, 274–276; records in, 263, 265–267, 269–271; reports from, 280–281; sorting records in, 272–274; suppliers database table example for, 271–272; suppliers form example for, 280; suppliers report example for, 281; tables in, 263, 265–267, 267–272
date and time, 132
date data type, spreadsheets and, 230
Delete key, 19, 70, 202
deleting files and folders, 30
delivery services, 152–154
departments, 42, 390
Dictation button, speech recognition and, 196
dictionary, speech recognition and, adding words to, 206–207
digital cameras, 6, 302
digital devices, 3
digital ink, Ink option, handwriting recognition and, 184–185

digital video disks (DVD), 24
direct dial calls, 366–367
directories, telephone communications and, 366
disaster recovery plans, 370–371
discrete speech, 198
discrimination, in hiring, 452–453
disk drive, 6
distribution, logistics careers, 222
diversity, 455
division, using numeric keypad, 125–126
Dollar Sign ($) key, 115–116
domain name system (DNS), 39–40
double clicking the mouse, 13
double spacing, 129
Dragon NaturallySpeaking, 190
Drawing toolbar, 166
Dreamweaver, 319
drug use, 448

E

E key, 75–76
ecommerce, 223
Edit menus, 14
edited copy, 106–107, 132–133
education and training careers, 60–61
electronic organizers, 353
electronic test package, x
electronic writing tablet, 174
e-mail, 37, 48, 52–56, 135; addresses for, 53; attachments to, 55; career building and, 416–417; client software for, 52; forwarding, 56; inbox for, 56; Internet Service Provider

(ISP) and, 52; managing messages in, 55–56; *Microsoft Outlook* and, 52; on-screen keyboard input for, 188; receiving, 56; replying to, 56; sending, 53–54; servers for, 52; viruses and, 54–55
employee handbook, 454
employee rights and obligations in 452–455
employee right-to-know laws, 453–454
employment agencies, 400, 403
End key, 18
engineering careers, 217–218
Enter (Return) key, 70, 73
entering text, 18–19
enunciation, 196, 360
envelopes, 141–142
Equal Sign (=) key, 118–119
ergonomics, 109
error correction, speech recognition and, 201–205
Escape key, 70
ethics, 65–66; codes of, 65–66; commitments, 103; computer crime, 9; confidential information, 159; copying paper records, 377; diversity in workplace, 455; gossip, 206; honesty in presentations, 291; intellectual property rights, copyright, 46–47, 338; job descriptions, 402; loyalty, 186; privacy issues, 277; resumes and honesty, 410; searching the Web, 59; stewardship of company funds, 245
evaluating presentations, 314–316

event flyer, 171
Excel (*See Microsoft Excel*)
Exclamation Point (!) key, 115–116
exercises, 5

F

F key, 71–74
favorites lists, 47, 48
faxes, 225
features of DigiTools, viii–ix
FedEx®, 152–154
fiber optic networks, 7
fields, databases and, 263, 268
file management, 24–31; arranging files in, 29; copying files and folders in, 28–29; deleting files and folders in, 30; devices used for, 24; file system for, 25–26; folder creation in, 26–27, 30–31; moving files and folders in, 28–29; naming files and folders in, 25; renaming files and folders in, 27; types of files in, 25; viewing details of files in, 29; *Windows Explorer* and, 25–26
File menu, 19, 20
file system, in *Windows*, 25–26
file transfer protocol (FTP), 432
filing procedures, records management and, 373–374
Fill, 168
fill and fill series, spreadsheets and, 249–251
filtering records, databases and, 273–274
finance careers, 460–461
Find and replace, 14
firewalls, 37
floppy disks, 24

INDEX

flowcharting, 221
flyers, 171
folder creation, 26–27, 30–31
folders (*See* file management)
follow-up letters, career building and, 449–450
font styles, 131
footers, 160; presentations and, 303–304; spreadsheets and, 254–255
formatting text, 130–131; HTML for, in Web site design, 324–326; speech recognition and, 211–216; spreadsheets and, 246
forms, databases and, 277–280
formulas, spreadsheet, 230, 236–242; absolute and relative cell referencing in, 239–240; creating, 236–238; functions in, 238–239; number formats and, 237
forwarding email, 56
frequently asked questions (FAQs), 44–45
FrontPage (*See* Microsoft FrontPage)
fuel costs, 241–242
function keys, 70
functions, spreadsheets and, 238–239

G

G key, 81–82
global marketplace, 223
goals, company, 241
GoLive, 319
Google, 57
gossip and ethics, 206
government and public administration careers, 458–460

grammar check, 137–138
graphics (*See also* visual aids for presentations), 165–171; AutoShapes for, 168; charts and graphs in, 257–261; clip art as, 165, 167–168; Drawing toolbar for, 166; file formats for, 303; Fill in, 168; HTML and handles in, 166, 336; inserting, in a report, 169–170; pixels in graphics/pictures of, 336; resizing and moving, 166–167; visual aids for presentations and, 295–301; wrapping text around, 166–167
Greater Than (>) key, 115–116

H

H key, 75–76
hackers, 9
handles, graphic images, 166
handouts, presentations, and, 294
Handwriting Options feature, 177
handwriting recognition, 4, 172, 174; Automatic Recognition feature in, 177–178; block print in, 180–181; Correction button for, 180; cursive writing style in, 182; digital ink, option in, 184–185; Handwriting Options feature in, 177; handwriting tools for, 174–175; Help for, 176; input to writing pad in, 178–181; Language bar or input panel for, 174–178;
hanging indents, 160
harassment, career building and, 454
hard drives, 24
hard returns, 72
hardware skills, vii–viii, 2, 3–7
headers, 160; presentations and, 303; spreadsheets and, 254–255
health care careers, 382–383
health science careers, 218–220
Help, 14; handwriting recognition and, 176;
help wanted ads, 402
hexadecimal values for colors, HTML and Web site design in, 334–335
hierarchical design, HTML and Web site design in, 320
highlighting (selecting) text, 129
hiring practices and discrimination in, 452–453
hits, during Web search, 57
home keys, 71–74
Home key, 18
homonym correction, speech recognition and, 202, 204–205
honesty, 291
hospitality and tourism careers, 64–65
hubs, 6

HTML and Web site design, 4, 34, 319–348; adding files to, 328–329; anchor tags in, 329–331; attributes and values in, 333–334; color schemes in, 332, 346; computer safety and ergonomics sample for, 340–341; converting file formats for use in, 342–345; copying and pasting Web information from, 45, 338; creating, simple style, 326–332; design issues in, 319, 321–323; general design tips for, 340; graphics in, 332, 336–338; hexadecimal values for colors in, 334–335; hierarchical design for, 320; HTML tags in, 323–326; hyperlinks in, 40–41, 331–332, 345–346; index page for, 331–332; information design in, 319, 339–346; interaction design in, 319, 320–323; linear design for, 321; Microsoft Word documents converted to, 339–341; navigating through, 40–41, 346; Notepad to create, 326–329; personal Web page for employment and, 423–424, 439; pixels in graphics/pictures of, 336; presentation of, 319; random access design for, 320; RSI survey Web page sample for, 344–345; workplace injury Web page sample for, 342–343; hubs, 6

human services careers, 382–383
hyperlinks, 40–41, 331–332, 345–346
hypertext markup language (*See* HTML and Web site design)
Hyphen (-) key, 104–105

I

I key, 77–78
IBM Via Voice, 190, 198
icons, 12
important records, 369
inbox, email, 56
income statements, 241
indentation, 160
Index of help, 23
index page, HTML and Web site design in, 331–332
indexing and coding records, 374
information design, HTML and Web site design in, 319, 339–346
information technology (IT) careers, 383–384
Ink option, handwriting recognition and, 184–185
input, 3
Input panel, handwriting recognition and, 174–178
input technology skills, vii
Insert date feature, 132
Insert key, 70
insertion point, 18
Instant Messaging (IM), 49
instructor's materials, x
intellectual property rights, copyright, 46–47, 338
interaction design, HTML and Web site design in, 319, 320–323

interfaces, computer (*See* computer interfaces)
international travel arrangements, 223–224
Internet (*See* networks, intranets, Internet)
Internet Explorer (*See also* browser software), 35, 324
Internet Service Provider (ISP), 52
Internet telephony, 49
InterNIC, 40
interviews, job, 440–449
intranet (*See* networks, intranets, Internet)
IP addresses, 39–40, 41
italic, 130–131, 211–212

J

J key, 71–74
job descriptions, 392–393, 396–397, 402, 418–419
job interviews, 440–449
job opportunities, evaluation of, 224
justified text, 136–137

K

K key, 71–74
keyboard, 6; on-screen, in mobile computing, 186–188
keyboarding, 4
keying position, 71, 109, 120

L

L key, 71–74
labels, 141–142
landscape page orientation, 167
Language bar; handwriting recognition and, 174–178; speech recognition and, 192–193

laptop (*See* portable/laptop computers)
law and public safety careers, 461–462
left aligned text, 132, 136–137, 213
Left Shift key, 83–84
Less Than (<) key, 115–116
letter of application, 420–423, 438
letterhead stationery, 136
letters, 135–143; block format, 138–139; business type, 140; envelopes for, 141–142; formats for, 136, 137–138; personal-business type, 138–139
letters of recommendation, 434–435
life cycle of records, 369–370
life style and work, 463
line breaks, speech recognition and, 198–200
line spacing; memos and, 129; speech recognition and, 213–214
linear design, in HTML and Web site design, 321
links (*See* hyperlinks)
Linux, 12
listener profile, presentations and, 284–285
lists, 157
login, 38–39
logistics careers, 222
long distance service, 366–368
Lotus SmartSuite, 15
loyalty and ethics, 186
Lycos, 57

M

M key, 91–92
Macintosh OS, 12

magnetic media storage, 371
mailing costs, 153
mailing labels, 141–142
managers, conflicts with, 67
manufacturing careers, 220–221
margins, memos and, 134; reports and, 157; spreadsheets and, 252; unbound reports, 155–156
markers and handles, tables and, 149
marketing careers, 380–381
math, using numeric keypad, 125–126
media selection, for presentations, 292
meeting rooms, for presentations, 309
memos, 127–135; confidential, 128; creating, 133–134; editing, 132–133; line spacing in, 129; margins for, 134; styles of, 128; text selection in, 129; updating, 130
menu bar, 13
merging table cells, 149–150
message of presentation, 286–291
message taking, telephone communications and, 362–363
messages, in communication, 3–4
microfiche, 372
microforms, 372
micrographics, 371
microphone position, speech recognition and, 191–192
microphone settings adjustment for speech recognition, 194

Microsoft FrontPage (See also HTML and Web site design); 227–261, 319
Microsoft Office, 16
Microsoft Excel (See also Microsoft Office), 16
Microsoft Access (See also Microsoft Office), 16
microchips, 2, 3
microprocessors or microchips, 2, 3
Microsoft Office, 16, 319
  applications in, 15–16, 198
  entering text in, 18–19; File menu in, 19, 20; Help in, 23; opening and closing documents in, 20–21; opening and closing programs in, 16–17; previewing and printing documents in, 21–22; saving a file in, 19–20
Microsoft Outlook (See also Microsoft Office: time management), 16, 52, 358
Microsoft PowerPoint (See also Microsoft Office: presentations), 16, 283–318
Microsoft Publisher (See also Microsoft Office), 16
Microsoft Word (See also HTML and Web site design), 15, 319; document conversion for, 339–341; spreadsheets and, 228
minimize/maximize buttons, 14–15
mission statements, 42
mobile computing, 172–173
modems, 37
monitor, 6
mouse, 6, 7, 13

N
N key, 81–82
  naming files and folders, 25
  natural resources careers, 62
  navigation: in databases, 264; in HTML and Web site design, 346; in spreadsheets, 230–231
  navigation bars, 43
  nepotism, 453
  netiquette, 48, 50–51
  Netscape Navigator *(See also* browser software), 35, 324
  networks, intranets, Internet, use policy (AUP) for, 9–10, 11; access points in, 7; accessing, 37–38; bookmarks in, 47–48; copying and pasting Web information from, 45; Domain Name System (DNS) and, 39–40; email and, 37, 48, 52–56; favorites lists in, 47, 48; fiber optic, 7; firewalls, 37; Frequently Asked Questions (FAQs) and, 44–45; hackers and, 9; hyperlinks and, 40–41; Instant Messaging (IM) and, 49; intellectual property rights, copyright and, 46–47, 338; Internet
  moving files and folders, 28–29
  moving graphic images, 166–167
  Mozilla *(See also* browser software), 35, 324
  multiplication, using numeric keypad, 124
  multitasking, 17
  My Computer icon, 25

O
O key, 79–80
  Occupational Outlook Handbook job searches, 397–398
  office suite applications, 15–16
  Office XP handwriting recognition *(See also* handwriting recognition), 174
  nonessential records, 369
  Northern Light, 57
  Notepad, 319
  notes, presentations and, 287
  Num Lock key, 70
  number data types, spreadsheets and, 229, 237
  number keys, 108–114
  numbered lists, 157
  numbers in text, 225
  numeric keypad, 70, 120–123

P
P key, 93–94
  packet data, 37
  page breaks, 161
  page orientation, 167
  Page up/down keys, 19
  pagers, 6
  Palm OS, 12
  Palm Pilot *(See* personal digital assistants)
  paper records, 371
  paragraphs, 160; Click and Type feature, 170; paragraph symbols in text, 130; speech recognition and, 198–200
  Parentheses (( )) keys, 117–118
  passwords, 8
  paste command, 134–135, 233
  online job search, 397–401
  on-screen keyboard input, 186–188
  opening documents, *Microsoft Office*, 20–21
  opening programs, *Windows* and *Microsoft Office*, 16–17
  opening remarks, for presentations, 309–310
  Opera *(See also* browser software), 35, 324
  operating systems, 4, 10–13
  optical media storage, 372
  organization and team work, 42–44
  orientation of page, 253
  outlining ideas for presentation, 286–291
  Outlook (See Microsoft Outlook)
  output, 4
  local access to, 36; login for, 38–39; modems, 37; navigating, 40–41, 43; netiquette and, 48, 50–51; packet data, 37; passwords for, 8; paths to addresses in, 41; registered users of, 41; remote access, 37; safety and security in, 8, 37–38; searching, 57–59; spam and, 10; streaming data/media and, 49–50; TCP/IP, 37; Top Level Domains (TLDs) and, 40; Uniform Resource Locators (URLs) in, 39–40, 41; usernames for, 38; video conferences and, 49–50; viruses and, 9; Web access to, 38–39
  telephony and, 49; IP addresses in, 39–40, 41;

paths and network addresses, 41
peer checks, 5
Percent (%) key, 115–116
Period (.) key, 83–84
person-to-person calls, 367
Personal Computers (PC), 6
Personal Digital Assistants (PDAs), 6, 12, 173
personal planners, 353
personal/laptop computers, 172–173
personal-business letter, 138–139; handwriting recognition and, practice in, 185; speech recognition and, 209–210
photocopiers, photocopying, 375–377
pie charts, 260–261
pixels in graphics/pictures, 336
placement centers and services, 403
plagiarism, 46, 338
Plus Sign (+) key, 117–118
Pocket PC (*See* personal digital assistants)
pointer (*See* cell pointer)
polite behavior, 442
portable/laptop computers, 6
portfolios, 428–433, 439
portrait page orientation, 167, 253
possessives, 225
postage meters, 142–143
POSTNET codes, 141
posture, 311–312
Pound Sign (#) key, 117–118
PowerPoint (*See* Microsoft PowerPoint)
prepaid phone cards, 367
Presentation Wizard, 307–308
presentation, in Web site design, 319

presentations, 4, 283–318; animation, motion and sound in, 304–307; audience response and, 310–312; audio in, 293–294, 304–307; body language and gestures in, 311–312; career building and, 435–436; closing remarks in, 314; delivering, 308–318; evaluation of, 314–316; graphics in, 299; handouts for, 294; headers and footers for, 303–304; honesty in, 291; HTML and Web site design in, document conversion for, 342–345; individual delivery of, 316–317; listener profile for, 284–285; media selection in, 292; message of, organizing and outlining ideas for, 286–291; Microsoft PowerPoint in, 283–318; notes for, 287; opening remarks for, 309–310; planning and preparation for, 283; preparing to present, meeting rooms and appearance, 309; Presentation Wizard for, 307–308; purpose of, 284; question/answer session in, 312–314; slides in, inserting, deleting, arranging, 298–301; storyboarding for, 286; team delivery of, 317–318; team work in, 294–295; visual aids in (*See* visual aids), 291–292, 295–301, 312
previewing documents, *Microsoft Office*, 21–22

printing, 21–22; headers and footers in, 254–255; page options in, 253–254; portrait page orientation in, 253; print area settings in, 253, 254; scaling in, 254; spreadsheets and, 251–256
prioritizing, 351–356
privacy issues, 277
private mail delivery services, 152–154
productivity, 5
professional organizations, 403
professional teams, 390
profile, presentations and, 284–285
programs (*See* applications; programs)
project management, 352, 463–465
pronouns, 465
proofreading, 68
Publisher (*See* Microsoft Publisher)
pull down menu, 13
punctuation, 386; spacing, 96; speech recognition and, 199–200

Q
Q key, 95–96
qualifications vs. employer needs, 410–411, 426–427
quality management, 384
querying databases, 274–276
Question Mark (?) key, 99–100
question/answer session, in presentations, 312–314
Quotation Mark ( ) key, 101

R
R key, 77–78
random access design, HTML and Web site, 320

range (*See* cell range)
rating scale, for job interviews, 446–447
receiving email, 56
records, databases and, 263, 265–267, 269–271
records management, 4, 368–377; copying paper records in, 375–377; disaster recovery plans and, 370–371; filing procedures for, 373–374; indexing and coding records in, 374; life cycle of records in, 369–370; retention schedules for, 369, 370; storage; equipment and supplies for, 373; storage media for (paper, magnetic, etc.), 371–372
recovery plans, 370–371
reference page, 154, 161–164; citing online sources, 67–68
references, job, 407–409, 415
registered users, 38
relative cell referencing, 239–240
reminder systems, 352–353
remote access, 37
renaming files and folders, 27
Repetitive Stress Injury (RSI), 108–109, 191
replying to email, 56
reports, 154–164; creating, 158–159; databases and, from database, 280–281; graphics inserted into, 169–170; headers and footers in, 160; indentation in, 160; lists in, 157; margins in, 157; page breaks in, 161; reference page in, 154, 161–164; speech recognition and,

S

S key, 71-74
Safari (See also browser software), 35
salaries and wages, 394-396
sales and service (marketing) careers, 380-381
Save, 20
Save As, 19
saving a file, 19-20
scaling, spreadsheets and, 254
scanners, 6, 302-303
scanning of electronic resumes, 425
SCANS competency skills, 458
scientific research and engineering careers, 217-218
Scratch That command, speech recognition and, 201-202
script copy, 106-107, 112
scroll bars, 13
scrolling, 76, 230
search engines, 57
searching the Web, 57-59
citing online sources from, 67-68
security, network, 8, 37-38
firewalls in, 37; Internet, 37-38
selecting in tables, 149
selecting text, 129, 132
self checks, 5
Semicolon (;) key, 71-74
sending email, 53-54
sentence structure, 225
servers, email, 52
shading table cells, 150
Shift keys, 70, 83-84, 87-88
show/hide command, 130, 133
shrinkage, 448
single spacing, 129
Slash (/) key, 115-116
slides, presentations and, inserting, deleting, arranging, 298-301
smoking, 448
software skills, vii-viii, 3, 4
sorting records, databases and, 272-274
Space bar, 70, 72
spam, 10
speaking skills, 360
speech recognition, 4, 190;
Backspace and Delete in, 202; capitalization and changing case in, 208-209;
character formats (in bold, italic, etc.), 211-212;
Correction command in, 202-204; Dictation button in, 196; dictionary of, adding words to, 206-207;
enunciation in, 196; error correction in, 201-205;
formatting text in, 211-216;
homonym correction in, 202, 204-205; Language bar for, 192-193; line spacing in, 213-214;
microphone position for, 191-192; microphone settings adjustment for, 194;
navigation in, 198-200; new line/new paragraph commands in, 198-200; origins of, 198; personal-business letter practice in, 209-210; punctuation in, 199-200; report practice in, 214-216; Scratch That command in, 201-202;
Speech Tools option in, substituting words in, 192-193;
text alignment in, 205; text in, dictating, 213;
training in, 192-193, 200, 206-207;
user profile for, 192-193; voice commands in, voice dictating, 194-196; voice protection during, 191
Speech Tools option, speech recognition and, 192-193
spell checking, 137-138
splitting table cells, 149-150
sponsored sites, in Web searches, 58
spreadsheets, 4, 227-261;
absolute and relative cell referencing in, 239-240;
active cells in, 229; budget example of, 256; cell pointer in, 229; cell range in, 229; centering on page, 252; charts and graphs in, clearing cells in, 257-261; columns, rows, and cells in, 229, 231, 243-251; creating, 231-233; currency data/signs in, 248; cut, copy, paste in, 233; data types in, 229-230, 237, 246; dates and times in, 230; deleting/inserting columns and rows in, 243-245; editing and formatting, 234-236, 243-251; entering data into, Enter/Delete/Cancel, 230; fill and fill series in, 249, 250-251; formatting data in, 246, 247-248; formulas in, (See also formulas), 230, 236-242; fuel cost example for, 241-242; functions in, 238-239; headers and footers in, 254-255; height

of rows in, 246–247; HTML and Web site design in, document conversion for, 342–345; inserting worksheet into, 249–250; margins for, 252; *Microsoft Excel* in, 227–228; *Microsoft Word* and, 228; naming worksheets in, 233; navigating within worksheets of, 230–231; numbers in, 229, 237; page options in, 253–254; portrait page orientation in, 253; printing, 251–256; scaling in, 254; tabs on worksheets in, 233; text in, 229; workbooks in, 228; worksheet basics in, 227–236

StarOffice, 15

starting a program/application, 13

starting your new career, 428–455

stewardship of company funds, 245

storage equipment and supplies, records management and, 373

storage media, records, (paper, magnetic, etc.), 371–372

storyboards for presentations, 286

straight copy, 106–107

strategic plans, 241

streaming data/media, 49–50

substituting words, speech recognition and, 205

subtraction, using numeric keypad, 124

switches, 6

symbol keys, 108

T

T key, 79–80

Tab key, 70, 101–103

tables, 143–152; AutoFit feature and, 147; changing structure of, 150–151; columns, rows, and cells in, 144, 146–148; creating, 144–145; databases and, 263, 265–267, 267–272; markers and handles in, 149; merging and splitting cells in, 149–150; modifying features of, 146–148; moving within, 145; selecting in, 149; shading cells in, 150

tablet PCs, 6, 173, 174

tabs, 131

tabs, on worksheets, 233

tags, HTML, 323–326

tangible vs. intangible rewards of work, 386

task lists, time management and, 353–358

Task panes in, 14

TCP/IP, 37

team work, 4, 42–44, 390; presentations and, 294–295; report writing, 164

telecommuting, 224

telephone communications, 359–368; costs and, controlling, 367–368; directories for, 366; first impressions and, 359–360; incoming calls in, 360–363; Internet telephony and, 49; long distance service and, 366–368; message taking in, 362–363; outgoing calls in, 364; role playing to improve, 362; speaking skills, enunciation, and attitude in, 360; time zone differences and, 365; toll free service in, 367; voice mail and, 364

templates, 437–438

temporary employment agencies, 400, 403

test package, electronic, x

text data type, spreadsheets and, 229

text entry, 18

text formatting (*See* formatting text)

text, speech recognition and, dictating, 196–197

textual citations, 154, 161–164

time data type, spreadsheets and, 230

time management, 4, 350–358; *Microsoft Outlook* in, 350–358; personal tasks list example in, 358; prioritizing in, 351–356; project management in, 352; reminder systems for, 352–353; task lists in, 353–358; updating tasks in *Outlook* for, 357–358

time zones, 365

title bar, 13

title page, 170–171

toll free service, telephone communications and, 367

toolbars, 13, 20

Top Level Domains (TLDs), 40

Total Quality Management (TQM), 384–385

training, speech recognition and, 192–193, 206–207

transportation, distribution, logistics careers, 222

TV commercial example, 387

two-line input method, handwriting recognition and, 183–184

U

U key, 85–86

U.S. Department of Education, 32

unbound reports, 154, 155–156

Underline (_) key, 118–119

underlined text, 130–131, 133, 211–212

Undo button, 132, 145

Uniform Resource Locators (URLs), 39–40, 41

United States Postal Service, 152–154

useful records, 369

user interface (*See* computer interfaces)

user profile, speech recognition and, 192–193

usernames, 38

V

V key, 93–94

values settings, HTML and Web site design in, 333–334

video conferences, 49–50

video phone, 2

View menus, 14

viewing details of files, 29

viruses, 9, 54–55

visual aids for presentations (*See also* graphics), 291–295, 301, 312; animation, motion and sound in, 304–307; audio in, 304–307; color choice in, 296–297; design strategies for, 295; headers and footers for, 303–304; slides in, inserting, deleting, arranging, 298–301; text in, 297–298; white space in, 295

vital records, 369

INDEX

V

voice commands, speech recognition and, dictating, 194–196
voice mail, 364

W

W key, 87–88
Web browsers, 4, 35–36, 324
Web hosting services and resumes, 417
Web sites (See HTML and Web site design; networks, intranets, Internet)
Web sites for employment opportunities in, 399–400
WebCrawler, 57
whistle-blower laws, 186
white space, 295
Windows, 12, 24–31
Windows Explorer (See also file management), 25–26
Windows key, 70
Windows Pocket PC, 12
Windows Tablet PC Edition, 12
Windows XP Tablet PC Edition, 190
wireless phones, 6
Word (See Microsoft Word)
word processing, 4, 127
work area arrangement, 71, 109
work ethics, 386
work related injuries, 342
work/life balance, 463
workbooks (See spreadsheets)
worker's compensation, 342
works cited page (See also reference page), 154
worksheets (See spreadsheets)
wrapping text around graphics, 166–167
writing pad, handwriting recognition and, 175–176, 178–181
writing skills, 225, 386, 465; citing online sources, 67–68; handwriting recognition and, 180
writing style and handwriting recognition, 178

X

X key, 91–92

Y

Y key, 89–90
Yahoo!, 57

Z

Z key, 97–98
Zero key, 110–112
ZIP codes, 141, 148–149